Tissue Engineering

INTERFACE SCIENCE AND TECHNOLOGY
Series Editor: ARTHUR HUBBARD

In this series:

Vol. 1: Clay Surfaces: Fundamentals and Applications
    Edited by F. Wypych and K.G. Satyanarayana

Vol. 2: Electrokinetics in Microfluidics
    By Dongqing Li

Vol. 3: Radiotracer Studies of Interfaces
    Edited by G. Horányi

Vol. 4: Emulsions: Structure Stability and Interactions
    Edited by D.N. Petsev

Vol. 5: Inhaled Particles
    By Chiu-sen Wang

Vol. 6: Heavy Metals in the Environment
    Edited by H.B. Bradl

Vol. 7: Activated Carbon Surfaces in Environmental Remediation
    Edited by T.J. Bandosz

Vol. 8: Tissue Engineering: Fundamentals and Applications
    By Y. Ikada

INTERFACE SCIENCE AND TECHNOLOGY – VOLUME 8

Tissue Engineering

Fundamentals and Applications

Yoshito Ikada

Suzuka University of Medical Science
Suzuka, Japan

Amsterdam • Boston • Heidelberg • London • New York • Oxford
Paris • San Diego • San Francisco • Singapore • Sydney • Tokyo

Academic Press is an imprint of Elsevier

ELSEVIER B.V.	ELSEVIER Inc.	**ELSEVIER Ltd**	ELSEVIER Ltd
Radarweg 29	525 B Street, Suite 1900	**The Boulevard, Langford Lane**	84 Theobalds Road
P.O. Box 211, 1000	San Diego, CA 92101-4495	**Kidlington, Oxford OX5 1GB**	London WC1X 8RR
AE Amsterdam	USA	**UK**	UK
The Netherlands			

First edition 2006

Library of Congress Cataloging in Publication Data
A catalog record is available from the Library of Congress.

British Library Cataloguing in Publication Data
A catalogue record is available from the British Library.

ISBN-13: 978-0-12-370582-2
ISBN-10: 0-12-370582-7
ISSN: 1573-4285

∞ The paper used in this publication meets the requirements of ANSI/NISO Z39.48-1992 (Permanence of Paper).
Printed in China by CTPS

Table of Contents

Preface xi

List of Abbreviations xvii

CHAPTER 1: SCOPE OF TISSUE ENGINEERING 1

1. Functions of Scaffold 1

2. Absorbable Biomaterials 4
   2.1. Natural Polymers 6
      2.1.1. Proteins 6
      2.1.2. Polysaccharides 10
      2.1.3. Natural Composite—ECM 14
   2.2. Synthetic Polymers 17
      2.2.1. Poly(α-hydroxyacid)s [Aliphatic α-polyesters
         or Poly(α-hydroxyester)s] 18
      2.2.2. Hydrogels 22
      2.2.3. Others 23
   2.3. Inorganic Materials—Calcium Phosphate 25
   2.4. Composite Materials 25

3. Pore Creation in Biomaterials 26
   3.1. Phase Separation (Freeze Drying) 27
   3.2. Porogen Leaching 28
   3.3. Fiber Bonding 29
   3.4. Gas Foaming 29
   3.5. Rapid Prototyping 29
   3.6. Electrospinning 30

4. Special Scaffolds 31
   4.1. Naturally Derived Scaffolds 31
      4.1.1. ECM-like Scaffolds 32
      4.1.2. Fibrin Gel 34
      4.1.3. Matrigel™ 34
      4.1.4. Marine Natural Scaffold 34
   4.2. Injectable Scaffolds 35
   4.3. Soft, Elastic Scaffolds 35
   4.4. Inorganic Scaffolds 35
   4.5. Composite Scaffolds 36

5. Surface Modifications 36
   5.1. Cell Interactions in Natural Tissues 36
   5.2. Artificial Surface in Biological Environment 38

6. Cell Expansion and Differentiation 41
   6.1. Monolayer (2-D) and 3-D Culture 42
   6.2. Cell Seeding 44
       6.2.1. Serum 46
       6.2.2. Cell Adhesion 47
       6.2.3. Seeding Efficiency 48
       6.2.4. Assessment of Cells in Scaffolds 49
       6.2.5. Gene Expression of Cells 51
   6.3. Bioreactors 51
       6.3.1. Spinner Flask 53
       6.3.2. Perfusion System 55
       6.3.3. Rotating Wall Reactor 56
       6.3.4. Kinetics 59
   6.4. Externally Applied Mechanical Stimulation 60
   6.5. Neovascularization 63

7. Growth Factors 65
   7.1. Representative Growth Factors 66
       7.1.1. BMPs 66
       7.1.2. FGFs 67
       7.1.3. VEGF 67
       7.1.4. TGF-β1 68
       7.1.5. PDGF 68
   7.2. Delivery of Growth Factors 69

8. Cell Sources 70
   8.1. Differentiated Cells 71
   8.2. Somatic (Adult) Stem Cells 74
       8.2.1. MSCs 77
       8.2.2. Adipose-Derived Stem Cells 81
       8.2.3. Umbilical Cord Blood-Derived Cells 81
   8.3. Cell Therapy 81
       8.3.1. Angiogenesis 82
       8.3.2. Cardiac Malfunction 82
   8.4. ES Cells 84
       8.4.1. Cell Expansion and Differentiation 85
       8.4.2. Somatic Cell Nuclear Transfer 86

References 87

CHAPTER 2: ANIMAL AND HUMAN TRIALS OF ENGINEERED TISSUES 91

1. Body Surface System 91
   1.1. Skin 91
      1.1.1. Without Cells 91
      1.1.2. Keratinocytes 91
      1.1.3. Keratinocytes on Acellular Dermis 93
      1.1.4. Keratinocytes + Fibroblasts 93
      1.1.5. Keratinocytes + Melanocytes 94
      1.1.6. Stem Cell Transplantation 95
   1.2. Auricular and Nasoseptal Cartilages 96
   1.3. Adipose Tissue 101

2. Musculoskeletal System 105
   2.1. Articular Cartilage 105
   2.2. Bones 119
   2.3. Tendon and Ligament 129
      2.3.1. Ligaments 131
      2.3.2. Tendons 135
   2.4. Rotator Cuff 138
   2.5. Skeletal Muscle 139
   2.6. Joints 140
      2.6.1. Large Joints 141
      2.6.2. Small Joints: Phalangeal Joints 142

3. Cardiovascular and Thoracic System 144
   3.1. Blood Vessels 145
      3.1.1. Large-Calibered Blood Vessels 145
      3.1.2. Coronary Artery 150
      3.1.3. Angiogenesis 153
      3.1.4. Neovascularization 156
   3.2. Heart Valves 158
   3.3. Myocardial Tissue 162
   3.4. Trachea 169

4. Nervous System 173
   4.1. Neuron 173
   4.2. Spinal Cord 173
   4.3. Peripheral Nerve 174

5. Maxillofacial System 188
   5.1. Alveolar Bone and Periodontium 189
   5.2. Temporomandibular Joint 195

5.3. Enamel and Dentin 198
5.4. Mandible 199
5.5. Orbital Floor 202

6. Gastrointestinal System 205
6.1. Esophagus 205
6.2. Liver 206
6.3. Bile Duct 209
6.4. Abdominal Wall 210
6.5. Small Intestine 210

7. Urogenital System 213
7.1. Bladder 213
7.2. Ureter 215
7.3. Urethra 215
7.4. Vaginal Tissue 216
7.5. Corporal Tissue 217

8. Others 218
8.1. Skull Base 218
8.2. Dura Mater 219
8.3. Cornea 220
8.4. Prenatal Tissues 221

References 223

CHAPTER 3: BASIC TECHNOLOGIES DEVELOPED FOR TISSUE
            ENGINEERING 235

1. Biomaterials 235
1.1. Naturally Occurring Polymers 235
1.1.1 Proteins 235
1.1.2. Polysaccharides 243
1.2. Synthetic Polymers 250
1.2.1. Poly(α-hydroxyacid)s 250
1.2.2. Hydrogels 263
1.2.3. Polyurethanes 265
1.2.4. Others 268
1.3. Calcium Phosphate 272
1.4. Composites 272

2. Fabrication of Porous Scaffolds 274
2.1. Freeze Drying 274
2.2. Porogen Leaching 278

2.3. Gas Foaming 279
2.4. Rapid Prototyping 282
2.5. Electrospinning 285
2.6. UV and Laser Irradiation 290

3. Novel Scaffolds 292
    3.1. Naturally Derived Scaffolds 292
        3.1.1. ECM-like Scaffolds 292
        3.1.2. Tissue-Derived Scaffolds 295
        3.1.3. Fibrin Gel 297
        3.1.4. Natural Sponge 298
    3.2. Injectable Scaffolds 299
    3.3. Elastic Scaffolds 299
    3.4. Inorganic Scaffolds 300
    3.5. Composite Scaffolds 301

4. Surface Modification of Biomaterials and Cell Interactions 303

5. Growth Factors and Carriers 309
    5.1. Growth Factor–like Polymers 309
    5.2. Carriers 311
    5.3. Combined and Sequential Release of Growth Factors 323
    5.4. Gene Transfer 325

6. Cell culture 328
    6.1. Cell Seeding 329
    6.2. Co-culture 336
    6.3. Bioreactors 338
        6.3.1. Spinner Flask Reactor 339
        6.3.2. Perfusion Reactor 339
        6.3.3. Rotating Reactor 342
    6.4. Kinetics 343
    6.5. Mechanical Stimulation 345
    6.6. Cell counting and distribution in scaffolds 353

7. Examples of Cell Culture 356
    7.1. Differentiated Cells 356
        7.1.1. Muscular Cells 356
        7.1.2. Fibroblasts 360
        7.1.3. Chondrocytes 363
        7.1.4. Bone Cells 368
        7.1.5. Vascular Cells 371
        7.1.6. Hepatocytes 377
        7.1.7. Oral Cells 377

7.1.8. Neuronal Cells 378
7.1.9. Retinal Cells 379
7.2. Stem Cells 379
7.2.1. MSCs 380
7.2.2. Adipose-Derived Stem Cells 396
7.2.3. NSCs 397
7.2.4. ES Cells 400

References 405

CHAPTER 4: CHALLENGES IN TISSUE ENGINEERING 423

1. Problems in Tissue Engineering 423

2. Sites for Neotissue Creation 424
2.1. *Ex vivo* Tissue Engineering 426
2.2. *In situ* Tissue Engineering 428

3. Autologous or Allogeneic cells 429
3.1. Allogeneic Cells 430
3.2. Autologous Cells 431

4. Cell Types 432

5. Risks at Cell Culture 436

6. Scaffolds for Large Animals and Human Trials 438
6.1. Mechanical Strength 439
6.2. Bioabsorption Rate 442
6.3. Tailoring of Ultrafine Structure 443

7. Importance of Neovascularization 446

8. Carriers for Growth Factors 448

9. Primary Roles of Each Player in the Tissue Engineering Arena 453
9.1. Scientists and Engineers 455
9.2. Clinicians 456
9.3. Manufacturers 457
9.4. Regulatory Agencies 459

References 461

Index 463

Preface

Current clinical technologies, especially donor transplants and artificial organs, have been excellent life-saving and life-extending therapies to treat patients who need to reconstitute diseased or devastated organs or tissues as a result of an accident, trauma, and cancer, or to correct congenital structural anomalies. For long, most scientists and clinicians believed that damaged or lost tissues could only be replaced by organ transplantation or with totally artificial parts. Although advances in surgical techniques, immunosuppression, and postoperative care have improved survival and quality of life, there are still problems associated with the use of biological grafts, such as donor site morbidity, donor scarcity, and tissue rejection. With regard to prostheses, a variety of synthetic and natural materials have been developed for replacement of lost tissues, but the results have not always been satisfactory. For instance, there is a great concern over the long-term performance of artificial devices. Treatments for end-stage renal failure by kidney dialyzers are based solely on unphysiological driving forces and are not able to mimic active molecular transport accomplished by renal tubular cells. Silicone for breast reconstruction after surgical mastectomy or lumpectomy for treatment of breast cancer may cause foreign body reactions and infection. Autologous adipose tissues have been clinically used to regenerate adipose tissues in depressed regions in the breast, but this therapy has problems of absorption and subsequent volume loss of transplanted adipose tissues. Such serious problems remaining unsolved and the need for improved treatments have motivated research aimed at alternative approaches creating new tissues. Tissue engineering emerged as a promising alternative in which organs or tissues can be repaired, replaced, or regenerated. The tissue engineering paradigm is to isolate specific cells through a small biopsy from a patient, to grow them on a three-dimensional scaffold under precisely controlled culture conditions, to deliver the construct to the desired site in the patient's body, and to direct new tissue formation into the scaffold that can be absorbed over time.

To avoid confusion, a brief explanation will be required for the relationship among tissue engineering, regenerative medicine, cell therapy, and embryonic stem (ES) cells. In recent years, ES cells have attracted surprisingly much attention because the pluripotent cells are anticipated to be able to differentiate into any cells responsible for formation of all kinds of tissues and organs present in the body.

For instance, Parkinson's disease and insulin-dependent diabetes might be cured using ES cells. This emerging therapy is called "regenerative medicine". In some limited cases, injection of cells to patients is sufficient for the medical treatment. This is termed "cell therapy" or "cellular therapy". However, in many other cases where lost tissues or organs have three-dimensional, bulky complex structure, cell injection alone is not effective as a cure because of quick scattering of injected cells from the site of injection. In such cases, we have to provide a guiding and scaffolding framework for cells to adhere to, expand, differentiate, and produce matrices for neotissue formation. This is the principle of tissue engineering. We can say therefore that regenerative medicine involves two concepts both of which make use of cells for therapies. One is cellular therapy (no use of scaffold) and the other is tissue engineering (assisted by scaffold). Obviously, the creation of three-dimensional complex tissues starting from ES cells also needs the technology of tissue engineering, but it will be several decades before advances of cell biology enable the widespread human use of ES cells.

It is worthwhile historically reflecting on what has happened in tissue engineering. In the spring of 1987, the Engineering Directorate of the National Science Foundation (NSF) of USA held a Panel discussion focusing on future directions in bioengineering. The target research areas appeared to overlap and the Panel coined the term "tissue engineering" to consolidate their efforts. It was this panel meeting in the spring of 1987 that produced the first documented use of the term "tissue engineering". On the basis of this initial Panel discussion, a Panel meeting on Tissue Engineering was held at the NSF in October 1987 [1]. The United States gave birth to the field of tissue engineering through pioneering efforts in reparative surgery and biomaterials engineering. In 1993, Langer and Vacanti presented an overview on tissue engineering showing how this interdisciplinary field had applied the principles of engineering and the life sciences to the development of biological substitutes that restore and improve tissue function [2]. As such, the goal of tissue engineering is to create cell–scaffold constructs to direct tissue regeneration and to restore function through the delivery of living elements, which become integrated into the patient. Since then, tissue engineering has attracted many scientists and surgeons with the hope of revolutionizing methods of healthcare treatment and dramatically improving the quality of life for millions of people throughout the world. Indeed, earlier work has successfully demonstrated creation of new tissues by using cells on biodegradable polymer scaffolds.

A review article by Lysaght and Reyes in 2001 demonstrated that at the beginning of 2001, tissue engineering R&D was being pursued by 3300 scientists and support staff in more than 70 companies with a combined annual expenditure of over $600 million [3]. Furthermore, the aggregate investment in the sector since 1990 exceeded $3.5 billion and the sector witnessed the entry of many new startup firms. As many as 16 startup firms focusing on this sector reached the milestone of initial public offerings (IPOs) and had a combined market capitalization of $2.6 billion. However, until the time of writing, tissue engineering has not yet delivered many products for better healthcare nor many successful companies making them, although the tissue engineering concepts have been around for 20 years, with serious activity for about 15 years. Lysaght and Hazlehurst wrote in a review published in 2004 [4]:

> . . . As recently as February 2003, the normally skeptical *Economist* reported, "these are exciting times for tissue engineers. The technology for growing human body parts is advancing rapidly. Already it is possible to cultivate sheets of human skin. And huge efforts are underway to develop even more complex structures, such as heart valves and whole organs such as the liver" (*The Economist*, February 1, 2003). Such highly favorable media treatment has its benefits, but research-minded professionals increasingly recognized a disconnect with the realities. And such disconnects rarely lead to happy endings. . . .

Technical knowledge and skill must develop if tissue engineering is to become a successful reality. Numerous research areas are critical for the success of tissue engineering. Many research centers of tissue engineering in the world have devoted much of their efforts to challenges in cell technologies. Engineered tissues are possibly produced from both autologous and allogeneic cells. Allogeneic products are amenable to large-scale manufacturing at single sites, while autologous therapies will likely lead to more of a service industry, with a heavy emphasis on local or regional cell banking/expansion. Probably because of this difference, two large American companies (*e.g.*, ATS and Organogenesis) focused on allogeneic products, but bankrupted in 2002. One of the major reasons for the bankruptcy may be their overinvestment in the overestimated market for their expensive tissue engineering products. In a cost-controlled healthcare environment, only those technologies capable of providing a major enhancement to quality of life and a reduction in expenditure will be driven forward.

In a review article, Breuer *et al.* described as follows [5]:

> . . . The holy grail of any tissue-engineering project would be the successful clinical application and use of the neotissue. Shin'oka *et al.* have applied the techniques used in creating a tissue-engineered heart valve to construct autologous vascular grafts for use as venous conduits in more than 40 children with varying forms of complex congenital heart disease. . . . They used a copolymer of either PGA and ε-caprolactone [P(CL/LA)] or P(CL/LA) reinforced with poly-L-lactide (PLLA) to construct their tubular grafts. . . . Shin'oka's first operation was performed in May 1999. Immediate postoperative results have been excellent, and there are now long-term follow-up results for these patients. Serial postoperative angiographic, computerized tomography, or magnetic resonance imaging examinations revealed no dilatation or rupture of grafts, and there have been no complications related to the tissue-engineered autografts. Although histological evaluation is not possible, no unwanted calcification or microcalcification has been found by current imaging studies in these patients. Shin'oka's results demonstrate the clinical utility and feasibility of tissue engineering in medicine. . . . Despite the excellent results of Shin'oka

et al. and Dohmen *et al.*, the clinical application of cardiovascular tissue-engineering techniques is premature by U.S. standards. The development of standards for preclinical trials to provide justification for establishing FDA clinical investigations of tissue-engineered products is in its infancy. The rapid development of cardiovascular tissue-engineering research has far outpaced the ability of regulatory agencies to develop policies to govern product development. The completion of preclinical studies that provide a firm foundation on which to base clinical trials is essential for the rational and responsible development of this promising technology. . . .

The motivation for writing this book was to address, as a collaborative engineer of the Shin'oka team, the reason why progress in tissue engineering has been so slow, with still so limited clinical applications of engineered tissues. This book is composed of four chapters. Chapter 1 provides an overview of contemporary tissue engineering research. This chapter will be helpful to readers new to the field who are very enthusiastic about tissue engineering. The most recent advances in animal experiments and human trials associated with tissue engineering are described in Chapter 2. This may help readers to understand current activities of tissue engineering applications, but readers will also learn how small numbers of engineered tissues have been applied to patients. This means that tissue engineering is still at an early stage in terms of clinical applications and needs much more of a contribution from different fields. Chapter 3 covers a large number of scientific papers published on basic technologies related to the tissue engineering area. However, the number of papers referred to in this book had to be drastically limited because of the extraordinarily vast number of publications in this field. The last chapter, the most important in this book, is devoted to demonstrating which technologies are the real bottlenecks that retard the clinical application of tissue engineering. Chapter 4 is therefore substantially different from Chapters 2 and 3 which are a brief compilation of literature on current tissue engineering research. In contrast, Chapter 4 involves thoughts and suggestions of the author of this book for developing the engineering systems needed to produce functional engineered tissues on the basis of his long-standing experience in research, including absorbable biomaterials, drug delivery, artificial organs, and tissue engineering. This writing style discriminates this work from other tissue engineering books that mostly consist of many chapters written by different contributors.

The author wishes that this book will serve as a base for directing future research of tissue engineering toward revolutionizing healthcare, especially repair of damaged and lost tissues as well as regeneration of neotissues with complex structures.

REFERENCES

1. C.W. Patrick Jr, A.G. Mikos, and L.V. McIntire, Eds, *Frontiers in Tissue Engineering*, Pergamon, 1998.
2. R. Langer and J.P. Vacanti, Tissue engineering, *Science*, **260**, 920 (1993).
3. M.J. Lysaght and J. Reyes, The growth of tissue engineering, *Tissue Engineering*, **7**, 485 (2001).

4. M.J. Lysaght and A.L. Hazlehurst, Tissue engineering: The end of the beginning, *Tissue Engineering*, **10**, 309 (2004).
5. C.K. Breuer, B.A. Mettler, T. Anthony, V.L. Sales, F.J. Schoen, and J.E. Mayer, Application of tissue-engineering principles toward the development of a semilunar heart valve substitute, *Tissue Engineering*, **10**, 1725 (2004).

Yoshito Ikada

List of Abbreviations

MATERIALS

Synthetic Polymers

PGA	poly(glycolide), poly(glycolic acid)
PLA	poly(lactide), poly(lactic acid)
PLLA	poly(L-lactide), poly(L-lactic acid)
PDLLA	poly(D,L-lactide), poly(D,L-lactic acid)
PGLA(PLGA)	glycolide-lactide copolymer(lactide-glycolide copolymer)
PCL	poly(ε-caprolactone)
P(LA/CL)	lactide-ε-caprolactone copolymer
PEG [PEO]	poly(ethylene glycol) [poly(ethylene oxide)]
PTFE	polytetrafluoroethylene
PTMC	poly(1,3-trimethylene carbonate)

Natural Polymers

HAc	hyaluronic acid or hyaluronate
CS	chondroitin sulfate
GAG	glycosaminoglycan
PHA	poly(β-hydroxyalakanoate)
FN	fibronectin
LN	laminin
FGF	fibroblast growth factor
bFGF	basic fibroblast growth factor
BMP	bone morphogenetic protein
EGF	epidermal growth factor
VEGF	vascular endothelial growth factor
TGF	transforming growth factor
KGF	keratinocyte growth factor
ALP	alkaline phosphatase
OP	osteopontin
OCN	osteocalcin
MMP	matrix metalloprotease
GFP	green fluorescent protein
EGFP	enhanced GFP
IL	interleukin
RGD	arginine-glycine-aspartic acid

Ceramics

HAp	hydroxyapatite
TCP	tricalcium phosphate

Low-molecular-weight molecules

GA	glutaraldehyde
WSC	water soluble carbodiimide
EDAC	1-ethyl-3-(3-dimethylaminopropyl)-carbodiimide
NHS	N-hydroxysuccinimide

TISSUES

ECM	extracellular matrix
CNS	central nervous system
PNS	peripheral nervous system
ACL	anterior crucial ligament
SIS	small intestine submucosa

CELLS

ES	embryonic stem
MSC	mesenchymal stem cell, marrow-derived stem cell
BMSC	bone marrow-derived stem cell, bone marrow-derived mesenchymal stem cell, bone-marrow stromal cell
ADAS	adipose-derived adult stem
NSC	neural stem cell
MNC	mononuclear cell
EPC	endothelial progenitor cell
EC	endothelial cell
HUVEC	human umbilical vein endothelial cell
UC	urothelial cell

CELL CULTURE

2-D	two dimensional
3-D	three dimensional
FCS (BFS)	fetal calf serum (bovine fetus serum)
PBS	phosphate buffered solution
DMEM	Dulbecco's modified Eagle's medium
MEM	Minimal essential medium
AA	ascorbic acid
PRP	platelet-rich plasma

RWV	rotating-wall vessel
RT–PCR	reverse transcription polymerase chain reaction

MISCELLANEOUS

BSE	bovine spongiform encephalitis
PERV	porcine endogenous retrovirus
GTR	guided tissue regeneration
GBR	guided bone regeneration
MW	molecular weight

Chapter 1

Scope of Tissue Engineering

In tissue engineering, a neotissue generally is regenerated from the cells seeded onto a bioabsorbable scaffold, occasionally incorporating growth factors: *cells + scaffold + growth factors → neotissue*. In theory, any tissue could be created using this basic principle of tissue engineering. However, in order to achieve successful regeneration of tissues or organs based on the tissue engineering concept, several critical elements should be deliberately considered including biomaterial scaffolds that serve as a mechanical support for cell growth, progenitor cells that can be differentiated into specific cell types, and inductive growth factors that can modulate cellular activities. The fundamentals of tissue engineering will be presented in this chapter.

1. FUNCTIONS OF SCAFFOLD

When a tissue is severely damaged or lost, not only large numbers of functional cells but also the matrix in tissue, generally called *extracellular matrix* (ECM), are lost. It is difficult to imagine how small-molecule drugs or even recombinant proteins would be able to restore the lost tissue and reverse the function. Because tissue represents a highly organized interplay of cells and matrices, the fabrication of replacement tissue may be facilitated by mimicking the spatial organization in tissue. To this end, we should provide an artificial or biologically derived ECM for cells to create a neotissue. Isolated cells have the capacity to form a tissue structure only to a limited degree when placed as a suspension on tissue, because they need a template that guides cell organization. In tissue engineering we designate the substitute of native ECM as "scaffold", "template", or "artificial matrix". Scaffold provides a three-dimensional (3-D) ECM analog which functions as a template required for infiltration and proliferation of cells into the targeted functional tissue or organ. If any assistance by scaffold is not required for cells, we call it "cell (or cellular) therapy" or "cell transplantation". Cell therapy avoids the complications of surgery, but allows replacement of only those cells that perform the biological functions including hormone secretion and enzyme synthesis. It would be therefore convenient to divide regenerative medicine into two subgroups, as shown in Fig. 1.1, depending on the scaffold requirement.

The primary function of scaffolds is to provide structure for organizing dissociated cells into appropriate tissue construction by creating an environment that enables 3-D cell growth and neotissue formation. When cells attached to a scaffold are implanted, they will be incorporated into the body. Cell attachment is the first critical element in initiating cell growth and neotissue development. Natural or

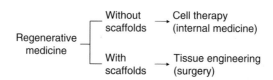

Fig. 1.1   Classification of regenerative medicine based on the use of scaffold.

synthetic biomaterials utilized for scaffold fabrication are mostly selected on the basis of their biocompatibility, bioabsorbability, and mechanical properties. Much of the secondary scaffold processing is performed to make the scaffold more porous for enhancement of cell infiltration and neotissue ingrowth. To promote cell attachment various cell adhesion molecules such as laminin (LN) have been used to coat the scaffold before cell seeding. The traditional method of seeding polymer scaffolds with cells has employed static cell culture techniques. For instance, a concentrated cell suspension is pipetted onto a collagen-coated polymer scaffold and left to incubate for variable periods of time for cells to adhere to the polymer. Dynamic cell seeding employs a method in which either the medium or the medium and scaffold are in constant motion during the incubation period.

Scaffolds should encourage the growth, migration, and organization of cells, providing support while the tissue is forming. Finally, as demonstrated in Fig. 1.2, the scaffolds will be replaced with host cells and a new ECM which in turn should provide functional and mechanical properties, similar to native tissue. The material and the 3-D structure of scaffolds have a significant effect on cellular activity. Depending on the tissue of interest and the specific application, the required scaffold material and its properties will be quite different. In general, a biologically active scaffold should provide the following characteristics: (1) a 3-D, well-defined

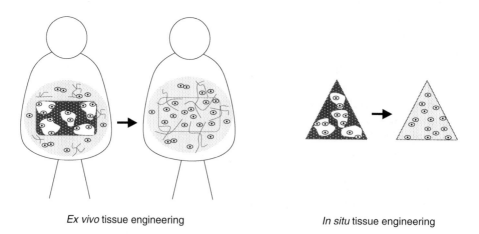

Ex vivo tissue engineering   *In situ* tissue engineering

Fig. 1.2   Role of scaffold in tissue engineering.

porous structure to make the surface-to-volume ratio high for seeding of cells as many as possible; (2) a physicochemical structure to support cell attachment, proliferation, differentiation, and ECM production to organize cells into a 3-D architecture; (3) an interconnected, permeable pore network to promote nutrient and waste exchange; (4) a non-toxic, bioabsorbable substrate with a controllable absorption rate to match cell and tissue growth *in vitro* or *in vivo*, eventually leaving no foreign materials within the replaced tissues; (5) a biological property to facilitate vasculature network formation in the scaffold; (6) mechanical properties to support or match those of the tissue at the site of implantation and occasionally to present stimuli which direct the growth and formation of a load-bearing tissue; (7) a mechanical architecture to temporarily provide the biomechanical structural characteristics for the replacement "tissue" until the cells produce their own ECM; (8) a good carrier to act as a delivery system for bioactive agents, such as growth factors; (9) a geometry which promotes formation of the desired, anisotropic tissue structure; (10) a processable and reproducible architecture of clinically relevant size and shape; (11) sterile and stable enough for shelf life, transportation, and production; and (12) economically viable and scaleable material production, purification, and processing.

Scaffolds initially fill a space otherwise occupied by natural tissue, and then provide a framework by which a tissue will be regenerated. The architecture of 3-D scaffold can also control vascularization and tissue ingrowth *in vivo*. In this capacity, the physical and biological properties of the material are inherent in the success of the scaffold. Selection and synthesis of the appropriate scaffold material is governed by the intended scaffold application and environment in which the scaffold will be placed. For example, a scaffold designed to encapsulate cells must be capable of being gelled without damaging the cells, must allow appropriate diffusion of nutrients and metabolites to and from the encapsulated cells and surrounding tissue, and stay at the site of implantation with sufficient mechanical integrity and strength. Scaffold heterogeneity has been shown to lead to variable cell adhesion and to affect the ability of the cells to produce a uniform distribution of ECM. Tissue synthesized in a scaffold with non-uniform pore architecture may show inferior biomechanical properties compared to tissue synthesized in a scaffold with a more uniform pore structure. In scaffolds with equiaxed pores, cells aggregate into spherical structures, while in scaffolds with a more elongated pore shape, cells align with the pore axis.

Once the scaffold is produced and placed, formation of tissues with desirable properties relies on scaffold mechanical properties on both the macroscopic and the microscopic levels. Macroscopically, the scaffold must bear loads to provide stability to tissues as it forms and to fulfill its volume maintenance function. On the microscopic level, cell growth and differentiation and ultimate tissue formation are dependent on mechanical input to the cells. As a consequence, the scaffold must be able to both withstand specific loads and transmit them in an appropriate manner to the surrounding cells and tissues. Specific mechanical properties of scaffolds include elasticity, compressibility, viscoelastic behavior, tensile strength, failure strain, and their time-dependent fatigue.

2. ABSORBABLE BIOMATERIALS

Biomaterials are a critical enabling technology in tissue engineering, because they serve in various ways as a substrate on which cell populations can attach and migrate, as a 3-D implant with a combination of specific cell types, as a cell delivery vehicle, as a drug carrier to activate specific cellular function in the localized region, as a mechanical structure to define the shape of regenerating tissue, and as a barrier membrane to provide space for tissue regeneration along with prevention of fibroblast ingrowth into the space. In many cases a biomaterial serves a dual role as scaffold and as delivery device. Degradation and absorption of biomaterials are essential in functional tissue regeneration, unless the application is aimed at long-term encapsulation of cells to be immunologically isolated. Materials that disappear from the body after they have fulfilled their function obviate concerns about long-term biocompatibility. The by-products of degradation must be non-toxic, similar to the starting material. For a biomaterial to be accepted in the medical system, its safety and efficacy must be proven with any therapy. Ideally, the rate of scaffold degradation should mirror the rate of new tissue formation or be adequate for the controlled release of bioactive molecules. Table 1.1 represents naturally occurring and synthetic biomaterials that possess hydrolysable bonds in the main chain. Their current medical applications are summarized in Table 1.2. Clinical applications of absorbable biomaterials have a long history similar to those of non-absorbable biomaterials. Absorbable sutures like catgut and hemostatic or sealing agents like collagen, oxidized cellulose, and α-cyanoacrylate polymers have been the front runners in the medical use of absorbable biomaterials. The largest clinical application of absorbable biomaterials at present is for suturing and ligature.

The technical term "degradable" that has been used for biomaterials has several synonyms including biodegradable, absorbable, bioabsorbable, and resorbable. In general, the term "degradable" is used when the molecular weight (MW) of the polymer constructing the material does decrease over time. When such degradation takes place only in a biological environment where enzymes exist, the term "biodegradable" is preferably used because the degradation of material is a result of enzymatic, biological action. When a material disappears by being taken into another body, we call the phenomenon "absorption". In this case a decrease in MW is not a necessary condition. A good example for this is alginate which is water soluble but becomes water-insoluble upon addition of divalent cations such as Ca^{2+}. This water-insoluble gel recovers to a water-soluble sol when a high concentration of monovalent, sodium ion is present. If this sol–gel transition takes place through ion exchange in the living body, one can say that the alginate–Ca^{2+} complex has been "absorbed" into the body without a decrease in MW of the starting alginate. As this example suggests, the term "absorbable" (or resorbable) seems to be more suitable than "degradable" so far as their medical use is concerned, because absorption includes both chemical degradation (either by passive hydrolytic or by enzymatic cleavage) and physicochemical absorption (through simple physical dissolution into aqueous media). Here the term "absorbable" (or "bioabsorbable") will be mostly

Table 1.1
Hydrolyzable bonds in bioabsorbable macromoecules and the representatives

Bond	Chemical Structure	Representative Polymers
Ester	$-\overset{\overset{\displaystyle O}{\|\|}}{C}-O-$	PGA, PLA, Poly(malic acid)
Peptide	$-\underset{\underset{\displaystyle H}{\|}}{N}-\overset{\overset{\displaystyle O}{\|\|}}{C}-$	Collagen, fibrin, synthetic polypeptides
Glycoside	$\overset{-O}{\underset{-C}{>}}C-O-C\overset{C-}{\underset{O-}{<}}$	HAc, alginate, starch, chitin, chitosan
Phosphate	$-\overset{\overset{\displaystyle O}{\|\|}}{\underset{\underset{\displaystyle O}{\|\|}}{P}}-O-$	Nucleic acid
Anhydride	$-\overset{\overset{\displaystyle O}{\|\|}}{C}-O-\overset{\overset{\displaystyle O}{\|\|}}{C}-$	$+C\overset{\overset{\displaystyle O}{\|\|}}{}-\langle\rangle-O-CH_2)_4-\overset{\overset{\displaystyle O}{\|\|}}{C}-O+_n$
Carbonate	$-O-\overset{\overset{\displaystyle O}{\|\|}}{C}-O-$	$+CH_2-CH_2-O-\overset{\overset{\displaystyle O}{\|\|}}{C}-O+_n$
Orthoester	$\overset{-C}{\underset{-O}{>}}C\overset{O-}{\underset{O-}{<}}$	
Carbon–carbon	$-CH_2-\underset{\underset{\displaystyle O-}{\overset{\displaystyle \|}{C=O}}}{\overset{\overset{\displaystyle CN}{\|}}{C}}-$	Poly(isobutyl cyanoacrylate)
Phosphazene	$-\overset{\|}{\underset{\|}{P}}=N-$	Polydiaminophosphazene

Table 1.2
Medical applications of bioabsorbable polymers

I. For surgical operation
   1. Suturing, stapling
   2. Adhesion, covering
   3. Hemostasis, sealing

II. For implantation
   1. Adhesion prevention
   2. Bone fixation
   3. Augmentation
   4. Embolization, stenting

III. For drug delivery
   1. Sustained release
   2. Drug targeting

used. The term "bioabsorbable" is used simply because the absorption proceeds in the biological environment.

Specific bulk and surface properties—including mechanical strength, absorption kinetics, wettability, and cell adhesion—are required for the biomaterials that are used for tissue engineering. Large numbers of both biologically derived and synthetic materials that meet these requirements have been extensively explored in tissue engineering. These biomaterials can be categorized according to several schemes. Here their categorization is based on their source.

2.1. Natural Polymers

The origin of naturally occurring polymers is human, animals, or plants. Materials from natural sources such as collagen derived from animal tissues have been considered to be advantageous because of their inherent properties of biological recognition, including presentation of receptor-binding ligands and susceptibility to cell-triggered proteolytic degradation and remodeling. However, the biologically derived materials have concerns, especially complexities associated with purification, sustainable production, immunogenicity, and pathogen transmission. Apart from this fact, medical applications of absorbable natural polymers are limited, because their mechanical strength is not strong enough when hydrated. One exception is chitin (and chitosan) that is a crystalline polymer. Most of natural polymers are soluble in aqueous media or hydrophilic. Because water-soluble polymers are not appropriate as a 3-D scaffold, they should be converted into water-insoluble materials by physical or chemical reactions.

2.1.1. Proteins

The major source of naturally derived proteins has been bovine or porcine connective tissues from peritoneum, blood vessels, heart valves, and intestine. These connective tissues have excellent mechanical properties; reconstructed collagen sponge is incomparable to natural collagenous tissues with respect to the tensile properties. Additional drawbacks of naturally derived materials include possible risks of prion such as bovine spongiform encephalitis (BSE), immunogenicity of eventually remaining cells, their remnants, and biopolymers themselves. However, many of natural polymers can promote cell attachment owing to the presence of cell adhesion sequence.

Collagens
Collagen is the most abundant protein within the ECM of connective tissues such as skin, bone, cartilage, and tendon. At least 20 distinct types of collagen have been identified. The primary structural collagen in mammalian tissues is type I collagen (or collagen I). This protein has been well characterized and is ubiquitous across the animal and plant kingdom. Collagen contains a large number of glycine (almost 1 in 3 residues, arranged every third residue), proline and 4-hydroxyproline residues.

A typical structure is -Ala-Gly-Pro-Arg-Gly-Glu-4Hyp-Gly-Pro-. Collagen is composed of triple helix of protein molecules which wrap around one another to form a three-stranded rope structure. The strands are held together by both hydrogen and covalent bonds, while collagen strands can self-aggregate to form stable fibers. Collagen is naturally degraded by metalloproteases, specifically collagenase, and serine proteases, allowing for its degradation to be locally controlled by cells present in tissues.

Allogeneic and xenogeneic, type I collagens have been long recognized as a useful scaffold source with low antigenic potential. Bovine type I collagen has perhaps been the biological scaffold most widely studied due to its abundant source and its history of successful use. Type I collagen is extracted from the bovine or porcine skin, bone, or tendon through alkaline or enzymatic procedures. Most of the telopeptide portion present at the end of collagen molecule with antigenic epitopes is removed during the extraction processes. The low mechanical stiffness and rapid biodegradation of the extracted but untreated collagen have been crucial problems that limit the use of this biomaterial as scaffold. Since crosslinking is an effective method to improve the biodegradation rate and the mechanical property of collagen, crosslinking treatments have become one of the most important issues for collagen technology. Two kinds of crosslinking methods are known for collagen: chemical and physical. The physical methods that do not introduce any potential cytotoxic chemical residues include photooxidation, UV irradiation, and dehydrothermal treatment (DHT). Chemical methods are applied generally when higher extents of crosslinking than those provided by the physical methods are needed. The reagents used in the chemical crosslinking include glutaraldehyde (GA), water-soluble carbodiimide (WSC) such as 1-ethyl-3-(3-dimethylaminopropyl)-carbodiimide (EDAC), hexamethylene diisocyanate, acyl azides, glycidyl ethers, polyepoxidic resins, and so on. A bifunctional reagent bridging amino groups between two adjacent polypeptide chains through Schiff base formation, GA, is the most predominant choice for collagen crosslinking because of its water solubility, high crosslinking efficiency, and low cost, although GA is a potentially cytotoxic aldehyde. Carbodiimides (CDIs) have been widely used for activation of carboxyl groups of natural polymers under the acidic conditions that are necessary for protonation of the CDI nitrogens, leading to nucleophilic attack of the carboxylate anion at the central carbon to form an initial *O*-acylisourea. The EDAC has been called a "zero-length crosslinker" because it catalyzes the intermolecular formation of peptide bonding in collagen without becoming incorporated. This crosslinking reaction results in formation of water-soluble urea as only one by-product. If both unreacted EDAC and urea are thoroughly rinsed from the material, the concern over the release of toxic residuals, commonly associated with other chemical crosslinking agents, will be reduced.

Crosslinking by UV and DHT does not introduce toxic agents into the material, but both the treatments partially give rise to denaturation of collagen. When collagen fiber scaffolds for ligament tissue engineering are crosslinked by DHT, in combination with CDI, approximately 50% of these implants rupture prior to neoligament formation, due possibly to the collagen denaturation caused by DHT. Although physical crosslinking can avoid introducing potential cytotoxic residues, most of the physical treatments do not yield high enough crosslinking to meet the demand for collagen as scaffold. Therefore, chemical treatments are applied in many cases with use of traditional GA, WSC, and other methods. Such chemical crosslinking has been shown to reduce biodegradation of collagen.

Traditional collagen crosslinking reagents may impart some degree of cytotoxicity, caused by the presence of unreacted functional groups or by the release of those groups during enzymatic degradation of the crosslinked protein. Furthermore, the chemical reaction that occurs between amine and/or carboxylic acid groups must be averted in the case of *in situ* crosslinking of cell-seeded gels. Methods have been developed that allow for collagen materials to directly crosslink without incorporation of crosslinking reagents. To date, the recognized mechanisms for strengthening collagen constructs in the presence of cells are nonenzymatic glycation and enzyme-mediated crosslinking techniques, thereby enhancing mechanical strength while remaining benign toward the cells.

A number of studies have shown that the major antigenic determinants of collagen are located within the terminal regions. In still other cases, evidence has been presented to suggest that central determinants also play a role in collagen–antibody interactions. Collagens are treated with proteolytic enzymes to remove the terminal telopeptides. However, in some cases, telopeptide remnants persisting following pepsin treatment (Fig. 1.3) [1] have been shown to be sufficiently large so that the antigenic activity of the pepsin-treated and native forms is almost indistinguishable. Further detailed study may be needed to characterize the human immune response to xenogeneic collagen.

Although several commercial skin products are based on bovine type I collagen which has been licensed for clinical use by the Food and Drug Administration (FDA), this would seem not to be a good long-term solution for clinical use. The possible risk of virus and prion tends to reduce the use of collagen in tissue engineering, but the use of collagen in tissue engineering likely still continues because of its excellent properties as scaffold. Development of good assay kits for virus check and reasonable consideration of risk/benefit balance are required for the safe application of collagen. A big challenge at the moment is to produce inexpensive human recombinant collagen that is completely free of any virus and prion.

Besides collagen, elastin plays a major role in determining the mechanical performance of some native tissues. Elastin fibers can extend 50–70% under physiological loads, and depending on the location of the vessel, elastin content can range from 33 to 200% that of collagen. In native tissues, elastin exists in stable fibers that resist both hydrolysis and enzymatic digestion.

Gelatin
Gelatin is obtained by denaturing collagen by heating animal tissues including bone, skin, and tendon. Alkaline pretreatment of collagen, which converts asparagine and glutamine residues to their respective acids, produces acidic gelatin with isoelectric

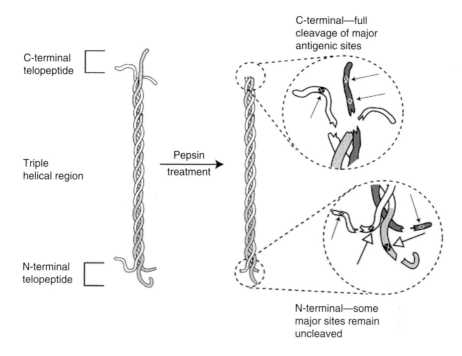

C-terminal—full
cleavage of major
antigenic sites

C-terminal
telopeptide

Triple
helical region

Pepsin
treatment

N-terminal
telopeptide

N-terminal—some
major sites remain
uncleaved

Fig. 1.3 Telopeptide removal from collagen via pepsin treatment.

points below 7, while extraction with diluted acid or enzymes yields basic gelatin with isoelectric points higher than 7. Gelatin is a heterogeneous mixture of single and multistranded polypeptides, each with extended left-handed proline helix conformations and containing between 300 and 4000 amino acids. The triple helix of type I collagen extracted from skin and bones is composed of two a1 (I) and one a2 (I) chains, each with MW of ~95 kDa, width of 1.4 nm, and length of 290 nm. Gelatin consists of mixtures of these strands together with their oligomers and breakdown (and other) polypeptides. Solutions undergo coil–helix transition followed by aggregation of the helices by the formation of collagen-like right-handed triple-helical proline/hydroxyproline (OHP)-rich junction zones. Higher levels of these pyrrolidines result in stronger gels. Each of the three strands in the triple helix requires 25 residues to complete one turn; typically there would be between one and two turns per junction zone. Gelatin films containing greater triple-helix content swell less in water and are consequently much stronger.

Gelatin has been used for a range of medical applications including adhesion prevention because of good processability, transparency, and bioabsorbability. Aqueous solution of gelatin sets to a gel through hydrogen bonding below room temperature and recovers to the sol state upon raising the temperature to destroy the hydrogen bonding. This reversible sol–gel transition facilitates the molding of gelatin into definite shapes such as block and microsphere, but chemical crosslinking is required when its dissolution in aqueous media at the body temperature should be avoided. When applied as a scaffold of cells or a carrier of growth factors, gelatin needs to be permanently crosslinked. Glutaraldehyde has most frequently been used for chemical crosslinking of gelatin to link lysine to lysine, similar to other proteins.

Fibrin
Fibrin is a product of partial hydrolysis of fibrinogen by the enzymatic action of thrombin. Upon crosslinking it converts to gel. This is called "fibrin glue" or "surgical adhesive". Human fibrin adhesives are approved and available in most major geographical regions of the world. Fibrin is applied to patients as a liquid and solidifies shortly thereafter *in situ*. Furthermore, fibrin gel can be readily infiltrated by cells, because most migrating cells locally activate the fibrinolytic cascade.

Silk Fibroin
Due to their high strength, native silk proteins from silkworm have been used in the medical field as suture material for centuries. Undesirable immunological problems attributed to the sericin protein of silk limited the use of silk in the last two decades. However, purified silk fibroin, which remains after removal of sericin, exhibits low immunogenicity and retains many of the attributes of native silk fibers. This has sparked a renewed interest in the use of silk fibroin as a biomaterial. Silk fibroin has unique properties that meet many of the demands for scaffolds. Silk exhibits high strength and flexibility and permeability to water and oxygen. In addition, silk fibroin can be molded into fibers, sponges, or membranes, making silk a good substrate for biomedical applications such as implant biomaterials, cell culture scaffolds, and cell carriers.

2.1.2. Polysaccharides

In addition to proteinous materials, naturally occurring polysaccharides and their derivatives have been employed for scaffold fabrication. Among them are hyaluronic acid (HAc), chitin, chitosan, alginate, and agarose. A prominent feature common to most of polysaccharides is the lack of cell-adhesive motifs in the molecules. This makes these biopolymers suitable as biomaterials for fabrication of a scaffold whose interaction with cells should be minimized. An exception is chitosan which has basic NH_2 groups.

Hyaluronic Acid
Industrially, HAc is obtained from animal tissues such as umbilical cord, cock's comb, vitreous body, and synovial fluid. Biotechnology also produces HAc on a large scale. HAc is an only non-sulfated glycosaminoglycan (GAG) that is present in all connective tissues as a major constituent of ECM and plays pivotal roles in wound healing. As shown in Fig. 1.4, this linear, non-adhesive polysaccharide consists of repeating disaccharide units (β-1,4-D-glucuronic acid and

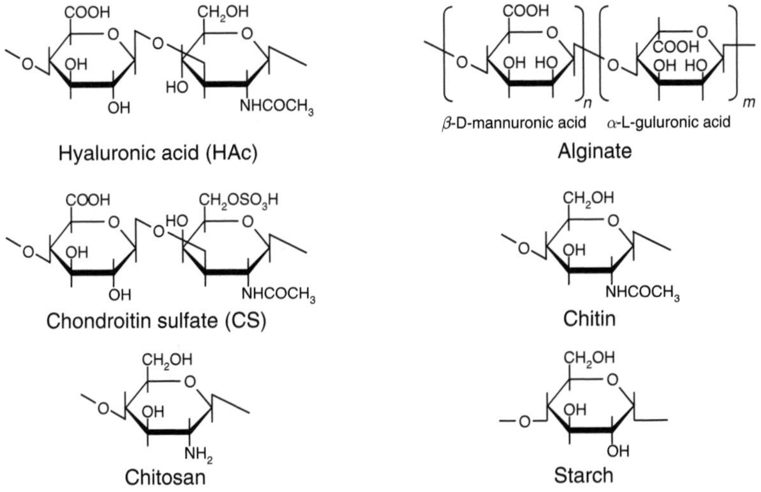

Fig. 1.4 Chemical structure of polysaccharides used for tissue engineering.

β-1,3-*N*-acetyl-D-glycosamine) with weight–average MWs (Mw) up to 10,000 kDa. This anionic polymer is also a major constituent of the vitreous (0.1–0.4 mg/g), synovial joint fluid (3–4 mg/ml) and hyaline cartilage, where it reaches approximately 1 mg/g wet weight. Serum HAc levels range from 10 to 100 μg/l, but are elevated during disease. Clearance of HAc from the systemic circulation results in a half-life of 2.5–5.5 min in plasma.

In solution, HAc assumes a stiffened helical configuration due to hydrogen bonding, and the ensuing coil structure traps approximately 1000-fold weight of water. The highly viscous aqueous solutions thus formed give HAc unique physicochemical and biological properties that make it possible to preserve tissue hydration, regulate tissue permeability through steric exclusion, and permit joint lubrication. In the ECM of connective tissues, HAc forms a natural scaffold for binding other large GAGs and proteoglycans (aggrecans), which are maintained through specific HAc–protein interactions. Consequently, HAc plays important roles in maintaining tissue morphologic organization, preserving extracellular space, and transporting ions, solutes, and nutrients. Along with ECM proteins, HAc binds to specific cell surface receptors such as CD44 and RHAMM. The resulting activation of intracellular signaling events leads to cartilage ECM stabilization, regulates cell adhesion and mobility, and promotes cell proliferation and differentiation. The HAc signaling takes place also during morphogenesis and embryonic development, modulation of inflammation, and in the stimulation of wound healing. In correspondence with these functions, HAc is a strong inducer of angiogenesis, although its biological activity in tissues has been shown to depend on the molecular size. High MW native-HAc (n-HAc) has been shown to inhibit angiogenesis, whereas degradation products of low MW stimulate endothelial cell proliferation and migration. Oligosaccharide HAc fragments (o-HAc) have been shown to induce angiogenesis in several animal models as well as within *in vitro* collagen gels. HAc is naturally hydrolyzed by hyaluronidase, allowing cells in the body to regulate the clearance of the material in a localized manner.

Owing to its unique physicochemical properties, unmodified HAc has been widely used in the field of visco-surgery, visco-supplementation, and wound healing. However, the poor mechanical properties of this water-soluble polymer and its rapid degradation *in vivo* have precluded many clinical applications. Therefore, in an attempt to obtain materials that are more mechanically and chemically robust, a variety of covalent crosslinking via hydroxyl or carboxyl groups, esterification, and annealing strategies have been explored to produce insoluble HAc hydrogels. For example, HAc-esterified materials, collectively called "Hyaff™", are prepared by alkylation of the tetrabutylammonium salt of HAc with an alkyl or benzyl halide in dimethyl formamide solution. Crosslinked HAc has been prepared using divinyl sulfone, 1,4-butanediol diglycidyl ether, GA, WSC, and a variety of other bifunctional crosslinkers. However, the crosslinking agents are often cytotoxic small molecules, and the resulting hydrogels have to be extracted or washed extensively to remove traces of unreacted reagents and by-products.

Alginate

Alginates are linear polysaccharide derived primarily from brown seaweed and bacteria. They are block copolymers composed of regions of sequential (1-4)-linked β-D-mannuronic acid monomers (M-blocks), regions of α-L-guluronic acid (G-blocks), and regions of interspersed M and G units (Fig. 1.4). The length of the M- and G-blocks and sequential distribution along the polymer chain varies depending on the source of alginates. These biopolymers undergo reversible gelation in aqueous solution under mild conditions through interaction with divalent cations including Ca^{2+}, Ba^{2+}, and Sr^{2+} that can cooperatively bind between the G-blocks of adjacent alginate chains creating ionic interchain bridges. This highly cooperative binding requires more than 20 G-monomers.

Gels can also be formed by covalently crosslinking alginate with adipic hydrazide and poly(ethylene glycol) (PEG) using standard CDI chemistry. Ionically crosslinked alginate hydrogels do not specifically degrade but undergo slow uncontrolled dissolution. Mass of the alginate-Ca^{2+} is lost through ion exchange of calcium followed by dissolution of individual chains, which results in loss of mechanical stiffness over time. Alginates are easily processed into any desired shape with the use of divalent cations. One possible disadvantage of using alginates is its low and uncontrollable *in vivo* degradation rate, mainly due to the sensitivity of the gels towards calcium chelating compounds (*e.g.*, phosphate, citrate, and lactate). Several *in vivo* studies have shown large variations in the degradation rate of calcium-crosslinked sodium alginates. Hydrolytically degradable form of alginate and an alginate derivative, polyguluronate, are oxidized alginate and poly(aldehyde guluronate), respectively.

Chondroitin Sulfate

Chondroitin sulfate (CS) is composed of repeating disaccharide units of glucuronic acid and *N*-acetylgalactosamine with a sulfate group and a carboxyl group on each disaccharide (Fig. 1.4). Chondroitin sulfate is a constituent of ECM, contributing to the functionality of the extracellular network. The cartilage ECM consists of type II collagen and proteoglycans including aggrecan, which are responsible for the tissue's compressive and tensile strength, respectively. Chondroitin sulfate forms the arms of the aggrecan molecule in cartilage.

Chitosan and Chitin

Chitosan is a linear polysaccharide of (1-4)-linked D-glucosamine and *N*-acetyl-D-glucosamine residues derived from chitin, which is found in arthropod exoskeletons (Fig. 1.4). The degree of *N*-deacetylation of chitin usually varies from 50 to 90% and determines the crystallinity, which is the greatest for 0 and 100% *N*-deacetylation. Chitosan is soluble in dilute acids which protonate the free amino groups. Once dissolved, chitosan can be gelled by increasing the pH or extruding the solution into a non-solvent. Chitosan derivatives and blends have also been gelled via GA crosslinking, UV irradiation, and thermal variation. Chitosan is degraded by lysozyme, and the kinetics of degradation is inversely related to the degree of crystallinity. Figure 1.5 shows the dependence of resorption of chitin on the hydrolysis extent when partially hydrolyzed chitin (or partially acetylated chitosan) is subcutaneously implanted in rat [2]. In contrast with 100% homopolymeric chitosan, partially hydrolyzed chitin or partially acetylated chitosan and chitin are absorbable and high in the tensile strength, but it seems that clear evidence has not yet been presented regarding its safety, especially when implanted in the human body.

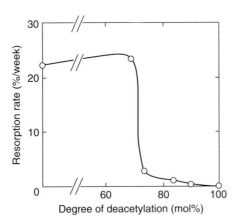

Fig. 1.5 Dependence of the initial resorption rate on films of chitin and its deacetylated derivatives on the degree of deacetylation.

2.1.3. *Natural Composite—ECM*

The native ECM provides a substrate containing adhesion proteins for cell adhesion, and regulates cellular growth and function by presenting different kinds of growth factors to the cells. The ECM is a complex structural protein-based entity surrounding cells within mammalian tissues. Most normal vertebrate cells cannot survive unless they are anchored to the ECM. In tissues and organs, major ECM components are structural and functional proteins, glycoproteins, and proteoglycans arranged in a unique, specific 3-D ultrastructure, as illustrated in Fig. 1.6. Each tissue or organ has its own unique set and content of these biomolecules. In skin, the collagen:elastin ratio is about 9:1, whereas in an artery this ratio is 1:1 averaging all artery layers, and 1:9 when considering the lamina elastica only. In ligaments, the collagen:elastin ratio is also 1:9, and in lung about 1:1. Likewise, the amount and type of GAGs, another major ECM component, varies from matrix to matrix. For instance, in cartilage CS is the major GAG making up 20% of the dry weight. In skin, dermatan sulfate is the most abundant (about 1% of the dry weight), whereas in the vitreous body of the eye the major GAG is hyaluronate.

Natural ECMs are gels composed of various protein fibrils and fibers interwoven within a hydrated network of GAG chains. In their most elemental function, ECMs thus provide a structural scaffold that, in combination with interstitial fluid, can resist tensile (via the fibrils) and compressive (via the hydrated network) stresses. In this context it is worth mentioning just how small a proportion of solid material is needed to build mechanically quite robust structures. Structural ECM proteins include collagens—some of which are long and stiff and thus serve structural functions whereas others serve connecting and recognition functions—and elastin, which forms an extensive crosslinked network of elastic fibers and sheets. The anisotropic fibrillar architecture of natural ECMs has apparent consequences

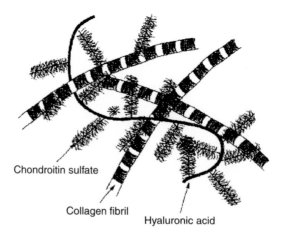

Fig. 1.6 Component arrangement in ECM (cartilage).

for cell behavior. Because of a tight connection between the cytoskeleton and the ECM through cell surface receptors, cells sense and respond to the mechanical properties of their environment by converting mechanical signals into chemical signals. Consequently, the biophysical properties of ECMs influence various cell functions, including adhesion and migration. Moreover, the fibrillar structure of matrix components brings about adhesion ligand clustering, which has been demonstrated to alter cell behavior. Structural ECM features, such as fibrils and pores, are often of a size compatible with cellular processes involved in migration, which may influence the strategy by which cells migrate through ECMs.

Natural ECMs modulate tissue dynamics through their ability to locally bind, store, and release soluble bioactive ECM effectors such as growth factors to direct them to the right place at the right time. When many growth factors bind to ECM molecules through, for example, electrostatic interactions to heparan sulfate proteoglycans, it raises their local concentration to levels appropriate for signaling, localizes their morphogenetic activity, protects them from enzymatic degradation, and in some cases may increase their biological activity by optimizing receptor–ligand interactions. Growth factors are required in only very tiny quantities to elicit a biological response.

The macromolecular components of natural ECMs are degraded by cell-secreted and cell-activated proteases, mainly by matrix metalloproteases (MMPs) and serine proteases. This creates a dynamic reciprocal response, with the ECM stimulating the cells within it and cellular proteases remodeling the ECM and releasing bioactive components from it.

With the discovery that ECM plays a role in the conversion of myoblasts to myotubes and that structural proteins such as collagen and GAGs are important in salivary gland morphogenesis it became obvious that ECM proteins serve many functions including the provision of structural support and tensile strength, attachment sites for cell surface receptors, and as a reservoir for signaling factors that modulate such diverse host processes as angiogenesis and vasculogenesis, cell migration, cell proliferation and orientation, inflammation, immune responsiveness, and wound healing. Stated differently, the ECM is a vital, dynamic, and indispensable component of all tissues and organs and is a nature's scaffold for tissue and organ morphogenesis, maintenance, and reconstruction following injury.

Until the mid 1960s the cell and its intracellular contents, rather than ECM, was the focus of attention for most cell biologists. However, ECM is much more than a passive bystander in the events of tissue and organ development and in the host response to injury. The distinction between structural and functional proteins is becoming increasingly blurred. Domain peptides of proteins originally thought to have purely structural properties have been identified and found to have significant and potent modulating effects upon cell behavior. For example, the RGD (R: arginine; G: glycine; D: aspartic acid) peptide that promotes adhesion of numerous cell types was first identified in the fibronectin (FN) molecule; a molecule originally described for its structural properties. Several other peptides have since been identified in "dual function" proteins including LN, entactin, fibrinogen, types I and VI collagen, and vitronectin. The discovery of cytokines, growth factors, and potent functional proteins that reside within the ECM characterized it as a virtual information highway between cells. The concept of "dynamic reciprocity" between the ECM and the intracellular cytoskeletal and nuclear elements has become widely accepted. The ECM is not static. The composition and the structure of the ECM are a function of location within tissues and organs, age of the host, and the physiologic requirements of the particular tissue. Organs rich in parenchymal cells, such as the kidney, have relatively little ECM. In contrast, tissues such as tendons and ligaments with primarily structural functions have large amounts of ECM relative to their cellular component. Submucosal and dermal forms of ECM reside subjacent to structures that are rich in epithelial cells (ECs) such as the mucosa of the small intestine and the epidermis of the skin. These forms of ECM tend to be well vascularized, contain primarily type I collagen and site-specific GAGs, and a wide variety of growth factors.

Collagen types other than type I exist in naturally occurring ECM, albeit in much lower quantities. These alternative collagen types each provide distinct mechanical and physical properties to the ECM and contribute to the utility of the intact ECM as a scaffold for tissue repair. Type IV collagen is present within the basement membrane of all vascular structures and is an important ligand for endothelial cells, while type VII collagen is an important component of the anchoring fibrils of keratinocytes to the underlying basement membrane of the epidermis. Type VI collagen functions as a "connector" of functional proteins and GAGs to larger structural proteins such as type I collagen, helping to provide a gel-like consistency to the ECM. Type III collagen exists within selected submucosal ECMs, such as the submucosal ECM of the urinary bladder, where less rigid structure is demanded for appropriate function. The relative concentrations and orientation of these collagens to each other provide an optimal environment for cell growth *in vivo*. This diversity of collagen within a single material is partially responsible for the distinctive biological activity of ECM scaffolds and is exemplary of the difficulty in re-creating such a composite *in vitro*, although the translation of the ECM functions to the therapeutic use of ECM as a scaffold for tissue engineering applications has been attempted.

The ECM of the basement membrane that resides immediately beneath ECs such as urothelial cells (UCs) of the urinary bladder, endothelial cells of blood vessels, and hepatocytes of the liver is comprised of distinctly different collections of proteins including LN, type IV collagen, and entactin. All ECMs share the common features of providing structural support and serving as a reservoir of growth factors and cytokines. The ECMs present these factors efficiently to resident cell surface receptors, protect the growth factors from degradation, and modulate their synthesis. In this manner, the ECM affects local concentrations and biological activity of growth factors and cytokines and makes the ECM an ideal scaffold for tissue repair and reconstruction.

The GAGs are also important components of ECM and play important roles in binding of growth factors and cytokines, water retention, and the gel properties of ECM. The heparin binding properties of numerous cell surface receptors and of many growth factors [*e.g.*, fibroblast growth factor (FGF) family, vascular endothelial growth factor (VEGF)] make the heparin-rich GAGs extremely desirable components of scaffolds for tissue repair. The GAG components of the small intestine submucosa (SIS) consist of the naturally occurring mixture of CSs A and B, heparin, heparin sulfate, and HAc. To date it will need time-consuming labor or much money to obtain a large amount of purified components of the natural ECMs and reconstruct an ECM from the purified components.

Since ECM plays an important role in a tissue's mechanical integrity, crosslinking of ECM may be an effective means of improving the mechanical properties of tissues. The ECM crosslinking can result from the enzymatic activity of lysyl oxidase (LO), tissue transglutaminase, or nonenzymatic glycation of protein by reducing sugars. The LO is a copper-dependent amine oxidase responsible for the formation of lysine-derived crosslinks in connective tissue, particularly in collagen and elastin. Desmosine, a product of LO-mediated crosslinking of elastin, commonly is used as a biochemical marker of ECM crosslinking. The LO-catalyzed crosslinks that are present in various connective tissues within the body—including bone, cartilage, skin, and lung—are believed to be a major source of mechanical strength in tissues. Additionally, the LO-mediated enzymatic reaction renders crosslinked fibers less susceptive to proteolytic degradation.

2.2. Synthetic Polymers

Before the prion shock, naturally derived materials had attracted much attention because of their natural origin which seemed to guarantee the biocompatibility. However, reports on the Creutsfeld-Jacobs disease due to the implanted sheets made from human dried dura mater diverted the focus of biomaterial scientists to non-biological materials such as synthetic polymers. Synthetic materials have long been applied for replacements of tissues and organs, fulfilling some auxiliary functions, especially improving comfort and the well-being of patients. Further possibilities exist now for synthetic materials to create tissues and organs with controlled mechanical properties and well-defined biological behavior. In the biomaterial area, there are two kinds of synthetic polymers, non-absorbable and absorbable. Non-absorbable polymers have been used as key materials for artificial organs, implants, and other medical devices. In most cases, absorbable polymers are not adequate as the major component of permanent devices, since absorption or degradation of materials in these applications has a meaning almost identical to the material deterioration which is an undesirable, negative concept for biomaterials in permanent use. Widespread clinical use of silicone, poly(ethylene terephthalate) (PET), polyethylene, polytetrafluoroethylene (PTFE), and poly(methyl methacrylate) (PMMA) as important components of artificial organs and tissues is owing to the excellent chemical stability or non-degradability in the body. If these materials undergo degradation more or less in the body, this will definitely raise a serious concern because one cannot deny that degradation by-products might evoke untoward reactions in the body. If these stable polymers exhibit deterioration over time, this might not always be due to hydrolysis but due to attack by active oxygens generated in the body as a result of inflammation.

For simplicity, bioabsorbable, synthetic polymers are here classified into three groups: poly(α-hydroxyacid)s, synthetic hydrogels, and others.

2.2.1. Poly(α-hydroxyacid)s [Aliphatic α-polyesters or Poly(α-hydroxyester)s]

The majority of bioabsorbable, synthetic polymers that are currently available is poly(α-hydroxyacid)s that have repeating units of–O-R-CO-(R: aliphatic) in the main chain. This is mainly because most of them have the potential to produce scaffolds with sufficient mechanical properties and some of them have been approved by the U.S. FDA for a variety of clinical applications as absorbable biomaterials with biosafe degradation by-products. By contrast, aromatic polyesters with phenyl groups in the main chain do not undergo any appreciable degradation in physiological conditions. The monomers used for synthesis of poly(α-hydroxyacid)s include glycolic acid (or glycolide), and L- and DL-lactic acid (or L- and DL-lactide) with a hydroxyl group on the α carbon. These monomers can yield not only homopolymers but also copolymers when polymerized together with other monomers such as ε-caprolactone (CL), p-dioxanone, and 1,3-trimethylene carbonate (TMC). Chemical structures of α-hydroxyacid polymers, copolymers, and their monomers are shown in Fig. 1.7.

Homopolymers
The most widely used absorbable sutures are made from polyglycolide (PGA) or poly(glycolide-co-lactide) (PGLA) with a glycolide (GA)/L-lactide (LLA) ratio of 90/10. This PGLA with the 90% content of GA is included in PGA here, because PGLA with a GA/LLA ratio of 90/10, which is commercially available as a multifilament suture and a Vicryl mesh (Vicryl, Ethicon, USA), exhibits properties quite similar to PGA (100% GA polymer). Poly(L-lactide) (PLLA) has been clinically used after molding into pin, screw, and mini-plate for fixation of fractured bones and maxillofacial defects of patients. Both PGA and PLLA are crystalline polymers which can provide medical devices with excellent mechanical properties, but PGA degrades mostly too quickly while PLLA degrades too slowly for use as scaffold. Nevertheless, both of them have primarily been chosen as polymers for scaffold fabrication in numerous studies worldwide.

Non-woven PGA fabrics have extensively been used as a scaffold material for cell growth in the effort to engineer many types of tissues. However, scaffolds fabricated from PGA fibers lack sufficient dimensional stability to allow molding into distinct shapes and degrade rapidly to disturb processing of this material after exposure to aqueous media. To overcome these problems, the PGA fabrics are often dipped in solution of polylactide (PLA), followed by evaporation of the solvent to deposit stiff PLA coating on the fabrics. In general, cell adhesion onto such blended materials is influenced by the polymer component existing at the

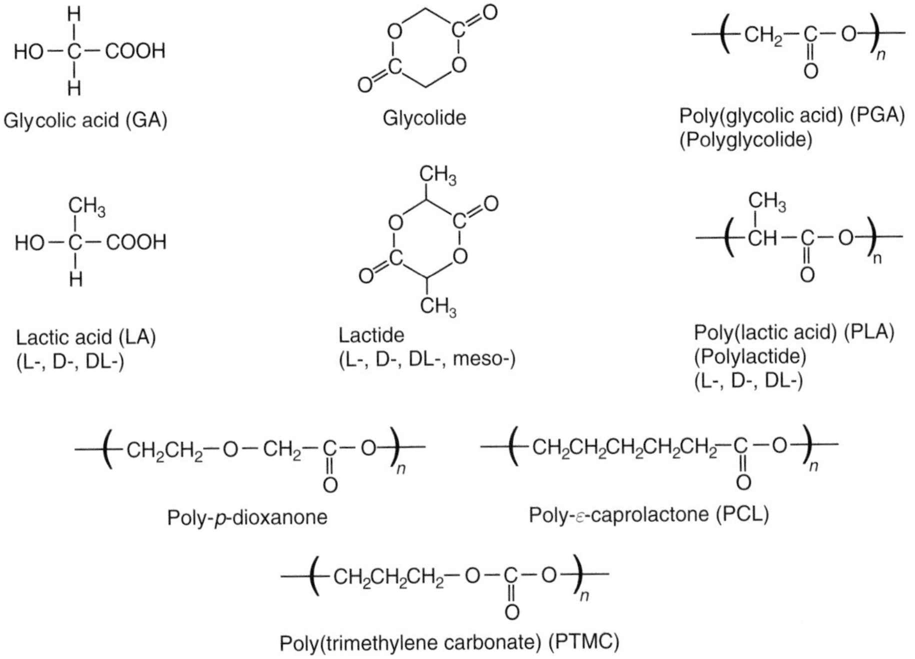

H
|
HO—C—COOH
|
H

Glycolic acid (GA)

Glycolide

$\left(CH_2-\underset{\underset{O}{\|}}{C}-O\right)_n$

Poly(glycolic acid) (PGA)
(Polyglycolide)

CH₃
|
HO—C—COOH
|
H

Lactic acid (LA)
(L-, D-, DL-)

Lactide
(L-, D-, DL-, meso-)

$\left(\underset{\underset{CH_3}{|}}{CH}-\underset{\underset{O}{\|}}{C}-O\right)_n$

Poly(lactic acid) (PLA)
(Polylactide)
(L-, D-, DL-)

$\left(CH_2CH_2-O-CH_2-\underset{\underset{O}{\|}}{C}-O\right)_n$

Poly-*p*-dioxanone

$\left(CH_2CH_2CH_2CH_2CH_2-\underset{\underset{O}{\|}}{C}-O\right)_n$

Poly-*ε*-caprolactone (PCL)

$\left(CH_2CH_2CH_2-O-\underset{\underset{O}{\|}}{C}-O\right)_n$

Poly(trimethylene carbonate) (PTMC)

Fig. 1.7   Chemical structure of *α*-hydroxyacid polymers, copolymers, and their monomers.

outermost surface, while the degradation kinetics is the simple addition of kinetics of each component degradation.

It takes a longer time than a few years for homopolymers of LLA and *ε*-CL to be completely absorbed, whereas TMC homopolymer degrades too quickly in the presence of water. Homopolymers of poly(D,L-lactide) (PDLLA) are bioabsorbed at a little higher rate than PLLA, because of the absence of crystalline regions. In case there is no need to distinguish between PLLA and PDLLA, the term PLA will be used below to include PLLA and PDLLA.

Copolymers
The aliphatic copolyester that has the largest clinical application is poly(LA-*co*-GA) (PLGA) mostly with an LLA/GA ratio around 50/50. This copolymer has clinically been used as a carrier of peptide drugs for their sustained release. In many cases, copolymers are preferred for scaffold fabrication because of their more versatile, physicochemical properties. Figure 1.8 shows the decrease in tensile strength in buffered phosphate solution (PBS) versus the hydrolysis time for various aliphatic polyesters (Table 1.3) in the fiber form [3]. Copolymerization of monomers A and B offers a great potential for modifications of polymers A or B, by controlling the physical and biological properties of bioabsorbable polymers, such as degradation rate, hydrophilicity, mechanical properties, and *in vivo* shrinkage.

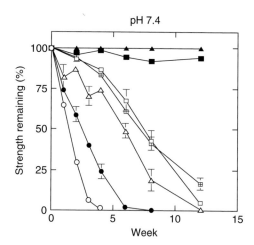

Fig. 1.8   Tensile strength change of monofilament sutures upon immersion in buffer solution of pH 7.4 at 37°C. Open circles, MONOCRYL; open squares, PDS II; open triangles, MAXON; solid circles, BIOSYN; crossed squares, P(LA/CL); solid squares, ETHILON; and solid triangles, PROLENE.

Assume that homopolymer A is absorbed while homopolymer B exhibits no or insignificant absorption in the body. Blending of homopolymers A and B does not change the degradation kinetics of each homopolymer, but copolymerization of monomer B with monomer A converts homopolymer B to a bioabsorbable polymer component. The scheme is illustrated in Fig. 1.9 for the equimolar copolymerization of monomers A and B. The average continuous sequence of each monomer in the copolymer chain is governed by polymerization conditions such as initiator and temperature. If the continuous sequence of monomer B in the copolymer chain is shorter than a critical length below which oligomers B are soluble or dispersible in aqueous media, the copolymer A–B becomes bioabsorbable owing to the degradation of monomer A unit. A typical example is

Table 1.3

Chemical structure of suture materials

Suture	Chemical Composition (wt Ratio)
MONOCRYL®	glycolide:ε-caprolactone = 75:25
MAXON®	glycolide:trimethylene carbonate = 67.5:22.5
BIOSYN®	glycolide:dioxanone:trimethylene carbonate = 60:14:26
PDS II®	p-dioxanone = 100
P(LA/CL)	L-lactide:ε-caprolactone = 80:20
ETHILON®	caprolactam = 100
PROLENE®	propylene = 100

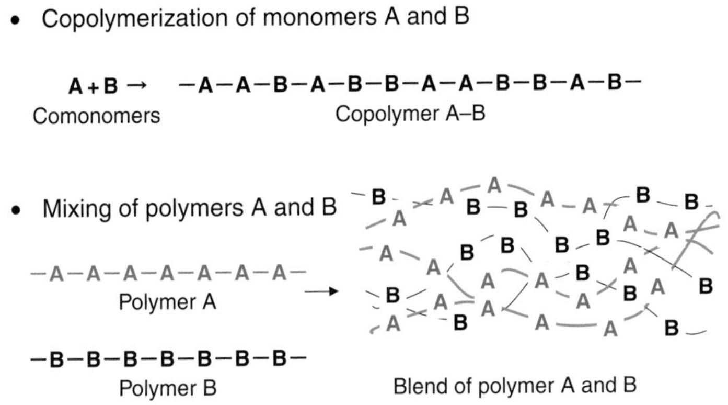

- Copolymerization of monomers A and B

A + B →   −A−A−B−A−B−B−A−A−B−B−A−B−
Comonomers   Copolymer A–B

- Mixing of polymers A and B

−A−A−A−A−A−A−A−
Polymer A

−B−B−B−B−B−B−B−
Polymer B   Blend of polymer A and B

Fig. 1.9   Difference between copolymer A–B and blend of polymer A and polymer B.

LA–CL copolymers that are absorbed in the body at rates higher than LA homopolymer and CL homopolymer that is virtually non-absorbable. Furthermore, this copolymerization converts the brittle LA homopolymer into much more rubber-like, tough polymer. Figure 1.10 shows how copolymerization of LLA with CL yields polymers with low Young's moduli [4] and high resorption rates. The DLLA copolymerization with CL will produce copolymers with properties different from those of LLA–CL copolymers, since long sequences dominating in the LLA chains will result in small crystallite formation by associating together, in marked contrast with DLLA sequences that do not have any potential to crystallize.

Copolymerization of DLLA or CL with TMC has also been attempted. The TMC homopolymer of high MW is an amorphous elastomer that shows good mechanical performance, combining high flexibility with high tensile strength, but degrades very slowly at pH 7.4 and 37°C. High MW copolymers of TMC and DLLA with 20–50 mol% of TMC are amorphous, relatively strong elastomers, can maintain mechanical properties up to 3 months at *in vitro* degradation, and are absorbed in less than a year. In contrast, copolymers of TMC and CL degrade more slowly than TMC–DLLA copolymers. The TMC–CL copolymers with high contents of CL are semi-crystalline, very flexible, and tough, so that they can maintain mechanical properties for more than 1 year when incubated in buffer solution at pH 7.4 and 37°C.

In addition to A–B type copolymers, A–B–A type triblock copolymers have been actively synthesized. For instance, PEG–PLA–PEG triblock copolymers can be synthesized as follows. In the first step, diblock copolymers of methoxy(Me).PEG–PLA are prepared by ring-opening polymerization of LA in the presence of Me.PEG–OH. Then, the resultant diblock copolymer is reacted with hexamethylene diisocyanate (HMDI) at a high temperature to connect the two diblock copolymer chains. The

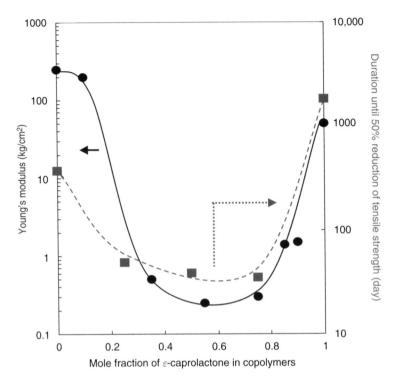

Fig. 1.10   Young's modulus and hydrolysis time until to 50% reduction of tensile strength for LLA-ε-CL copolymer plotted against the mole fraction of ε-CL comonomer.

reasoning for block copolymerization of LA with PEG is to reduce the surface hydrophobicity of PLA scaffolds or make suspension in aqueous media.

Many bioabsorbable polymers like PLA and PGA lack functional groups which facilitate further functionalization. Therefore, poly(LA-*co*-lysine) was synthesized to immobilize peptides on the pendant amino groups onto the lysine side chains, while carboxyl-functionalized PLA was prepared by copolymerizing side-chain-protected β-alkyl-α-malate and subsequent deprotection of the carboxyl groups by hydrogenation or alkaline treatment.

2.2.2. Hydrogels

Crosslinking of water-soluble polymer chains produces water-insoluble 3-D networks. The product is called "hydrogel". Several distinctive features, including tissue-like viscoelasticity, diffusive transport, and interstitial flow characteristics, make synthetic hydrogels excellent physicochemical mimetics of natural ECMs and candidates for soft tissue scaffolds. Indeed, hydrogels have a potential to efficiently encapsulate cells and high water contents to allow for nutrient and waste transport. Soft tissues of our body are also like a hydrogel because of their high water contents ranging between 70 and 90%. The structural integrity of hydrogels depends on crosslinks formed between polymer chains via physical, ionic, or covalent interactions. Synthetic hydrogels can be processed under relatively mild conditions, have structural properties similar to ECM, and can be delivered in the body in a minimally invasive manner, particularly by injection. A variety of synthetic materials including PEG, poly(vinyl alcohol) (PVA), and poly(acrylic acid) may be used to form hydrogels. These polymers are not biodegradable but water soluble unless crosslinks are introduced. Thus, once the introduced crosslinks are broken, the resulting polymer chains become again water soluble, being absorbed in the body, followed by excretion from kidney or bile duct, depending on the molecular size. Naturally derived polymers including collagen, gelatin, fibrinogen, albumin, polypeptides, HAc, CS, agarose, alginate, and chitosan can also provide hydrogels.

Poly(ethylene glycol) or poly(ethylene oxide) (PEO) with the same chemical structure as PEG is one of the synthetic polymers that is most commonly used for hydrogel fabrication in tissue engineering and is currently approved by FDA for several applications. Each end of PEG chains can be modified with either acrylates or methacrylates to facilitate photocrosslinking. When the modified PEG is mixed with an appropriate photoinitiator and crosslinked via UV exposure hydrogels are formed. This photo-induced method to produce hydrogels facilitates diverse, minimally invasive applications via arthroscopy/endoscopy or subcutaneous injection for tissue replacement or augmentation. Thermally reversible hydrogels have also been created from block copolymers of PEG and PLLA.

Although *in situ* forming hydrogels offer advantages as cell delivery carriers, cells do not usually adhere to highly hydrated gels because of no built-in cell-adhesive ligands and limited protein adsorption, so that the cells encapsulated inside gels are present in a "blank" environment wherein there is little to no interaction of integrins and other cell surface receptors with the gels.

2.2.3. Others

Besides aliphatic polyesters and hydrogels, a number of synthetic polymers have been used for scaffolds. The chemical structures of these polymers are shown in Fig. 1.11. Only polyurethanes (PUs) and polyphosphazens will be described here, because these two polymers have been studied by different research groups.

Polyurethanes
A considerable majority of tissue engineering literature has utilized a narrow array of polyester scaffolds although their mechanical properties are limited with respect to high strain, elastic capabilities. As a result, interest has been paid on developing bioabsorbable elastomers with high strain, elastic capabilities, and mechanical strength. The impact of mechanical forces on tissue development is increasingly appreciated in efforts to engineer load-bearing and mechanically responsive tissues. The polymers applicable for these tissues should address the requirement

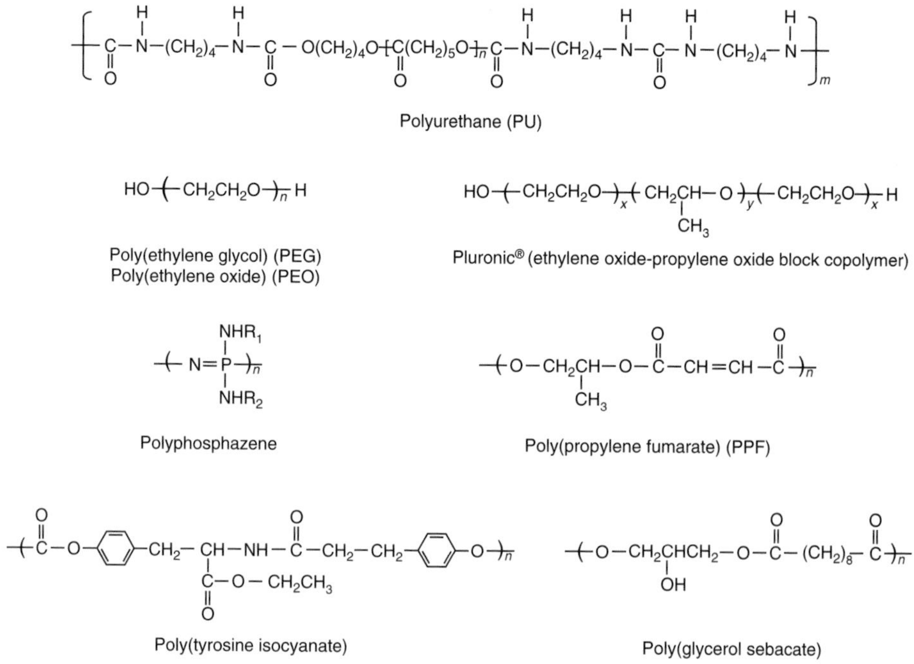

Polyurethane (PU)

Poly(ethylene glycol) (PEG)
Poly(ethylene oxide) (PEO)

Pluronic® (ethylene oxide-propylene oxide block copolymer)

Polyphosphazene

Poly(propylene fumarate) (PPF)

Poly(tyrosine isocyanate)

Poly(glycerol sebacate)

Fig. 1.11 Chemical structure of synthetic, absorbable polymers used for tissue engineering (except for α-hydroxyacid polymers).

to transmit mechanical cues to the tissues over the course of tissue development, regardless of *in vitro* mechanical training or dynamic *in vivo* movement. Segmented PU elastomers potentially meet these requirements, because these polymers have been used for elastic devices such as indwelling catheters, intra-aortic balloons, and left ventricular assist devices.

Segmented PUs are basically synthesized from a polymer with a terminal hydroxyl group at both chain ends (HO–P–OH) and an excess of diisocyanate (OCN–R–NCO). Upon mixing them under an excess of diisocyanate, a prepolymer having terminal isocyanate groups is produced under formation of urea bonds between –OH and –NCO. Addition of a diamine chain extender to this prepolymer increases the chain length to form segmented PU. The P portion associates to yield a soft segment while the R portion forms a hard segment. If the P portion comprises hydrolysable units, the resultant PU exhibits bioabsorption.

Polyphosphazenes
Polyphosphazenes comprise a large class of macromolecules with alternating phosphorus and nitrogen atoms in the backbone and a wide variety of organic, inorganic, or organometallic side groups. Some linear polyphosphazenes undergo rapid hydrolytic degradation, but exhibit poor mechanical properties.

2.3. Inorganic Materials—Calcium Phosphate

Biological apatite found naturally in bone comprises a range of minerals and has osteoconductive and osteophilic properties. Inorganic minerals have been used for cell scaffolding as well. Among them are hydroxyapatite (HAp) [$Ca_{10}(PO_4)_6(OH)_2$] and β-tricalcium phosphate (β-TCP) [β-$Ca_3(PO_4)_2$]. Synthetic HAp is very slowly absorbed in the body by attack of osteoblasts, whereas β-TCP undergoes absorption probably both with and without such biological assist. A precursor of biological apatite in bone tissue is octacalcium phosphate (OCP) [$Ca_8H_2(PO_4)_65H_2O$]. The OCP is absorbed at a higher rate than HAp and β-TCP. A large advantage of HAp as scaffold is that this is a natural component of bones and bioactive (or bioconductive). The inherent brittleness of calcium phosphates limits their use to the exclusive graft material in non-load-bearing defects.

Coral with a chemical composition of calcium carbonate ($CaCO_3$) has an advantage of good communication between pores and is readily converted to HAp, resulting in a high affinity for cells with tissue-regenerating ability. A porous material prepared from coral skeletons is an optimal scaffold for bone regeneration, but the mechanical properties of the composite in the hydrate state are much inferior to those of the natural bone. In addition, coral plays an important role in the marine environment and has now become difficult to obtain, because its use is controlled by environmental regulations such as the Washington Treaty.

2.4. Composite Materials

Inorganic scaffolds are too brittle and not amenable to trimming at operation table, when prefabricated as porous structure. This issue has led some investigators to fabrication of composites from inorganic and organic materials. A well-known example is a composite from HAp and soft collagen, which is a biomimetic from a compositional point of view. Owing to its pliability one can trim the porous composite with scissors into a desired shape.

The PLA, PGA, and some other synthetic polymers have an established safety record in humans, but are not osteoconductive. Apatite-coated polymer composites combine the osteoconductive property of bioceramics and the mechanical resilience of organic polymers. Polymeric scaffolds impregnated with apatite promote *in vitro* cellular attachment and bone nodule formation, as well as bone formation *in vivo*, providing an appropriate osteogenic environment for tissue engineering. Such bioabsorbable and bioactive composites have been developed, combining resorbable polymers with calcium phosphates, bioactive glasses, or glass-ceramics in various scaffold architectures. Besides imparting bioactivity to a polymer scaffold, the addition of bioactive phases to bioabsorbable polymer may alter polymer degradation behavior. Bioactive phases in bioabsorbable polymer allow rapid exchange of protons in water for alkali in the glass or ceramic, which may provide a pH buffering effect at the polymer surface, reducing acceleration of acidic degradation products from the polymer. A further advantage of using bioactive inorganic particles for composite preparation with absorbable polymers is increased mechanical properties owing to the filler effect.

3. PORE CREATION IN BIOMATERIALS

The 3-D architecture of scaffolds with proper pore size, pore shape, alignment, and interconnectivity can have significant effects on regulating the tissue-specific morphogenesis of cultured cells. The rate of tissue ingrowth increases as the porosity and the pore size of the implanted devices increase. The transport of molecules through pores is a function of their size, connectivity, and tortuosity. Thus, the architecture of scaffolds will significantly affect nutrient and oxygen transport within the 3-D matrix, which may directly affect the cell motility during tissue regeneration. The porous structure of scaffold is characterized by two parameters, pore size and porosity. The average pore diameter must be large enough for cells to migrate through the pores and small enough to retain a critical total surface area for appropriate cell binding. To allow for transport of cells and metabolites the scaffold must have a high specific surface and large pore volume fraction in addition to an interconnected pore network. Scaffold pore size has been shown to influence cell adhesion, growth, and phenotype. The optimal scaffold pore size that allows maximal entry of cells as well as cell adhesion and matrix deposition varies with different cell types. Generally, the optimal pore size ranges from 100 to 500 μm and the optimal porosity is above 90%. Various bone engineering groups have noted that pore sizes of greater than 300 μm have a greater penetration of mineralized tissue in comparison with smaller pore sizes, while at pore sizes of 75 μm, hardly any mineralized tissue is found within the scaffold. High porosity enables maximal conversion of cells and tissue invasion necessary for construction, together with conduits suitable for blood vessel formation.

Scaffolds of hydrogel type such as collagen gel and fibrin glue have no geometrical pores recognizable with SEM, but nutrients and oxygen are delivered to cells via physical diffusion through the aqueous medium. The term "channel" may be more appropriate for characterizing the pore structure, because interconnected pores form a channel. However, fabrication of scaffolds with typical channel structures is extremely tedious [5].

A variety of techniques have been developed to fabricate porous scaffolds, of which some include woven and non-woven fiber-based fabrics, solvent casting, salt (or particulate) leaching, temperature-induced phase separation, gas foaming with pressurized carbon dioxide, melt molding, high-pressure processing, membrane lamination, forging, injection molding, pressing, inkjet printing, fused deposition modeling, electrospinning, rapid prototyping, among others. A simple method is to use absorbable fibers as a starting material, because fibers can yield a variety of "porous" products including woven or knitted cloth, non-woven fabric, web, mesh, felt, and fleece. Conventional spinning can produce either multi-filaments or monofilaments. A representative example is products from PGA which has very few specific solvents but readily yields fibers by melt spinning. To produce aggregates of PGA fine fibers by means of electrospinning, we need expensive solvents such as trifluoropropyl alcohol for preparing the stock solution for electrospinning of PGA.

Another simple method available in small labs for porous scaffold fabrication is to use the solution of absorbable polymers. One can readily obtain a porous sheet or a 3-D block when a polymer solution is subjected to freeze drying. When a polymer solution is mixed with porogens such as NaCl particulates followed by drying and removal of the porogens from the dried material, we obtain a porous scaffold with the pore size determined by the particulate size. Supercritical liquids have also been applied to fabrication of porous materials. More sophisticated methods applied for scaffold formation are imprinting of polymers and photo-polymerization of monomers along with computer-aided machines. The 3-D printing is a novel technique and the application of this technique for tissue engineering is still limited. These modern technologies, together with computer programming, allow production of scaffolds with very regular and ultrafine structures. The control over scaffold architecture using these fabrication techniques is highly process driven and not design driven.

3.1. Phase Separation (Freeze Drying)

A widely used method for preparation of porous scaffolds is the thermally induced phase separation, in which the solution temperature is lowered to induce phase separation of the homogeneous polymer solution. The phase separation mechanism may be liquid–liquid demixing, which generates polymer-poor and polymer-rich liquid phases. The subsequent growth and coalescence of the polymer-poor phase would develop to form pores in scaffolds. On the other hand, when the temperature is low enough to allow freeze of the solution, the phase separation mechanism would be solid–liquid demixing, which forms frozen solvent and concentrated polymer phases. By adjusting the polymer concentration, using different solvents, or varying the cooling rate, phase separation could occur via different mechanisms, resulting in the formation of scaffolds with various morphologies. "Freeze drying" is one of the most extensively used methods that produce matrices with porosity greater than 90%. The pore sizes depend on the growth rate of ice crystals during the freeze-drying process.

After removal of the liquid or frozen solvent contained in the demixed solution, the space originally occupied by the solvent would become pores in the prepared scaffolds. Obviously, in the stage of solvent removal, the porous structure contained in the solution needs to be carefully retained. Without freeze drying, a rise in temperature during the drying stage could result in remixing of the phase-separated solution or remelting of the frozen solution, leading to destruction of the porous structure. Thus, the reason of using freeze drying to remove solvent is quite obvious: to keep the temperature low enough that the polymer-rich region would not redissolve and possesses enough mechanical strength to prevent pore collapse during drying. Although freeze drying is a widely used method to prepare porous scaffolds, it is time consuming and energy consuming. For instance, it often takes 4 days to remove solvent by freeze drying. During this 4-day period, a lot of energy is consumed to keep vacuum and to maintain the low temperature needed for drying. Another problem encountered in the application of freeze drying is the occurrence of surface skin. During the freeze-drying stage, if the temperature is not controlled low enough, the polymer matrix in the demixed solution would not be rigid enough to resist the interfacial tension caused by the evaporation of solvent. Thus, the porous structure collapses and dense skin layers occur in the prepared scaffold.

In the case of hydrogels, the freeze-dry processing does not require additional chemicals, relying on the water already present in hydrogels to form ice crystals that can be sublimated from the polymer, creating a particular micro-architecture. Because the direction of growth and the size of ice crystals are a function of the temperature gradient, linear, radial, and/or random pore directions and sizes can be produced with this methodology.

For the freeze-drying process, the solvent vapor pressure at the drying temperature (usually very low) needs to be high enough to allow its removal. Dimethylcarbonate can be also used as a solvent because it exhibits high vapor pressure and melting point around $0°C$ which makes it suitable for sublimation. Because of a solubility parameter close to that of dioxane, this solvent might be convenient as an alternative to dioxane (potentially carcinogen) for freeze drying of poly(α-hydroxyacid)s. Dimethyl sulfoxide (DMSO) cannot be used as the solvent for PLGA for the freeze-drying process due to its low vapor pressure. This limitation of choosing solvent may be lifted when the scaffolds are prepared by a freeze-extraction method. Freeze extraction and freeze gelation are of a type of phase separation. These methods also fix the porous structure under freezing condition, but in the subsequent drying stage the freeze-drying process is not needed. The principle of the freeze-extraction method is to remove the solvent by extraction with a non-solvent. After the removal of solvent, the space originally occupied by the solvent is taken by the non-solvent and the polymer is then surrounded with the non-solvent. Under this circumstance, even at room temperature, the polymer would not dissolve. Hence, drying can be carried out at room temperature to remove the non-solvent, leaving space that becomes pores in the scaffold. For the freeze-extraction process, DMSO can be used as the solvent for preparation of PLGA scaffolds, because it can be easily extracted out by ethanol aqueous solution.

3.2. Porogen Leaching

The porogen leaching method also has been widely used to prepare porous scaffolds, because of easy operation and accurately controlling pore size and porosity. The salt leaching technique consists of adding salt particulates to a polymer solution. The overall porosity and level of pore connectivity are regulated by the ratio of polymer/salt particulates and the size of the salt particulates.

3.3. Fiber Bonding

Fiber mesh consists of individual fibers either woven or knitted into 3-D patterns of variable pore sizes. They can be obtained by deposition of a polymer solution over a non-woven mesh of another polymer, and the subsequent evaporation. The advantages of fiber meshes are the large surface area for cell attachments and the rapid diffusion of nutrients, but they are not suitable for the fine control of porosity.

3.4. Gas Foaming

The gas-foaming technique uses high-pressure CO_2 gas processing. The porosity and pore structure depend on the amount of gas dissolved in the polymer, the rate and type of gas nucleation, and the diffusion rate of gas molecules through the polymer to the pore nuclei. However, this technique often produces a scaffold that is too compact for the cells. This may be improved by associating the gas foaming with salt leaching. Another technique derived from improvement of gas foaming consists of the substitution of high-pressure CO_2 with ammonium bicarbonate, which acts also as porogen. In this case the porosity depends only on the amount of salt particulates added, whereas the pore diameter is due to the size of the salt crystals.

3.5. Rapid Prototyping

Rapid prototyping (RP) or solid free-form fabrication (SFF) refers to a group of technologies that build a physical, 3-D object in a layer-by-layer fashion. The RP is a subset of mechanical processing techniques which allow highly complex structures to be built as a series of thin two-dimensional (2-D) slices using computer-aided design (CAD) and computer-aided manufacturing (CAM) programs. These techniques essentially allow researchers to predefine properties such as porosity, interconnectivity, and pore size. The RP methodologies include stereolithography (SLA), selective laser sintering, ballistic particle manufacturing, and 3-D printing. Examples of RP applied to generate tissue engineering scaffolds include laser sintering to fabricate Nylon-6 scaffolds, fused deposition molding of poly(ε-caprolactone) (PCL) scaffolds, and SLA of PU patterns. Direct and indirect SLA have been used to make ceramic scaffolds and cancellous bone structure models. Few studies address the generation of scaffolds by RP techniques for soft tissue engineering such as 3-D hydrogel scaffolds. SLA, one of the most common types of RP, operates by selectively shining a laser beam onto a vat of liquid photopolymer. This method has been employed in numerous biomedical applications such as building models of biological structures, bone substrate scaffolds, and heart valve scaffolds. Using an RP technique to pattern not only the scaffold but also cells, will accelerate and improve tissue assembly. On the basis of this idea, various RP methods have emerged, but they involve time-consuming sequential writing processes with a narrow processing window and relatively low resolution. The SFF fabrication technologies create 3-D structures in laminated fashion from numerical models. Commercially available SFF technologies, such as fused deposition modeling, SLA, and selective laser sintering have been utilized to fabricate scaffolds. Computer-aided engineering has also emerged to fabricate multifunctional scaffolds with control over scaffold composition, porosity, macroarchitecture, and mechanical properties based on optimization models. To print tissue engineering scaffolds, inkjet solvent binder is printed onto a powder bed of porogens and polymer particles. The solvent will dissolve the polymer and evaporate, and the polymer will re-precipitate to form solid structures. The final porosity is achieved after particulate leaching and solvent removal. Figure 1.12 shows a custom-designed fiber-deposition device [6]. Hollister reviewed how integration of computational topology design and SFF has made scaffolds with designed characteristics possible [7].

3.6. Electrospinning

Electrospinning provides a mechanism to produce nanofibrous scaffolds from synthetic and natural polymers, with high porosity, a wide distribution of pore diameter, high surface-area-to-volume ratio, and morphological similarities to natural collagen

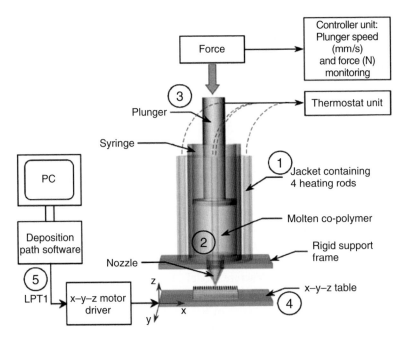

Fig. 1.12 The 3-D deposition device consisting of five main components: (1) a thermostatically controlled heating jacket; (2) a molton copolymer dispensing unit consisting of a syringe and nozzle; (3) a force-controlled plunger to regulate flow of molten copolymer; (4) a stepper motor driven x–y–z table; and (5) a positional control unit consisting of stepper-motor drivers linked to a personal computer containing software for generating fiber deposition paths.

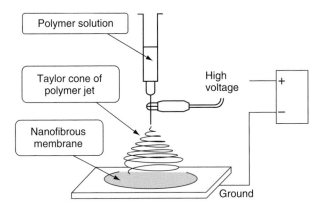

Fig. 1.13 Electrospinning apparatus of polymer nano-fibrous membranes.

fibrils. Fiber diameters are in the range from several micrometers down to less than 100 nm. The electrospinning process is based on a fiber spinning technique driven by a high-voltage electrostatic field using a polymeric solution or liquid. Figure 1.13 represents an electrospinning apparatus [8]. The underlying physics of this technique relies on the application of an electrical force, especially when at the polymer droplet surface it overcomes the surface tension force, and a charged jet is ejected. As the solvent evaporates, the charge density increases on the fibers, resulting in an unstable jet, which stretches the fibers over about one million fold. The variables controlling the behavior of the electrified fluid jet during electrospinning can be divided into fluid properties and operating parameters. The relevant fluid properties are viscosity, conductivity, dielectric constant, boiling point, and surface tension. The operating parameters include flow rate, applied electric potential, and distance between the tip and the collector called "air gap".

The final product of electrospinning generally consists of randomly interconnected webs of sub-micron size fibers. Nanofibrous scaffolds formed by electrospinning, by virtue of structural similarity to natural ECM, may represent promising structures for tissue engineering applications.

4. SPECIAL SCAFFOLDS

4.1. Naturally Derived Scaffolds

Connective tissues are typically tough and pliable in contrast with synthetic, hydrophilic polymers. The high toughness and pliability of naturally derived materials in spite of their high water contents such as 70% are due largely to ordered orientation of collagen fibrils and the presence of crosslinked elastin in the connective tissue. Some researchers have attempted to fabricate scaffolds from collagen molecules for tendon and ligament tissue engineering, but the reconstructed scaffolds do not provide such good mechanical properties as natural ones because of difficulty in assembling the collagen fibers to specific orientations. When scaffolds are derived from natural tissues, they do not need the pore formation process and can provide the optimal surface for cell adhesion because of the presence of cell adhesion sites. Matrices for tissue engineering are being derived by extraction or partial purification of whole tissue, removing some components and leaving much of the 3-D matrix structure intact, likely with growth factors as well. Acellular biological tissues have therefore long been proposed to be used as scaffolds for tissue repair and tissue regeneration. The guiding principle behind such acellular grafts is that the immunogenic response associated with allografts is sufficiently reduced through removal of the cells. Once implanted, the acellular graft serves as a natural scaffold into which surrounding cells readily migrate, forming the foundation for new tissue. These biomaterials are composed of ECM proteins and polysaccharides that are conserved among different species and can serve as scaffolds for cell attachment, migration, and proliferation. In addition to the inherent cell compatibility, biological tissues primarily maintain desired shapes and the strength of the tissues from which the materials have been derived. This can be an advantage over synthetic materials in regard to materials processing. In addition, acellular tissues may provide a natural microenvironment for host cell migration and proliferation to accelerate tissue regeneration.

4.1.1. *ECM-like Scaffolds*

Tissues are basically constructed from cells and ECM, and the cells will lose their shape and function when taken out from the native environment. An artificial 3-D polymeric scaffold must thus not only serve as a physical structural support, but also play an active important role in cell migration, growth, and vascularization throughout the 3-D architecture in both *in vitro* cell culture and *in vivo* tissue regeneration. The design and manufacture of 3-D polymeric scaffolds that mimic native ECM are thus thought to be crucial to the success of tissue engineering. The ECM exists in all tissues and organs, but can be harvested for use as a therapeutic scaffold from relatively few sources. Dermis of the skin, submucosa of the small intestine and urinary bladder, pericardium, basement membrane and stroma of the decellularized liver, and the decellularized Achilles tendon are all potential sources of ECM.

The host response to ECM-derived scaffolds is largely dependent upon the methods used to process the material. Chemical and non-chemical crosslinking of ECM proteins have been utilized extensively in an effort to modify the physical, mechanical, or immunogenic properties of naturally derived scaffolds. Chemical crosslinking generally involves aldehyde or CDI, and photochemical protein crosslinking has also been investigated. Although these crosslinking methods provide certain desirable physical or mechanical properties, the end result is the modification of a biologically interactive material into a relatively inert scaffold. There is abundant literature on the use of modified ECM scaffolds, especially chemically crosslinked biological scaffolds, for tissue repair and replacement. Porcine heart valves, decellularized and crosslinked human dermis (Alloderm™), and chemically crosslinked purified bovine type I collagen (Contigen™) are examples of such products currently available for use in humans. Similarly, modified ECM scaffolds have been used for the reconstitution of the cornea, skin, cartilage, bones, and nervous system.

The scaffold modifications typically result in inhibition of the scaffold degradation and cellular infiltration into the scaffold. Although there may be clinical uses for such modified biomaterials, these properties are counterintuitive to many approaches in the field of tissue engineering; especially, those approaches in which cells are seeded upon scaffolds prior to or at the time of implantation. In contrast, ECM scaffolds that remain essentially unchanged from the native ECM elicit a host response that promotes cell infiltration and rapid scaffold degradation, deposition of host-derived neomatrix, and eventually constructive tissue remodeling with minimum formation of scar tissue. The native ECM represents a fundamentally different material than the ECM scaffold that has been chemically or otherwise modified. In the native ECM, cells clear a path by secreting and activating proteases, such as MMPs, serine proteases, and hyaluronidase, that specifically degrade protein or proteoglycan components of the pericellular matrix. Degradation is highly localized because of the involvement of membrane-bound proteases, complexation of soluble proteases to cell surface receptors, and a tightly regulated balance between active proteases and their natural inhibitors. On the other hand, amoeboid migration is driven by cell shape adaptation (that is, squeezing through pre-existing matrix pores) and deformation of the ECM network.

Porcine-derived ECM scaffolds that have not been modified, except for the decellularization process and terminal sterilization, have been used for the repair of numerous body tissues. Immediately following implantation *in vivo*, there is an intense cellular infiltrate consisting of equal numbers of polymorphonuclear leukocytes and mononuclear cells (MNCs). By 3 days postimplantation, the infiltrate is almost entirely MNCs in appearance with early evidence for neovascularization. Between day 3 and 14, the number of MNCs increases, vascularization becomes intense, and there is a progressive degradation of the xenogeneic scaffold with associated deposition of host-derived neomatrix. Following day 14, the MNC infiltrate diminishes and there is the appearance of site-specific parenchymal cells that orient along lines of stress. These parenchymal cells consist of fibroblasts, smooth muscle cells, skeletal muscle cells, and endothelial cells (ECs) depending upon the site in which the scaffold has been placed. A devitalized collagen-based scaffold from small intestine submucosa (SIS) has been shown to induce site-specific regeneration in numerous tissues, including blood vessels, tendon, abdominal wall, ligaments, skin, urinary bladder, musculoskeletal repair, and dural substitute [9].

The characteristic of the intact ECM that distinguishes it from other scaffold materials is not only its diversity of structural proteins and associated bioactive molecules, but also their unique spatial distribution that makes the ECM an uncomparably tough material. As illustrated in Fig. 1.14 [10], fibrous components of connective tissues are not randomly oriented but align in specific manners. The axis of alignment of collagen fibers in native myocardium varies across

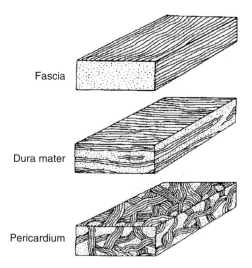

Fascia

Dura mater

Pericardium

Fig. 1.14 Orientation of collagen fibrils in connective tissues.

the thickness of the heart wall. Such mechanical anisotropy is required also for engineered heart tissue.

4.1.2. Fibrin Gel

The advantage of fibrin gels over the other gels is that it can be obtained autologous. Furthermore, cells entrapped in the fibrin gels were reported to produce more collagen and elastin than those entrapped in collagen gel. Fibrin degrades within several days by cell-associated enzymatic activities when no degradation inhibitors are used. In tissue engineering applications using cells encapsulated in fibrin gel, degradation inhibitors are often used to preserve the scaffold function of the fibrin. The effect of the inhibitor concentration on collagen formation in the gels differs among studies.

4.1.3. Matrigel™

Matrigel is commercially available from BD Biosciences (San Jose, CA, USA). This gel is a basement membrane preparation extracted from Engelbreth-Holm-Swarm mouse sarcoma and solubilized in Dulbecco's modified Eagle's medium (DMEM). This contains several components of basement membranes enriched with LN. Matrigel is liquid at 2–8°C and sets to a gel rapidly at 22–35°C.

4.1.4. Marine Natural Scaffold

Some natural biomaterials including coral, sea urchins, and marine sponges are much less expensive than the natural ECMs and appear to provide affordable, readily available scaffolds with a number of unique and suitable properties such as open interconnected channels.

4.2. Injectable Scaffolds

Large-scale, prefabricated scaffolds require invasive surgery for implantation, are difficult to contour to the defect shape, have predetermined mechanical properties, and induce poor vascularization. Therefore, injectable biomaterials such as collagen, fibrin, gelatin, alginate, HAc, chitosan, and PEG gels emerged as candidates for scaffolds in the replacement of large-size tissue defects. Injectable hydrogel materials are advantageous because they require a minimally invasive procedure and conform to irregular shapes. However, they are mechanically weaker than soft tissues and many cells perform poorly when suspended within a bioinert hydrogel.

4.3. Soft, Elastic Scaffolds

For soft tissue engineering, synthetic elastomeric materials with tunable degradation properties would be preferable. For instance, for cardiovascular applications, elastomeric behavior with low modulus would allow the transmission of stresses to seeded or infiltrating cells early in the implant or culture period. Such mechanical training has been shown to be important in developing mechanically robust tissues with appropriate cellular orientation. The PGA, PLA, and their copolymers (PLGA) are relatively stiff and nonelastic and are not ideally suited for engineering of soft tissues under a mechanically demanding environment such as cardiovascular, urological, and gastrointestinal tissue, unless specifically processed. Mechanical signals are thought to be necessary for developing the cell alignment that leads to tissue structure with correct biomechanical properties and functions. Elastic scaffolds may be essential for the engineering of any tissue under conditions of cyclic mechanical strain. This is the reason for synthesis of absorbable and pliable PUs.

4.4. Inorganic Scaffolds

The most abundant bioceramic HAp can be used as a scaffold for bone regeneration because HAp will be integrated in the regenerated bone so far the shape and size is acceptable, although the absorption rate is quite low unless osteoblast frequently attacks. However, HAp scaffolds whose pores are not interconnected but independent are not adequate for tissue engineering. Continuous pore interconnection is a prerequisite as scaffold. Modification of marine coral into HAp possessing porous structure suitable for tissue engineering has been applied for scaffold fabrication, but the use of coral-derived HAp should be refrained if the coral harvest violates the Washington Treaty. In contrast to HAp, β-tricalcium phosphate (β-TCP) exhibits rapid degradation and hence is a good candidate for inorganic scaffolds. To serve as the scaffold in which bone ingrowth and vascularization take place, β-TCP should be highly porous, but this inorganic material tends to produce a porous block with brittle and fragile structure. This drawback may be overcome by formation of composites from β-TCP and flexible polymers, which must be osteoconductive if portion of β-TCP is present on the surface of scaffold.

4.5. Composite Scaffolds

Ceramics including dense and porous HAp, TCP ceramics, bioactive glasses, and glass-ceramics have been combined with a number of polymers including collagen, chitosan, and poly(α-hydroxyacid)s. The combination of such polymers with a bioactive component takes advantage of the osteoconducting properties (bioactivity) of HAp and bioactive glasses and of their strengthening effect on polymer matrices. The composite is expected to have superior mechanical properties than the neat (unreinforced) polymer and to improve structural integrity and flexibility over brittle glasses and ceramics for eventual load-bearing applications. Composite fabrication research has focused on developing polymer/ceramics blends, precipitating ceramic onto polymer templates, coating polymers onto ceramics, or ceramics onto polymers. Polymer processing techniques including combined solvent-casting and salt-leaching, phase separation and freeze drying, and immersion-precipitation have been used for the preparation of highly porous PLA/HAp scaffolds. *In situ* apatite formation can also be induced by a biomimetic process in which polymer foams are incubated in a simulated body fluid.

5. SURFACE MODIFICATIONS

Both bulk (*e.g.*, strength and degradability) and surface properties of biomaterials are important in tissue engineering. Generally, bulk and surface properties are interwound. Often, materials are chosen for their favorable bulk properties, and their unfavorable biological interactions are improved by surface modifications. The success of tissue engineering depends on interactions at the cell–scaffold interface, including the cell adhesion and proliferation, expression and activity of regulatory signaling molecules, and biomechanical stimuli.

5.1. Cell Interactions in Natural Tissues

A highly dynamic and complex array of biophysical and biochemical signals are transmitted from the outside of a cell by various cell surface receptors and integrated by intracellular signaling pathways. The signals converge to regulate gene expression and ultimately establish cell phenotype. This indicates that the ultimate decision of a cell to differentiate, proliferate, migrate, apoptose, or perform other specific functions is a coordinated response to the molecular interactions with ECM effectors. This flow of information between cells and their ECM is bidirectional.

In multicellular organisms contacts of cells with neighboring cells and the surrounding ECM are mediated by cell adhesion receptors. Among them the integrin family comprises the most numerous and versatile groups. They are a large family of heterodimeric, cell surface molecules, and are the most prominent ECM adhesion receptors of animal cells for many of the ECM adhesion molecules. They play not only a major role in linking the macromolecules of the ECM with the cell's cytoskeleton, cell–cell adhesion, and binding to proteases, but are also important in processes like embryogenesis, cell differentiation, immune response, wound healing, and hemostasis.

Integrins consist of two non-covalently associated transmembrane subunits, termed α and β. To date 18α and 8β subunits are known, which form 24 different heterodimers. The combination of particular α and β subunits determines the ligand specificity of the integrin. Some integrins, however, are highly promiscuous, *e.g.*, the $\alpha v\beta 3$ integrin binds to vitronectin, FN, von Willebrand factor, osteopontin (OP), tenascin, bone sialoprotein, and thrombospondin. Vice versa, ECM molecules like FN are ligands for several integrins. Fibronectin is a disulfide-linked dimeric glycoprotein prominent in many ECMs and present at about 300 μg/ml in plasma. It interacts with collagen, heparin, fibrin, and cell surface receptors of the integrin family. In 1984, the tripeptide motif RGD (Fig. 1.15) [11] was identified by Pierschbacher and Rouslahti as a minimal essential cell adhesion peptide sequence in FN. The RGD peptides inhibit cell adhesion to FN on the one hand, and promote cell adhesion when they are immobilized on surfaces on the other hand. Since then, cell-adhesive RGD sites were identified in many other ECM proteins, including vitronectin, fibrinogen, von Willebrand factor, collagen, LN, OP, tenascin, and bone sialoprotein as well as in membrane proteins, in viral and bacterial proteins, and in snake venoms. About half of the 24 integrins have been shown to bind to ECM molecules in an RGD-dependent manner. Since other important adhesive motifs have been also identified, the RGD sequence is not the "universal cell recognition motif", but it is unique with respect to its broad distribution and usage. The conformation of the RGD-containing loop and its flanking amino acids in the respective proteins are mainly responsible for their different integrin affinity. Other factors that contribute to integrin-ligand binding affinity include the activation of integrins by divalent cations and cytoplasmatic proteins.

The process of integrin-mediated cell adhesion comprises a cascade of four different partly overlapping events: cell attachment, cell spreading, organization of actin cytoskeleton, and formation of focal adhesions. First, in the initial attachment step the cell contacts the surface and some ligand binding occurs that allows the cell to withstand gentle shear forces. Secondly, the cell body begins to flatten and its plasma membrane spreads over the substratum. Thirdly, this leads to actin organization into microfilament bundles, referred to as stress fibers. In the fourth step the

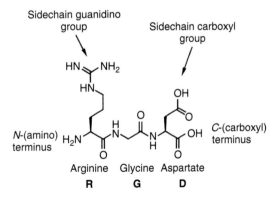

Fig. 1.15   The molecular formula and nomenclature of RGD.

formation of focal adhesions occurs, which link the ECM to molecules of the actin cytoskeleton. Focal adhesions consist of clustered integrins and more than 50 other transmembrane-associated and other cytosolic molecules. During the four steps of cell adhesion integrins are employed in physical anchoring processes as well as in signal transduction through the cell membrane.

It is well established that integrin-mediated cell spreading and focal adhesion formation trigger survival and proliferation of anchorage-dependent cells. In this context the expression of the anti-apoptotic protein Bcl-2, induced by $\alpha5\beta1$ integrins, and the suppression of the p53 pathway by focal adhesive kinase (FAK) are discussed. In contrast, loss of attachment causes apoptosis in many cell types, referred to as "anoikis" (a Greek word meaning homelessness). Anoikis can even be induced in the presence of immobilized ECM molecules when non-immobilized soluble ligands like RGD peptides are added, as illustrated in Fig. 1.16 [11]. Stable linking of RGD peptides to a surface is essential to promote strong cell adhesion, because formation of focal adhesions only occurs if the ligands withstand the cell's contractile forces. These forces are able to redistribute weakly adsorbed ligands on a surface, which leads only to weak fibrillar adhesions later on. Furthermore, cells can remove mobile integrin ligands by internalization.

5.2. Artificial Surface in Biological Environment

Because common poly(α-hydroxyester)s like PLA and PGA are hydrophobic, porous scaffolds fabricated with these polymers are floating in cell culture medium. When cells in culture medium are plated on the top of the porous scaffold

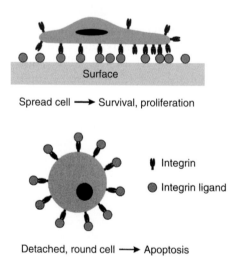

Fig. 1.16 Opposite effects of integrin ligands. Immobilized ligands act as agonists of the ECM, leading to cell adhesion and cell survival, while non-immobilized ligands act as antagonists, leading to cell deattachment, a round cell shape, and apoptosis.

or injected into its interior for seeding, the majority of its pores remain empty since the scaffold does not absorb the culture medium. Because it is crucial to obtain a uniform distribution of initial seeded cells throughout the scaffold volume for the creation of a tissue with homogeneous cellularity, some approaches have been attempted to improve hydrophilicity of polymer scaffolds and thus to ensure uniform and dense cell seeding. They include treatments of the scaffolds by prewetting with ethanol, hydrolysis with NaOH, oxidation with perchloric acid solution, oxygen or ammonia plasma discharge treatment, physical or chemical coating with some hydrophilic polymers or cell-adhesive proteins, and blending with hydrophilic polymers such as Pluronics (ethylene glycol-propylene glycol block copolymers).

When foreign materials are exposed to a biological environment, ECM proteins are non-specifically adsorbed on the surface of nearly all the materials, masking its specific surface properties, and then cells indirectly interact with the material surface through the adsorbed ECM proteins [12]. A range of proteins from the culture medium come adsorbed to the substrate surface, reflecting the relative abundance of proteins in the culture medium. However, over time, weakly adsorbed proteins will leave the surface, while those bound more strongly will remain. Other protein molecules will come in and replace those that have been desorbed. The implication of this series of events is that the final nature of the attached protein layer reflects both the surface chemistry of the underlying substrate and the composition of the culture medium. This highlights the importance of optimizing the surface chemistry of the substrate to give the best performance of the cells in a particular medium. The nature of the protein layer that is adsorbed to the underlying substrate has a profound effect on the attachment and subsequent development of the cells in culture. Not only is the mass of bound protein an issue but, more importantly, the orientation of the protein with respect to the surface may be crucial. Complexity of the cell adhesion process that spans a broad range of time and length scales no doubt contributes to some of the differences among reports of cell interaction with materials. Adhesion of anchorage-dependent mammalian cells is traditionally viewed as occurring through at least four major steps that precede proliferation: protein adsorption, cell–substratum contact, cell–substratum attachment, and cell adhesion/spreading. Interactions between cells and polymers in serum are mediated by proteins that have been either preimmobilized on the material surface, absorbed from the surrounding medium, or secreted by cells during culture. Cell contact and attachment involves gravitation/sedimentation to within 50 nm or so of a surface whereupon physico-chemical forces conspire to close the cell–surface distance gap. Attached cells then spread over the surface, depending on the compatibility with the surface typically within hours.

Functional scaffolds should be capable of eliciting specific cellular responses and directing new tissue formation. This is mediated by biomolecular recognition instead of by non-specifically adsorbed proteins, similar to the native ECM. Extensive studies have been performed to render scaffolds biomimetic by surface modification, as this is the simplest way to make biomimetic scaffolds. Scaffold with an informational function, *e.g.*, with the RGD sequence which facilitates cell attachment, must be better than non-informational synthetic polymers. These biomimetic materials potentially mimic many roles of ECM in tissues. For example, incorporation of peptide sequences renders the surface of biomaterials cell adhesive that inherently have been non-adhesive to cells. The design of biomimetic materials that are able to interact with surrounding tissues by biomolecular recognition is an attempt to make the materials such that they are capable of directing new tissue formation mediated by specific interactions, which can be manipulated by altering design parameters instead of by non-specifically adsorbed ECM proteins. For mimicking the native ECM, either long chains of ECM proteins such as FN, vitronectin, and LN or short peptide fragments composed of several amino acids along the long chain of ECM proteins can be the candidates for surface modification.

The early work on the surface modification of biomaterials with bioactive molecules has used long chains of ECM proteins for surface modification. Biomaterials can be also coated with these proteins, which usually have promoted cell adhesion and proliferation. Since the finding of the presence of signaling domains that are composed of several amino acids along the long chain of ECM proteins and primarily interact with cell membrane receptors, the short peptide fragments have been used for surface modification in numerous studies. Particularly, the use of a short peptide for surface modification is advantageous over the use of the long chain of native ECM proteins. The native ECM protein tends to be randomly folded upon adsorption to the biomaterial surface such that the receptor binding domains are not always sterically available. However, the short peptide sequences are relatively more stable during the modification process than long chain proteins such that nearly all peptides modified with spacers are available for cell binding. In addition, short peptide sequences can be massively synthesized in laboratories more economically. The biomimetic material modified with these bioactive molecules can be used as a tissue engineering scaffold that potentially serves as artificial ECM providing suitable biological cues to guide new tissue formation.

The most commonly used peptide for surface modification is RGD. Additionally, other peptide sequences such as Tyr-Ile-Gly-Ser-Arg (YIGSR), Arg-Glu-Asp-Val (REDV), and Ile-Lys-Val-Ala-Val (IKVAV) have been immobilized on various model substrates. In order to provide stable linking, RGD peptides should be covalently attached to the polymer via functional groups like hydroxyl, amino, or carboxyl groups. Simple adsorption of small RGD peptides only leads to poor cell attachment. When a polymer does not have functional groups on its surface, these have to be introduced by blending, copolymerization, or chemical or physical treatments. Different coupling techniques have been employed to ensure covalent binding of the peptides to the surface of the materials. In most cases, peptides are linked to polymers by reacting an activated surface carboxylic acid group with the nucleophilic *N*-terminus of the peptide, as shown in

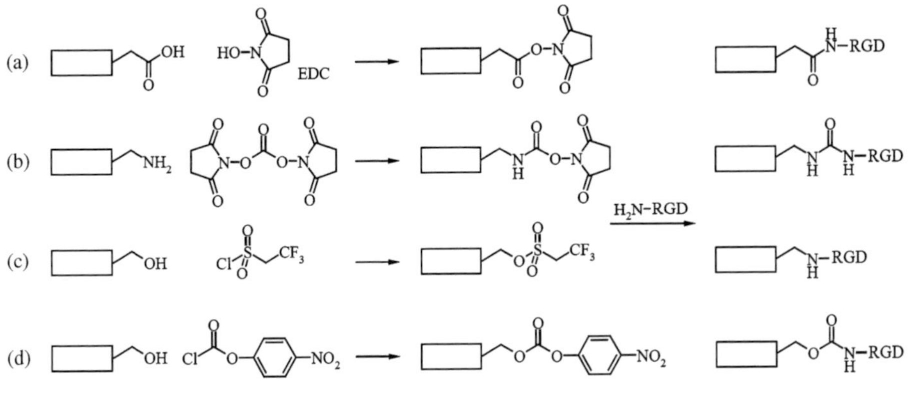

Fig. 1.17 RGD peptides react via the *N*-terminus with different groups on polymers: (a) carboxyl groups, preactivated with a carbodiimide and NHS to generate an active ester; (b) amino groups, preactivated with DSC; (c) hydroxyl groups, preactivated as tresilate; and (d) hydroxyl groups, preactivated as *p*-nitrophenyl carbonate.

Fig. 1.17a [11]. First, the surface carboxyl group is converted to an active ester, that is less prone to hydrolysis, *e.g.*, *N*-hydroxysuccinimide (NHS) ester, and, second, this is coupled to the peptide in water. Polymers that contain surface amino groups can be treated with succinic anhydride to generate surface carboxyl groups, which can be reacted with RGD peptides as described above. Amino groups can directly be converted into preactivated carboxyl groups by using an excess of bisactivated moieties like *N,N'*-disuccinimidyl carbonate (DSC), as shown in Fig. 1.17b. Surface containing hydroxyl groups can simply be preactivated with, for instance, tresyl chloride (Fig. 1.17c) or *p*-nitrophenyl chlorocarbonate (Fig. 1.17d).

A bifunctional crosslinker that has a long spacer arm can be used for the immobilization of peptides to the surface, which can enable the immobilized peptides to move flexibly in the biological environment. For polymer substrates lacking appropriate functional groups for a coupling reaction, a photochemical immobilization method has been utilized to graft cell-binding peptides. In order to examine that any cellular responses to the modified substrates are mediated solely by the immobilized peptides, the experiment is performed under serum-free conditions.

6. CELL EXPANSION AND DIFFERENTIATION

For the clinical use of tissue engineering with isolated cells, a small number of cells are initially isolated from a small biopsy from the specific body part of patients or others and then expanded in number in conventional monolayer culture before they are seeded into scaffolds. In many cases the number of harvested cells is not large enough for repairing the lost or damaged tissues, especially when used clinically in the reconstruction of large defects. Therefore, it is necessary to multiply the isolated cells for tissue engineering with a sufficient number of cells. Generally, the smaller the cell amount, the ECM formation is less. Moreover, if the density of cells seeded into scaffolds is low—in other words, the distance between the neighboring cells is long—the production of ECM such as collagen and GAG from the cells will be poor because of insufficient communication between the cells. Although the promise of tissue engineering is tremendous, it has only seldom been accomplished in humans, largely because many cells are needed to generate even small amounts of tissue. It will be often necessary to generate large amounts of tissue, starting with very few cells. To circumvent this problem, cell culture is performed to multiply cells under retention of their phenotypic characteristics either before or after seeding them into scaffold.

In cell culture the modulation of cell phenotype between the synthetic and quiescent states is important. The modulation is based on biochemical or environmental cues. A central issue in blood vessel culture is to balance the competing goals of smooth-muscle cell proliferation and ECM deposition (synthetic state or dedifferentiation), and the contractile phenotype associated with differentiation and maturation. For culture of engineered vessels *de novo*, an increased synthetic state is required, whereas at the conclusion of vessel culture, a minimally proliferative, quiescent phenotype is desired. Since cells would dedifferentiate when seeded into polymeric structures *in vitro*, in depth investigations are necessary to find out how dedifferentiation of seeded cells can be prevented.

To treat traumatic or congenital cartilage defects with tissue engineering techniques, a relatively small number of donor cells, either chondrocytes or progenitor cells, are expanded *in vitro* until sufficient cells are obtained. However, *in vitro* multiplication of chondrocytes in monolayer results in dedifferentiation of these cells. Expansion in high seeding density cultures often fails to produce sufficient chondrocytes, even after several passages. Lower seeding densities may increase cell yield, but bear the risk of decreased redifferentiation capacity. Differentiation of stem cells to target cells *in vitro* needs specific culture media.

6.1. Monolayer (2-D) and 3-D Culture

Advances in cell culture techniques have culminated in the field of tissue engineering. The most common method to increase the number of cells is 2-D monolayer cell culture on a flat substrate. The 2-D culture is excellent for cell expansion but sometimes induces the loss of native functional natures of cells. In monolayer cultures, cells are forced to grow in one plane under space-limiting conditions. This results in an obviously artificial growth environment, in contrast to development *in vivo*. Conventional monolayer cultures of chondrocytes have the disadvantage of producing matrix that differs from that produced *in vivo*, losing their typical phenotype within several days or weeks as they "dedifferentiate". Freshly isolated articular chondrocytes express cartilage-specific type II collagen and hyaline cartilage markers, aggrecan, but during prolonged culture and serial subculture these cells lose their spherical shape, begin to dedifferentiate to a fibroblast-like phenotype, and produce predominantly unspecific type I collagen. The phenotypic stability of adult human chondrocytes is lost quickly on expansion in serial monolayer cultures than that of cells of juvenile humans. This loss of phenotype in monolayer culture is reversible if chondrocytes are cultured in 3-D culture systems embedded in solid support matrices, such as collagen, agarose, or alginate gels. However, there are disadvantages to these systems, including the slow rate of proliferation and the substantial decline of matrix production in suspension cultures. Cardiomyocytes cultured in monolayers also exhibit properties different from native heart tissue, because of structural differences between 2-D and native environments, and because of the effects of cell isolation and *in vitro* cultivation.

The 3-D culture has often been stated to be preferable to 2-D culture, because most of the cells in our body are present in 3-D environments [12]. Many cellular processes including morphogenesis and organogenesis have been demonstrated to occur exclusively when cells are organized in a 3-D fashion. The fundamental process underlying most tissue engineering methodologies is 3-D culture at high cell density to enhance cell–cell interactions favorable for ECM production. The 3-D ECM culture systems have been developed to simulate natural interactions between cells and the extracellular environment. The 3-D culture maintains the cell phenotype but is poor for cell expansion. Cells in 3-D culture are surrounded with a substrate not only on one side of the cells but on many sides of the cells. Generally, the so-called "3-D scaffolds" having distinct pores of sizes much larger than the cells come in contact with cells only by their one side, as illustrated in Fig. 1.18a. Cells in right 3-D culture should be in contact

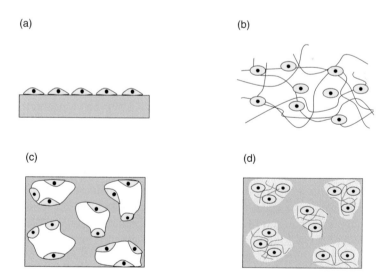

Fig. 1.18 Cells at 2-D and 3-D culture, and seeded on 3-D porous scaffold: (a) cells on flat dish (2-D culture); (b) cells in gel (3-D culture); (c) cells adsorbed on 3-D porous scaffold (2-D culture); and (d) cells entrapped in biosynthesized ECM.

with a substrate from multiple directions, as shown in Fig. 1.18b. However, the 2-D culture may change to 3-D culture once the cells begin to be surrounded by the matrix produced by the cells themselves, as shown in Fig. 1.18d. In culture of chondrocytes for cartilage repair the isolated cells have been expanded in monolayer culture, dedifferentiated, and then redifferentiated in a 3-D cell arrangement for new cartilage formation. A typical sign of dedifferentition of chondrocytes is the switch from type II collagen to type I collagen synthesis. Endothelial cells grown in 2-D systems vary from 3-D model systems. A wide variety of cell types exhibit enhanced maintenance of their differentiated phenotype if cultured in 3-D systems instead of monolayers, which is attributed to associated differences in cell shape and/or increases in intercellular communication. It was shown that neonatal rat ventricular myocytes in confluent monolayers couple on average with six cells, whereas the same cells in native ventricles couple on average with nine cells [13].

One major constraint in the use of 3-D scaffolds has been the limitation of cell migration and tissue ingrowth within these structures. Bovine aortic endothelial cells can survive or proliferate in a 2-D model in the absence of angiogenic factors; however, they die in a 3-D collagen lattice in the absence of angiogenic factors [14]. Because cells located in the interior scaffold receive nutrients only through diffusion from the surrounding media in static culture, high cell density on the exterior of the scaffold may deplete nutrient supply before these nutrients can diffuse to the scaffold interior to support tissue growth. In addition, diffusive limitations may inhibit the efflux of cytotoxic degradation products from the scaffold and metabolic wastes produced in the scaffold interior. Some examples of possible diffusive limitations of high cell density 3-D culture are summarized in Table 1.4 [15]. In a series of related studies, maximum penetration depth of osseous tissues was reported to be in the range of 200–300 μm within porous 3-D PLGA scaffold after 2 months of static culture. Although several attempts have been made to alter scaffold geometry to provide adequate diffusion within 3-D constructs with some success, in-growth limitations within 3-D scaffolds remain a pervasive problem in tissue engineering.

Some of the factors known to influence the phenotype of cells are growth factors, cell–material interactions, and mechanical stimulation, but the basic steps of cell-culturing processing, regardless of 2-D and 3-D cultures, involve sufficient nutrient transfer throughout the culture medium and sufficient removal of waste products from around the cells. Other important parameters that must be monitored are oxygen concentration and pH of the culture medium, as changing these parameters can have a positive or negative effect on the cell growth.

6.2. Cell Seeding

Seeding cells into porous scaffolds using autologous or xenogeneic serum is generally the first step of tissue engineering. Achieving the optimal cell density and

Table 1.4
Representative examples of high-density bone cell cultivation within 3-D scaffolds

Cell Type	Pore Size (mm)	Pore Volume (%)	Scaffold Thickness (mm)	Cell Density (cells/ml)	Observation
Rat marrow	300–500	90	1.9	a	Maximum osseous penetration of 240 ± 82 μm (day 56)
Rat calvaria	150–300	90	1.9	a	Maximum osseous penetration of 220 ± 40 μm (day 56)
	500–710	90	1.9	a	Maximum osseous penetration of 190 ± 40 μm at (day 56)
SaOS-2	187	30	2.5	9.5×10^7	Preferrential cell growth on scaffold exterior (day 7)
Rat marrow	300–500[b]	78	6	$5.8^b \times 10^{6\ c}$	Preferrential cell growth on scaffold exterior (day 7)

Notes
Refer to the original work for the references.
[a] Values of cell density within the scaffold could not be determined from the reference.
[b] An average pore diameter of 400 mm was assumed to facilitate model calculations.
[c] Reported values for cell and scaffold void volume of 3.5×10^6 and 0.60 cm² were used to determine cell density (cells/ml).

desired cell distribution in scaffolds is a major goal of cell seeding technologies in tissue engineering. General seeding requirements for 3-D scaffolds include (a) high yield to maximize the utilization of donor cells, (b) high kinetic rate to minimize time in suspension for anchorage-dependent cells, (c) specially uniform distribution of attached cells to provide basis for uniform tissue regeneration, (d) high initial construct cellularity to enhance the rate of tissue development, and (e) appropriate nutrient and oxygen supply to maintain cell viability during the seeding procedure. Efficiency of cell seeding is usually low and in many cases cells are found close to the surface of the scaffold, but not in the interior. In addition, time-consuming, costly, and physiologically stressful cell seeding and tissue assembly techniques would become limiting factors in the development of clinically useful tissue-engineered products.

The spatial organization of cells provides cell–cell adhesion cues that are important in directing numerous biological functions, such as embryonic tissue development, organ formation, and tissue regeneration. This spatial organization is necessary for the preservation of vital cell–cell interactions as well as cell phenotype.

6.2.1. Serum

As mentioned above, reaction of tissue-engineered constructs requires the *in vitro* propagation of human cells to increase cell numbers. The *in vitro* cell growth can be stimulated by supplementation of the growth medium with serum. So far, culture techniques have required the use of xenobiotic material—fetal calf serum (FCS). The bovine serum has been used as a rich source of mitogens for cell proliferation. However, there is a risk of viral and/or prion disease transmission when bovine serum is used in clinical cell culture. To reduce the risk of disease transmission, it is preferable to culture cells under completely defined culture conditions, without any xenobiotic or donor cell material. The ideal medium is one that contains no xenobiotic products and in which any mitogens present are recombinant.

The choice of medium for the culture of autologous cells for clinical use may be at the discretion of the consultant medical doctors in many countries. Green's medium that has been in use for some 20 years for culture of skin cells for patients with burn injuries contains xenobiotic materials. Serum for clinical use must be sourced from countries such as New Zealand or Australia, where BSE has not been found to occur. In terms of risk management, regulatory authorities would prefer that cells cultured for clinical use avoid the use of bovine and other animal-derived products, to reduce the risk of disease transmission. Thus, a "holy grail" of cell culture for clinical use is to develop an entirely defined culture system. The use of autologous human serum in tissue engineering is increasing.

Cell growth medium often contains supplements, growth factors, and serum that may have an unknown influence on cell adhesion and proliferation. For example, the inclusion of the fibroblast feeder layer for keratinocyte expansion comes from the well-documented dependence of ECs on mesenchymal cell interactions between epidermal and dermal cells via soluble factors providing important signals in regulating the re-epithelialization of wound skin. Keratinocytes regulate the expression of keratinocyte growth factor (KGF) in fibroblasts through the release of interleukin-1β (IL-1β), a proinflammatory molecule that increases fibroblast proliferation and ECM production. The effect of solid substrate on the paracrine relationship between keratinocytes and fibroblasts as modulated by KGF and IL-1β is unclear.

Although the significance of breathing ambient air (containing 21% oxygen) to our survival is obvious, physiologic "normoxia" is much lower. Therefore, the traditional paradigm of culturing cells in humidified ambient air may not be optimal for maintaining certain cell types, including stem cells. Oxygen reduction to 3–6% promotes the survival of both peripheral and central nervous system stem cells, and can influence their fate by enhancing catecholaminergic differentiation. Similarly, human hematopoietic stem cells demonstrate increased self-renewal and bone marrow repopulating capability after hypoxic treatment. Furthermore, because cartilage is a relatively avascular tissue, chondrocytes are bathed in a naturally hypoxic milieu and rely on hypoxia-induced signaling for survival. Murine marrow-derived mesenchymal stem cells (MSCs) undergo enhanced osteochondrogenic differentiation in the setting of chronic hypoxia, perhaps by returning them to a more "natural" oxygen environment. In contrast, preadipocytes do not thrive under hypoxia, and the absence of oxygen triggers a hypoxia inducible factor (HIF)-1α response that represses genes essential to adipogenic differentiation. Thus, although reduced-oxygen incubation may promote the *in vitro* expansion and/or differentiation of many cell types by mimicking *in vivo* oxygen levels, it is possible to inhibit the differentiation of others by exposing them to reduced oxygen.

6.2.2. Cell Adhesion

Cell adhesion can be divided into three grades of adhesiveness: (1) *early adhesiveness*, meaning that a cell is attached but has not spread, (2) *intermediate adhesiveness* characterized by cell spreading, but which lacks stress fibers and focal adhesions, and (3) *late adhesiveness* indicating cell spreading with stress fibers and focal adhesions. Early adhesiveness is important since it is the first step of attachment to substitutes, which results in cell growth, differentiation, viability, and spreading. Figure 1.19 diagrams the general form and characteristics of attachment and proliferation rate assays, identifying quantitative parameters extracted by statistical fitting to data. Here, $\%I_{max}$ measures the maximum number of cells that attach to a surface from a cell suspension, expressed as percent of inoculum.

Integrins that mainly mediate the biological cell adhesion are involved in processes such as development, wound healing, tumor invasion, and inflammation. Many cells appear to be capable of attaching to implant materials through integrins, and a considerable number of proteins contain the requisite RGD and (G)RGD(S) (Gly-Arg-Asp-Ser) sequences which are recognized by integrins. Serum contains abundant proteins with unknown activities in terms of cell adhesion. A high concentration of FN in serum can serve as an immediate attachment protein and may anchor cells to implants. Cell adhesion to a substrate is dependent on the chemical properties such as material composition and wettability, and physical parameters such as porosity and roughness. With the passage of time the cell can produce the ECM, and the ECM proteins in turn will attach to the substitute surface.

Cells can be also cultured successfully on a surface coated with type I collagen, but there is a concern throughout Europe that BSE cannot be detected by any *in vitro* tests and therefore it is impossible to be confident that bovine material is BSE free unless it comes from herds that have never been exposed to BSE.

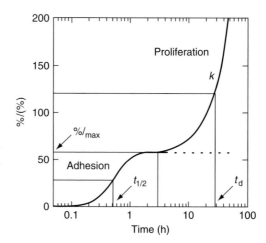

Fig. 1.19   Schematic illustrating cell adhesion and proliferation identifying quantitative parameters extracted from the variation of percentage of a cell inoculum ($\%I$) with time. $\%I_{max}$ is the maximum percentage of a cell inoculum that adheres to a surface from a sessile cell suspension and $t_{1/2}$ measures half-time to $\%I_{max}$. The proliferation rate (k) and cell-number doubling time (t_d) measure viability of attached cells.

6.2.3. Seeding Efficiency

Although there are some techniques that successfully introduce cells into biomaterials, seeding efficiencies are not yet at optimal levels, especially at low cell concentrations. Lower seeding densities affect the amount of time and resources required to obtain scaffolds ready for implantation. The systems developed for the seeding of cells onto scaffolds include from simple techniques such as static seeding, where cells and scaffolds are brought into direct contact and allowed to sit, relatively undisturbed, with the intention of cellular attachment and migration into the scaffolds, to more elaborate techniques such as pulsatile perfusion wherein medium flows under oscillatory pressures to try to mimic the natural environments. Dynamic seeding which induces medium flow within the scaffold pores and shear stress on cells seems to be the most applicable method when relatively thin (<1 mm) scaffolds with a large 2-D surface need to be seeded.

In addition to seeding a greater number of cells into scaffolds, it is important to achieve homogeneity in cellular distribution. Achieving a confluent coating on scaffold exteriors may not be ideal because of potential problems with vascular ingrowth and nutrient/waste transport. Such an exterior cellular coating is also undesirable because of cell migration: If there is no subsequent cell migration into the scaffold interior, a bioabsorbable scaffold would ultimately collapse and the goal of quicker tissue regeneration will not be achieved.

6.2.4. Assessment of Cells in Scaffolds

Strategies for investigating cell growth in scaffolds include cell viability, proliferation, and metabolic active assays. In most cases, it is unknown how many cells remain viable over time in scaffolds *in vitro* and *in vivo* without assays in which the viability of the samples must be compromised to perform the specific assessment. Various assays are available for assessing cultured cell proliferation. These include (1) mitochondrial enzyme reduction of tetrazolium salts into their respective formazon by-products [MTT (3-(4,5-dimethylthiazol-2-yl)-2,5-diphenyl tetrazo-zolium bromide) and MTS]; (2) cellular redox indicators (Alamar blue); (3) ATP quantification through biolumines-cence; (4) S-phase incorporation of radioactively labeled DNA precursors such as [^{3}H]thymidine and bromodeoxyuridine (BrDU); and (5) co-staining with a fluorescent DNA-specific dye for live cells (Hoechst 33258 and PicoGreen), and physical counting (hemocytometer). These assays are well established in characterizing cell proliferation in 2-D monolayer cultures of low cell densities.

The most common methods to visualize cells on biomaterials have been limited to light microscopy or use of dyes that compromises the viability of the cells. Conventional techniques, such as histomorphometry, electron microscopy, and Fourier transform infrared (FT-IR) imaging, are capable of giving information on tissue development, but they require that scaffolds be destructively sectioned. Other problems include the dissolution of polymeric scaffolds by the organic solvents used in the preparation of thin sections or the disruption of early mineral deposits by aqueous solvents. Some investigators reported on the successful application of confocal microscopy to study cells growing in the pores of scaffolds, but the depth penetration of this technique is limited and it cannot be applied to scaffolds that are optically opaque. What is required is a nondestructive technique that can provide spatially resolved, chemically specific, tissue-level information about the tissue formed within the pores of scaffolds.

DNA binding dyes coupled with confocal microscopy have been used to demonstrate the coverage of cells on materials. In a similar method, 3-D cell distribution on a material can be demonstrated with the fluorescence of enhanced green fluorescent protein (EGFP) and confocal microscopy without the need for special preparation of the cells. This approach facilitates visual assessment of cells as they exist in tissues and is not dependent on cell markers. Since EGFP is stably produced by the cells, no staining or cellular manipulation is required. Reporter genes have been utilized for a variety of applications ranging from gene expression and regulation to determination of efficiency of gene vector delivery. The technique allows the tracking of stem cells as they differentiate or become specialized. A reporter gene is inserted into a stem cell. This gene is only activated or "reports" when cells are undifferentiated and is turned off once they become specialized. Once activated, the gene directs the stem cells to produce a protein that fluoresces in a brilliant green color. Two commonly used reporter genes are EGFP and luciferase. Figure 1.20 illustrates the genetical modification of cell to express both EGFP and luciferase [16]. Both the fluorescent and luminescent signals from the cells follow a linear relationship as a function of cell

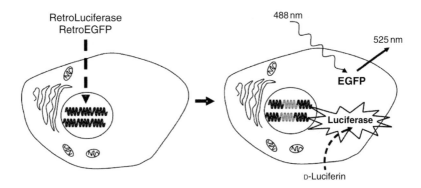

Fig. 1.20 Schematic depiction of cell genetically modified by retroviral vectors to express the reporter genes EGFP and luciferase. EGFP when excited with at ~488 nm will emit at ~525 nm. Luciferase in the presence of the substrate, D-luciferin, will emit light.

number. These relationships provide two opportunities for quantifying cellularity from simple extracts. Both genes, when expressed in mammalian cells, will produce a molecule that can be detected by different techniques. A chromophore-containing protein, EGFP, emits fluorescent light when excited with light at a wavelength of approximately 488 nm. Qualitatively, EGFP under excitation provides an opportunity to visualize only cells expressing this protein. In contrast, the luciferase enzyme hydrolyzes its substrate, D-luciferin, in the presence of ATP and oxygen to produce photons of the emission spectra ranging between 400 and 620 nm. This stable retroviral transfection of cells with green fluorescent protein (GFP) offers a pathway to study tissue development, with emphasis on distinguishing between cellular components initially seeded onto a construct and those occurring as a result of cell ingrowth from surrounding tissue.

Various seeding techniques can be assessed using a DNA assay to estimate the total number of cells within a scaffold. This approach, however, yields very little information about the distribution of the cells throughout the scaffold. For destructive cellularity assays which quantify cellular DNA, the presence of ECM and/or material particulates in the sample extracts can interfere with these analyzes by making it difficult to extract the DNA. A critical issue in tissue engineering concerns whether the cellular components of tissue-engineered structures are derived from cells harvested and seeded onto an acellular scaffold or from cells originating from surrounding tissue (*e.g.*, proximal and distal anastomosis in the case of cardiovascular tissue engineering), or from circulating pluripotent stem cells. To clarify this issue, some studies have utilized fluorescent carbocyanine dyes. It was possible to identify cells labeled by this technique *in vivo* for up to 6 weeks after transplantation, but more stable long-term labeling and tracing without adverse effects, such as cell toxicity, are required. Current histomorphometric techniques evaluating rates of tissue ingrowth tend either to measure the overall tissue content in an entire sample or to depend on the user to indicate a front of tissue ingrowth. Neither method is particularly suitable for the assessment of tissue ingrowth rates, as these methods either lack the sensitivity required or are problematic when there is a tissue ingrowth gradient rather than an obvious tissue ingrowth front.

Cells interacting with scaffolds may exhibit distinct patterns of gene expression depending on the molecular nature of the surface they are contacting. Many cell types grown in 3-D culture exhibit phenotypes drastically different from those of their plate-grown counterparts. In these cases gene expression studies are beneficial.

6.2.5. Gene Expression of Cells

Expression of specific genes is enhanced within cell-seeded constructs or engineered tissue. For instance, in the bone tissue engineering, Runx 2, alkaline phosphatase (ALP), and osteocalcin (OCN) that are important markers of different stages of bone matrix production are expressed. Runx 2 is a transcriptional activator essential for initial osteoblast differentiation and subsequent bone formation. The ALP is expressed by pre-osteoblasts and osteoblasts before the expression of OCN. Finally, OCN is a late marker that binds HAp and is produced by osteoblasts just before and during matrix mineralization.

6.3. Bioreactors

When 3-D cellular constructs are grown in static culture, cells on the outer surface of the constructs are typically viable and proliferate readily while cells within the constructs may be less active or necrotic. In the absence of a vascular blood supply *in vitro*, nutrient delivery to cells throughout 3-D tissue engineered constructs grown in static culture must occur by diffusion. As a result, thin tissues (*e.g.*, skin) and tissues that are naturally avascular (e.g., cartilage) have been more readily grown *in vitro* than thicker, vascular tissues such as bone. The engineering of tissues *ex vivo* in a bioreactor offers several benefits, such as better understanding of tissue development and mechanisms of disease, off-the-shelf revision of transplantable tissues, and possible scale-up for commercial production of engineered tissue. The bioreactors in tissue engineering are utilized for cell expansion on a large scale and production of 3-D tissues *in vitro*. Bioreactors offer several advantages over culture in plates and flasks. Bioreactors can be custom designed to engineer tissues with complicated 3-D geometry containing multiple cell types. More significantly, bioreactors can impart appropriate biochemical and mechanical stimuli in a controllable environment to promote cell growth, maturation, and tissue differentiation. Artificial tissue development in bioreactors generally involves less handling than static culture methods to significantly reduce contamination risk and is amenable to scale-up for the generation of tissue products. Bioreactors can serve as tissue growth systems as well as packaging and shipping units that can be delivered directly to surgeons.

Bioreactors have been widely used from drug production to beer brewing to create products in an efficient manner. The basic concept of bioreactors that take advantage of microorganisms is to create an environment that is advantageous for microorganisms to the creation of a desired product, whether it be penicillin, alcohol, or ECM. Nutrient, waste, and oxygen levels must be carefully controlled to prevent the organisms from dying. There are three types for such bioreactors: batch, fed-batch, and continuous systems. In a batch system, organisms are combined with the required nutrients in a single-step process. A fed-batch process is similar to the batch reaction in that nothing is removed during the reaction. However, in contrast to the batch system, additional nutrients are added over a portion of the process to keep the reaction rate at its maximum. A continuous bioreactor adds nutrients and removes waste and products over the entire course of the reaction. Continuous reactors are the most efficient and cost-effective once the equipment is running.

The bioreactors used in engineering tissues are primarily batch processes and there is no need to run a large-scale operation. As custom design of engineered tissues is often necessary because of immunological concerns, a continuous reactor is not necessary. The bioreactors for engineering tissues, as opposed to bioprocessing, provide an environment in which cells continuously reside throughout the culturing period. Mass transfer becomes the main concern because nutrients and oxygen must be provided in sufficient amounts to tissues to grow to a usable size. The stimulus from a mechanical force can still be beneficial. Articular cartilage in the knee experiences shearing stresses during normal movement. On the surface of cartilage, a thin layer of synovial fluid provides lubrication, which reduces friction, but shearing of the tissue still occurs because of the solid-on-solid contact. Shear forces alter the phenotype of the cells seeded on a scaffold to one that is close to native cartilage. These forces can be emulated using a flowing fluid either across or through the cell-seeded scaffold. Microcarrier beads may effectively provide surface area for attachment of anchorage-dependent cells. In addition, agitation of microcarrier cultures leads to homogeneous culture conditions and improved gas-liquid oxygen transfer.

If constructs must be produced using patient-specific cells that do not produce an immune response, then continuous bioreactors are not required but large-scale bioreactors that resemble a batch process will have to be created. This makes the task much more difficult because the samples must be separated at all times and the equipment must be sterilized after each batch. A mechanical force must be applied to each scaffold in a controlled manner without any mixing of cell batches.

Tissue culture systems that provide dynamic medium flow conditions around or within tissue-engineered constructs should be designed to enhance nutrient exchange and cell growth *in vitro*. Such tissue culture systems are required as bioreactors to engineer thicker, more uniform tissues for transplantation. In addition to enhancing mass transport, bioreactor systems may be useful to deliver controlled mechanical stimuli such as flow-mediated shear stress, matrix strains, or hydrostatic pressures to tissue constructs. This general approach has been used to culture a variety of 3-D constructs including bone, cartilage, muscle, and blood vessels. Tissue culture systems that incorporate dynamic medium flow conditions for developing 3-D tissue constructs include spinner flasks, perfusion systems, and rotary cell bioreactors. In general, improved cell viability, proliferation, and ECM production have been demonstrated in dynamic systems relative to static controls. Internal fluid flow in each system is achieved in different ways. In spinner flasks, stirred culture medium is moved past the scaffolds at fixed positions within the vessel. In rotating bioreactors, the scaffolds are not fixed, but in continuous free fall within the culture medium during bioreactor rotation. In perfusion cultures, culture medium is directly perfused using a uniform fluid pressure head applied to the scaffolds and fluid. In spinner flasks, flow and mixing of culture medium is associated with turbulent shear at construct surfaces. Mass transport between the constructs and culture medium is enhanced by convection, whereas mass transport within the constructs remains controlled by diffusion. In perfused reactors, interstitial flow of culture medium enhances mass transport throughout the construct volume, and exposes all cells to laminar shear.

6.3.1. Spinner Flask

One of the simplest bioreactors is the mechanically stirred flask and the most common mechanically stirred bioreactor is the spinner flask. Scaffolds seeded with cells are attached to needles hanging from the cover of the flask and suspended within a stirred suspension of cells with addition of sufficient medium to cover the scaffolds, as shown in Fig. 1.21 [17]. This method has produced favorable results, but potential weaknesses of the technique include the amount of time required for seeding (typically 24 h), low efficiency at low cell concentrations, and undesirable side effects associated with mechanically stirred bioreactors of high shear rates. The degree of shear stress depends on stirring speed and the morphology of the scaffold. Cell damage has been observed at 150–300 rpm in microcarrier cultures, and although there is no apparent physical cell damage at 50 rpm, a fibrous capsule does form on the construct surface. The local shear

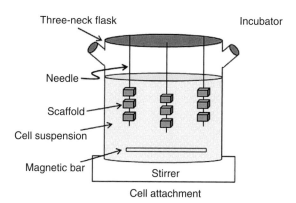

Fig. 1.21 Schematic view of a spinner flask designed to assist initial cell attachment. The cell-attached scaffolds are transferred onto 24-well plates for subsequent cell proliferation and differentiation assays.

force experienced by the cells is produced by eddies created by the turbulent flow of the impeller. The smallest turbulent eddies are on the order of several hundreds of micron with velocities of approximately 0.5 cm/sec. Cell flattening and proliferation, and formation of an outer capsule, are caused by the pressure and velocity fluctuations associated with turbulent mixing. Table 1.5 represents the high-shear bioreactor summary [18].

Using impellers is popular in basic cell culturing because it increases the rate of mass transfer to the cells, but impellers create many problems. Non-uniform mass transfer rates, nutrient and pH gradients, and shear gradients impart a non-uniform mechanical stimulus over the construct, resulting in the formation of tissue that is inferior to that produced by other culturing techniques. The shear force at the surface of the impeller is reported to be 10 times higher than anywhere else within the bioreactor. Because of this, cells located closer to the impeller will exhibit more of an injury response because of the high levels of shear. Cells farther away will not produce a fibrous capsule as long as the rotation rate is low enough, but they will receive fewer nutrients and experience a higher pH because of the lessened mixing.

Table 1.5
High-shear bioreactor summary

First Author	Proteoglycan	Collagen	Additional Notes
Gooch (construct, 2- to 4-week bovine knee, 18×10^6 cells/cm^3 PGA, 6 weeks)	36% decrease in GAG composition; 2- to 3-fold increase in GAG synthesis	80% increase in collagen composition	Constructs retained low levels of GAG after synthesis
Vunjak-Novakovic (construct, 2- to 3-week bovine knee, 25×10^6 cells/cm^3 PGA, 8 weeks)	60% increase in GAG composition	125% increase in collagen composition	Fibrous capsule formed
Smith (monolayer, adult bovine wrist, 1.5×10^5 cells/cm^2, 3 days)	2-fold increase in [^{35}S]-sulfate incorporation	Not evaluated	High levels of prostaglandin E$_2$ and metalloproteinase released
Freed (construct, 2- to 3-week bovine knee, 29×10^6 cells/cm^3 PGA, 6 weeks)	No significant change	50% increase in collagen composition	Primarily type I collagen
Dunkelman (construct, 4- to 8-month rabbit joints, P2, 25×10^6 cells/cm^3 PGA, 4 weeks)	25% dry weight GAG (15–30% in native cartilage)	15% dry weight collagen (50–73% in native cartilage)	Direct perfusion; more tissue formed on periphery than middle

Refer to the original work for the references.

6.3.2. Perfusion System

Mechanically stirred bioreactors are used in scaffold-seeding experiments but they might not be optimal for producing tissues such as cartilage. The most successful results in any bioreactor are based on the use of scaffolds seeded dynamically and at high densities. However, 3-D hydrogels that include cells during gel formation do not need to be seeded in this manner and cells can be distributed evenly at high densities without the use of a mechanically stirred bioreactor. For 3-D porous scaffolds, a few systems that force medium through the scaffold give the most thickness-independent results. This type of system is called a "direct perfusion bioreactor", as illustrated in Fig. 1.22 [19]. This type of bioreactor uses a pump to perfuse medium continuously through the interconnected porous network of the scaffold, rather than around the edges. The constant availability of fresh medium, the mechanical action of shear stress, and the ability to transport nutrients through an increasingly dense mass of cells and ECM material have favored its use.

Cells within the bulk of a construct feel the shear force from the fluid flowing through the pores and may produce ECM in response to it. A modification that can be made to direct perfusion bioreactors is to mix a portion of the old medium with the fresh medium. By recycling some of the medium, beneficial proteins such as growth factors and interleukin 4 are kept in the system. Otherwise, chemical signals would be lost from the developing tissue. The tissue then develops in a medium that contains all the signal proteins that are produced, the cell-secreted collagen, and proteoglycans that are flushed out of the scaffold but are still in the medium.

One of the problems that becomes apparent when using direct perfusion bioreactors is non-uniform cellular secretions through the thickness of the construct as well as damage to some of the cells. If fluid is flowing from one side of a scaffold to the other, then the front surface will have a greater mechanical stress exerted on it because of the oncoming flow. Conversely, the back surface does not feel the force except for inside of the pores. This will cause a thicker matrix to form on the front side of the scaffold as compared with the rest of the construct. For high-shear direct

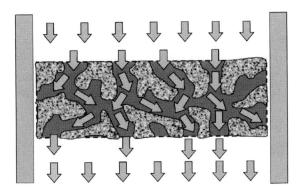

Fig. 1.22 Direct perfusion bioreactor.

perfusion devices, the fibrous response is similar to the capsule formed in spinner flasks except that it occurs only on one side of the construct. The tissue ends up having inferior mechanical characteristics compared with native tissue. The direct perfusion bioreactor must use low flow rates to fix this problem, which would nullify the effectiveness of this particular bioreactor system.

6.3.3. Rotating Wall Reactor

The conventional bioreactor has the disadvantage that high shear forces are generated, which damage the cells and hinder proper tissue-specific differentiation. Decreasing the stirring rate while increasing the viscosity of the culture medium might partially reduce the hydrodynamic damage, but aggregates formed under these conditions still exhibit necrotic centers. The ability of the direct perfusion bioreactors to provide a nutrient-rich environment for the cells and stimulate them simultaneously was carried over into the design of the more sophisticated rotating-wall device. The major change is mainly in the application of a mechanical force, because high or even moderate levels of them are undesirable in the formation of some tissues. A rotating fluid environment was found to be the best way to produce a low-shear bioreactor. Efforts at tissue reconstruction would benefit from a venue which promotes cell–cell association while avoiding the detrimental effects of high shear stress. Such a venue might be provided by microgravity, because it was observed many years ago that cells in suspension tended to aggregate when exposed to microgravity in space [20]. In an effort to derive the potential beneficial effects of microgravity and low fluid shear for cell culture here on the earth, scientists at the National Aeronautics and Space Agency (NASA) introduced the rotary wall vessel (RWV) bioreactor. Probably, the best overall results for cell culturing currently come from a modified version of the direct perfusion bioreactor called the "rotating-wall bioreactor" [18]. The NASA originally created it as a "microgravity" environment for cell culture, but later successfully as a low-shear, high-diffusion bioreactor for many cell types.

The horizontally rotating culture vessel was initially designed to simulate some aspects of microgravity and has proved valuable in the generation of 3-D cultures of a variety of transformed and non-transformed cells. The system allows anchorage-dependent cells to grow in 3-D and multiple layers within the framework. The NASA-developed bioreactor allows for the formation of 3-D aggregates and promotes co-spacial localization between similar and dissimilar cells under conditions of controlled access to O_2 and nutrients. As illustrated in Fig. 1.23, the most popular type of the bioreactor is composed of two cylinders, where microcarriers or scaffolds are placed in the annular space between the two cylinders. Gas exchange is allowed through the stationary inner cylinder while the outer cylinder is impermeable and rotates in a controlled fashion. Under carefully selected rotational rates the free falling of the scaffolds inside the bioreactor due to gravity can be balanced by the centrifugal forces due to the rotation of the outer cylinder, thus establishing microgravity-like culturing constitutions.

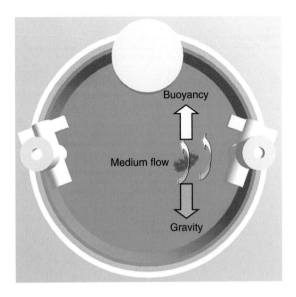

Fig. 1.23 Rotating-wall vessel bioreactor.

Rotating-wall bioreactors are horizontally rotated with fluid-filled culture vessels equipped with membrane diffusion gas exchange to optimize gas/oxygen supply. The initial rotational speed is adjusted so that the culture medium, the individual cells, and pre-aggregated cell constructs or tissue fragments rotate synchronously with the vessel, thus providing for an efficient low-shear mass transfer of nutrients and wastes. As the cell aggregates grow in size and exhibit increasing sedimentation velocities, the shear stress can approach up to 0.5 dyn/cm^2. These simulated microgravity conditions facilitate spatial co-location and 3-D assembly of individual cells into large aggregates. There is laminar flow of medium over the surface of constructs with fairly uniform shear stress distribution. For example, the maximum shear stress on the surface of constructs is approximately 0.3 dyn/cm^2 for a 120-ml rotating vessel containing only one construct and operated at rotational speeds of $\sim$25 rpm. The low-shear bioreactor summary is given in Table 1.6 [18]. The time-averaged gravitational vector acting on these cellular assemblies is reduced to about 10^{-2} g. In a typical experiment on board a space craft or the space station cycling in near-earth orbit, the gravitational force is approximately 10^{-4}–10^{-6} g.

Cell culture conditions in the "simulated microgravity" environment of rotating-wall bioreactors combine two beneficial factors: low shear stress, which promotes close apposition (spatial co-location) of the cells; and randomized gravitational vectors. Close apposition of the cells in the absence of shear forces presumably promotes cell–cell contacts and the initiation of differentiative cellular signaling via specialized cell adhesion molecules. This process then might lead to the rapid establishment and expansion of aggregate cultures which, unlike in conventional bioreactors, are not disrupted by shear forces. In addition, the low-shear environment, in concert with

Table 1.6

Low-shear bioreactor summary

First Author	Proteoglycan	Collagen	Additional Notes
Gooch (construct, 2- to 4-week bovine knee, 76×10^6 cells/cm^3 PGA, 4 weeks)	2.9-fold increase in GAG composition	1.6-fold increase in collagen composition	Supplemented with IGF-I
Martin (construct, 2- to 3-week bovine knee, 127×10^6 cells/cm^3 PGA, 6 weeks)	75% increase in GAG composition	39% increase in collagen composition	Equilibrium modulus and GAG composition reach native cartilage levels after 7 months *in vitro*
Freed (construct, 2- to 3-week bovine knee, 127×10^6 cells/cm^3 PGA, 6 weeks)	68% of native GAG levels per gram wet weight	33% of native type II collagen levels per gram wet weight	Type II collagen crosslinked in constructs

Refer to the original work for the references.

randomized gravitational vectors, might restrict diffusion of mitogenic and differentiative growth factors, which are secreted by the cells. These autocrine/paracrine feedback mechanisms might further enhance the aggregation and differentiation, and contribute to the observed capability of this environment to maintain high density cell cultures. Cultures in a 3-D matrix or scaffolds such as collagen and agarose gel largely maintain the chondrocyte phenotype, but the use of RWVs as scaffold-free bioreactors also leads to successful differentiation and the hyaline production. The most significant advantages of the RWV for a 3-D tissue culture are low shear forces, the reduced risk of cell damage, and the increased opportunity for cell–cell interaction. Thus, primarily dissociated cells can reassemble to form a 3-D tissue-like mass. The spatial orientation may be critical to the differentiation of growing aggregate cells and to the regeneration of a functional ECM.

Cell aggregates generated *in vitro* (in either suspension cultures or stirred bioreactors) that exceed about one mm in size invariably develop necrotic cores. Culture conditions in RWV bioreactors are unique in that the fluid dynamics of the system allow for efficient mass transfer of nutrients and oxygen diffusion. In this environment, dissociated cells can assemble into macroscopic tissue aggregates several mm in size, which are largely devoid of such necrotic cores. With the possible exception of avascular tissue of low cellularity and slow metabolism, such as cartilage, all tissues, including those growing in RWV bioreactors, will eventually require internal, blood-vessel-like conduits for the delivery of oxygen and nutrients as well as for the removal of waste products. Attempts to generate 3-D constructs as replacement tissues will necessitate combining RWV technology with innovative methods for creating bioengineered blood conduits such as growing endothelial cells on the inside and outside of tubular scaffolds.

6.3.4. Kinetics

The success of scaffolds for tissue engineering is typically coupled to the appropriate transport of gases, nutrients, proteins, cells, and waste products into, out of, and/or within the scaffold. To obtain fluid flow and nutrient flux in porous, 3-D scaffolds, dynamic culture methods (Fig. 1.24) [15] have been employed. However, the primary mass transport property of interest, at least initially, is diffusion. In a scaffold, the rate and distance a molecule diffuses depend on both the material and the molecule characteristics and interactions. As a consequence, diffusion rates will be affected by the MW and size of the diffusion species (defined by Stokes radii) compared to the pore of scaffolds. For example, molecules such as glucose, oxygen, and vitamin B_{12}, with MWs less than 1300 Da and Stokes radii less than 1 nm, are able to freely diffuse into and from ionically crosslinked hydrogels. Gel properties such as polymer fraction, polymer size, and crosslinker concentration determine the gel nanoporous structure. However, higher MW molecules such as albumin and fibrinogen are not able to freely diffuse, and their rate of diffusion is further decreased by increases in polymer concentration, in crosslinker concentration, and/or in the extent of gelation.

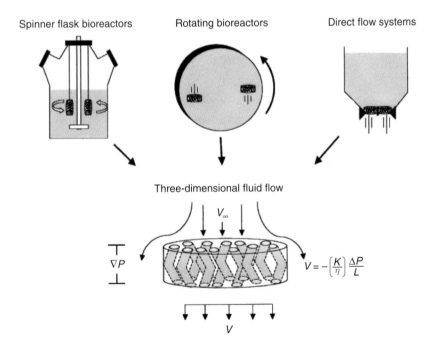

Fig. 1.24 Representative overview of dynamic culture methods used to obtain fluid flow and nutrient flux in porous 3-D scaffolds. In spinner flask bioreactors, culture medium is stirred around scaffolds at fixed positions within the vessel. In rotating bioreactors, scaffolds are maintained in continuous free fall through the culture medium. In a direct flow system, scaffolds are fixed in position and perfused with culture medium by a direct application of a uniform fluid pressure head.

Ultimately, diffusion requirements and subsequent material choice depend on the scaffold application.

The mass transfer of nutrients and waste between a scaffold and the surrounding medium follows the equation below:

$$J_s = -DS \frac{dC}{dx}$$

where J_s is the diffusion of a solute, D is the diffusivity of the solute, S is the surface area normal to the direction of solute diffusion, and dC/dx is the concentration gradient. However, diffusion is just one of many parameters that are important in the cell-culturing process.

To evaluate the role of nutrient diffusion and consumption within cell-seeded scaffolds in the absence of flow, Botchwey *et al.* developed a 1-D glucose diffusion model, as will be described in Chapter 2. Such a quantitative analysis will provide a basis for development of new dynamic culture methodologies to overcome the limitations of passive nutrient diffusion in 3-D cell–scaffold composite systems. However, many factors still remain to be quantified, such as the transport of oxygen, other nutrients, cell–cell signaling molecules, growth factors, metabolic wastes, and cell chemotactic factors in dynamic culture, all of which may have an effect on cell function and tissue synthesis.

6.4. Externally Applied Mechanical Stimulation

Some tissues exist in a mechanically dynamic environment. Blood vessels are continuously exposed to mechanical forces that lead to adaptive remodeling. Although there have been many studies characterizing the responses of vascular cells to mechanical stimuli, the precise mechanical characteristics of the forces applied to cells to elicit these responses are not clear. Soft musculoskeletal tissues also adapt to immobilization and realize strengthening in response to exercise. Tendons are in a continuous state of dynamic remodeling. Cells suspended within a 3-D network of ECM respond to external changes via receptor–ligand interactions that relay signals from outside the cell to the cytoskeletal domain and thereby influence subsequent cellular function such as attachment, migration, differentiation, and apoptosis. As natural tissues, especially those which should resist against mechanical loading and pressure—such as cartilages, bones, ligaments, cardiac muscle, and blood vessels— are subjected to mechanical stimuli during development, it will be reasonable to assume that mechanical stimuli, pulsed or unpulsed, should be given to cell–scaffold constructs aimed at least at *in vitro* tissue engineering of musculoskeletal and cardiovascular tissues. Indeed, numerous studies have shown local mechanical signals to be a key factor directing the development, growth, repair, and maintenance of bone and cartilage. Since some cells can sense their mechanical environment such as endogenously generated tension, exogenous stimulus may be used as a conditioning modality to influence the efficacy of tissue engineered replacements for load-bearing tissues. Thus, researchers have employed functional tissue engineering approaches that place special focus on using physical stimuli to encourage the development of a biomechanically functional tissue. A premise of *in vitro* tissue engineering is that one may enhance the rate and quality of tissue growth by re-creating *in vitro* some of the same conditions that the tissue experiences *in vivo*. Articular cartilage is amenable to such an approach, as it is a tissue defined not only by its cellular constituents but also by its dynamic physical environment. Therefore, appropriate mechanical stimulation has been applied to cells during their culture, especially in case cells in the human body live in an environment heavily influenced by mechanical forces such as in load-bearing tissues. This is the major reason why chondrocyte has been very often selected as a cell model for tissue engineering under mechanical stimulation. Another reason is that the cartilage in which chondrocytes live is an avascular tissue and receives oxygen and nutrients from the synovial fluid. Chondrocyte is known to be one of the most robust cells and to develop differently on the basis of what culturing processes are used. The presence of mechanical forces such as hydrostatic pressure or direct compression stimulates chondrocytes to secrete more ECM as compared with static culture.

There are two major methods for mechanically stimulating cells outside of their culturing environment to enhance their growth. One is the bioprocessing which uses mechanical stimuli only at intervals during the culturing period. The other is the bioreactor system which uses a constant mechanical force to stimulate the cells. The advantage of these approaches is the introduction of a mechanical stimulus and an increase in diffusion through the porous scaffold. In the bioprocess, mechanical stimuli are given to cells via either hydrostatic pressure or direct compression. When cartilage located in articular joints is loaded during walking, running, or shifting weight while standing, the force is transmitted throughout the tissue which contains water by 75–80 wt%, being absorbed primarily by the fluid. The pressure produced by the compressed fluid acts uniformly on the chondrocytes within ECM. This hydrostatic pressure ranges between 7 and 10 MPa during normal activity. One technique that uses hydrostatic pressure as a mechanical stimulus is to employ a two-step process that separates culturing from force application. The cells are kept mostly in static culture medium, where they are nourished. At prescribed times, the cells are moved to a hydrostatic chamber where a specified load is placed on them. Another technique uses a semicontinuous perfusion system that feeds the cells and applies hydrostatic pressure in the same device. In this case the cell cultures do not have to be moved as much, which reduces the possibility of contamination, and the process can be automated to run for long periods of time without any need for manual labor. The production of proteoglycans and type II collagen is currently the main experimental indicator of positive mechanical stimulation, although a few studies also report DNA synthesis or aggregate moduli. Table 1.7 summarizes representative results of bioprocesses which use hydrostatic pressure as a mechanical stimulus [18]. The duration and magnitude of hydrostatic loading vary widely between studies, but the most successful studies dynamically load cartilage for longer periods of time.

In the case of engineering blood vessels, mechanical stimulation can be provided via radial distension of the silicone tube around which a tubular scaffold is

Table 1.7
Hydrostatic pressure summary

First Author	Proteoglycan	Type II Collagen	Additional Notes
Smith (monolayer, adult bovine wrist, 10^5 cells/cm^2, 4 days)	21-fold increase in mRNA synthesis	9-fold increase in mRNA synthesis	10 MPa, 1 Hz
Carver (construct, 1-week bovine knee, 2×10^6 cells/cm^3 PGA, 5 weeks)	2-fold increase in concentration	No significant increase	3.5 MPa, 5/15 s (on/off) for 20 min every 4 h
Smith (monolayer, adult bovine wrist, 10^5 cells/cm^2, 4 h)	31% increase in mRNA synthesis	36% increase in mRNA synthesis	1% fetal bovine serum 10 MPa, 1 Hz

Refer to the original works for the references.

sewn, as demonstrated in Fig. 1.25 [21]. *In vitro* fluid-induced shear stress, as occurs in agitated microcarrier cultures, can be used as a model system for mechanical stimulation of articular chondrocytes. Culture of chondrocytes on microcarriers may have beneficial effects for cartilage tissue engineering applications. Efficient expression of chondrocytes, without the loss of their ability to express the differentiated phenotype

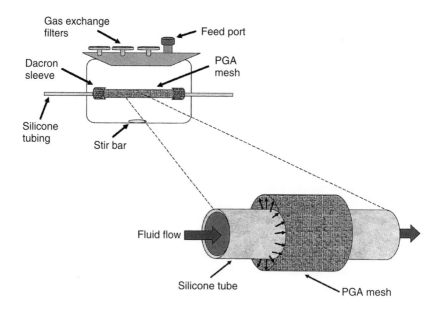

Fig. 1.25 Vessels made of PGA are sewn around a silicone tube and are attached to the sides of a glass bioreactor with a nondegradable Dacron sleeve. Mechanical stimulation of the construct is provided via radial distension of the silicone tube as a result of fluid flow.

would be highly beneficial for the engineering of cartilage for clinical purposes. In particular, such expression would exclude costly and time-consuming techniques of redifferentiation.

The thickness of constructs is significantly limited by the diffusion properties associated with porous scaffolds. Dynamic compression helps alleviate these diffusion limitations through pressure gradients within the scaffold as well as by a secondary mixing effect on the surrounding medium. In a healthy person, articular cartilage undergoes direct compression thousands of times a day without any long-term injury to the tissue. Unlike hydrostatic pressure, there is an actual solid-on-solid contact that takes place under direct compression. Mass transfer for constructs under direct compression is expected to be better than for those cultured under hydrostatic pressure or by the static culture approach. The main parameters that must be set when using dynamic compression are the frequency of the applied load, the strain or force used, and the duration of the experiment.

New bioreactors can be developed that take advantage of several different mechanical stimuli, as well as a good culturing environment, all in one package. For example, hydrostatic pressure could be combined with a rotating bioreactor to create a stimulating environment that is self-contained. Because scaffolds are already cultured in a fluid medium, hydrostatic pressure could be applied without removing anything from the sterile environment. This reduces the chance of contamination as well as limiting the amount of time needed to transfer the scaffolds between a stimulation device and a culturing environment. If the cells are fed in sufficient amounts and are surrounded by a good environment, they will act like native cells and secrete ECM into the scaffold pores. The addition of bioactive peptides and growth factors in the scaffolds will help matrix synthesis, especially in conjunction with mechanical forces.

Most of engineered tissues are histologically similar to natural tissues, but differ greatly in mechanical strength, stiffness, and functional properties. The mechanical properties of many tissues engineered to date are inferior to those of native tissues. Skeletal muscle constructs were demonstrated to have reasonable initial histology but force generation an order of magnitude below the native tissue [22]. Generally, such poor results are observed when cells seeded onto scaffolds are cultured under static conditions. Both morphology and force generation of engineered heart tissue may be modified by cyclic loading during construct development. It is necessary to investigate the effect of mechanical conditioning of cells on cell–scaffold constructs and to quantify their mechanical properties as well as their active force capabilities. Fiber arrays made from absorbable, rubber-like materials have been chosen for such an approach. The scaffold materials will guide cell orientation and provide mechanical support, and then degrade within a certain time frame but not before the graft attained sufficient mechanical strength.

6.5. Neovascularization

Mammalian cells require oxygen and nutrients for their survival. Tissues in the body, except for cartilage, overcome issues of sufficient nutrient and waste exchange with their surroundings by containing closely spaced capillaries that provide conduits for convective transport of nutrients and waste products to and from the tissues. Mammalian cells are located within 100–200 μm of blood vessels. This distance is the diffusion limit for oxygen. Figure 1.26 shows the relationship between the distance of tumor cells from nearby vessels and their degree of hypoxia and acidosis [23]. One of the dominant factors currently limiting clinical applications of tissue engineering is the inability of the scaffold to construct *de novo* a microvasculature. Tissue engineered constructs also require a capillary network for cell maintenance and function excluding those less than 100–200 μm thick, which may be oxygenated by diffusion. Few clinical trials in cardiac tissue engineering are due to unavailability of scaffolds with large size and excellent capability of vasculature. The lack of capillary network connected to the host tissue and the resulting poor, oxygen transport limit the thickness of generated tissue constructs to ∼0.1 mm, which is generally too thin for clinical application. The mass transport issue limits the size of engineered tissues to a millimeter scale at the largest, which is clinically insufficient if a large mass of tissue or a whole organ should be replaced. In some cases, the limitation of this mass transport of nutrients leads to loss of more than 95% of transplanted cell types. Fat tissue represents a highly vascularized tissue, and a volume-persistent culture of adipose tissue can be successful only via early vascularization of the cell–scaffold construct. A simple solution to avoid considerable cell death *in vivo*, especially in the central area, is to utilize native tissues with rich vascularity. Among them is the omentum which is a fold of the peritoneum anchored to the stomach and transverse colon that drapes over the small intestine. The omentum is highly vascular, contains a relatively large surface area, and is recruited to sites of intra-abdominal abscess, perhaps to "wall off" the area and prevent diffuse peritoneal contamination. The mesentery of the small intestine that anchors the bowel to the body is also vascular and supplies blood to the small intestine.

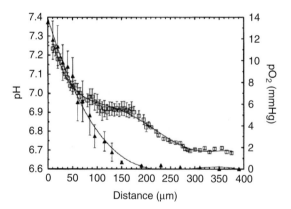

Fig. 1.26 Mean interstitial pH and partial pressure of oxygen (pO$_2$) profiles of 27-day-old tumors taken as one moves away from the nearest blood vessel. Open symbols, pH; closed symbols, pO$_2$.

7. GROWTH FACTORS

The term "growth factors" used here include differentiation factors and angiogenic factors in addition to growth factors along with bone morphogenetic proteins. It must be recognized that the redundancy found in most biological structures is such that precise characterization of a growth factor as falling in just one category above is misleading. Many growth factors may provide a host of functions and may modulate cell attachment, cell growth (or apoptosis), cell differentiation, cell migration, neovascularization, etc., and indeed may do so differently according to the biochemical, cellular, and biomechanical context into which they are placed. Both growth factors and differentiation factors are likely essential to establishing a sufficient number and architecture of appropriately functioning cells. Specific factors favor certain cell lineage, possibly through an inhibition of certain lineage rather than promotion of a specific one.

Growth factors may be exogenously added, or the cells themselves may be induced to synthesize them in response to chemical and/or physical stresses. Although cytokines and growth factors are present within ECM in vanishingly small quantities, they act as potent modulators of cell behavior. The list of these growth factors is given in Table 1.8. These factors tend to exist in multiple isoforms, each with its specific biological activity. Purified forms of growth factors and biological peptides have been investigated as therapeutic means of encouraging blood vessel formation (VEGF), inhibiting blood vessel formation (angiostatin), stimulating deposition of granulation

Table 1.8

Growth factors used for tissue engineering

Growth Factor	Abbreviation	Effects
Basic fibroblast growth factor	bFGF, FGF-2	Angiogenesis; Fibroblast and osteoblast mitogen
Bone morphogenetic proteins	BMP-2 BMP-7 (OP-1)	Growth and development of some tissues; Osteogenesis
Transforming growth factor-$\beta1$	TGF-$\beta1$	Proliferation and differentiation of bone forming cells; Fibroblast matrix synthesis
Vascular endothelial growth factor	VEGF	Angiogenesis; Proliferation and migration of endothelial cells
Platelet-derived growth factor	PDGF	Proliferation of smooth muscle cells; Fibroblast mitogen and matrix synthesis
Hepatocyte growth factor	HGF	Hepatocyte mitogen; Motogen and anti-apoptotic factor of cells; Angiogenesis
Keratinocyte growth factor	KGF	Epithelization of wounds
Epidermal growth factor	EGF	Proliferation of epithelial; Mesenchymal and fibroblast cells
Insulin-like growth factor	IGF-1	Cartilage development and homeostasis; Bone formation

tissue (platelet-derived growth factor, PDGF), and encouraging epithelialization of wounds (KGF). However, this therapeutic approach has struggled with determination of optimal dose, sustained and localized release at the desired site, and the inability to turn the factor "on" and "off" as needed during the course of tissue repair. An advantage of utilizing the ECM in its native state as a scaffold for tissue repair is the presence of all of the attendant growth factors (and their inhibitors) in the relative amounts that exist in nature and, perhaps most importantly, in their native 3-D ultrastructure.

Growth factors must greatly contribute also to tissue engineering at various stages of cell proliferation and differentiation. Thus, numerous studies have been performed using growth factors in the field of tissue engineering, and some growth factors have produced promising results in a variety of preclinical and clinical models.

7.1. Representative Growth Factors

7.1.1. BMPs

The BMPs represent a family of related osteoinductive peptides akin to differentiation factors. Research involving BMPs found its beginnings more than 35 years ago, when Urist observed that demineralized bovine bone matrix was capable of inducing endochondral bone formation when implanted ectopically into soft tissues of experimental animals [24]. Urist and colleagues subsequently discovered that low MW glycoproteins isolated from bone were responsible for the osteogenic activity observed earlier and were capable of inducing bone formation when delivered to ectopic or orthotopic locations. Wozney *et al.* and Celeste *et al.* subsequently cloned the first bone morphogenetic proteins: BMP-2, BMP-3, and BMP-4. Ozkaynak *et al.*, using similar techniques, cloned BMP-7 and BMP-8. At least 15 BMPs have been identified, many of which can induce chondro-osteogenesis in various mammalian tissues. The BMPs are homodimeric molecules (MW: 25–30 kDa) that regulate various cellular functions such as bone induction, morphogenesis, chemotaxis, mitosis, hematopoiesis, cell survival, and apoptosis. With the exception of BMP-1, which is a protease that possesses the carboxy-terminal procollagen peptide, BMPs are part of the transforming growth factor (TGF)-β superfamily and play a major role in the growth and development of several organ systems, including the brain, eyes, heart, kidney, gonads, liver, skeleton, skeletal muscle, ligaments, tendons, and skin. Although the sequences of members of the TGF-β superfamily vary considerably, all are structurally very similar. The BMPs are further divided into subfamilies based on phylogenetic analysis and sequence similarities: BMP-2 and 4-(dpp subfamily); BMP-3; BMP-5, -6, -7, -8A, and -8B (60A subfamily). And BMP-6 is characteristically expressed in prehypertrophic chondrocytes during embryogenesis. Several investigators have reported that BMP-6 plays a role in chondrocyte differentiation both *in vivo* and *in vitro*. When involved in osteoinduction, BMPs have three main functions. First, BMPs act as a chemotactic agent, initiating the recruitment of progenitor and stem cells toward the area of bone injury. Second, BMPs function as growth factors, stimulating angiogenesis and proliferation of stem cells from surrounding mesenchymal tissues. Third, BMPs function as differentiation factors, promoting maturation of stem cells into chondrocytes, osteoblasts, and osteocytes.

7.1.2. FGFs

The function of FGFs is not restricted to cell growth. Although some of the FGFs induce fibroblast proliferation, the original FGF (FGF-2 or basic FGF) is now known also to induce proliferation of endothelial cells, chondrocytes, smooth muscle cells, melanocytes, as well as other cells. It can also promote adipocyte differentiation, induce macrophage and fibroblast IL-6 production, stimulate astrocyte migration, and prolong neural survival. The FGFs are potent modulators of cell proliferation, motility, differentiation, and survival, and play an important role in normal regeneration processes *in vivo*, such as embryonic development, angiogenesis, osteogenesis, chondrogenesis, and wound repair. The FGF superfamily consists of 23 members, all of which contain a conserved 120-amino-acid core region that contains 6 identical, interspersed amino acids. The superfamily members act extracellularly through four tyrosine kinase FGF receptors, with multiple specificities noted for almost all FGFs. The FGFs are considered to play substantial roles in development, angiogenesis, hematopoiesis, and tumorigenesis. Human FGF-2, otherwise known as basic FGF, HBGF-2, and EDGF, is an 18 kDa, non-glycosylated polypeptide that shows both intracellular and extracellular activity. The FGFs are stored in various sites of the body under interactions with GAGs such as heparin and heparan sulfate in the ECM. And FGF-2 binds to heparin and heparan sulfate with a high affinity. These GAGs stabilize FGF-2 by protecting from inactivation by acid and heat as well as from enzymatic degradation. Also, heparin enhances the mitogenic activity of FGF-2 and serves as a cofactor to promote binding of FGF-2 to high affinity receptors.

7.1.3. VEGF

Of many unknown angiogenic factors, VEGF is unique in that it is the only known cytokine with mitogenic effects primarily confined to endothelial cells. The VEGF is produced by a variety of normal and tumor cells. Its expression correlates with periods of capillary growth during embryologic development, wound healing, and the female reproductive cycle, as well as with tumor expansion. In addition, suppression of VEGF expression in adult mice or neutralization of VEGF receptors suppresses tumor growth, while VEGF$^{\pm}$ mice are mortal *in utero*. Consequently, VEGF is thought to be a major promoter of both physiologic and pathologic angiogenesis. Furthermore, VEGF has been shown to be an anti-apoptosis survival factor for endothelial cells even during periods of microvessel stasis. In addition, VEGF can enhance tissue secretion of a variety of pro-angiogenic proteases, including uPA, MMP-1, and MMP-2. It has also been shown that inhibition of VEGF or the VEGF-R2 receptor can suppress expression of MMP-2 and MMP-3.

The development of blood vessels includes two processes. Vasculogenesis is the embryonic formation of blood islands, the earliest vascular system, by the differentiation from mesoderm of angioblasts and hematopoietic precursor cells (HPC). Angiogenesis is the sprouting of pre-existing vessels to form the vascular tree. In addition to a key role in embryonic development, angiogenesis is essential in such things as wound healing and tumor growth. While many growth factors exhibit angiogenic activity (FGFs, PDGF, TGF-α, HGF, and P/GF), most evidence points to a special role for VEGF. The VEGF is a dimeric glycoprotein that stimulates endothelial cells, induces angiogenesis and increases vascular permeability. There are four alternatively sliced variants of 121, 165, 189, and 206 amino acid residues. The receptors for VEGF are VEGFR-1 and VEGFR-2. Homozygous mutants of the VEGF receptors led to lack of vasculogenesis and death of mouse embryos on day 8, indicating that VEGF receptors are essential for the formation of a normal vasculature. The FGFs tend to be potent yet relatively non-specific growth factors with some angiogenic activity while the VEGF group trends to be relatively more specific to angiogenesis but with relative endothelial cell-specific, yet weaker, endothelial cell mitogenicity.

7.1.4. TGF-β1

The TGF-β1 is the superfamily of growth and differentiating factors of which BMP is a member. The MW of TGF-β1 is 25 kDa. This protein is synthesized in platelets and macrophages, as well as in some other cell types. When released by platelet degranulation or actively secreted by macrophages, TGF-β1 acts as a paracrine growth factor (*i.e.*, growth factor secreted by one cell exerting its effect on an adjacent second cell), affecting mainly fibroblasts, marrow stem cells, and osteoblast precursors. The TGF-β1 stimulates chemotaxis and mitogenesis (increase in the cell populations of healing cells) of osteoblast precursors, promotes differentiation toward mature osteoblasts, stimulates deposition of collagen matrix, and inhibits osteoclast formation and bone resorption. Therefore TGF-β represents a mechanism for sustaining a long-term healing and bone regeneration module.

7.1.5. PDGF

The PDGF is a glycoprotein existing mostly as a dimer of two chains of about equal size and MW (14–17 kDa). The protein that seems to be the first growth factor present in a wound is synthesized not only in platelet but also in macrophages and endothelium. It initiates connective tissue healing, including bone regeneration and repair. At the time of injury PDGF emerges from degranulating platelets and activates cell membrane receptors on target cells. The most important specific activities of PDGF include mitogenesis, angiogenesis (endothelial mitoses into functioning capillaries), and macrophage activation (debridement of the wound site and a second-phase source of growth factors for continued repair and bone formation). The PDGF is also stored in the bone matrix and released upon activation of osteoblasts, resulting in an increase of new bone formation. The PDGF is a potent stimulator of fibroblast cell migration, mitogenesis, proliferation, and matrix synthesis important in wound healing.

The four PDGF isoforms (A, B, C, and D) are characterized by a highly conserved eight-cysteine domain termed the PDGF/VEGF homology domain. The PDGF isoforms exist as disulfide-linked homodimers and heterodimers and differently bind homodimer and heterodimer combinations of two receptor tyrosine kinases, PDGF Rα and PDGF Rβ. The PDGF-AA is widely expressed in fibroblasts, osteoblasts, platelets, macrophages, smooth muscle cells, endothelial cells, and Langerhans cells. The PDGF-AA activity is ubiquitous, but dependent on cell expression of PDGF-Rα. The PDGF-AA plays key roles in protein synthesis, chemotaxis inhibition, embryonic neuron fiber development, and bronchial lung development. The PDGF-AB demonstrates mitogenic activity for vascular smooth muscle cells (VSMCs) as well as stimulating angiogenesis in the heart. Parallel to PDGF-AA and PDGF-BB expression, PDGF-AB is important in a wide variety of cellular processes of the immune, nervous, and cardiovascular systems. The PDGF-BB is mainly expressed in endothelial cells and participates in angiogenesis and arterialization of early organ, respiratory, and neuronal development. The PDGF-BB is also observed in platelets, neurons, macrophages, and fetal fibroblasts. The PDGF-BB, via PDGF Rβ, is involved in cellular proliferation and TGF-like activities. The PDGF-CC and PDGF–DD form a novel subgroup of the PDGF family distinguished by structural differences that include an *N*-terminal CUB domain. Widely expressed in multiple embryonic and adult cell, tissue, and organ types, PDGF-CC appears to be important for angiogenesis, cardiovasculature development, and tumorigenesis.

7.2. Delivery of Growth Factors

The use of growth factors has not always been achieved successfully *in vivo*. A major reason for this is the high diffusibility and short half-life time of growth factors *in vivo* to effectively retain their biological activities. Topical delivery of proteins remains in the site administered for a limited duration because of protein diffusion, proteolytic cleavage, and the early bioabsorption of carriers when carriers such as fibrin glue are applied. Application of growth factors in tissue engineering requires enhancement of their activities *in vivo* by means of adequate delivery systems.

The delivery methods include bolus injection; release of protein adsorbed directly on scaffold surfaces; in collagen sponge or in porous coatings; constantly delivery via osmotic pumps; and controlled release of protein trapped in an absorbable polymer. There have been reported numerous studies on the formulation of protein growth factors within absorbable polymers for use as drug delivery vehicles. Although trapping in absorbable polymers seems to yield formulations that can deliver active proteins, there is ample literature demonstrating that organic solvents can have a negative effect on protein association and function. By extension from normal tissue formation and repair, important variables in formulations for delivery systems include the concentration, timing, and sequence in which the growth factors are introduced.

To circumvent the difficulty of sustained release, high cost, and low commercial availability of growth factors, several methods have been attempted for the sustained

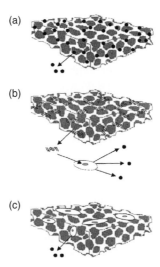

(a)

(b)

(c)

Fig. 1.27   Three primary polymeric growth factor delivery strategies: (a) growth factors are embedded within the polymer matrix and released; (b) genes encoding a growth factor are embedded within the polymer matrix and released, followed by cellular uptake and expression of the gene to produce growth factor; (c) growth factor is released from cells seeded on the polymer matrix that secrete the factor.

delivery of growth factors. As shown in Fig. 1.27 [25], the methods used to deliver growth factor molecules include the development of systems to deliver the protein itself, genes encoding the growth factor, or cells secreting the growth factor. Injection of recombinant plasmids and transplantation of gene-manipulated cells have been performed with an expectation that growth factors will be continuously produced from the modified cells for a certain period of time. The DNA most widely employed for such studies is those which are responsible for the biosynthesis of VEGF that induces vascularization. Plasmid vectors are relatively safe but vulnerable to nuclease attack and consequent inefficiency and expense. Viral vectors including adenoviruses, retroviruses, lentoviruses, etc., increase gene transfer efficiency but themselves have limitations including possible toxicologic and immunologic responses. Retroviruses are expressed only in proliferating cells and permanently integrate into genomic DNA.

8. CELL SOURCES

A human body consists of approximately 60 trillion cells. Human cells differentiate from stem cells into progenitor, precursor, and mature cells, forming a tree-like hierarchy structure as represented in Fig. 1.28. The starting cell for human body development is a fertilized ovum, which then forms an embryo as a result of repeated cleavages. The modura formed after 3–4 day and the blastocyst formed after 5–7 day cleavage contains pluripotent stem cells, named "embryonic stem" (ES) cells. Circulating blood cells survive for only a short time ranging from days to months. Throughout the entire

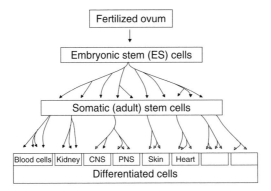

Fig. 1.28 From undifferentiated cells to differentiated cells.

life they are replenished by hematopoietic stem cells in bone marrow which provide a continuous source of progenitors for red cells, platelets, monocytes, granulocytes, and lymphocytes. In addition to hematopoietic stem cells, adult bone marrow contains also non-hematopoietic stem cells. The stem cells for non-hematopoietic tissues are referred to either as MSCs, because of their ability to differentiate into cells that can be roughly defined as mesenchymal, or as bone marrow stromal cells (BMSCs), because they appear to arise from the complex array of supporting stromal structures found in marrow. It has often been stated that a key factor in tissue engineering is lineage-committed precursor cells, especially multilineage stem cells, but almost differentiated cells also have large applications in the current tissue engineering.

Many different types of stem cell exist, but they all are found in very small populations in the human body; in some cases 1 stem cell in 100,000 cells in circulating blood. To identify these rare types of cells found in many different cells and tissues, scientists use stem cell markers. Each cell type has a certain combination of receptors on their surface that makes them distinguishable from other kinds of cells. In many cases, a combination of multiple markers is used to identify a particular stem cell type. Table 1.9 lists some of the markers commonly used to identify stem cells and to characterize differentiated cell types.

8.1. Differentiated Cells

Previous thinking was that many differentiated cells of the adult human have a limited capacity to divide, but some almost differentiated cells have widely been used in clinical tissue engineering. Among them are notably fibroblasts, keratinocytes, osteoblasts, endothelial cells, chondrocytes, preadipocytes, adipocytes, and tenocytes. In the following, we will focus on chondrocytes as representative of differentiated cells.

Despite a large number of studies achieved using chondrocytes, isolation of autologous chondrocytes for human use is invasive, requiring a biopsy from a

Table 1.9
Commonly used markers for stem cells and differentiated cell types

Tissue	Marker	Cell Type	Significance
Bone marrow	CD34$^+$ Scal$^+$ Lin$^-$ profile	Mesenchymal stem cell	Into adipocyte, osteocyte, chondrocyte, and myocyte
	CD29, CD44, CD105	Mesenchyme	A type of cell adhesion molecule
Bone	RunX		Transcriptional activatori
	ALP	Preosteoblast, osteoblast	Before the expression of OCN
	OCN	Osteoblast	Before and during matrix mineralization
Cartilage	Type II and IV collagen	Chondrocyte	Synthesized specifically by chondrocyte
	Sulfated proteoglycan	Chondrocyte	Synthesized by chondrocyte
Fat	ALBP	Adipocyte	Adipocyte lipid-binding protein
Nervous system	Nestin	Neural progenitor	A marker of neural precursors
	β–Tubulin–III	Neuron	Indicative of neuronal differentiation
Muscle	MyoD and Pax7	Myoblast, myocyte	Secondary transcriptional factors
	Myosin heavy chain	Cardiomyocyte	A component of contractile protein
	Myosin and MR4	Skeletal myocyte	Secondary transcriptional factors
Blood vessel	CD34	Endothelial progenitor	Cell surface protein
	Flkl	Endothelial progenitor, endothelial cell	Cell surface receptor protein
	VE-cadherin	Smooth muscle cell	Cell adhesion molecule

non-weight-bearing surface of a joint or a painful rib biopsy. In addition, the *ex vivo* expansion of a clinically required number of chondrocytes from a small biopsy specimen, which may itself be diseased, is hindered by deleterious phenotypic changes in the chondrocyte. Bone marrow–derived MSCs have been reported to differentiate into multiple cell types of mesenchymal origin. Bone marrow is a reservoir of both hematopoietic and non-hematopoietic stem cells. The MSCs produce tissues such as cartilage, bone, fat, and tendon. These tissues are used daily by plastic and orthopedic surgeons for the repair and augmentation of tissue defects. For these tissues or organ repair or replacement, tissue engineers seek to manipulate cell biology and cellular environments. However, to date, few tissue-engineered systems provide an autologous, minimally invasive, and easily customizable solution for the repair or augmentation of cartilage defects using MSCs.

The activity and function of articular chondrocytes during skeletal growth differ from those found after completion of growth. Chondrogenesis begins in the central core of the developing limb end. First, cartilage is formed from undifferentiated mesenchymal cells that cluster together and synthesize cartilage collagen, proteoglycans, and noncollagenous proteins. Type II collagen forms the primary component of the cross-banded collagen fibrils. The organization of these fibrils, into a tight meshwork that extends throughout the tissue, provides the tensile stiffness and strength of articular cartilage, and contributes to the cohesiveness of the tissue by mechanically entrapping the large proteoglycans. The tissue becomes recognizable as cartilage under light microscopy, when an accumulation of matrix separates the cells and they assume spherical shape. In growing individuals, the chondrocytes produce new tissue to expand and remodel the articular surface. With skeletal maturation, the rates of cell metabolic activity, matrix synthesis, and cell division decline. After completion of skeletal growth, most chondrocytes probably never divide, but rather continue to synthesize collagens, proteoglycans, and noncollagenous proteins. This continued synthetic activity suggests that the maintenance of articular cartilage requires substantial ongoing remodeling of the macromolecular framework of the matrix, and the replacement of degraded matrix macromolecules. With aging, the capacity of the cells to synthesize some types of proteoglycans and their response to stimuli, including growth factors, decrease. These age-related changes may limit the ability of the cells to maintain the tissue, and thereby contribute to the development of degeneration of the articular cartilage.

To promote restoration of normal biomechanical function and long-term integrity of the articular cartilage, integration with surrounding cartilage and subchondral bone is important. Otherwise, the osteochondral repair construct would delaminate or subside. A lack of the integration is the reason in part why tissue engineering approaches to osteochondral defect repair have had limited success. In an attempt to enhance the integration, tissue engineering strategies have utilized heterogeneous constructs in which the upper region promotes cartilage repair while the lower region is specifically designed to encourage bone integration. The ability of chondrocyte-seeded scaffolds to promote repair of the subchondral bone has been variable, due in part to differences in the chondrocyte phenotype within tissue-engineered cartilage constructs. The cellular component of an osteochondral repair strategy may provide critical signals for enhancing bone repair that are intrinsically lacking in an acellular or devitalized implant. During growth and endochondral bone repair, chondrocytes progress to a hypertrophic phenotype and play an important role in angiogenesis, osteoblast recruitment, and mineralized matrix formation. Conversely, articular chondrocytes do not express osteoinductive factors and may inhibit bone formation by producing antiangiogenesis factors such as tissue inhibitors of metalloproteases and troponin. Therefore, chondrocytes within a tissue-engineered cartilage construct have the potential to influence adjacent bone repair response.

Chondrocytes *in vivo* respond to chemical stimuli as well. Two such biochemical regulators of matrix biosynthesis found in articular joints are TGF-β1 and IGF-I. The IGF-I enhances matrix biosynthesis and mitotic activity in chondrocytes, decreases matrix catabolism, and can enhance the tissue properties of cartilage explants in long-term culture. The TGF-β1 exerts a similar influence on matrix biosynthesis, directing production toward increased amounts of larger, more anionic proteoglycan species.

The phenotype change is well known for chondrocytes which are responsible for the hyaline cartilage consisting of type II collagen in the normal articular cartilage, but under irregular culture conditions the same cells produce the fibrous cartilage made from type I collagen which is inferior to the hyaline cartilage with respect to mechanical properties required for articular cartilage. This means that the cells lose the chondrocyte phenotype and become fibroblast-like cells.

8.2. Somatic (Adult) Stem Cells

Mature cells, when allowed to multiply in an incubator, ultimately lose their effectiveness. Consequently, scientists are turning to other cell types. To be effective, cells must be easily procured and readily available; they must multiply well without losing their potential to generate new functional tissue; they should not be rejected by the recipient and not turn into cancer; and they must have the ability to survive in the low-oxygen environment normally associated with surgical implantation. Mature adult cells fail to meet many of these criteria. The oxygen demand of cells increases with their metabolic activity. After being expanded in the incubator for significant periods of time, they have a relatively high oxygen requirement and do not perform normally. A hepatocyte, for example, requires about 50 times more oxygen than a cell such as a chondrocyte and much attention has turned to progenitor cells and stem cells. True stem cells can turn into any type of cell, while progenitor cells are more or less committed to becoming cell types of a particular tissue or organ. Somatic adult stem cells may actually represent progenitor cells in that they may turn into all the cells of a specific tissue but not into any cell type. Somatic stem cells can be procured from the individual needing the new tissue and thus not be rejected. Since these cells are immature, they will survive a low-oxygen environment. Somatic stem cells normally reside within specific extracellular regulatory microenvironments—stem cell niches—consisting of a complex mixture of soluble and insoluble, short- and long-range ECM signals, which regulate their behavior. These multiple, local environmental cues are integrated by cells that respond by choosing self-renewal or a pathway of differentiation. Outside of their niche, adult stem cells lose their developmental potential quickly.

Somatic stem cells had been claimed to possess an unexpectedly broad differentiation potential that could be induced by exposing stem cells to the extracellular developmental signals of other lineages in mixed-cell cultures. This stem cell plasticity was thus thought to form the foundation for one of the multiple prospective uses of adult stem cells in regenerative medicine. However, experimental evidence supporting the existence of stem-cell plasticity has been refuted because stem cells have been shown to adopt the functional features of other features by means of cell-fusion-mediated acquisition of lineage-specific determinants (chromosomal DNA) rather than by signal-mediated differentiation.

Data demonstrating that stem cells could fuse with and subsequently adopt the phenotypes of other cell types indicated that the very co-culture assays originally interpreted to support plasticity instead were artifacts of cell fusion.

Two types of stem cell are available: ES cells and somatic or adult autologous stem cells. The clinical use of ES cells has critical problems of cell regulation including malignant potential, allogeneic immune response, and ethical issues concerning the cell source, while the problem with autologous stem cells concerns their cell source. Bone marrow is the most suitable cell source, because it involves not only hematopoietic stem cells but also MSCs. The potential advantages of using bone marrow MSCs include low cell numbers required at the initial culture, relative simple procedure for bone marrow harvest, and the cell maintenance of high biological activity from older donors. However, the yield of MSCs obtained from aspirated bone marrow blood is too low (approximately 1 per 10^5 adherent stromal cells) to use them as clinical cell source for tissue regeneration and, therefore, an *ex vivo* cell expansion will be necessary. The frequency of long-term repopulating cells is 1 in 35,000 total epidermal cells, or in the order of 1 in 10^4 basal epidermal cells, similar to that of hematopoietic stem cells in the bone marrow [26]. Hematopoietic stem cell frequency in the bone marrow was determined to be 1 in 10,000.

Stem cells have not taken on the identity of any specific cell type and are not yet committed to any dedicated function; they can divide indefinitely and may be induced to give rise to one or more specialized cell types. It seems very likely that each tissue or organ has one somatic stem cell even in adults. A well-known somatic stem cell is the MSC that is able to differentiate into a variety of tissues including skin, cartilage, bone, muscle, and fat, as illustrated in Fig. 1.29 [27]. It is clearly seen how versatile this adult stem cell is for clinical application. Stem cells of embryonic as well as adult tissue origin undergo the differentiation process and eventually reach functional maturity. In the use of stem cells as part of tissue engineering, cellular behaviors including the differentiation process must be carefully monitored.

Although the potential of ES cells is enormous, the use of embryonal sources of stem cells is controversial and major ethical and political issues impede their use. Issues surrounding the rights of the unborn fetus, and subsequent government regulation and limitations on availability and applicability of embryonic tissue, have put the brakes on what appeared to be a rapidly approaching clinical reality. In contrast, recent interest has emerged in the use of bone marrow–derived stem cells for tissue engineering applications. Adult-derived precursors potentially provide ample quantities of an autologous source of regenerative tissue without these ethical and political issues. Although demonstrations of bone marrow–derived MSC plasticity have been reported and debated, widespread use of adult-derived tissue will likely require a relatively painless, convenient, and safe procurement method. Some have suggested that the skin fulfills this role, while other reports have begun to emerge suggesting that adipose tissue—which is electively aspirated in large quantities—provides a readily available autologous source. Like bone marrow, adipose is supported by a stroma whose isolation yields a significant amount of cells capable of

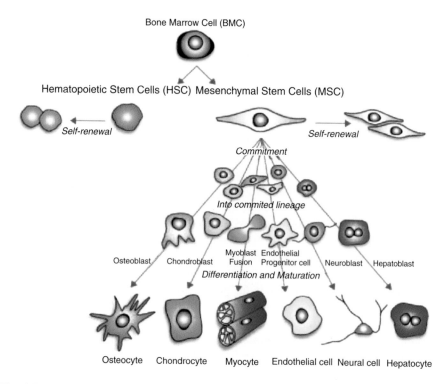

Fig. 1.29 MSCs differentiation cascade. There are at least two types of stem cells in bone marrow, namely hematopoietic stem cells and MSCs. One type of MSC repeats self-renewal, whereas the other type is committed to a specific cell lineage and goes through a lineage process. MSCs are reported to differentiate into a variety of cellular types, such as osteocytes, chondrocytes, myocytes, endothelial cells, neurons, hepatocytes, etc.

osteogenic, adipogenic, neurogenic, myogenic, and chondrogenic differentiation. Given that adipose-derived mesenchymal precursors can be harvested in abundant quantities under local anesthesia with little patient discomfort, they may emerge as an important source for cell-based therapy. One can envision a scenario in which a mesenchymal cell fraction is purified from a patient's bone marrow or liposuction aspirate, is exposed to oxygen and other environmental conditions optimized for differentiation along a certain lineage, and is ultimately returned to the same patient to fill a tissue defect.

Some other tissues have been shown to be sources of somatic stem cells. Periosteum, as well as adipose tissue and peripheral blood, is a good source that removes the confounding effect of hematopoietic stem cells. Periosteum has been shown to be an effective source of cells in the repair of osteochondral defects in an animal model but again suffers from painful procurement procedures and low cell yield. On the basis of the currently available therapeutic options, the ideal reconstitutive measure would cause insignificant donor morbidity, regenerate quickly the harvested tissue, have no size limitations, be readily available, have no issues of immunogenicity, and be of low cost. Good cell volume would obviate the need for *ex vivo* expansion and easy availability would allow for expensive preliminary *in vitro* human cell testing before clinical trials.

8.2.1. MSCs

Circulating blood cells survive for only a few days or months. This means that hematopoietic stem cells in bone marrow must provide a continuous source of progenitors for blood cells. Bone marrow also contains cells that meet the criteria for stem cells of non-hematopoietic tissues. The stem-like cells for non-hematopoietic tissues are referred to as MSCs, because they can be differentiated in culture into osteoblasts, chondrocytes, adipocytes, and myoblasts. The marrow stromal cells can be isolated from other cells in marrow by their tendency to adhere to tissue culture plastic.

The presence of stem cells for non-hematopoietic cells in bone marrow was first suggested by a German pathologist Cohnheim 140 years ago [28]. He studied wound repair by injecting an insoluble dye into the veins of animals and then looking for the appearance of dye-containing cells in wounds he created at a distant site. He concluded that most of the cells appearing in the wounds came from the bloodstream and, by implication, from bone marrow. This work raised the possibility that bone marrow might be the source of fibroblasts that deposit collagen fibers as part of the normal process of wound repair.

Evidence that bone marrow contains cells that can differentiate into fibroblasts as well as other mesenchymal cells has been available since the pioneering work of Friedenstein *et al.* [29]. They placed samples of whole bone marrow in plastic culture dishes, and, after 4 h, poured off the cells that were nonadherent. In effect, they discarded most of the hematopoietic stem cells. The small number of adherent cells were heterogenous in appearance, but the most tightly adherent cells were spindle shaped and formed foci of two to four cells. The cells in the foci remained dormant for 2–4 days and then began to multiply rapidly. After passage several times in culture, the adherent cells became more uniformly spindle shaped in appearance. The cells had the ability to differentiate into colonies that resembled small deposits of bone or cartilage.

Adult or somatic MSCs, commonly referred to as BMSCs, are stem cells originated from embryonic mesoderm and are a unique class of multipotent cells that are noncommitted and remain in an undifferentiated state. When induced by the appropriate biological cues including right chemicals, hormones, and growth factors, BMSCs are capable of extensive proliferation and differentiation into several phenotypes including fibrous tissues (*e.g.*, tendon), and hematopoiesis-supporting reticular stroma, and represent a heterogeneous cell population likely containing a range of progenitor cells. Of most importance from a tissue engineering standpoint is the fact that stem cells that have been transformed to osteoblasts are much like bone cells of a developing organism, and are capable of secreting large amounts of

ECM. Moreover, the ease of isolation of MSCs makes them very attractive for tissue engineering applications; aspiration of bone marrow is only slightly more complicated than a blood donation, and MSCs can be enriched to obtain a relatively pure population of cells. The technique using bone marrow cells (BMCs) does not require cell culture with serum from other species, which might be associated with a risk of infection. In addition, as large numbers of cells can be obtained from BMCs without culture, fewer steps in the preparation of tissues would mean a lower risk of contamination and less work and time. Fewer materials for culture would mean cost benefits, and the technique can be applied even in emergency cases in a clinical setting, as cell seeding requires only a few hours. In culture of cells generated from suspensions of marrow, colonies form from a single precursor cell termed the "colony-forming-unit fibroblast" (CFU-F). The progeny of these CFU-Fs is what have been defined as BMSCs. During prolonged cultivation *ex vivo*, adult BMSCs undergo two possible interdependent procedures: replicative aging and a decline in differentiation potential.

One strategy for tissue engineering is to use these MSCs or BMSCs harvested from marrow stroma. The MSCs have attracted extensive interest due to the surprising finding that they can also commit to neural cell lineages in both animal models and cell cultures. Under certain culture conditions, MSCs were induced to assume morphology and express protein markers typical of neurons. Adult MSCs were found to migrate into the brain and develop into astrocytes. These results suggest that MSCs may be useful also in neural tissue repair and regeneration. The use of MSCs has significant advantages over other cell types. First, the use of adult MSCs means that autologous transplantation is possible, thus avoiding detrimental immune responses often caused by allogeneic transplantation. Second, ethical concerns associated with the use of ES cells or fetal tissues are eliminated. Last, unlike neural stem cells (NSCs) or other neural precursors that need to be obtained via surgery, MSCs are relatively easy not only to obtain from a small aspirate of bone marrow, but also to expand in culture under conditions in which they retain some of their potential to differentiate into multiple cell lineages. For clinical use, it might be desirable to expand and differentiate MSCs into the cell types of choice *in vitro* before transplantation, because this strategy allows reproducible and reliable generation of well-defined transplants in a precisely controlled environment. Besides making tissue-engineered constructs of a single tissue type, an important potential advantage in using MSCs is the ability to design multifunctional or composite tissue constructs, such as osteochondral grafts, from a single cell source. Interestingly, there is evidence to suggest that certain types of bone marrow–derived cells can themselves produce an angiogenic mediator VEGF. This observation raises the possibility that these mesenchymal cells could potentially regulate angiogenesis.

Several clinical trials with human MSCs, or closely related cells, have started. For most of the human trials and animal experiments, MSCs are prepared with a standard protocol in which mononucleated cells are isolated from a bone marrow aspirate with a density gradient, and then both enriched and expanded in the presence of FCS by their tight adherence to plastic tissue culture dishes. For instance, bone marrow aspirates are mixed with Hanks' balanced salt solution (HBSS) containing heparin. The same amount of high-density Ficoll solution is carefully placed beneath the HBSS mixture followed by centrifugation. The nucleated cell layer existing at the interface between the HBSS and the Ficoll is removed and placed on a dish. After removal of non-adherent cells and the subsequent continuation of cell culture, cells form colonies from which MSCs are obtained. Cultures of human MSCs, unlike murine cells, become free of hematopoietic precursors after one or two passages and can be extensively expanded before they senesce. However, when the cells are expanded under standard culture conditions, they lose their proliferative capacity and their potential to be differentiated into lineages such as adipocytes, tenocytes, and chondrocytes. Moreover, cultures of human MSCs are morphologically heterogeneous, even when cloned from single-cell-derived colonies.

Studies on human MSCs become complicated by the fact that there is no consensus as to the characteristic surface epitopes that can be used to identify the cells. A series of antibodies to surface epitopes have been employed by several investigators, but none have been into general use. Sekiya *et al.* screened over 200 antibodies but did not find any that efficiently distinguish rapidly self-renewing cells from mature MSCs [30]. It is therefore difficult to compare the results that different research groups obtained either in animal models for diseases or in clinical trials. It follows that several parameters must be considered in preparing frozen stocks of human MSCs: (1) variations in the quality and number of MSCs obtained from different bone marrow aspirates, (2) the yield of cells required as frozen stocks, (3) the quality of the cultures in terms of their content of early progenitor cells that replicate most rapidly and have the greatest potential for multilineage differentiation, and (4) the number of cell doublings the cells have undergone before they are harvested and frozen.

Autologous MSCs have advantages over ES cells that may lead to teratocarcinoma formation. However, compared with ES cells, which have an unlimited proliferative life span (period before the cells reach growth arrest in culture) and consistently high telomerase activity, MSCs have very poor replicative capacity and short proliferative longevity. Thus, an important challenge in tissue engineering is to improve the replicative capacity of MSCs, thereby to obtain a number of MSC sufficient to repair large defects. Forced expression of telomerase in MSCs markedly increased their proliferative life span and MSCs with a high telomerase activity showed osteogenic potential.

Cell Expansion

The most common source of adult-derived stem cells is the bone marrow. MSC can be obtained easily and repeatedly by bone marrow aspiration, but isolation of marrow aspirates in great volume causes damage and pain, and it is difficult to isolate from the bone marrow 10^7–10^8 MSCs that are required for regeneration of large injured tissues. In addition, the heterogeneous nature of bone marrow with both hematopoietic and MSCs confounds the result of various therapies using bone marrow. The scarcity of MSC in bone marrow of older donors may impose additional requirements with respect to cell expansion and differentiation. Furthermore, a short life span of MSC and a reduction in their differentiation potential in culture have limited their clinical application. Thus, the ability to rapidly expand MSCs in culture is of obvious importance in using the cells for tissue engineering. The expansion of MSCs *in vitro* is a prerequisite for autologous cell transplantation. In other words, identification of growth factors that stimulate the proliferation of MSCs and support their multilineage differentiation potential is a critical step towards the clinical application of MSCs.

Cell Differentiation

The forces driving stem cell differentiation, or maintaining stem cells in a state of suspended undifferentiation, include secreted and bound messengers or homing signals. Specific chemicals and hormones that cause the transformation of MSCs to osteoblasts, chondrocytes, and adipocytes have been elucidated. Table 1.10 represents the commonly used *in vitro* environments for differentiations. For instance, osteogenic differentiation occurs when MSCs are treated with dexamethasone, β-glycerophosphate, and ascorbic acid (AA), and this differentiation to osteoblasts is characterized by gene expression of osteopontin (OP) and alkaline phosphatase (ALP). Just as important are the cellular environmental sensors sensitive to oxygen, temperature, chemical gradients, mechanical forces, and others cues in the microenvironment. The notion of microenvironments affecting stem cell division and function is not new. Schofield

Table 1.10

Lineage specific differentiation induced by media supplementation

Medium	Media	Serum	Supplementation
Osteogenic	Dulbecco's minimal essential medium (DMEM)	10% Fetal calf serum (FCS)	50 μM ascorbic acid 2-phosphate, 10 mM β-glycerophosphate, 100 nM dexamethasone
Chondrogenic	High-glucose DMEM		10 ng/ml transforming growth factor (TGF)-β3,100 nM dexamethasone, 6 μg/ml insulin, 100 μM ascorbic acid 2-phosphate, 1 mM sodium pyruvate, 6 μg/ml transferrin, 0.35 mM proline, 1.25 mg/ml bovine serum albumin
Adipogenic	DMEM	10% FCS	1 μM dexamethasone, 0.2 mM indomethacin, 10 μg/ml insulin, 0.5 mM 3-isobutyl-l-methylxanthine
Myogenic	DMEM	10% FCS, 5% horse serum	10 μM 5-azacytidine, 50 μM hydrocortisone
Cardiac	Isocove's modified DMEM	20% FCS	3 μM 5-azacytidine

dubbed these "niches" with respect to hematopoietic stem cells, and subsequent reports have described their presence in numerous tissues including neural, germline, skin, intestinal, and others [31].

8.2.2. Adipose-Derived Stem Cells

A second large stromal compartment found in human subcutaneous adipose tissue has received attention because of the presence of multipotent cells named adipose-derived adult stem (ADAS) cells. Under lineage-specific biochemical and environmental conditions, ADAS cells will differentiate into osteogenic, chondrogenic, myogenic, adipogenic, and even neuronal pathways. Although it remains to be determined whether ADAS cells meet the definition of stem cells, they are multipotential, are available in large numbers, are easily accessible, and attach and proliferate rapidly in culture, making them an attractive cell source for tissue engineering. Moreover, ADAS cells demonstrate a substantial *in vitro* bone formation capacity, equal to that of bone marrow, but are much easier to culture. It has been empirically shown that several growth factors and hormones, in combination with cellular condensation and rounded cell morphology, may promote the chondrogenic differentiation of ADAS cells.

8.2.3. Umbilical Cord Blood-Derived Cells

Human umbilical cord blood-derived cells may be alternative autologous or allogeneic cell source. Umbilical cord blood cells contain multipotent stem cells and these cells have been used to generate tissue-engineered pulmonary artery conduits in a pulsatile bioreactor [32]. Cells from umbilical cord artery, umbilical cord vein, whole umbilical cord, and saphenous vein segments were compared for their potential as cell sources for tissue-engineered vascular grafts [33]. All four cell sources generated viable myofibroblast-like cells with ECM formation including types I and III collagen and elastin. There were also CD34$^-$ umbilical cord cells which have the capacity to generate endothelial cells.

8.3. Cell Therapy

Stem cell therapy is an emerging field, but current clinical experience is limited. One limitation of cell therapy is that donor cells often cannot functionally replace the impaired cells immediately following transplantation and this delay may impact patient survival. In the field of adult stem cell therapy, it is estimated that there are currently over 80 therapies and around 300 clinical trials underway using such cells [34]. Hematopoietic stem cell transplants are routine clinical practice and more than 300 patients with type I diabetes have now undergone transplants of islet cells using the so-called "Edmonton protocol", with a significant proportion staying off insulin injections for several years. But the field remains beset by problems of reproducibility. Often, the level of therapeutic benefit is minimal as implanted cells die because of immune attack or other problems. It is commonly not clear whether cells are expanding, fusing with recipient cells, or exerting an effect through secreted growth factors. Unraveling these problems will require progress on several fronts.

8.3.1. Angiogenesis

Adult bone marrow is a rich reservoir of tissue-specific stem and progenitor cells. Among them is a scarce population of cells known as "endothelial progenitor cells" (EPCs) that can be mobilized to the circulation and contribute to the neoangiogenic processes. Circulating endothelial progenitor cells (CEPs) have been detected in the circulation either after vascular injury or during tumor growth. The CEPs primarily originate from EPCs within the bone marrow and differ from sloughed mature, circulating endothelial cells (CECs) that randomly enter the circulation as a result of blunt vascular injury. Preclinical studies have shown that introduction of bone marrow–derived endothelial and hematopoietic progenitors can restore tissue vascularization after ischemic events in limbs, retina, and myocardium [33]. Co-recruitment of angiocompetent hematopoietic cells delivering specific angiogenic factors facilitates incorporation of EPCs into newly sprouting blood vessels. Given the morbidity associated with limb ischemia, vascular stem cell therapy provides promising adjunct therapy to current bypass surgical approaches. In preclinical studies, introduction of bone marrow–derived EPCs effectively improves collateral vessel formation, thereby minimizing limb ischemia. In patients suffering from peripheral arterial disease, placement of autologous whole bone-marrow MNCs into ischemic gastrocnemius muscle resulted in restoration of limb function. Because MNCs contain both EPCs and myeloid cells, it remains to be determined whether the improvement in these studies was due in part to the introduction of myelomonocytic cells.

Given the complexity of organ-specific microenvironmental cues essential for functional incorporation of transplanted EPCs and CEPs, numerous hurdles have to be overcome for successful stem cell therapy for tissue vascularization. As damaged tissue may lose anatomical cues for functional organ neovascularization, *in vitro* manipulation of stem cells may be essential to facilitate *in vivo* incorporation. Identification of organ-specific cytokines—including the appropriate combinations of VEGFs, FGFs, PDGFs, IGFs, angiopoietins, other as-yet-unrecognized factors and ECM components for optimal culture conditions—will provide the platform for the differentiation of stem cells, allowing delivery of a large number of angiocompetent cells.

8.3.2. Cardiac Malfunction

Myocardial infarctions most commonly result from coronary occlusions, due to a thrombus overlying an atherosclerotic plaque. Because of its high metabolic rate, myocardium (cardiac muscle) begins to undergo irreversible injury within 20 minutes of ischemia, and a wavefront of cell death subsequently sweeps from the inner layers toward the outer layers of myocardium over a 3- to 6-h period. Although cardiomyocytes are the most vulnerable population, ischemia also kills vascular cells, fibroblasts, and nerves in the tissue. Myocardial necrosis elicits a vigorous inflammatory response. Hundreds of millions of marrow-derived leukocytes, initially composed of neutrophils and later of macrophages, enter the infarct. The macrophages phagocytose the necrotic cell debris and likely direct subsequent phases of wound healing. Concomitant with removal of the dead tissue, a hydrophilic provisional wound repair tissue rich in proliferating fibroblasts and endothelial cells—termed "granulation tissue"—invades the infarct zone from the surrounding tissue. Over time, granulation tissue remodels to form a densely collagenous scar tissue. In most human infarcts, this repair process requires 2 months to complete. At the organ level, myocardial infarction results in thinning of the injured wall and dilation of the ventricular cavity. These structural changes markedly increase mechanical stress on the ventricular wall and promote progressive contractile dysfunction. The extent of heart failure after a myocardial infarction is directly related to the amount of myocardium lost.

Related to myocardial repair following injury, the limited proliferative capacity of mature cardiomyocytes is the fundamental reason that numerous investigators have evaluated alternative cell sources. Cellular cardiomyoplasty to replace damaged myocardial cells has been attempted using a variety of cell types including bone-marrow precursor cells, skeletal myoblasts, satellite cells, muscle-derived stem cells, smooth muscle cells, late embryonic/fetal and neonatal cardiac cells, and ES cells. Umbilical cord blood cells have the capacity to form endothelial cells and myofibroblasts. Given this level of excitement, it is hardly surprising that the theory-to-therapy approach to the use of cells would be thrust forward into translational human studies at the earliest possible stage [34]. A prompt action came in 2001 with the publication of a paper in Nature by Orlic and colleagues, suggesting that stem cells derived from bone marrow can replace heart muscle lost as a result of heart attack, and can improve cardiac function. Injecting bone-marrow stem cells into an injured heart potentially represented a new therapy, triggering the launch of numerous clinical studies to investigate the effect of directly injecting these cells into the damaged heart muscle of patients following a heart attack. A major barrier to the long-term success of cellular cardiomyoplasty is the survival of the transplanted cells.

Recent reports suggest that hematopoietic stem cells can transdifferentiate into unexpected phenotypes such as skeletal muscle, hepatocytes, ECs, neurons, endothelial cells, and cardiomyocytes, in response to tissue injury or placement in a new environment. Although most studies suggest that transdifferentiation is extremely rare under physiological conditions, extensive regeneration of myocardial infarcts was reported after direct stem cell injection, promoting several clinical trials. Under conditions of tissue injury, myocardial replication and regeneration have been reported and a growing number of investigators have implicated adult bone marrow in this process, suggesting that marrow serves as a reservoir for cardiac precursor cells. It remains unclear which BMCs can contribute to myocardium, and whether they do so by transdifferentiation or cell fusion.

Independent studies by Murry *et al.* [35] and Balsam *et al.* [36] seriously challenged Orlic and colleagues' initial observations and the scientific underpinnings of the ongoing human studies [37]. Two strategies were used to show that bone-marrow stem cells do not take on the role of damaged heart cells. Murry *et al.* isolated and purified genetically modified bone-marrow stem cells from mice. The modification "tagged" the cells (with LacZ), enabling them to be detected in the recipient mouse heart, into which the cells were directly injected. Closer inspection of the recipient heart showed that the label could not be detected in heart muscle cells. Similar results were shown by Balsam *et al.*, although the approach was slightly different. Donor bone-marrow stem cells were transfused directly into the circulation of recipient. Again, the tag (GFP) could not be detected in heart muscle cells of the donor; indeed, the BMCs continued to differentiate into blood cells while in the heart. So, scientists are asking why there are wide discrepancies between the earlier report and the current investigations. As Murry *et al.* suggest, the differences may arise from the difficulty of tracking the *in vivo* fate of transplanted cells within an intact organ. Orlic *et al.* mainly relied on detecting unique protein constituents of bone-marrow stem cells using fluorescently tagged antibodies. Murry *et al.* and Balsam *et al.*, however, created intrinsic genetic markers that can be easily recognized without antibody staining. Owing to its high density of muscle-specific contractile proteins, intact heart muscle tends to have high inherent background fluorescence, and can also display non-specific antibody binding to the abundant muscle proteins. This makes it difficult, even for the most experienced labs using the most specialized microscopes, to track cell fate by simply using techniques that rely on fluorescent antibody staining of cardiac proteins. Various experiments using several types of stem cells support the view that transdifferentiation occurs rarely, if at all, in many organ systems, including heart muscle. Less than 2% of the transplanted or injected cells take on the *in vivo* fate of heart cells. If this is true, then the improvement in cardiac function seen by Orlic *et al.* might have arisen not because the stem cells transdifferentiated, but because new blood vessels were encouraged to grow around the injected area. Such growth of new blood vessels has been consistently found in transplantation studies of diverse cell types in the heart. Studies in large-animal models of transplanted bone-marrow-derived stem cells in the injured heart also failed to document cardiac regeneration [38]. Again, the implication is that any functional improvement seen may not be related to an increase in functioning heart muscle *per se*. A recent clinical study, in which bone-marrow stem cells were transplanted into injured hearts, was terminated because of serious cardiac side effects that threatened the blood flow to the heart. This again suggests that further experimental testing is warranted in large-animal model systems.

8.4. ES Cells

The first report of a stable ES cell line derived from a human blastocyst in 1998 was by Thomson *et al.* [39]. This led to a surge of interest in ES cells as a potential cell source for tissue engineering. The ES cells are derived from the inner cell mass of the preimplantation blastocyts. They are pluripotent and can be maintained and expanded in culture in an undifferentiated state. Markers such as Oct-4, SSEA-4, and Nanog have all been used to characterize and assess the pluripotent capacity of

ES lines when grown *in vitro*. However, the precise mechanisms by which the culture methods routinely employed in laboratories enable ES cells to remain pluripotent are still not fully understood and it is likely that further key genes that prevent the cell from proceeding with differentiation remain to be identified.

Human ES cell lines (hES) are pluripotent diploid cells that can proliferate in culture indefinitely and provide a unique system for studying the events in human embryonic development. The hES cells have the potential to generate all embryonic cell lineages when they undergo differentiation. Differentiation of hES can be induced in monolayer culture or by removing the cells from their feeder layer and growing them in suspension. This differentiation in suspension results in aggregation of the cells and formation of embryoid bodies (EBs), where successive differentiation steps occur.

From human ES cells we might be able to develop new transplantation therapies to replace diseased or aged cells or tissues. To this end, researchers need to develop methods with which they can derive from human ES cells their required cell types, such as cardiomyocytes or hematopoietic cells. Chemical cues provided directly by growth factors or indirectly by feeder cells can induce ES cell differentiation toward specific lineages. While ES cells show great promise for treating many diseases, such as heart disease, diabetes, and Parkinson's disease, non-matching ES cells would be rejected by patients' immune systems unless they take immunosuppressant drugs. The HLA (human histocompatibility leukocyte antigen) system has a central role in the initiation and development of immune rejection. However, hES cells and their differentiated progeny express highly polymorphic MHC (major histocompatibility complex) molecules that serve as major graft rejection antigens to the immune system of allogeneic hosts. To achieve sustained engraftment of donor cells, strategies must be developed to overcome graft rejection without broadly suppressing host immunity. One approach entails induction of donor-specific immune tolerance by establishing chimeric engraftment in hosts with hematopoietic cells derived from an existing hES cell line. To achieve best possible MHC matching we could establish large banks of HLA-defined and highly diversified hES cell lines, but this strategy might not be sufficient since minor rejection antigens are still present and difficult to define. Immunosuppressive drugs such as cyclosporine are administered to transplant recipients to prevent acute and chronic immune-mediated rejection of allogeneic bone marrow and organ transplants even with best possible MHC matching. Polymorphisms in many non-HLA histocompatibility antigens, including highly polymorphic mitochondrial and H-Y gene products, result in rejection even in HLA-matched individuals.

8.4.1. Cell Expansion and Differentiation

Routine propagation of mouse ES cells in an undifferentiated state can be achieved by culture upon mitotically inactivated mouse embryonic fibroblasts (MEFs) or upon gelatin-coated dishes in the presence of the interleukin-6 family member cytokine leukemia inhibitory factor (LIF). The LIF stimulates ES cell self-renewal following binding of the LIF receptor β/gp130 heterodimer and activation of the JAK/Stat3 signaling passway. Although the mechanism by which gelatin aids in the maintenance of ES cells in an undifferentiated state is not known, it is clear that surface properties of specific substrates can exert powerful effects upon cell growth and behavior.

By controlling the culture conditions under which ES cells are allowed to differentiate, it is possible to generate cultures that are enriched for lineage-specific precursors. To this end, stem cells require an additional ability to control growth and differentiation into useful cell types. The effects of biomaterials on the behavior of stem cells have not been studied in great detail. This is due in part to the potential diversity of biomaterials and the difficulty of large-scale hES cell production,

8.4.2. Somatic Cell Nuclear Transfer

The isolation of pluripotent hES cells and breakthroughs in somatic cell nuclear transfer (SCNT) in mammals have raised the possibility of performing human SCNT and generated potentially unlimited sources of undifferentiated cells for use in research, with potential applications in tissue repair and transplantation medicine. The SCNT concept, known as "therapeutic cloning", refers to the transfer of the nucleus of a somatic cell into an enucleated donor oocyte. In theory, the oocyte's cytoplasm would reprogram the transferred nucleus by silencing all the somatic cell genes and activating the embryonic ones. The ES cells would be isolated from the inner cell mass of the cloned preimplantation embryo. When applied in a therapeutic setting, these cells would carry the nuclear genome of the patient; therefore, after directed cell differentiation, the cells could be transplanted without immune rejection to treat degenerative disorders such as diabetes, osteoarthritis (OA), and Parkinson's disease among others.

A team led by veterinary cloning expert Woo Suk Hwang and gynecologist Shin Yong Moon of Seoul National University in South Korea showed that the cloning technique can work in humans [40]. The researchers described how they created a human ES cell line by inserting the nucleus of a human cumulus cell into a human egg from which the nucleus had been removed. (Cumulus cells surround the developing eggs in an ovary, and in mice and cattle they are particularly efficient nucleus donors for cloning.) After using chemicals to prompt the reconstructed egg to start dividing, the team allowed it to develop for a week to the blastocyst stage, when the embryo forms a hollow ball of cells. They then removed the cells that in a normal embryo are destined to become the placenta, leaving the so-called "inner-cell mass" that would develop into the fetus. When these cells are grown in culture, they can become ES cells, which reproduce indefinitely and retain the ability to form all the cell types in the body. The ES cell line the team derived seems to form bone, muscle, and immature brain cells, for example. The South Korean scientists suspect that their method of removing the egg's nucleus might have been one of the secrets of their success. Instead of sucking the nucleus out with a pipette, which in past work seemed to damage the protein machinery that controls cell division, the team nicked a small hole in the egg's membrane and gently squeezed out the genetic material. Perhaps the Korean scientists' most important advantage was the whopping 242 eggs they had to work with. The team obtained oocytes and donor cells from 16 healthy women, who underwent hormone treatment to stimulate their ovaries to overproduce maturing eggs. The women who donated specifically for the experiments were not compensated, and were informed that they would not personally benefit from the research.

In this study the adult cell (donor cell) and the egg (oocyte) came from the same person. This made difficult to prove that the embryo really was cloned. The process was also very inefficient, taking 242 eggs to create just one ES cell line. However, in 2005 the Hwang's team created 11 more ES cell lines from cloned embryos in an impressive study that answers all the criticisms of their original study. They have also greatly increased the efficiency of the process: the 11 lines came from just 185 fresh eggs donated by 18 unpaid volunteers, meaning an average of only 17 eggs was needed per ES cell line [41]. The donor "adult" cells came from patients aged 2–56, with a variety of conditions ranging from spinal injuries to an inherited immune condition. The work proves that matching ES cells can be derived via therapeutic cloning from donors of any age and sex.

REFERENCES

1. A.K. Lynn, I.V. Yannas, and W. Bonfield, Antigenicity and immunogenicity of collagen, *J. Biomed. Mater. Res. B: Appl. Biomater.*, **71B**, 343 (2004).
2. K. Tomihata and Y. Ikada, *In vitro* and *in vivo* degradation of films of chitin and its deacetylated derivatives, *Biomaterials*, **18**, 567 (1997).
3. K. Tomihata, M. Suzuki, and Y. Ikada, The pH dependence of monofilament sutures on hydrolytic degradation, *J. Biomed. Mater. Res.*, **58**, 511 (2001).
4. S.-H. Hyon, W.-J. Cha, T. Nakamura *et al.*, Synthesis and properties of elastomeric lactide-caprolactone copolymers, *J. Japanese Soc. Biomater.* (in Japanese), **14**, 216 (1996).
5. V. Karageorgiou and D. Kaplan, Porosity of 3D biomaterial scaffolds and osteogenesis, *Biomaterials*, **26**, 5474 (2005).
6. T.B.F. Woodfield, J. Malda, J. de Wijn *et al.*, Design of porous scaffolds for cartilage tissue engineering using a three-dimensional fiber-deposition technique, *Biomaterials*, **25**, 4149 (2004).
7. S.J. Hollister, Porous scaffold design for tissue engineering, *Nat. Mater.*, **4**, 518 (2005).
8. K. Fujihara, M. Kotaki, and S. Ramakrishna, Guided bone regeneration membrane made of polycaprolactone/calcium carbonate composite nano-fibers, *Biomaterials*, **26**, 4139 (2005).
9. S.F. Badylak, The extracellular matrix as a scaffold for tissue reconstruction, *Semin. Cell Dev. Biol.*, **13**, 377 (2002).
10. J. Parizek, P. Mericka, J. Spacek *et al.*, Xenogeneic pericardium as a dural substitute in reconstruction of suboccipital dura mater in children, *J. Neurosurg.*, **70**, 905 (1989).
11. U. Hersel, C. Dahmen, and H. Kessler, RGD modified polymers: Biomaterials for stimulated cell adhesion and beyond, *Biomaterials*, **24**, 4385 (2003).
12. A. Abbott, Cell culture: Biology's new dimension, *Nature*, **424**, 870 (2003).
13. V. Fast and A. Kleber, Microscopic conduction in cultured strands of neonatal rat heart cells measured with voltage-sensitive dyes, *Circ. Res.*, **73**, 914 (1993).

14. M. Kuzuya, S. Satake, M.A. Ramos *et al.*, Induction of apoptotic cell death in vascular endothelial cells cultured in three-dimensional collagen lattice, *Exp. Cell Res.*, **248**, 498 (1999).

15. E.A. Botchwey, M.A. Dupree, S.R. Pollack *et al.*, Tissue engineered bone: Measurement of nutrient transport in three-dimensional matrices, *J. Biomed. Mater. Res.*, **67A**, 357 (2003).

16. J.S. Blum, J.S. Temenoff, H. Park *et al.*, Development and characterization of enhanced green fluorescent protein and luciferase expressing cell line for non-destructive evaluation of tissue engineering constructs, *Biomaterials*, **25**, 5809 (2004).

17. H.-W. Kim, H.-E. Kim, and V. Salih, Stimulation of osteoblast responses to biomimetic nanocomposites of gelatin-hydroxyapatite for tissue engineering scaffolds, *Biomaterials*, **26**, 5221 (2005).

18. E.M. Darling and K.A. Athanasiou, Articular cartilage bioreactors and bioprocesses, *Tissue Engineering*, **9**, 9 (2003).

19. G.N. Bancroft, V.I. Sikavitsas, and A.G. Mikos, Design of a flow perfusion bioreactor system for bone tissue-engineering applications, *Tissue Engineering*, **9**, 549 (2003).

20. W.C. Hymer *et al.*, Feeding frequency affects cultured rat pituitary cells in low gravity, *J. Biotechnol.*, **47**, 289 (1996).

21. A. Solan, S. Mitchell, M. Moses *et al.*, Effect of pulse rate on collagen deposition in the tissue-engineered blood vessel, *Tissue Engineering*, **9**, 579 (2003).

22. R.G. Dennis, P.E. Kosnik, M.E. Gilbert *et al.*, Excitability and contractility of skeletal muscle engineered from primary cultures and cell lines, *Am. J. Physiol. Cell Physiol.*, **280**, C288 (2000).

23. G. Helmlinger, F. Yuan, M. Dellian *et al.*, Interstitial pH and pO2 gradients in solid tumors *in vivo*: High-resolution measurements reveal a lack of correlation, *Nat. Med.*, **3**, 177 (1997).

24. M.R. Urist, Bone: Formation by autoinduction, *Science*, **150**, 893 (1965).

25. R.R. Chen and D.J. Mooney, Polymeric growth factor delivery strategies for tissue engineering, *Pharm. Res.*, **20**, 1103 (2003).

26. T.E. Schneider, C. Barland, A.M. Alex *et al.*, Measuring stem cell frequency in epidermis: A quantitative *in vivo* functional assay for long-term repopulating cells, *Proc. Nat. Acad. Sci. USA*, **100**, 11412 (2003).

27. N. Kotobuki, M. Hirose, Y. Takakura *et al.*, Cultured autologous human cells for hard tissue regeneration: Preparation and characterization of mesenchymal stem cells from bone marrow, *Artif. Organs*, **28**, 33 (2004).

28. J. Cohnheim, Ueber Entzueundung und Eiterung, *Arch. Path. Anat. Physiol. Klin. Med.*, **40**, 1 (1867).

29. A.J. Friedenstein, U. Gorskaja, and N.N. Kulagina, Fibroblast precursors in normal and irradiated mouse hematopoietic organs, *Exp. Hematol.*, **4**, 267 (1976).

30. I. Sekiya, B.L. Larson, J.R. Smith, *et al.*, Expansion of human adult stem cells from bone marrow stroma: conditions that maximize the yields of early progenitors and evaluate their quality, *Stem Cells*, **20**, 530 (2002).

31. R. Schofield, The relationship between the spleen colony-forming cell and the haemopoietic stem cell, *Blood Cells*, **4**, 7 (1978).

32. S.P. Hoerstrup, A. Kadner, C. Breymann, *et al.*, Living, autologous pulmonary artery conduits tissue engineered from human umbilical cord cells, *Ann. Thorac. Surg.*, **74**, 46 (2002).

33. A. Kadner, G. Zund, C. Maures, *et al.*, Human umbilical cord cells for cardiovascular tissue engineering: A comparative study, *Eur. J. Cardiothorac. Surg.*, **25**, 635 (2004).

34. Editorial, Focus on cell therapies. Proceed with caution, *Nat. Biotechnol.*, **23**, 763 (2005).

35. C. Murry, M.H. Soonpaa, H. Reinecke *et al.*, Haematopoietic stem cells do not transdifferentiate into cardiac myocytes in myocardial infarcts, *Nature*, **428**, 664 (2004).
36. L.B. Balsam, A.J. Wagers, J.L. Christensen *et al.*, Haematopoietic stem cells adopt mature haematopoietic fates in ischaemic myocardium, *Nature*, **428**, 668 (2004).
37. D. Orlic, J. Kajstura, S. Chimenti *et al.*, Bone marrow cells regenerate infarcted myocardium, *Nature*, **410**, 701 (2001).
38. K. Wollert, G.P. Meyer, J. Lotz, *et al.*, Intracoronary autologous bone-marrow cell transfer after myocardial infarction: The BOOST randomised controlled clinical trial, *Lancet*, **364**, 141 (2004).
39. J.A. Thomson, J. Itskovitz-Eldor, S.S. Shapiro *et al.*, Embryonic stem cell lines derived from human blastocysts, *Science*, **282**, 1145 (1998).
40. W.S. Hwang, Y.J. Ryu, J.H. Park *et al.*, Evidence of a pluripotent human embryonic stem cell line derived from a cloned blastocyst, *Science*, **303**, 1669 (2004).
41. W.S. Hwang, S.I. Roh, B.C. Lee *et al.*, Patient-specific embryonic stem cells derived from human SCNT blastocysts, *Science*, **308**, 1777 (2005).

NOTE ADDED IN PROOF:

References 40 and 41 have been retracted as the results in them are deemed to be invalid. This retraction by Donald Kennedy was originally published in *Science Express* on 12 January 2006 and in *Science* 20 January 2006: Vol. 311, no. 5759, p. 335; DOI: 0.1126/Science.1124926.

Chapter 2

Animal and Human Trials of Engineered Tissues

It has frequently been stated that the field of tissue engineering has achieved brilliant progress and still is developing at a rapid rate. Investigators have attempted to engineer virtually every mammalian tissue, but unexpectedly in authorized international journals few results with respect to clinical trials of tissue engineering have been published. Even the reports describing tissue engineering research performed with large animals like dog and sheep have not been published in a large number. This chapter will present recent results on animal and human trials related to tissue engineering.

1. BODY SURFACE SYSTEM

1.1. Skin

Skin is composed of the epidermis and dermis. Epidermis or epithelium is formed by stratified keratinocytes. The upper layer of the epidermis is cornified and protects invasion by various foreign substances. Dermis is composed of an ECM such as collagen fibers and protects inside organs.

1.1.1. Without Cells

A bilayered artificial skin without cultured cells, as illustrated in Fig. 2.1, was first developed by Yannas *et al.* [1]. It is composed of an inner sponge layer of collagen and chondroitin-6-sulfate, a GAG, and an outer silicone layer. When placed on wounds, the inner sponge layer characteristically turns into dermis-like connective tissue. This kind of bilayered artificial skins, both wet and dry types, is commercially available and patients with massive deep burns have successfully been treated with the material.

1.1.2. Keratinocytes

Since Rheinwald and Green described the culture of keratinocytes using a 3T3 feeder layer in 1975 [2], clinical application of cultured keratinocytes has been performed for a permanent use. O'Connor *et al.* first reported the use of cultured autologous epithelial sheets in patient care [3].

The current clinical "gold standard" for laboratory expansion of keratinocytes as an augmentation of conventional split-thickness skin grafts is expansion of the cells with DMEM containing 10% FCS on a lethally irradiated layer of 3T3 mouse

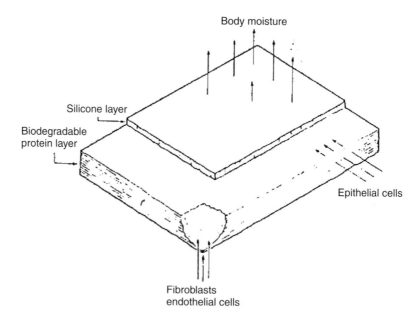

Body moisture

Silicone layer

Biodegradable
protein layer

Epithelial cells

Fibroblasts
endothelial cells

Fig. 2.1   Schematic representation of a standard version of the noncellular bilayer polymeric membrane (stage I).

fibroblasts and then detachment of these cells using dispase as an integrated sheet of cells. In the presence of the fibroblast feeder layer, keratinocytes expand rapidly and maintain their colony-forming potential, whereas keratinocytes quickly begin to undergo terminal differentiation in the absence of fibroblasts. With fibroblasts present, keratinocytes and fibroblasts produce basement membrane proteins, whereas, in the absence of fibroblasts, keratinocytes are unable to produce a basement membrane, poorly organized, and fail to properly differentiate. In the US, autologous cultured epithelium (Epithel™) has been commercially produced.

Boyce and Ham developed serum-free keratinocyte growth medium (MCDB 153 medium) [4]. To develop for the clinical benefit of patients a culture system in which keratinocytes can be cultured in the presence of fibroblasts under serum-free conditions, Higham *et al.* developed a new surface by plasma polymerization of acrylic acid and octa-1,7-diene monomers [5]. Keratinocytes expanded rapidly on a poly(acrylic acid) surface containing 9.2% carboxylate groups under serum-free culture conditions if a feeder layer of fibroblasts was present.

Allogeneic cultured epithelium which can be produced beforehand and stored has been used for the treatment of severe burns. This cultured epithelium is temporary and used in combination with autologous skin graft. It is also used for treatment of chronic skin ulcers. Cytokines and growth factors released from the allogeneic keratinocytes may accelerate wound healing. However, the cultured epithelium grafted onto the full thickness skin defect of patients often fails, probably because it has no dermal component. In addition, the graft can often shrink and be easily blistered and ulcerated. Another disadvantage is the 3- to 4-week period required to produce the cultured epithelium with the 3T3 feeder layer method, and 4–5 weeks with the serum-free keratinocyte growth medium method. Because wound treatment is required within 2 weeks for burn patients with large skin defects, new approaches are needed to achieve effective wound closure in patients with extensive skin loss or chronic ulcers.

1.1.3. Keratinocytes on Acellular Dermis

Cultured autografts of keratinocytes have persistent problems, including blistering, wound contracture, susceptibility to infection, and varying graft "take" rate, probably due to the lack of a dermal component and of a mature dermoepidermal junction. One dermal analog used for treatment of full-thickness wounds is an acellular human cadaver dermis, such as AlloDerm™ (LifeCell, NJ, USA), which retains ECM proteins and an intact basement membrane structure. Izumi *et al.* generated human *ex vivo*–produced oral mucosa equivalents (EVPOMEs) in a serum-free culture system without the use of irradiated xenogeneic feeder layer by seeding human oral keratinocytes onto AlloDerm™ [6]. They harvested oral keratinocytes from tripsinized discarded human oral mucosa. Harvested and expanded keratinocytes, cultured in a serum-free medium, were seeded onto rehydrated AlloDerm coated with type IV collagen. The keratinocyte–AlloDerm composites were cultured in medium for 4 days and then at an air–liquid interface for an additional 7 days or 14 days to enhance stratification of the epithelial layer of the EVPOMEs before grafting onto a subcutaneous pouch of severe combined immunodeficient (SCID) mice. The AlloDerm™ portion of the EVPOME showed no evidence of an adverse inflammatory reaction or rapid remodeling reaction in the early stage after grafting, but rather appeared to act as a biomimetic template that promoted the intradermal migration and infiltration of fibroblasts and host cells. The microvessel density of AlloDerm without epithelium was less than that of EVPOMEs with an epithelial layer. The EVPOMEs co-cultured at an air–liquid interface for 7 days had the optimal balance of neoangiogenesis and epithelial differentiation necessary for *in vivo* grafting. The presence of an intact and viable epithelial layer influenced secondary remodeling within the dermis of the EVPOME, most likely by its synthesis and release of cytokines, enzymes, and growth factors.

1.1.4. Keratinocytes + Fibroblasts

Bell *et al.* developed living skin-equivalent grafts consisting of fibroblasts cast in collagen lattices and seeded with ECs [7]. Liu *et al.* harvested autologous full-thickness skin, isolated keratinocytes and dermal fibroblasts, and separately seeded on a sheet of unwoven PGA fibers with Pluronic (ethylene oxide/propylene oxide copolymer) hydrogel [8]. In a porcine model, two full-thickness wounds deep to the fascia were created at the dorsal aspect of pigs and a titanium chamber was inserted to prevent tissue ingrowth from adjacent skin. The wounds were repaired either with a composite skin construct (experimental group) or with PGA alone (control group).

When examined grossly, neoskin formation was observed at 2 weeks postrepair in the experimental group. Although the wound was totally covered by engineered skin in the experimental group at 4 weeks, mature full-thickness skin was formed only in part of the wound area. However, the maturation completed uniformly at 8 weeks, and resembled native porcine skin. In contrast, only granulation tissue was observed in control wounds. Histologically, double layer skin was formed as early as 1 week postrepair. The rete ridges of the epidermis became enlarged and started to migrate deeply into the dermal part of the engineered skin at 2 weeks and had reached deeper at 4 weeks. At 8 weeks, these outreaching rete ridges had retreated to where they had originally resided and the histology appeared to be similar to that of normal skin. Furthermore, histological and TEM examination revealed the presence of a basement membrane structure in the tissue-engineered skin. Likewise, only granulation tissue was observed histologically in the control group.

To investigate the viability and distribution of autologous and allogeneic fibroblasts after implantation into skin, Morimoto *et al.* isolated fibroblasts from the skin of a guinea pig and used them as allogeneic fibroblasts [9]. Three full-thickness wounds were created on the backs of guinea pigs and an acellular collagen sponge, a collagen sponge seeded with autologous fibroblasts, and a collagen sponge seeded with allogeneic fibroblasts were implanted. Before implantation, fibroblasts were labeled with PKH26, a red fluorescent dye that stains the membrane of viable cells and distributed between cells when mitosis occurs. Three weeks after implantation, the PKH26-labeled autologous and allogeneic fibroblasts remained viable. In the wounds covered with the autologous fibroblast-seeded collagen sponge, the epithelization was the fastest, and the percent wound contraction was the smallest. In contrast, in the wounds covered with allogeneic fibroblasts, the epithelization was the slowest and the percent contraction was the largest.

To study the effects of progressive serum reduction on epidermal differentiation, quality of dermal and dermal–epidermal junctions, and expression of ECM proteins, Black *et al.* used a co-culture model including both fibroblasts and keratinocytes [10]. The cells were successively added to a dermal substrate composed of collagen, GAGs, and chitosan. Control skin equivalents (SEs) were cultured with medium containing 10% serum throughout the production process. Serum content was reduced to 1 and 0% at the air–liquid interface and compared with control SEs. First, they demonstrated that serum deprivation at the air–liquid interface improved keratinocyte terminal differentiation. Second, in the absence of serum, the specific characteristics of the SE were maintained, including epidermal and dermal ultrastructure, the expression of major dermal ECM components (human collagen types I, III, and V, FN, elastin, and fibrillin 1), and the dermal–epidermal junction (LN, human type IV collagen, and α_6 integrin).

1.1.5. Keratinocytes + Melanocytes

Vitiligo is common (1–2% of the population) and causes considerable distress to sufferers. The mainstay therapy is UV therapy with psoralens, but the majority of sufferers resort to camouflage cosmetics. This has led to the introduction of various grafting strategies to take functioning melanocytes from elsewhere in the body to graft depigmented areas. The keratinocytes with the melanocytes that are present after isolation of the cells from skin, referred to as "passenger" melanocytes, seem to ensure rapid healing with no scarring and in theory the passenger melanocytes should pigment the transplanted areas. However, Beck *et al.* found that it was difficult to maintain a reasonable percentage of melanocytes within the keratinocyte–melanocyte integrated sheets [11]. To develop an alternative strategy for transferring keratinocytes and melanocytes to the grafting site, they studied several culture systems with a view to find a combination of medium and culture surface suitable for the co-culture of primary keratinocytes and melanocytes immediately after isolation from skin. To achieve this combination of cells, they compared the ability of three culture surfaces and three media. The three culture surfaces were prepared by coating glass coverslips with type I collagen, nitrogen-containing plasma polymer (PlP), and none. Type I collagen provided a good surface for keratinocyte culture, while nitrogen-containing polymer coating was found to promote melanocytes to thrive alongside keratinocytes [12]. Major components of the three media are shown in Table 2.1 [11]. Green's medium (developed for keratinocyte culture) has been in clinical use for more than 20 years. The GIBCO keratinocyte defined medium was developed for keratinocyte culture and is serum free. The MCDB-153 medium was developed for *in vitro* melanocyte culture and contains bovine pituitary extract and therefore would not be viewed as appropriate for clinical use because of the risk of BSE. Keratinocytes with passenger melanocytes present were obtained from full-thickness skin of patients undergoing elective abdominoplasty and breast reductions. In Green's medium, the number of melanocytes increased until 3 days in culture, whether cells were cultured on type I collagen, plasma polymer, or glass. The melanocyte number then stabilized between days 3 and 7 on all three surfaces. Cells grown on type I collagen and glass achieved a higher number of melanocytes than did cells cultured on the PlP surface. The trends in melanocyte growth in the other two media were similar in that there was generally an increase in melanocyte number up to 3 days and then rarely any increase beyond this point. When cells were cultured in defined keratinocyte medium, melanocyte proliferation was the best on glass and PlP and the poorest on type I collagen. When cells were cultured in MCDB-153, good melanocyte numbers were achieved on glass, but worse for plasma polymer and poor for type I collagen.

1.1.6. Stem Cell Transplantation

To examine the contribution of bone marrow–derived cells to skin cells, Deng *et al.* transplanted chloromethyl-dialkyl carbocyanine (CM-DiI) fluorescence-labeled Flk-1$^+$ bone marrow mesenchymal stem cells (bMSCs) into BALB/c mice (*H-2Kd*, white) and into lethally irradiated C57BL/6 mice (*H-2Kb*, black) [13]. Fluorescence tracing revealed that donor cells could migrate and take residency in the skin, which was confirmed by Y chromosome–specific polymerase chain reaction (PCR) and Southern blot. The recipient mice grew white hairs about 40 days later and white

Table 2.1

Comparison of major components and Ca^{2+} concentrations of three media

Green's Medium	Defined Keratinocyte Serum-free Medium (Supplemented)	MCDB-153 Medium (Supplemented)
Basal Medium		
DMEM (ICN Biomedicals) (6%)[a]	Defined keratinocyte-SFM (GIBCO-BRL)[b]	MCDB 153 medium (Sigma)[a]
Ham's F12 (GIBCO-BRL) (20%)[a]	—	—
Supplements		
FCS (10%)[a]	—	Chelexed FCS (2%)
Insulin (5 µg/ml)	Insulin	Insulin (10 µg/ml)
Epidermal growth factor (10 ng/ml)	Epidermal growth factor	—
—	Fibroblast growth factor	—
Transferrin (5 µg/ml)	—	Transferrin (10 µg/ml)
Hydrocortisone (0.4 µg/ml)	—	Hydrocortisone (2.8 µg/ml)
L-Glutamine (2 mM)	—	L-Glutamine (2 mM)
Cholera toxin (8.5 ng/ml)	—	Cholera toxin (100 ng/ml)
Adenine (20 µg/ml)	—	—
—	—	Vitamin E (1 µg/ml)
—	—	Phorbol myristate acetate (10 nM)
—	—	Bovine pituitary extract (50 ng/ml)
Approximate [Ca^{2+}][a]		
250 mg/l	<4.0 mg/l	4.4 mg/l (in base medium)

[a] These items are the main source of calcium in the media and the approximate [Ca^{2+}] was calculated based on this.
[b] The concentrations of these items are considered proprietary information by the supplier (GIBCO-BRL, Life Technologies).

hairs could spread over the body. Immunochemistry staining and reverse transcription (RT)-PCR demonstrated that skin tissue within the white hair regions was largely composed of donor-derived *H-2 Kd* cells, including stem cells and committed cells. Furthermore, most skin cells cultured from white hair skin originated from the donor. These findings provide direct evidence that bone marrow–derived cells can give rise to functional skin cells and regenerate skin tissue.

1.2. Auricular and Nasoseptal Cartilages

Ear pathology includes loss of cartilage due to trauma, and congenital defects such as microtia, a congenital defect characterized by a small, abnormally shaped or absent outer ear that typically requires multiple stages of reconstructive surgery. The auricle of ear has a layered structure composed of skin, perichondrium, and cartilage. The overlying skin is very thin (800 μm), making ear reconstruction very challenging in efforts to match the ear's natural dermal thinness, 3-D shape, and contour. The ECM of auricular cartilage, classified as "elastic cartilage", is composed mainly of proteoglycans, type II collagen, and a large network of elastin. Elastic cartilage is quite different from articular hyaline cartilage. The elastic fibers allow the ear to undergo large deformation bending, giving the ear its characteristic flexibility. The functional purpose of auricular cartilage is to mechanically support the ear, allowing for large-scale deformation. Auricular cartilage is surrounded on both sides by perichondrium, a connective tissue that is the interface between the cartilage and the skin. In native tissue, the perichondrium contains the vascular supply, which is important for the growth and maintenance of the avascular auricular cartilage.

Microtia with an occurrence rate of 0.025% presents a very challenging surgical reconstruction when compared to most other congenital deformities. Traditional repair methods have included some sort of biomaterials for prosthetic implants, such as alloplastic polyethylene implants or autologous costal cartilage grafts. Clinicians have focused on surgical approaches for the regeneration of congenitally malformed or traumatized cartilages in auricle, nose, eyelids, and trachea. However, auricular reconstruction for cartilage defects, such as congenital microtia, still remains one of the most difficult challenges in reconstructive surgery. The current standard approach for ear reconstruction is to use autologous costal cartilage grafts fabricated piece by piece into a structure mimicking the natural ear shape. This reconstructive method often fails due to unfavorable thickness and rigidity of the reconstructed ears. Several problems are apparent in this reconstructive approach: (1) the elastic character, 3-D architecture, and final ear size are variable and can change through gradual resorption of the transplanted costal cartilage as the young patient grows; (2) chest deformities occur with a 60% or greater incidence in patients under 10 years of age; and (3) more than three surgical procedures are often needed to effect a satisfactory result. This early effort using autologous costal cartilage is clearly in need of modification to achieve more favorable outcomes.

Cao *et al.* were the first to report on a regenerative concept using a non-woven PGA mesh in combination with bovine chondrocytes in order to tissue engineer a 3-D, elastic cartilage graft [14]. The textile construct was shaped into a human infant ear by using a plaster mold in combination with thermal treatment. Dipping the fragile scaffold in a PLA solution stabilized the 3-D geometry. However, the highly porous construct had to be stented externally for 4 weeks because the scaffold was not able to resist the contraction forces of healing. Britt and Park described autologous chondrocyte seeding onto PGA–PLLA ear-shaped templates in a rabbit model [15]. The fine details of the auricle graft were lost after long-term follow-up. Saim *et al.* demonstrated that injection of chondrocytes in a biodegradable hydrogel of Pluronic F-127 resulted in the formation of the cartilage tissue in the shape of human ear helix, but not a whole complex ear, in an immunocompetent autologous porcine model [16]. Haisch *et al.* described the manufacture of auricular-shaped PGLA–PLLA construct by a silicone cylinder, and subsequently seeded with human nasoseptal chondrocytes [17].

To form elastic cartilage in regular-shaped, mechanically stable, and biomimetic constructs, Hutmacher *et al.* used a 3-D synthetic framework in combination with a hydrogel, within an immunocompetent animal model [18]. They manufactured scaffolds with porosity of 65% using poly (ε-caprolactone) (PCL) filaments and fused deposition modeling. Lay-down pattern of 0/60/120° head spead, fill gap, and raster angle for every layer were programmed through a software. Chondrocytes were harvested from the ear cartilage of adult Yorkshire pigs and mixed with type I collagen. The chondrocyte/collagen hydrogel mixture was injected into scaffolds, which were then placed in a fibrin glue solution to form a capsule of approximately 0.6 mm thickness. Following culturing for a period of 1 week, seeded and control scaffolds were placed subcutaneously in the paravertebral fascia in two pigs. In the rabbit model, ear cartilage of white rabbit was used to harvest chondrocytes. Scaffolds were divided into 4 groups (A: scaffold + cells + allogeneic fibrin glue; B: scaffold + cells; C: scaffold + allogeneic fibrin glue; and D: scaffold only). The scaffold/hydrogel/chondrocytes constructs showed islets of cartilage and mineralized tissue formation within the cell-seeded specimens in both pig and rabbit models. Specimens with no cells seeded showed only vascularized fibrous tissue ingrowth. The 3-D PCL scaffolds fabricated with a computer-aided machine were found to withstand the wound contraction forces, which resulted in supporting the regeneration of islets of elastic cartilage in both autologous rabbit and pig models. However, even with the use of collagen or fibrin gel as a biomimetic cell carrier, the mechanically stable scaffolds did not allow the regeneration of a large mass of structural and functional cartilage when loaded with dedifferentiated chondrocytes and implanted in a non-chondrogenic host environment.

Shieh *et al.* chose chondrocytes from adult sheep cartilage as the cell source [19]. Three types of auricular polymeric scaffolds were used: PGA, PCL, and P4HB. For fabrication of the PGA-ear, a non-woven mesh of PGA fibers was treated with 3% PLLA chloroform solution. Following PLLA treatment, the fabric was made into the shape of a human ear using a negative mold. For the PCL and P4HB ears, a sucrose ear template was prepared beforehand. The granular sucrose was broken into small particles using a grinder, and was sieved to a particle size of 250–500 μm. The selected granular sucrose was mixed with a minimal amount of water, and poured into the negative ear mold to form into an ear-shaped sucrose template. The ear scaffold was then constructed from the sucrose ear template by solvent casting/particulate leaching. During *in vitro* propagation, chondrocytes tended to undergo dedifferentiation characterized by a decrease in the synthesis of cartilage matrix molecules, type II collagen and GAG, and upregulation of type I collagen, which could be reversed by subsequent 3-D culture system, such as agarose or alginate. Shieh *et al.* used TGF-β2 and IGF-1 to induce and improve the capacity of redifferentiated chondrocytes to generate a properly assembled matrix for the 3-D ear scaffold and decided the optimal seeding density with 50 million cells per cubic centimeter and cultured in rotating bioreactors under dynamic condition for a growth period of 8 weeks for the ear constructs based on the above evidence. In addition, type I collagen gel was used to encapsulate chondrocytes and subsequently seeded onto 3-D scaffold. Under all these controlled *in vitro* conditions, they found that cells started to undergo chondrogenesis at the construct periphery and proceeded appositionally from periphery toward its center with deposition of ECM around chondrocytes themselves. The DNA content, related to the cell number, remained stable after cell seeding to the various types of scaffolds. After 8 weeks of *in vitro* cultivation, the GAG content significantly increased in PGA and PCL groups compared with that of 4 weeks, and the value of PGA group was even more than two-fold than the native cartilage. In PCL and poly-4-hydroxybutyrate (P4HB) groups, the GAG content also reached the level similar to the native's after 8 weeks of cultivation. The PGA group produced much more type II collagen than PCL and P4HB groups after 8 weeks of *in vitro* cultivation. All engineered constructs retained a significantly lower amount of elastin than the controls. Substantial neocartilage was formed in all the ear constructs after 10 months *in vivo* remolding in the nude mice except the infected one in P4HB group. The tissue-engineered PCL ear preserved a better gross architecture than PGA and P4HB ears.

According to the mechanism of cartilage formation, majority of the chondrocytes mature and induce cartilage, which is gradually replaced by mineralized tissue through endochondral ossification by apoptosis-associated mineralization. Some chondrocytes, however, remain as permanent chondrocytes and compose such lifelong cartilage as the trachea, auricle, joint, and nasal septum. This evoked an interest in Isogai *et al.* to produce tissue-engineered constructs with other sources of chondrocytes [20]. Common locations for cartilage harvests in clinical settings are the outer ear, rib, and nasal septum. To test a hypothesis that chondrocytes store site-directed property whose morphology and function are different in relation to the location, Isogai *et al.* investigated chondrogenic potential of the chondrocytes harvested from newborn calves at four different locations—(1) nasoseptal, (2) articular, (3) costal, and (4) auricle *in vitro*—and elucidated the difference in chondrogenesis by examining morphology and function of the neocartilage in the shape of human auricle. In experiment 1 (*in vitro*), four types of chondrocytes were cultured to compare ability of cell proliferation. The ECM production was also studied by a presence of metachromasia and by phenotypic expressions of type II collagen and aggrecan mRNA using RT-PCR assay. In experiment 2 (*in vivo*), cells were then seeded onto P(LA/CL) (50:50) scaffolds which were designed in two shapes: disc and human auricle. These cell/polymer constructs were cultured for one week and subsequently implanted in the subcutaneous space of nude mice for 20 weeks. The result of proliferation of four chondrocytes is given in Fig. 2.2, which demonstrated increased nasoseptal and costal chondrocyte proliferation by four-fold in 3 weeks of culture. The two chondrocytes also showed higher accumulation of cartilaginous ECM than articular chondrocytes. Auricular chondrocytes showed moderate cell proliferation and markedly lower expression of mRNA in type II collagen and aggrecan. Figure 2.3 shows the transitional change of the dimensions (diameter and thickness) in the disc-shaped constructs (original dimension: 15 mm in diameter and 0.5 mm in thickness). As can be seen, neocartilages generated from hyaline cartilage origin (nasoseptum, articular, and costal chondrocytes) showed a persistent increase in thickness over time without changes in diameter. Inversed growth direction was

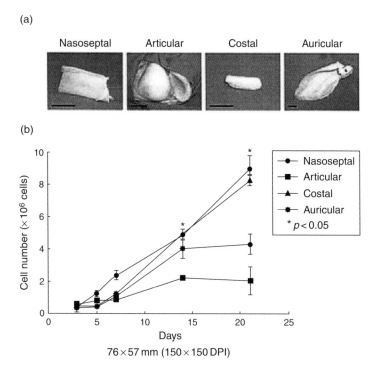

Fig. 2.2   Gross morphology of the sites for bovine cartilage harvested (a) and the *in vitro* proliferation of chondrocytes isolated from nasoseptal, articular, costal, and auricular tissues of newborn calves (b).

observed in neocartilages generated from elastic cartilage origin (auricular chondrocytes), which showed a slight increase in diameter without changes in thickness. It seems that growth direction in neocartilage is different and closely associated with the type of chondrocytes. Phenotypic expression in auricular chondrocytes was upregulated *in vivo*, which resulted in increased accumulation of cartilaginous ECM with the presence of elastic fiber. An analysis on chondrocytes at four harvest sites thus elicited significantly different chondrogenic responses. These results indicate that auricular cartilage is a potent harvest site for chondrocytes to induce the improved composition, 3-D contour, thickness, and controlled growth for engineering auricular cartilage.

Silicone has been the major material for nasal augmentation, although it often causes problems after implantation. To overcome the limitations in the use of silicone, Yanaga *et al.* developed a technique that uses cultured autologous auricular chondrocytes as a graft material for nasal augmentation [21]. First, approximately 1 cm^2 of auricular cartilage was collected and isolated chondrocytes were cultured in flasks to produce a gel-type mass consisting of the immature matrix. This mass, comprising a large number of cultured chondrocytes, was injected into a pocket created by dissection of the area for augmentation. The gel hardened over a period of 7–10 days after

(a)

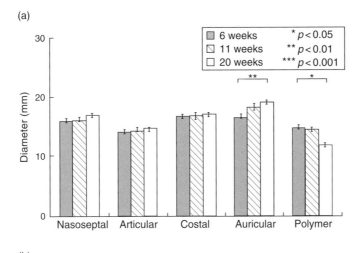

(b)

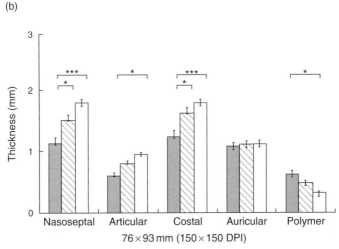

76×93 mm (150×150 DPI)

Fig. 2.3 Transitional changes of diameter (a) and thickness (b) dimensions of disk-shaped constructs retrieved at 5–20 weeks after implantation in nude mice.

injection as it became mature cartilage. This technique was applied to nine lesions of eight patients, as shown in Table 2.2. After surgery, the shape and contours at the graft sites were satisfactory. The follow-up period for the patients ranged from 6 months to 2 years, and neither change in shape nor cartilage resorption was noticed. This method was applicable to chin augmentation and deformed cranium as well.

1.3. Adipose Tissue

The correction of malformations and volume deficits of soft tissue still represents a challenge in plastic surgery. The use of synthetic materials, for example, in breast

Table 2.2

Clinical results for treatments with cultured autologous human auricular chondrocytes

Case	Age (years)	Gender	Site and Symptom	Complication	Resorption	Follow-up Period (months)
1	23	F	Nose (exposure of silicone implant)	—	—	18
2[a]			Chin (bone resorption)	—	—	6
3	35	F	Saddle nose (trauma)	—	—	24
4	21	F	Nose (primary augmentation)	—	—	10
5	29	F	Nose (exposure of silicone implant)	—	—	12
6	37	F	Nose (deviation of silicone implant)	—	—	8
7	23	F	Nose (convexoconcave deformity)	—	—	14
8	25	F	Nose (primary augmentation)	—	—	8
9	44	F	Nose, forehead, and temporal lesion (fibrous dysplasia)	—	—	16

[a] Case 2 involves the same person as Case 1.

surgery may lead to infections, tissue reactions such as foreign-body reactions, and capsular contracture. To control the local volume of adipose tissues, physicians simply remove or aspirate adipose tissue for a reduction or autograft a fat pad for an increment. The trials attempting to transplant autologous minced adipose tissues to depressed regions or scars in the breast and facial areas have not always been successful. Autologous adipose tissue loses up to 60% of its volume after transplantation by necrosis due to insufficient initial blood supply after cell transplantation. This probably occurs for two reasons. First, mature adipocytes have a high metabolic rate and high sensitivity to hypoxia. Second, in the first few days after transplantation these transplants are nourished only by diffusion from the surrounding tissue, which is cell sufficient only over a distance of ~150 μm. Vessels from the surrounding tissues grow after a delay of up to 8 days and even then they nourish only the periphery of the implant at that time. Since the transplanted adipose tissues are absorbed in this way and replaced by fibrous tissue and oil cysts, *de novo* regeneration of adipose tissue seems to be more promising for the reconstruction of adipose tissue than the autograft of native adipose tissue.

To achieve early vascularization after cell transplantation, Borges *et al.* attempted to differentiate ECs into capillary-like structures immediately after cell transplantation in the receiving organism by the cotransplantation of autologous preadipocytes in a fibrin matrix [22]. They added ECs in spheroids to fibrin, because it was reported that ECs suspended in fibrin underwent apoptosis whereas EC aggregates showed a significantly higher degree of differentiation and survival rate [23]. More resistant preadipocytes were added as a single-cell suspension, so that an EC spheroid–preadipocyte suspension in fibrin was created. They used a highly vascularized chorioallantoic membrane of the chicken embryo which served as the vessel-forming host tissue. For this investigation the classic chorioallantoic membrane model was modified so that a model especially suitable for *in vivo* investigations in the field of tissue engineering was established, as illustrated in Fig. 2.4. Human adipose tissue, free of mature preadipocytes, and human dermal microvascular ECs (HDMVECs), was obtained by plastic surgery procedures. The HDMVECs (2200) were seeded onto non-adherent, round-bottom 96-well plates to generate a single EC spheroid. When HDMVEC spheroids were brought into fibrin glue without preadipocytes (group 1), no cells could be found in the composite after explantation. After transferring HDMVECs and preadipocytes as a single-cell suspension in fibrin (group 2), both CD31-negative preadipocytes and CD31-positive HDMVECs could be found in four constructs. The HDMVECs formed only a few CD31-positive tube-like structures with vessel lumena; nevertheless, branching in the sense of capillary formation was not found. However, in the groups containing HDMVEC spheroids with the preadipocyte single-cell suspension in fibrin (group 3), CD31-negative preadipocytes and CD31-positive HDMVECs could be detected, and by the intercellular connections of these capillary-like structures a large branched network was formed. On day 7 after transplantation, lumen-forming human ECs with intralumenally located erythrocytes could be demonstrated immunohistochemically.

Yamashiro *et al.* considered adipose tissue–derived vascular stromal (VS) cells to be a promising candidate as the precursor cell for adipose tissue engineering and

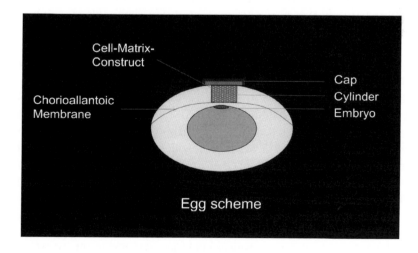

Fig. 2.4 Chorioallantoic membrane cylinder model.

attempted to expand human adipose tissue–derived VS cells and differentiate them into adipose [24]. For this purpose, human VS cells obtained from mature adipose tissue were transfected with adenovirus vector carrying the bFGF gene. The bFGF protein was observed in VS cell nuclei 24 h after transfection and in the cytoplasm and extracellular space 72 h after transfection. Naïve VS cells were almost static *in vitro* and proliferated in a dose-dependent manner on stimulation with recombinant bFGF(rbFGF), but bFGF-transfected VS cells proliferated spontaneously to the same extent as naïve VS cells when stimulated with rbFGF at 100 ng/ml. The former cell started to proliferate on day 3 after transfection and the proliferation pattern was similar to that of the latter cells, although only a slight amount of bFGF was detected in the culture medium when the bFGF-transfected cells started to proliferate. The proliferation of bFGF-transfected VS cells was completely inhibited by bFGF neutralizing antibody, which also completely inhibited the proliferation of naïve VS cells stimulated with rbFGF. Under conditions favoring differentiation to adipocytes, bFGF-transfected VS cells stopped proliferation and started to accumulate lipid in the cytoplasm.

A technique was developed to regenerate volume-stable adipose tissues *in vivo* [25]. Dome-shaped support structures fabricated from PGA fiber-based matrix reinforced with PLLA were utilized to provide a mechanically stable space for regeneration of volume-stable adipose tissues. The mechanical support structures would withstand the *in vivo* compressive forces and allow for the formation of volume-stable adipose tissues. Human preadipocytes mixed with injectable fibrin matrix, a space-filling matrix, were injected into the space under the dome-shaped support structures placed in subcutaneous pockets of athymic mice (group I). Injection of either fibrin matrix without preadipocytes under the support structures (group II) or fibrin matrix containing preadipocytes into subcutaneous spaces with no support structures (group III) served as controls. Six weeks after implantation, the original implant volume was maintained approximately in groups I and II, whereas group III showed significant implant shrinkage. The compressive modulus of the mechanical support structures did not change significantly over the 6-week incubation in PBS at 37°C. Histological analyses of the implants showed regeneration of adipose tissues in group I. In contrast, groups II and III did not show extensive adipose tissue formation.

Choi *et al.* evaluated *in vivo* adipose tissue formation using MSCs and injectable PLGA spheres [26]. The MSCs and adipo-MSCs (MSCs cultured in adipogenic medium for 7 days) were attached to PLGA spheres and cultured for 7 days, followed by injection into nude mice for 2 weeks. Before the *in vivo* study, two groups of cell-attached PLGA spheres were examined *in vitro* to assess the differentiation capacity and cell proliferation characteristics of MSCs on the PLGA sphere hybrids. The MPLGA group consisted of complete medium-cultured MSCs attached to PLGA spheres, and then cultured in adipogenic medium for 7 days. The APLGA group consisted of adipo-MSCs attached to PLGA spheres, and then cultured in adipogenic medium for 7 days. The difference between lipid accumulation in adipo-MSCs at 1 and 7 days was much higher *in vitro* than in the MSCs. Two weeks after injection, a massive amount of new tissue was formed in the APLGA group, whereas only a small amount was formed in the MPLGA group. The researchers verified via GFP testing that the newly formed tissue originated from the injected MSCs, and confirmed by oil red O staining and Western blot that the created tissue was actual adipose tissue. The primary MSCs were transfected with GFP using GFP Fusion TOPO™ TA expression kit and FuGENE 6 transfection reagent.

To engineer adipose tissue constructs, Alhadlaq *et al.* encapsulated adult stem cell–derived adipogenic cells in a photopolymerizable PEG diacrylate (PEGDA) hydrogel system [27]. Bone marrow–derived adult human mesenchymal stem cells (hMSCs) were preconditioned by 1 week of exposure to adipogenic-inducing supplement followed by photoencapsulation in PEGDA hydrogel in predefined shape and dimensions. Complementary *in vitro* and *in vivo* approaches were designed to demonstrate not only the differentiation of hMSCs into adipogenic cells, but also the volume stability of tissue-engineered constructs after 4 weeks of *in vitro* incubation and *in vivo* implantation. In two parallel experiments, the resulting hMSC-derived adipogenic cell–polymer constructs were either incubated *in vitro* in adipogenic medium or implanted *in vivo* in the dorsum of immunodeficient mice for 4 weeks. Tissue-engineered adipogenic constructs demonstrated positive reaction to oil red O staining both *in vitro* and *in vivo*, and expressed PPAR-γ2 adipogenic gene marker *in vivo*. By contrast, control PEGDA hydrogel constructs encapsulating undifferentiated hMSCs failed to demonstrate the adipogenic gene marker and were negative for oil red O staining. Recovered *in vitro* and *in vivo* constructs maintained their predefined physical shape and dimensions.

2. MUSCULOSKELETAL SYSTEM

2.1. Articular Cartilage

Cartilage plays an important role in defining facial structure and aesthetics, as well as covering diarthrodial joints with nearly frictionless gliding surfaces, while mediating the transfer of load within the joint to the underlying subchondral bone. In the general cartilage composition, chondrocytes occupy lacunae in the matrix, and produce cartilaginous ECM, which consists of type II collagen (13%), proteoglycans (7%), and water (80%). Cartilage is traditionally classified into 3 types based on the kind of fiber present; hyaline, elastic, and fibrocartilage. Hyaline cartilage is found at nasoseptal, articular, costal, trachea, and larynx. Elastic cartilage is hyaline cartilage with increased elasticity and found at ear and epiglottis. Elastic fibers are densely found around chondrocytes, spread throughout the tissue, and connected to perichondrium in elastic cartilage. Fibrocartilage is a dense fibrous tissue with small regions of hyaline cartilage and found at discs, and attachment of ligaments and tendons to bone. These cartilages except for fibrocartilage are surrounded by perichondrium, a vascular fibrous capsule. Nutrient vessels generally penetrate through perichondrium, which supplies nutrition to chondrocytes principally by the mechanism of diffusion through matrix. Articular chondrocyte, however, is dependent on its nutrition by synovial fluid. Blood vessels and lymphatics are not therefore present within cartilage.

Hyaline cartilage is the dense white tissue that covers the articulating surfaces of bones and it functions to transmit high stresses generated in diarthrodial joints, up to 5–18 MPa in the hip joint, by virtue of its unique material properties: compressive equilibrium aggregate modulus of 0.3–1.0 MPa, tensile equilibrium modulus of 15–40 MPa, and permeability of 0.5–5 $\times$ 10^{-15} m^4/(N.S)$^{1.5}$. The dense ECM that generates unique material properties is composed of GAG and collagen (primarily type II). Whereas its singular structure allows cartilage to function in a harsh loading environment, its dense composition and lack of innervations and vascularization severely limit its self-repair and remodeling capabilities after significant trauma or disease, in comparison with other musculoskeletal tissues such as bone and muscle. After completion of skeletal growth, most chondrocytes probably never divide, but rather continue to synthesize collagens, proteoglycans, and noncollagenous proteins. This continued synthetic activity suggests that the maintenance of articular cartilage requires substantial ongoing remodeling of the macromolecular framework of the matrix, and the replacement of degraded matrix macromolecules.

Cartilage defects resulting from aging, joint injury, and developmental disorders cause joint pain and loss of mobility. Perhaps due to its complex structure and composition, articular cartilage has been known to be one of the most common sources of pain and suffering in the human body. Natural articular cartilage is composed of a charged solid matrix phase consisting of charged proteoglycan macromolecules and collagen fibers, an interstitial fluid phase, and an ion phase. The proteoglycan matrix is mainly responsible for the equilibrium compressive stiffness of cartilage, the collagen fibers influence the instantaneous compressive and tensile response of the tissue, and the interstitial fluid phase affects the transient response of cartilage during compression. The ability of cartilage to regenerate or "heal" itself decreases with age, because, with age, chondrocyte (cartilage-synthesizing cell) numbers decrease, structural organization of cartilage becomes altered, and a well-defined calcified zone arises in cartilage.

Although articular cartilage is a metabolically active tissue, the chondrocytes in the matrix have a relatively slow rate of turnover. Because of the limited capacity for spontaneous repair, articular cartilage may lead to progressive damage and degeneration. In older patients with diffuse articular cartilage injury, total joint arthroplasty is the treatment of choice. In young patients with focal articular cartilage injury, autologous chondrocyte implantation, mosaicplasty, or marrow-stimulating techniques, such as microfracture or multiple drilling, are better alternatives than joint replacement therapy, but the results are variable and the techniques have some limitations. Autologous chondrocyte transplantation with a periosteal graft has shown encouraging results, but the predictability and reliability of hyaline or fibrocartilage formation are still questionable. Marrow-stimulating techniques result in fibrocartilage formation, which has a lower mechanical strength than hyaline cartilage and only limited repair capacity. For mosaicplasty, the major problems are limited availability of autologous tissue and donor site morbidity; this procedure also involves destroying healthy non-weight-bearing tissue to treat diseased tissue, and both the donor site and the treated area would be expected to degenerate. These limitations have stimulated the development of tissue-engineered cartilages that mimic the function of native tissue and that could be used in the repair of large cartilage defects.

Dorotka *et al.* determined the benefit of implanting a 3-D collagen matrix in addition to microfracture for the treatment of chondral defects in a sheep model, and whether there is any need for autologous chondrocytes in the matrices [28]. To this end, they evaluated the behavior of ovine chondrocytes and BMSC on a matrix comprising types I, II, and III collagen *in vitro*, and the healing of chondral defects in an ovine model treated with the matrix, either unseeded or seeded with autologous chondrocytes, combined with microfracture treatment. The animal study included 22 chondral defects in 11 sheep, divided into four treatment groups. Group A: microfracture and collagen matrix seeded with chondrocytes; B: microfracture and unseeded matrices; C: microfracture; and D: untreated defects. All animals were sacrificed 16 weeks after implantation, and two defects, 1 and 2.5 cm distal from the intercondylar notch, were introduced in the medial condyle of the femur. The defects were outlined with the punch down to the subchondral bone and cartilage was removed with small curettes. All treatment groups of the animal model showed better defect filling compared to untreated knees. Table 2.3 gives the histomorphometric results. The cell-seeded group had the greatest quantity of repair tissue and the largest quantity of hyaline-like tissue. Although the collagen matrix was an adequate environment for BMSC *in vitro*, the additionally implanted unseeded collagen matrix did not increase the repair response after microfracture in chondral defects. Only the

Table 2.3
Histomorphometric evaluation of the four treament groups (mean $\pm$ s.e.m.)

	Microfracture and Cell-seeded Matrix (Group A)	Microfracture and Unseeded Matrix (Group B)	Microfracture (Group C)	Untreated (Group D)
Total defect fill (%)	65 $\pm$ 8	43 $\pm$ 8	54 $\pm$ 7	22 $\pm$ 10
Fibrous tissue (% of defect area)	2 $\pm$ 2	11 $\pm$ 2	8 $\pm$ 4	14 $\pm$ 8
Transitional tissue (% of defect area)	8 $\pm$ 4	23 $\pm$ 7	17 $\pm$ 5	5 $\pm$ 2
Hyaline cartilage (% of defect area)	54 $\pm$ 8	6 $\pm$ 3	21 $\pm$ 7	3 $\pm$ 1
Articular cartilage (% of defect area)	1 $\pm$ 1	3 $\pm$ 3	8 $\pm$ 5	0
% Bonded to subchondral bone	16 $\pm$ 9	62 $\pm$ 10	63 $\pm$ 13	18 $\pm$ 5
% Bonded to adjacent cartilage	31 $\pm$ 18	46 $\pm$ 10	49 $\pm$ 7	27 $\pm$ 6
% Intact calcified layer	54 $\pm$ 4	65 $\pm$ 10	79 $\pm$ 4	99 $\pm$ 0

matrices seeded with autologous cells in combination with microfracture were able to facilitate hyaline-like cartilage regeneration. Using a rabbit model of osteochondral injury, Willers *et al.* inoculated autologous chondrocytes onto a type I/III collagen scaffold and implanted into 3-mm osteochondral knee defects [29]. All untreated defect histology showed inferior fibrocartilage and/or fibrous tissue repair, whereas autologous chondrocyte implantation (ACI) with type I/III collagen membrane regenerated cartilage with healthy osteochondral architecture in osteochondral defects at 6 weeks. At 12 weeks, articular cartilage regeneration was maintained, with reduced thickness and proteoglycan compared with the adjacent cartilage. Both 6-week and 12-week ACI with collagen membrane revealed significant improvement as compared with untreated controls. To further examine the efficacy of cartilage regeneration by ACI, they conducted a dose–response study using chondrocytes at various cell densities between 10^4 and 10^6 cells/cm^2. The results showed that cell density had no effect on outcome histology.

Guo *et al.* investigated the possibility of using implants of autologous MSCs seeded onto porous bioceramic β-tricalcium phosphate (β-TCP) to repair articular cartilage defects in a sheep model [30]. Figure 2.5 shows the SEM micrograph of the ceramic. As can be seen, the porous β-TCP bioceramic had homogeneous spherical pores with a porosity of 70%, a pore diameter of 450 $\pm$ 50 μm, and interconnections of 150 $\pm$ 50 μm in diameter. Culture-expanded autologous MSCs were seeded onto the bioceramic scaffold. A 4-cm incision was made over the right knee of each animal by a medial longitudinal parapatellar incision. The patella was dislocated laterally and the knee was flexed 90° to expose the weight-bearing area of the medial femoral condyle. One defect, 8 mm in diameter and 4 mm in depth, was created mechanically and the depth of penetration was limited to 4 mm. In the experimental group, the defects were filled with β-TCP seeded with autologous MSCs. In control

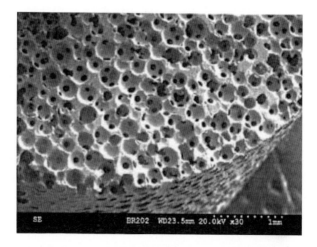

Fig. 2.5   ESM micrograph of β-TCP ceramics.

group 1, the defects were filled only with β-TCP. In control group 2, defects were generated on the left side of the rear limbs and then on the right side 12 weeks later. Twenty-four weeks later, the defects were resurfaced with hyaline-like tissue, and an interface between the engineered cartilage, the adjacent normal cartilage, and the underlying bone was observed. From 12 to 24 weeks postimplantation, modification of neocartilage was obvious in the rearrangement of surface cartilage and the increase in GAG level. Guo *et al.* further investigated the repair of sheep articular defects with bioceramic-chondrocyte constructs, prepared by seeding of autologous chondrocytes into the β-TCP bioceramic scaffolds used in the study above, and evaluated their efficacy for cartilaginous tissue regeneration and potential for human clinical application [31]. Neocartilage tissue completely resurfaced the cartilage defects after 24 weeks. Typical hyaline cartilage structure was generated in the engineered cartilage. Biodegradation of bioceramic was notable, leading to bioceramic fragmentation and particle formation. Numerous ceramic particles and numerous macrophages were visible in the neocartilage tissue.

To repair osteochondral defects in articular cartilage, Tanaka *et al.* fabricated plugs made of chondrocytes in collagen gel overlying a resorbable porous β-TCP block, which were implanted into defects in rabbit knees [32]. Eight weeks after implantation of the biphasic construct, histologic examination showed hyaline-like cartilage formation that was positive for safranin-O and type II collagen. At 12 weeks, most of the β-TCP was replaced by bone, with a small amount remaining in the underlying cartilage. The average composition of engineered cartilage is shown in Fig. 2.6. In the cell-seeded layer, the newly formed middle and deep cartilage adjacent to the subchondral bone stained with safranin-O, but no staining was observed in the superficial layer. In addition, cell morphology was distinctly different from the deep levels of the reparative cartilage, with hypertrophic cells at the bottom of the cartilaginous layer. At 30 weeks, β-TCP had completely been resorbed and a tidemark was observed in some areas. In contrast, controls (defects filled with a β-TCP block alone) showed no cartilage formation but instead had subchondral bone formation.

Johnson *et al.* undertook a study (1) to demonstrate the ability of isolated articular, auricular, and costal chondrocytes suspended in fibrin glue to form neocartilage in an *in vivo* model, (2) to demonstrate the ability of the tissue-engineered cartilage to integrate into living native cartilage, and (3) to evaluate the quality of the adhesive bonds created between tissue-engineered cartilage and native cartilage by biomechanical testing [33]. Discs of articular cartilage and articular, auricular, and costal chondrocytes were harvested from swine. Articular, auricular, or costal chondrocytes suspended in fibrin glue (experimental), or fibrin glue alone (control), were placed between discs of articular cartilage, forming trilayer constructs, and implanted subcutaneously into nude mice for 6 and 12 weeks. Specimens were evaluated for neocartilage production and integration into native cartilage with histological and biomechanical analysis. New matrix was formed in all experimental samples, consisting mostly of neocartilage integrating with the cartilage discs. Control samples developed fibrous

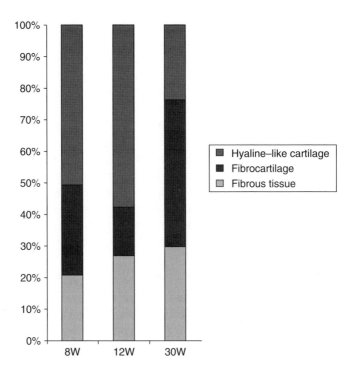

Fig. 2.6   The average composition of reparative cartilage from each set of six rabbits over time.

tissue without evidence of neocartilage. Ultimate tensile strength (UTS) values for experimental samples significantly increased than those of controls, as shown in Fig. 2.7.

   To investigate the effect of basic fibroblast growth factor (bFGF) on the healing of full-thickness articular cartilage defects, Tanaka *et al.* mixed bFGF with collagen gel and implanted the mixture into full-thickness articular cartilage defects drilled into rabbit knees [34]. At 4 weeks, treatment with 100 ng of bFGF had greatly stimulated cartilage repair in comparison with defects filled only with plain collagen gel. The average scores on the histological grading scale were significantly better for the defects treated with bFGF than for the untreated defects. Immunohistochemical staining for type II collagen showed that this cartilage-specific collagen was diffusely distributed in the repaired tissue at 50 weeks. Nishida *et al.* used two strategies to investigate the reparative effect of CTGF/CCN2 on articular cartilage *in vivo* [35]. The CTGF/CCN2 is a connective growth factor that stimulates the proliferation and differentiation of articular chondrocytes *in vitro*. They produced monoiodoacetic acid (MIA)-induced experimental osteoarthritis (OA) model in rats. To clarify the direct effect of CTGF/CCN2 on the repair of articular cartilage, they investigated whether full-thickness defects

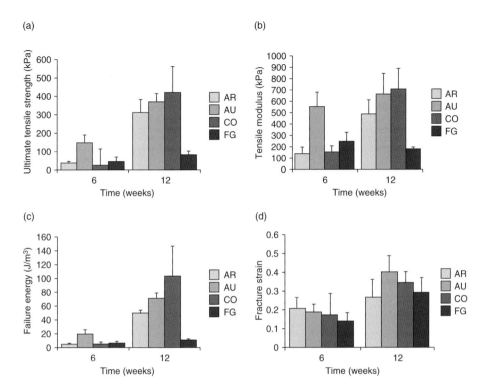

Fig. 2.7 Time course of changes in ultimate tensile strength over total 12 weeks *in vivo* for constructs formed with articular (AR), auricular (AU), and costal (CO) chondrocytes suspended in fibrin glue, or fibrin glue (FG) alone. Five to eight samples were tested per group per time point.

made in rat articular cartilage could be re-filled with hyaline-like tissue after implantation of CTGF/CCN2. A single injection of CTGF/CNN2 incorporated in gelatin hydrogel (CTGF/CNN2-hydrogel) into the joint cavity of MIA-induced OA model rats repaired their articular cartilage to the extent that it became histologically similar to normal articular cartilage. To examine the effect of CTGF/CCN2 on the repair of articular cartilage, they created defects (2 mm in diameter) on the surface of articular cartilage *in situ* and implanted CTGF/CCN2-hydrogel or PBS-hydrogel therein with collagen sponge. In the group implanted with CTGF/CNN2-hydrogel collagen, new cartilage filled the defect 4 weeks postoperatively. In contrast, only soft tissue repair occurred when the PBS-hydrogel collagen was implanted.

Rotter *et al.* examined the generation of tissue-engineered cartilage using PGLA (Ethisorb; Ethicon) in an autologous immunocompetent pig model [36]. The goals of the study were to determine the role of interleukin 1α (IL-1α) in this system and to assess the effect of serum treatment on tissue generation. Porcine auricular chondrocytes were seeded onto Ethisorb discs cultured for 1 week in medium supplemented with either FCS or serum-free insulin–transferrin–selenium supplement. Specimens were implanted autologously in pigs with unseeded scaffolds as controls. Histology revealed acute inflammation surrounding degrading scaffold. Cartilage formation was observed as early as 1 week after implantation and continued to increase with time; however, homogeneous matrix synthesis was not present in any of the specimens. Strong IL-1α expression was detected in chondrocytes at the implant periphery and in cells in the vicinity of degrading polymer. Histologically there was no significant difference between the experimental groups with respect to the amount of matrix synthesis or inflammatory infiltration. The GAG content was significantly higher in the serum-free group. Case *et al.* evaluated the effects of chondrocyte viability and mechanical loading on bone formation resulting from implantation of tissue-engineered cartilage constructs into a well-vascularized bone defect using a lapine hydraulic bone chamber (HBC) model shown in Fig. 2.8 (see p. 352 in Chapter 3 for the detail of this chamber model) [37]. Chondrocytes were isolated from the cartilage harvested from the patellar groove and femoral condyles of rabbits. The articular chondrocytes were seeded onto PGA felt discs, cultured for 4 weeks *in vitro*, and then transferred to empty bone chambers previously implanted into rabbit femoral metaphyses. Histological analysis of the 4-week cartilage constructs revealed that cellular attachment and matrix deposition occurred throughout the PGA felt. The implantation of cartilage constructs into the HBC model supported formation of a complex repair environment containing bone, cartilage, and fibrous tissue. Viable chondrocytes within the tissue-engineered cartilage constructs significantly enhanced bone formation, as compared with constructs devitalized by a freeze–thaw procedure and implanted in

Fig. 2.8   Dissembled hydraulic bone chamber implants: non-loaded (left) and loaded (right).

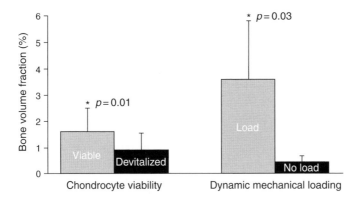

Fig. 2.9 Bone volume fraction quantified using micro-CT images of chamber biopsies. Left: Chondrocyte viability experiment ($n = 8$) comparing implantation of viable constructs with that of devitalized constructs after 4 weeks *in vivo*. Right: Dynamic mechanical loading experiment ($n = 5$) comparing chamber tissue loaded for 4 weeks with tissue under no-load control conditions after 8.5 weeks *in vivo*.

contralateral control chambers. The application of an intermittent cyclic mechanical load was found to increase the bone volume fraction of the chamber tissue from 0.4 to 3.6% as compared with no load control biopsies, as shown in Fig. 2.9.

Osteochondritis dissecans (OCD), in which an osteochondral fragment separates from a normal vascular bone bed, is quite different from osteonecrosis, in which the fragment separates from an avascular bone bed. Many etiologies have been proposed, including trauma or repetitive microtrauma and impingement of the tibial spine, stress fracture with no specific injury, and vascular insult. Brittberg *et al.* previously reported on the early successful results of autologous chondrocyte transplantation for the treatment of focal cartilage defects [38]. Slivers of cartilage (300–500 mg) were obtained from the upper minor load-bearing area of the medial femoral condyle of the injured knee. The samples were transferred to the cell-culture laboratory. The cells were incubated in tissue-culture flasks. Implantation was performed 14–21 days after the biopsy. Peterson *et al.* assessed the intermediate to long-term results of this technique in a large group of patients with OCD [39]. Fifty-eight patients with radiographically documented OCD of the knee underwent treatment with autologous chondrocyte transplantation between 1987 and 2000 and were assessed clinically with the use of standard rating scales. Twenty-two patients consented to arthroscopic second-look evaluation of graft integrity. The defect was located on the medial femoral condyle in 39 patients and on the lateral femoral condyle in 19. The mean lesion size was 5.7 cm^2, and the mean defect depth was 7.8 mm. After a mean duration of follow-up of 5.6 years, 91% of the patients had a good or excellent overall rating on the basis of a clinical evaluation and 93% had improvement on a patient self-assessment questionnaire. Two patients had a failure of treatment in the early postoperative period. Only one patient who had had a good or excellent rating at two years had a decline in clinical status at the time of the last follow-up. Thus, treatment of OCD lesions of the knee with autologous chondrocyte transplantation produced an integrated repair tissue with a successful clinical result in >90% of patients (Tables 2.4 and 2.5).

Ochi *et al.* presented the clinical outcome of transplanting autologous chondrocytes, cultured in collagen gel, for the treatment of full-thickness defects of cartilage in 28 knees (26 patients) over a minimum period of 25 months [40]. For the medical treatment they obtained specimens of cartilage weighing 300–500 mg through an anteromedial or anterolateral approach either from a detached cartilage fragment or from an unloaded area of either the medial or the lateral condyle. The isolated cells were mixed in a collagen–medium mixture and the mixture was placed in a culture dish. When cultured for 3–4 weeks, the collagen gel including chondrocytes became

Table 2.4
Clinical outcomes for 58 patients

Scoring System	
Brittberg clinical rating (*no. of patients*)	
Preoperative	
Excellent	0
Good	0
Fair	2
Poor	56
Postoperative	
Excellent	31
Good	22
Fair	4
Poor	1
Postoperative self-assessment (*no. of patients*)	
Improved	54 (93%)
Same or worse	4 (7%)
Cincinnati rating (*points*)	
Preoperative	2.0
2 years postoperative	8.9
2–10 years postoperative	9.8
Tegner–Wallgren score (*points*)	
Preoperative	6.3
2 years postoperative	8.3
2–10 years postoperative	10.2
Lysholm score (*points*)	
Preoperative	44.3
2 years postoperative	89.3
2–10 years postoperative	92.4
Brittberg–Peterson VAS score (*points*)	
Preoperative	80.2
2 years postoperative	31.2
2–10 years postoperative	26.7

Table 2.5
Radiographic findings, magnetic resonance imaging findings,
and macroscopic integrity

Variable	
Radiographic findings ($n = 27$) (*no. of patients*)	
Osteophyte formation	15
Joint-space narrowing	8
Cyst formation	3
Sclerotic changes	12
Bone reconstruction	2
Subchondral bone flattening	4
Tibial spine prominence "sharpening"	9
Magnetic resonance imaging findings ($n = 15$) (*no. of patients*)	
Bone remodeling	7
Cyst formation	2
Cartilage-like repair tissue	13
Graft integrity score ($n = 22$) (*points*)*	11.2 (4–12)

* The data are given as the mean, with the range in parentheses.

opaque and acquired a jelly-like hardness. Gel transplantation eliminated locking of the knee and reduced pain and swelling in all patients. Arthroscopic assessment indicated that 26 knees (93%) had a good or excellent outcome (Table 2.6). There were few adverse features, except for marked hypertrophy of the graft in 3 knees, partial detachment of the periosteum in 3 and partial ossification of the graft in 1 knee. Biomechanical tests revealed hard transplants, similar to the surrounding cartilage.

Wakitani *et al.* applied cell transplantation to repair human articular cartilage defects in osteoarthritis (OA) knee joints [41]. The study group comprised 24 knees of 24 patients with knee OA who underwent a high tibial osteotomy. Adherent cells in bone-marrow aspirates were expanded by culture, embedded in collagen gel, transplanted into the articular cartilage defect in the medial femoral condyle, and covered with autologous periosteum at the time of 12 high tibial osteotomies. The other 12 subjects served as cell-free controls. In the cell-transplanted group, as early as 6.3 weeks after transplantation the defects were covered with white to pink soft tissue, in which metachromasia was partially observed. Forty-two weeks after transplantation, the defects were covered with white soft tissue, in which metachromasia was partially observed in almost all areas of the sampled tissue and hyaline cartilage-like tissue was partially observed. Although the clinical improvement was not significantly different, the arthroscopic and histological grading score was better in the cell-transplanted group than in the cell-free control group. The result is shown in Table 2.7. Higher grading score indicates better tissue repair.

To verify the usefulness of a photoacoustic measurement method for evaluation of the viscoelastic properties of engineered cartilage, Ishihara *et al.* used cartilage

Table 2.6

Details of the lesions, transplants, and outcome for the 28 knees (26 patients) treated with chondrocytes transplanted in atelocollagen gel

Case	Size of Lesion (cm²)	Cell Number (×10⁶)	Harvest Site*	Lysholm Score Preop	Lysholm Score Postop	Arthroscopic Grade (ICRS)	Graft Stiffness (Centre) at 2 years (% of Control)	Graft Stiffness (Margin) at 2 years (% of Control)
1	2.4	42.0	Free body	68	85	7		
2	3.0	1.9	NWA	75	95	12		
3	1.4	0.88	NWA	81	97	12	93.2	101.0
4	2.2	0.7	Free body	71	92	10		
5	16.0	4.3	NWA	65	96	12		
6	2.3	0.58	NWA	74	98	12		
7	2.0	3.7	NWA + free body	73	100	11		
8	2.3	0.53	NWA + DC	85	100	11	64.1	75.8
9	3.8	3.0	NWA + DC	61	100	12	73.4	74.7
10	2.9	0.53	NWA + free body	78	100	12		
11	2.4	0.3	NWA + DC	63	100	12		
12	3.8	1.4	NWA + DC	56	98	11	103.5	101.3
13	4.2	0.55	NWA + free body	61	86	9		
14	3.5	1.6	NWA	75	100	9	92.7	94.7
15	2.9	1.4	NWA	73	98	9	54.6	133.4
16	2.8	1.2	NWA + DC	76	96	12	96.1	98.2
17	2.0	0.38	NWA + DC	74	96	11	91.6	93.2
18	4.0	0.9	NWA	50	95	9	83.0	99.7
19	2.5	0.23	NWA	54	86	7	93.0	98.0
20	1.7	4.6	NWA	53	96	9	78.1	97.9
21	0.7	1.1	NWA + DC	64	100	12	86.0	90.0
22	0.8	0.4	NWA	72	100	11	97.3	100.9
23	2.3	0.8	NWA	76	97	9	95.0	104.1
24	2.6	0.15	NWA	90	100	9	64.9	127.1
25	1.0	0.85	NWA + DC	89	100	9	100.8	103.2
26	3.4	1.3	NWA + DC	91	100	11	154.8	101.3
27	1.1	1.2	NWA + DC	72	100	11		
28	2.0	0.4	NWA + DC	59	95	9		

* NWA, non-weight-bearing area; DC, detached cartilage.

Table 2.7

Arthroscopic and histological grading score at first and second look operation

	First Look			Second Look		
		Cell−			Cell−	
	Cell+	per	ab + per	Cell+	per	ab + per
Patient no.	12	6	6	9	3	3
Weeks after surgery	6.3	7.2	6.6	42	46	40
Arthroscopic						
A. Width	2.5 ± 0.7	2.3 ± 0.8	2.3 ± 0.8	2.9 ± 0.3	2.3 ± 1.2	2.3 ± 0.6
B. Thickness	2.8 ± 0.5	2.7 ± 0.5	2.8 ± 0.4	2.8 ± 0.4	2.3 ± 1.2	2.0 ± 0.0
C. Surface regularity	2.2 ± 0.8	1.8 ± 0.8	2.0 ± 0.9	2.8 ± 0.4	1.7 ± 0.6	2.0 ± 0.0
D. Stiffness	0.3 ± 0.5	0.2 ± 0.4	0.2 ± 0.4	1.0 ± 0.5	0.7 ± 0.6	1.0 ± 0.0
E. Color	0.8 ± 0.4	0.2 ± 0.4	0.2 ± 0.4	1.0 ± 0.0	0.7 ± 0.6	1.0 ± 0.0
Subtotal	8.6 ± 1.7	7.2 ± 2.0	7.5 ± 1.5	10.4 ± 1.2	7.7 ± 4.0	8.3 ± 0.6*
Histological						
F. Cell morphology	0.6 ± 0.7	0.2 ± 0.4	0.2 ± 0.4	1.6 ± 0.7	1.0 ± 1.0	1.0 ± 1.0
G. Matrix staining	0.7 ± 0.9	0.2 ± 0.4	0.3 ± 0.5	3.4 ± 0.7	1.3 ± 1.2	2.0 ± 1.7
Subtotal	1.3 ± 1.4	0.3 ± 0.8	0.5 ± 0.8	5.0 ± 1.2	2.3 ± 2.1*	3.0 ± 2.7
Total	9.8 ± 2.0	7.5 ± 2.2*	8.0 ± 0.9*	15.4 ± 1.4	10.0 ± 6.1*	11.3 ± 2.3*

* Significantly different between cell+ and cell− (Mann–Whitney U-test).
  Cell+, cell-transplanted; cell−, cell-free.
  per, multiple perforation; ab + per, abrasion with multiple perforation.

tissues obtained by culture for various periods up to 12 weeks as samples [42]. The relaxation times measured by the photoacoustic method agreed well with the intrinsic viscoelastic parameters with a correlation coefficient of 0.98. Dickinson *et al.* developed quantitative biochemical outcome measures for the repair of articular cartilage [43]. They described the production of a new anti-peptide antibody and the development and validation of a method for the extraction and immunoassay of type I collagen. The assay was used, in conjunction with existing assays for type II collagen and proteoglycans, to measure levels of the matrix components in repair tissue biopsies obtained from patients treated with a tissue engineering therapy Hyalograft® C. Frozen sections cut from the same biopsies were stained for proteoglycans, and immunohistochemical analysis was used to assess types I and II collagen staining. They found that histological grading of matrix protein abundance to classify repair cartilage as hyaline or fibrocartilagenous was often misleading. In addition, Dickinson *et al.* demonstrated the ability to measure collagen crosslinks in repair tissue biopsies. Figure 2.10 shows the relative proportions of mature and immature collagen crosslinks. In 61% of the biopsies, the ratio of mature to immature crosslinks fell below the minimal ratio measured in seven natural adult human cartilages, indicating that the collagen was still turning over rapidly.

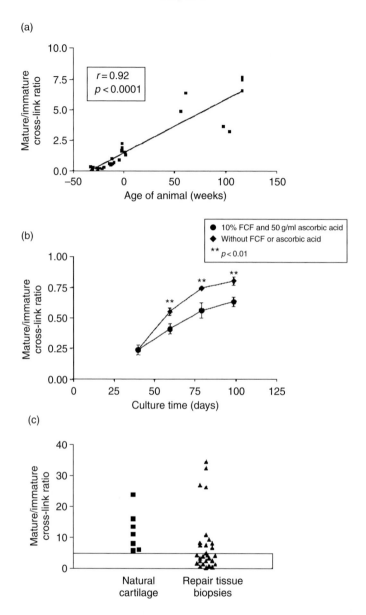

Fig. 2.10 Mature and immature collagen crosslinks as a function of age of animal (a), culture time (b), and tissue origin (c).

Periosteum has two distinct layers: a thick outer fibrous layer and a thin inner cambium layer which is adjacent to bone and contains specific chondrocyte precursor cells. The MSCs are known to persist in the cambium layer of periosteum. Periosteum-derived cells are easy to obtain and to handle, and contain a population of undifferentiated cells that can undergo osteogenic or chondrogenic differentiation, depending on the environmental conditions. Previous studies have reported the critical nature in which periosteum should be harvested and handled in order to preserve periosteum viability. This stems from the fact that the mesenchymal cells of the cambium layer are only lightly adherent to the periosteum and may be left on the bone surface if harvested inappropriately. In the resting state, the cambium layer of the periosteum contains spindle-shaped undifferentiated mesenchymal cells, three or four cell layers in thickness. Culturing rabbit periosteum in agarose gel in the presence of TGF-β1 promoted chondrogenesis, probably due to the proliferation and differentiation of the chondrogenic precursor cells located in the cambium layer of the periosteum [44, 45]. However, Stevens *et al.* found that *in vitro* cell culture of chondrocytes in the alginate gel yielded alginate/cell constructs that lacked the continuous, interconnected collagen/proteoglycan network of hyaline cartilage [46]. In addition, to demonstrate that this gel system was capable of supporting periosteum-derived chondrogenesis, they harvested periosteum using a technique that yielded large viable explants with no noticeable loss to the cambium layer. After 10 days *in vitro* culture in an alginate gel system in the presence of TGF-β1, the cellularity in this layer increased resulting in a twofold increase in thickness of the cambium layer. Periosteal chondrogenesis was evident at 3 weeks. After 6 weeks of *in vitro* culture, a significant amount of cartilage was generated with 54% of the total area of the explant staining positive for safranin-O. The presence of type II collagen was also confirmed in the safranin-O stained areas, indicating that the newly generated cartilage was not fibrocartilaginous in nature. To evaluate the reactive tissue formed after damaging the periosteum, Emans *et al.* bilaterally dissected periosteum from the proximal medial tibia of rabbits [47]. Reactive periosteal tissue was harvested 10, 20, and 40 days postsurgery and analyzed for expression of types I, II, and X collagen, aggrecan, osteopontin, glyceralaldehyde-3-phosphate dehydrogenase (GAPDH), and osteonectin by RT-PCR. The average mRNA expression (expressed as a ratio of β-actin) is shown in Fig. 2.11. Both β-actin and GAPDH were considered as internal control. Reactive tissue, appearing first as a thickening, in time turned from soft to hard tissue and was observed in 93% of the rabbits, as shown in Fig. 2.12. Histologically, this tissue consisted of hyaline cartilage at follow-up days 10 and 20. Expression of type II collagen and aggrecan was present at 10 and 20 days postsurgery. Highest expression was at 10 days. Immunohistochemistry confirmed the presence of cartilage, which was positive for types I and II collagen at 10 days and only for type II collagen at 20 days. At 20 days postsurgery the onset of bone formation was also observed. At 40 days postsurgery, the reactive tissue had almost completely turned into bone.

2.2. Bones

Autologous and allogeneic transplantations are frequently performed to fill bone defects and gaps in multifragment or comminuted fractures, cysts, or in patients suffering from inflammatory or tumorous skeletal diseases. Resection of cancerous bone leaves voids that must be filled by either tissue grafts or implants.

Chapter 2

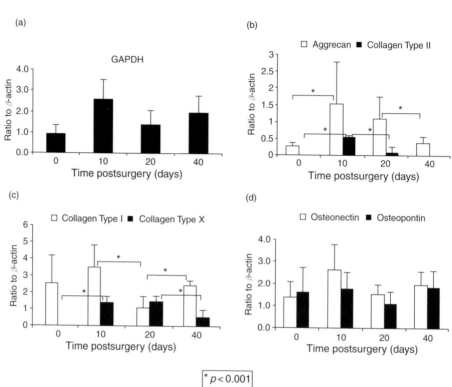

(a)

GAPDH

(b)

☐ Aggrecan ■ Collagen Type II

(c)

☐ Collagen Type I ■ Collagen Type X

(d)

☐ Osteonectin ■ Osteopontin

$^*p < 0.001$

Fig. 2.11 Histograms presenting relative mRNA levels compared with the level of β-actin mRNA during chondrogenic and osteogenic differentiation of reactive tissue produced by dissection of periosteum. (a) Expression of GAPDH; (b) expression of aggrecan and collagen type II; (C) expression of collagen type I and collagen type X; and (d) expression of osteopontin and osteonectin.

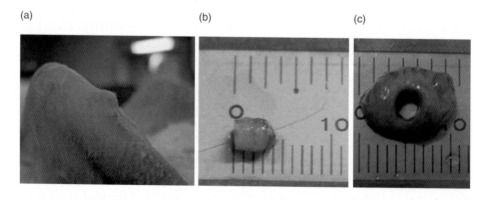

(a) (b) (c)

Fig. 2.12 (a) *In vitro* generation of cartilage from periosteum at the upper medial part of the right tibia of a rabbit, 10 days postsurgery. (b) A cartilage construct (3 mm thick and 3 mm in diameter) cored out of (c) A large sample, resected 10 days postsurgery.

Autologous bone graft derived from iliac crest has long been considered the "gold standard" biological graft material used most frequently and necessary for skeletal reconstruction, but sufficient autologous bone material may not be available from children. The limited tissue supply makes large defect repairs problematic. In addition, biomechanical and histologic studies revealed that autologous spongiosa in a transplant model did not show optimal mechanical stability for at least 2 years and could result in persistent impairment of skeletal stability. Clinical alternatives to autografts include allogeneic bone (bank bone), ceramics, demineralized bone matrix, bone marrow, and composite grafts (consisting of combinations of materials), but none of these materials has been met with widespread acceptance. Cadaver-derived bones taken from bone-bank, which can be obtained in sufficient quantities and have been used for some time with good results, are, however, associated with the risk of immune rejection in some patients and the potential for infection transmitted from the donor, whereas artificial bones do not possess regenerative ability. Patient's own bone is ideal for grafting from the aspect of avoiding rejection and maintaining osteogenic potential, and also produces the best clinical results. However, given the damage to the healthy portion of the bone and soft tissue at the donor site, the amount of bone that can be harvested is limited.

In response to these problems, biomedical engineers have been developing tissue-engineered bone graft alternatives. Since the early 1990s, numerous reports have been published on bone formation ectopically or in critical-sized defects. The concept studied most extensively for bone tissue engineering is the combination of BMSC with a biocompatible porous scaffold. Formation of the vertebrate skeleton through endochondral bone formation begins with the migration of undifferentiated mesenchymal cells from the lateral plate mesoderm to sites destined to become bones. The cells undergo a condensation step and then form a cartilaginous scaffold or "mold" that defines the morphology of the bone. Cells at the center synthesize an ECM that is rich in type II collagen, proteoglycans, and related macromolecules. They then hypertrophy and undergo apoptosis as the matrices are gradually replaced by invasion of blood vessels, followed by appearance of osteoblasts that synthesize bone matrix.

A massive bone defect with a clinically relevant volume was reconstructed by transplanting an engineered bone in which MSCs expanded in autologous serum were combined with a porous scaffold [48]. Constructs in which MSCs were combined with a porous coralline-based HAp (CHA) scaffold having the same architecture as natural coral but a lower resorption rate were prepared. After implantation, these constructs were found to have the same osteogenic potential as autologous bone grafts in terms of the amount of newly formed bone present at 4 months and to have been completely replaced by newly formed, structurally competent bone within 14 months. Although the rate of bone healing was improved when CHA–MSC constructs were used (five of seven animals healed) as compared with the coral–MSC construct (one of seven healed), it was still less satisfactory than that obtained with autografts (five of five healed). Nishikawa *et al.*

developed fully opened interconnected porous HAp having two different pore sizes [49]. One had pores with an average size of 150 μm in diameter, an average 40-μm interconnecting pore diameter, and 75% porosity (HA 150), while the other had pores with an average size of 300 μm in diameter, an average 60–100-μm interconnecting pore diameter, and 75% porosity (HA 300). These ceramics were combined with rat marrow mesenchymal cells and cultured for 2 weeks in the presence of dexamethasone. The cultured ceramics were then implanted into subcutaneous sites in syngeneic rats and harvested 2–8 weeks after implantation. All the implants showed bone formation inside the pore areas as evidenced by decalcified histological sections and microcomputed tomography images. The bone volume increased over time. At 8 weeks after implantation, extensive bone volume was detected not only in the surface pore areas but also in the center pore areas of the implants. As shown in Fig. 2.13, a high degree of ALP activity with a peak at 2 weeks and a high level of osteocalcin (OCN) with a gradual increase over time were detected in the implants. The levels of these biochemical parameters were higher in HA 150 than in HA 300. Yoshikawa *et al.* examined the possibility of clinical application of BMCs for bone reconstruction surgery [50]. A 3-ml sample of bone marrow was collected from the iliac bones of 27 orthopedic patients, followed by culture in minimal essential medium (MEM) containing 15% FCS (Fig. 2.14). In all 7 patients randomly selected from these 27 patients, significant *in vitro* osteogenic ability of marrow mesenchymal cells was demonstrated by SEM and biochemical analyses. To investigate the *in vivo* osteogenic potential of this human cultured bone, porous ceramics were impregnated with marrow cells and subcultured in osteogenic culture medium (standard medium supplemented with sodium β-glycerophosphate, vitamin C phosphate, and dexamethasone). After 3 weeks of subculture, the cultured bone/porous ceramics were grafted into the abdominal cavity of nude mice. Histological and biochemical (ALP and human OCN) examination indicated that the bone-regenerating ability of human marrow cells may not depend on age, as shown in Fig. 2.15, and that cultured artificial bone is useful for bone regeneration treatment if appropriate cultured marrow cells could be successfully prepared.

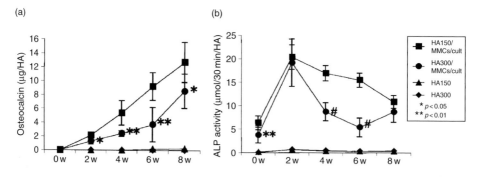

Fig. 2.13 *In vivo* ALP activities (b) and osteocalcin contents (a) of implants.

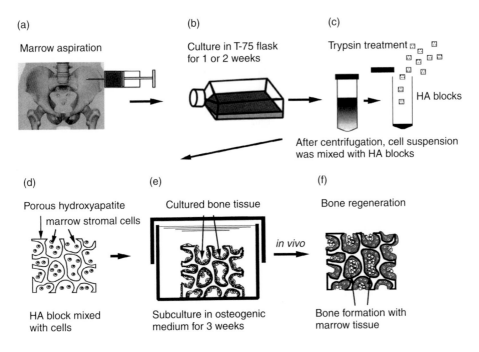

(a)

Marrow aspiration

(b)

Culture in T-75 flask
for 1 or 2 weeks

(c)

Trypsin treatment

HA blocks

After centrifugation, cell suspension
was mixed with HA blocks

(d)

Porous hydroxyapatite
| marrow stromal cells

HA block mixed
with cells

(e)

Cultured bone tissue

in vivo

Subculture in osteogenic
medium for 3 weeks

(f)

Bone regeneration

Bone formation with
marrow tissue

Fig. 2.14   Preparation of cultured artificial bone.

Bone regeneration by recombinant human (rh) bone morphogenetic protein-2 (BMP-2) in a block copolymer composed of poly(D,L-lactide) (PDLLA) with randomly inserted *p*-dioxanone (DX) and PEG (PLA–DX–PEG) and porous β-TCP blocks was studied in a critical-sized rabbit bone defect model [51]. When a composite of PLA–DX–PEG (250 mg) and β-TCP (300 mg) loaded with or without rhBMP-2 (50 μg) was implanted into a 1.5 cm intercalated bone defect created in a rabbit femur,

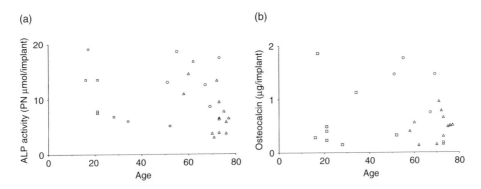

(a)

(b)

Fig. 2.15   (a) ALP activity 2 months after grafting of human bioartificial bone; (b) Human osteocalcin level 2 months after grafting of human bioartificial bone. (□) Trauma; (▵) osteoarthrosis; (○) rheumatoid arthritis.

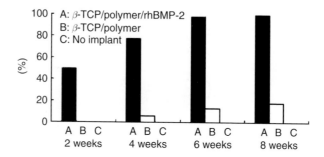

Fig. 2.16 BMP-loaded group promoting a time-dependent increase in callus.

the bony union of the defect was recognized only in the BMP-loaded group. As shown in Fig. 2.16, the BMP-loaded group showed a time-dependent increase in callus of nearly 100% at 6 weeks, but the group with β-TCP and polymer without rhBMP-2 showed bone formation only to less than 20%. No bone formation was recognized in the control group. When the BMP-loaded composites group was followed up to 24 weeks, all defects were completely repaired without residual traces of implants.

Schmoekel *et al*. designed a tripartite BMP-2 fusion protein, denoted as TG-pl-BMP-2, and produced it recombinantly [52]. An *N*-terminal transglutaminase substrate (TG) domain provides covalent attachment to fibrin during coagulation under the influence of the blood transglutaminase factor XIIIa. A central plasmin substrate (pl) domain provides a cleavage site for local release of the attached growth factor from the fibrin matrix under the influence of cell-activated plasmin. A C-terminal human BMP-2 domain provides osteogenic activity. The TG-pl-BMP-2 in fibrin was evaluated *in vivo* in critical-sized craniotomy defects in rats. As shown in Fig. 2.17, the fusion protein induced 76% more defect healing with bone at 3 weeks with a dose of 1 µg/defect than wildtype BMP-2 in fibrin. After a dosing study in rabbits,

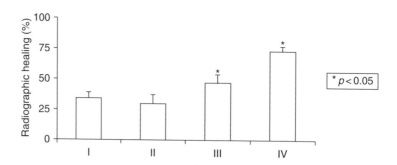

Fig. 2.17 Healing of calvarial defects in the rabbit. The dose-sensitivity of healing induced by TG-pl-BMP-2 incorporated in fibrin was examined in 6 mm defects in the rabbit, 4 per animal. Empty defects (I), fibrin alone (II), fibrin with 5 µg TG-pl-BMP-2 (III), and fibrin with 15 µg TG-pl-BMP-2 (IV) were tested ($n = 8$ for each group).

the engineered growth factor in fibrin was evaluated for pancarpal fusion in dogs, where it induced statistically faster and more extensive bone bridging than equivalent treatment with cancellous bone autograft. Peterson *et al.* assessed the efficacy of BMP-2-producing human adipose–derived MSCs to heal a critical-sized femoral defect in a nude rat model [53]. Human adipose tissue was obtained from healthy donors. Cells were grown in culture and infected with a BMP-2-carrying adenovirus. Five million cells were applied to a collagen-ceramic carrier and implanted into femoral defects. Eleven of the 12 femora in the group treated with human processed lipoaspirate (HPLA) cells genetically modified to overexpress BMP-2 had healed at 8 weeks. All eight of the femora treated with the rhBMP-2-impregnated collagen-ceramic carrier healed. No statistically significant difference was detected between these two groups. Cowan *et al.* investigated the ability of adipose-derived stromal (ADS) cells and bone marrow-derived stromal (BMS) cells to accelerate *in vivo* osteogenesis on *ex vivo* recombinant human BMP-2 and retinoic acid stimulation [54]. Mouse osteoblasts, ADS cells, and BMS cells were seeded onto apatite-coated PLGA scaffolds, stimulated with rhBMP-2 and retinoic acid *ex vivo* for 4 weeks, and subsequently implanted into critical-sized (4 mm) calvarial defects. Areas of complete bony bridging were noted as early as 2 weeks *in vivo*; however, osteoclasts were attracted to the scaffold as identified by calcitonin receptor staining and tartrate-resistant acid phosphatase activity staining. Hokugo *et al.* applied gelatin hydrogel incorporating platelet-rich plasma (PRP) to a bone defect of rabbit ulna to evaluate bone formation [55]. Successful bone regeneration was observed at the bone defect treated with the gelatin hydrogel incorporating PRP, in marked contrast to the control groups. When in contact with gelatin, growth factors, such as PDGF and TGF-β_1, were released from the PRP. The growth factors in PRP were likely immobilized in the hydrogel through physicochemical interaction with gelatin molecules and released from the hydrogel in concert with hydrogel degradation.

Stevens *et al.* have shown that large volumes of bone can be engineered without the need for cell transplantation and growth factor administration [56]. The approach is based on the manipulation of a deliberately created space within the body, such that it serves as an "*in vivo* bioreactor", wherein the engineering of the neotissue is achieved by invocation of a healing response within the bioreactor space. Specifically, in the context of bone engineering, they hypothesized that by creating this "space" between the surface of a long bone and the membrane rich in pluripotent cells that covers it, namely the periosteum, the cell population and biomolecular signals necessary for the formation of bone could be locally derived. They created such bioreactors in the tibia of rabbits and then provided volume to this space by injecting a calcium-alginate gel that crosslinked *in situ*. After 4 weeks, the bioreactor space was reconstituted by functional living bone, as shown in Fig. 2.18. When that bone was removed and transplanted to damaged bone sites within the same animals, the new bone integrated seamlessly.

Kasten *et al.* examined ectopic *in vivo* bone formation with and without MSC on a resorbable calcium-deficient HAp (CDHA), β-TCP, and demineralized bone matrix (DMB) [57]. Before implantation in the back of SCID mice, carriers were freshly

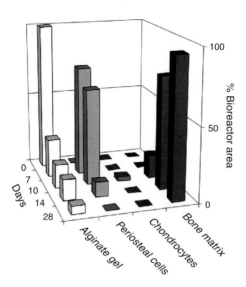

Fig. 2.18   Time course of tissue progression and gel resorption.

loaded with 2×10^5 expanded human MSC or loaded with cells and kept under osteogenic conditions for two weeks *in vitro*. Deposits of osteoid at the margins of ceramic pores occurred independent of osteogenic pre-induction, contained human cells, and appeared in MSC/CDHA composites (2/8) compared to MSC/β-TCP (1/8). There appeared to be a trend toward improved *in vivo* bone formation after osteogenic induction of MSC for 2 weeks in CDHA (3/8) in comparison to β-TCP (0/8). Bone was of human origin and no hematopoietic tissue or cartilage was found, although metachromasia in toluidine blue staining indicated proteoglycan deposition in several MSC/CDHA and MSC/β-TCP samples. The ALP activity was significantly higher in samples with MSC versus empty controls. Furthermore, ALP was significantly higher for all ceramics when compared to the DMB matrix. Hui *et al.* evaluated the ability of MSCs isolated from different origins—bone marrow, periosteum, or fat—to treat partial growth arrest in immature (6-week-old) rabbits [58]. Up to 50% of the medial half of the proximal physis of the tibia was excised in these rabbits. Three weeks later, the bony bridge was excised, and fibrin glue with and without MSCs was transferred into the physeal defect of different rabbits. Contralateral tibias, without undergoing operation, served as self-control. Each of four groups was injected separately with bone marrow–derived MSCs (group I), periosteum-derived MSCs (group II), fat-derived MSCs (group III), and fibrin glue alone (control, group IV). Similar proliferative rates for three MSCs were demonstrated on days 4, 7, and 11 of primary culture. However, MSCs derived from bone marrow and periosteum appeared to be more homogeneous than that from fat. All MSCs demonstrated chondrogenic and osteogenic differentiation potentials *in vitro*. The tibias in groups I and II showed significant correction of varus angulation at 16 weeks. However, the varus angulation in group III remained significantly obvious when compared with group I ($p < 0.05$). The length discrepancies between operated and normal tibiae in groups I, II, and III were significantly corrected compared with control.

Cheng *et al.* proposed a large animal model for the analysis of tissue-engineered bone fabrication [59]. The wool on the left chest wall of skeletally mature sheep (*n* = 20) was sheared and the rib cage was exposed through two incisions oriented parallel to the ribs (Fig. 2.19). Animals were each implanted with three rectangular (1 × 1 × 4 cm³), hollow tissue-molding chambers that were empty (control) or filled with equal weights (6.71–6.78 g) of particulate autologous bone graft (MBG) or bone graft that was autoclaved to denature stored growth factors (DeMBG). The MBG provided scaffold and bioactive factors and DeMBG provided only scaffold. The chambers were enclosed on five sides and securely implanted so that the open face was apposed to the osteogenic (*i.e.*, cambium) layer of the rib periosteum for 3, 6, 9, 12, or 24 weeks, after which the chambers were harvested and the contents analyzed. Each chamber contained osseous and fibrovascular tissue. The MBG-containing chambers had the best maintenance of tissue volume compared with DeMBG-containing or empty chambers, but it still decreased steadily over time. Despite this, the MBG-containing chambers showed continuous active bone formation. There was increasing calcified tissue with penetration of osteogenesis up to a mean of 0.75 ± 0.15 cm from the periosteum by 9 weeks, and the osteogenic area peaked at 0.59 ± 0.13 cm² by 12 weeks.

Vacanti *et al.* reported the use of a tissue-engineered distal phalanx to replace the bone in a patient who had a partial avulsion of the thumb [60]. Eight sections of periosteum, each 1 cm², were harvested from the distal part of the left radius of the

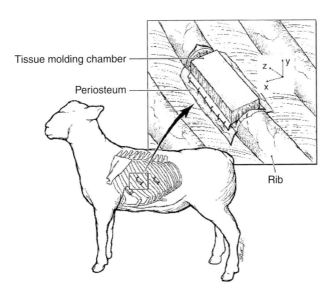

Fig. 2.19 Implantation of tissue-molding chambers at three alternating rib levels after removal of a bone segment.

patient and placed in tissue culture flasks containing tissue culture medium with 10% FCS and others. The flasks were incubated for the next 9 weeks. After 7 weeks, the FCS was replaced by serum from the patient. During the 9-week incubation period, cells shed from the periosteum multiplied to form a monolayer on the bottom of each flask. Twelve weeks after the injury, the skin graft on the dorsum of the thumb was incised longitudinally and a pocket was created beneath the flap of the thumb for placement of a cell–scaffold complex. To create the cell–scaffold complex, a block of specially treated natural coral (porous HAp) was carved into the approximate size and shape of the distal phalanx of the thumb. Twenty million periosteal cells were suspended in 1% alginic acid in saline. The porous coral implant was placed into the pocket and injected with the cell suspension, which was then fixed in position by applying a calcium chloride solution onto the surface of the implant. This step resulted in the formation of a stable alginate hydrogel that encapsulated the cells within the gel that saturated the coral implant. Twenty-eight months after the implantation, the patient had a thumb of normal length and strength, with some sensation. Radiographs obtained before the implantation and 6 weeks and 28 months afterwards showed some dorsal subluxation but no loss of volume or fragmentation of the implant. The mineral density of the implant was 0.903 g/cm^3 immediately after the implantation and 0.690 and 0.481 g/cm^3 at 5 and 10 months, respectively, after the implantation, as compared with 0.382 g/cm^3 in the contralateral phalanx.

To prevent failure of the bone–prosthesis interface, Ohgushi *et al.* attempted a tissue engineering approach using mesenchymal cells of patients [61]. They collected a small amount of fresh marrow cells from the patient's iliac crest and expanded the number of mesenchymal cells. They applied the cells to a ceramic ankle prosthesis and cultured them to form an osteoblasts/bone matrix on the prosthesis, as shown in Fig. 2.20. They used the tissue-engineered prostheses on three patients suffering from ankle arthritis and followed their progress for at least 2 years. X-ray examinations revealed early radiodense appearance (bone formation) around the cell-seeded areas of the prostheses about 2 months after operation, after which a stable host bone–prosthesis interface was established. All patients showed high clinical scores after operation and did not exhibit inflammatory reactions.

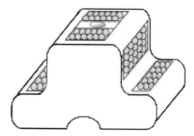

Fig. 2.20   Schematic presentation of tibia alumina component. Some surface areas consist of clusters of ceramic beads (represented in gray). Thus, the surface is porous and cells can be easily applied.

Aseptic nontraumatic osteonecrosis of the femoral head is a disorder that can lead to femoral head collapse. Core decompression of the hip with an 8-mm trephine is the most common procedure used to treat the early stages of osteonecrosis of the femoral head. Since osteonecrosis may be a disease of mesenchymal cells or bone cells, there is a possibility that bone marrow containing osteogenic precursors implanted into a necrotic lesion of the femoral head may be of benefit in the treatment of this condition. For this reason, Gangji and Hauzeur studied the implantation of autologous bone-marrow mononuclear cells in a necrotic lesion of the femoral head to determine the effect on the clinical symptoms and the stage and volume of osteonecrosis [62]. They studied 13 patients (18 hips) with stage-I or stage-II osteonecrosis of the femoral head. The hips were allocated to a program of either core decompression (the control group) or core decompression and implantation of autologous bone-marrow mononuclear cells (the bone-marrow-graft group). After 24 months, there was a significant reduction in pain and in joint symptoms within the bone-marrow-graft group. At 24 months, five of the eight hips in the control group had deteriorated to stage III, whereas only one of the ten hips in the bone-marrow-graft group had progressed to this stage. Survival analysis showed a significant difference in the time to collapse between the two groups. Implantation of BM-MNCs was associated with only minor side effects.

To develop tissue engineering techniques in posterior spine fusion as an alternative to autologous bone, van Gaalen *et al.* cultured syngeneic BMCs in the presence of bone differentiation factors and seeded them on porous HAp particles [63]. Seven rats underwent a posterior surgical approach, in which scaffolds with or without cells were placed on both sides of the lumbar spine. After 4 weeks, all rats were killed and examined radiographically, by manual palpation of the excised spine, and histologically for signs of bone formation or spine fusion. All rats that received cell-seeded scaffolds showed bone formation. Autologous BMSC–calcium-phosphate ceramic composites were constructed *in vitro* and implanted as a bone graft substitute for lumbar anterior interbody fusion in rhesus monkeys to determine the osteogenic capacity of the composites [64]. Nine adult rhesus monkeys underwent lumbar L3–L4 and L5–L6 discectomy and interbody fusion via an anterior retroperitoneal approach. Two fusion sites in each animal were randomly assigned to two of three treatments: autologous tricortical iliac crest bone graft (autograft group), cell-free ceramic graft (ceramic group), or BMSC–ceramic composite graft (BMSC group). Autologous BMSCs were expanded in culture and stimulated with osteogenic supplement. The BMSC group achieved lumbar interbody fusion superior to that of the ceramic group, both biomechanically and histologically. The BMSC group and the autograft group showed equivalent biomechanical stiffness. Ceramic residues were significantly greater in the ceramic group versus the BMSC group.

2.3. Tendon and Ligament

Tendon and ligament are dense fibrous connective tissues that attach muscles to bones, and bones to bones, respectively. They possess high tensile strength that is crucial in mediating the normal movement and stability of joints. Tendons are complex composite materials, composed primarily of water (55% of wet weight), pro-teoglycans (<1%), cells, and type I collagen (85% of dry weight). The primary struc-ture in the structural hierarchy of tendons is collagen polypeptides characterized by the presence of glycine at every third amino acid. Collagen monomers are assembled into fibrils which are then grouped into fibers. The bundles of collagen fibers along with fibroblasts are grouped into fascicles. Ligaments are morphologically and micro-scopically similar to tendons, but there are biochemical differences between the two. Ligaments are metabolically more active than tendons. Ligaments have more cellular nuclei, a higher DNA content, and greater amounts of reducible crosslink between collagen fibers. Types I and III collagens comprise approximately 90 and 10% of the collagen in ligaments, respectively. The water content in the wet weight of ligaments is 60% or more. Tendons and ligaments transmit load with minimal energy loss and deformation, but they are not an entirely non-extendable cable that directly transfers the length change or force of a contracting muscle to the bone. Tendons and ligaments are perfectly elastic as long as the strain does not exceed 4%, after which the viscous range commences and then macroscopic failure occurs at strains around 8%.

When tendon injury leaves a large gap, it is usually difficult to bridge. If the tendon is completely missing, then a graft or replacement device is usually necessary. Ligaments, on the other hand, have poor healing ability after any injury. The current "gold standard" for surgical repair of tendon or ligament injuries is to use autologous tendon. However, the mechanical strength and structural characteristics of the host tissue are permanently altered during repair. During anterior cruciate ligament (ACL) reconstruction, often with the use of patellar tendon (PT), an initial loss of strength in the host tissue is typically observed from the time of transplantation. A gradual increase in strength may occur, but typically never reaches the original magnitude. The alternative was to use acellular synthetic materials, including Gore Tex prosthesis (polytetrafluoroethylene, PTFE), Stryker-Dacron ligament PET, and Kennedy liga-ment augmentation device (LAD) (polypropylene). Although these synthetic grafts exhibited excellent short-term results, decades of ligamentoplasty with these synthetic ligaments did not yield successful long-term results, due to mechanical mismatch, poor abrasion resistance, high incidence of fatigue failures, and limited integration between the graft and the host tissue. Most of the synthetic materials did not approxi-mate the material properties of tendon and ligament, resulting in stress shielding in the natural tissue. Moreover, wear debris resulted in an immunological response that ultimately led to implant failure and additional surgery. Therefore, an ideal treatment has yet to be defined as a long-term solution for tendon or ligament repair.

In view of these problems, it is logical to consider bioabsorbable augmentation devices in combination with biologicals for tendon and ligament repair and regeneration. More desirable means for the repair of tendon and ligament may rely on a tissue engineering approach. However, this approach for load-bearing tissues such as tendon and ligament encounters a large problem because regeneration of such tissues is limited under static culture conditions and will need mechanical stimulation for its enhancement, seemingly similar to the native ligament growth. Therefore, a number of studies have attempted to apply cyclic stress to scaffold–cell constructs *in vitro*

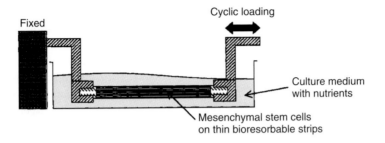

Fig. 2.21   Schematic side view of the culture chamber of a cyclic traction device.

culture to mimic physiological loading conditions. For instance, Goh *et al.* designed a cyclic stretching machine, as illustrated in Fig. 2.21, for tendon/ligament tissue engineering *ex vivo* [65]. Several other research groups also reported that the application of multidimensional mechanical strains to cells seeded onto absorbable, fibrous scaffolds upregulated the ligament fibroblast markers and resulted in significant cell alignment, density, and the formation of oriented collagen fibers. Nevertheless, any animal studies that guarantee clinical trials to patients have not yet been reported. As the scaffold for tendon and ligament repair, bioabsorbable materials such as collagen, chitin, PGA, and PLA have been used. When a scaffold is used in a shape of braided fibers, new tissue ingrowth among the braided fibers is poor because of the limited internal space of braided fiber scaffolds. Collagen gel shows little ability to hold its 3-D structure and resist mechanical forces.

2.3.1. Ligaments

The native ACL is an intraarticular ligament that controls normal motion and acts as a joint stabilizer. It connects the femur to the tibia and is completely enveloped by synovium. The ACL consists of a large number of fiber bundles arranged into three areas: anteromedial, posterolateral, and intermedial, accommodating low levels of friction tension during a wide range of motion. The human ACL has an average length of 27–32 mm and a cross-sectional area of 44–58 mm². The collagen fibrils in ligaments display a periodic change, called a "crimp pattern". In ACL, this crimp pattern repeats every 45–60 μm. The ligament is surrounded by a sheath of vascularized epiligament. Rupture of ACL is one of the most common sports-related injuries and ruptured ACL fails to heal spontaneously due to the ACL's intrinsically poor healing potential and limited vascularization. Optimal healing is achieved when the continuity of collagen fibers is maintained. Due to the lack of tissue organization and difference in crimp pattern between the new and the old ECM, there is a difference in the mechanical properties of the new scar tissue and the original ligament leading to a reduction in mechanical properties. When injury causes a full rupture of the ligament midsubstance or detachment of the ligament, an insertion point surgery is required. Untreated ruptures are associated with long-term complications such as pain, joint swelling, repeated episodes of instability,

Chapter 2

and OA. Currently treatment modalities utilizing autologous grafts such as bone–patellar tendon (PT)–bone and hamstring tendon have demonstrated clinically functional outcomes providing adequate functional stability. The PT graft material is usually removed with a piece of bone from the patella and from the insertion point at the tibia. The "bone–PT–bone" graft is then fed through a tunnel drilled through the tibia, drawn across the knee, and anchored into a tunnel drilled through the femur. However, the choice of autologous grafts in revision surgery grafts is limited by donor site-related problems such as harvest site infection, nerve injury, and patellar fracture. Allografts are restricted in use due to the potential for infectious disease transfer and unreliable graft incorporation.

These drawbacks have led to a tissue engineering approach to induce ACL repair via gradual replacement of the resorbable scaffold by host fibrous tissue. The scaffold must withstand repetitive tensile loading and an enzymatic environment without significant degradation during the initial period of tissue ingrowth. Postoperatively, an ACL reconstruction device experiences loads and enzymes simultaneously. For collagen implants, each of these variables is dependent on the method and extent of collagen crosslinking. Collagen denaturation may cause ACL reconstruction scaffolds to degrade too rapidly when exposed to the cyclic loads and harsh enzymatic environment of the postsurgical knee joint.

van Eijk *et al.* compared ACL and skin fibroblasts and BMSCs with regard to cell proliferation and differentiation (matrix synthesis) on 3-D biopolymer scaffolds to study the effect of initial seeding density on proliferation and matrix synthesis [66]. The scaffolds were made of PGA 11-μm fibers, braided in bundles to a multifilament and coated with 90% CL and 10% GA copolymer, as shown in Fig. 2.22. Five strands were braided together and pieces approximately 1 cm in length were knotted at the ends with PGA suture. All tested cell types attached to the suture material, proliferated, and synthesized ECM rich in type I collagen. On day 12 the scaffolds seeded with BMSCs showed the highest DNA content and the highest collagen production, whereas scaffolds seeded with ACL fibroblasts showed the lowest DNA content and collagen production. Figure 2.23 shows the result.

Fig. 2.22 Scanning electron micrograph of scaffolds seeded and cultured for 12 days with ACL fibroblasts.

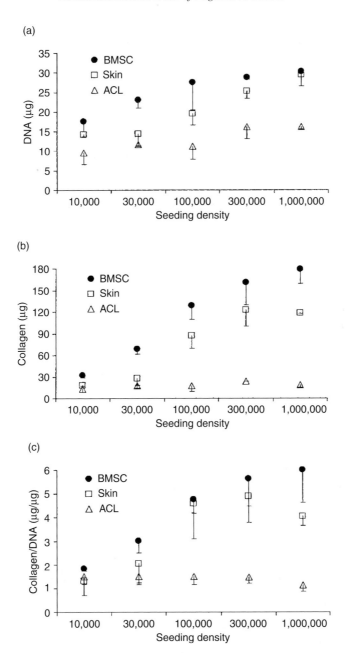

Fig. 2.23 Effect of seeding density on DNA content (a), collagen production (b), and collagen/DNA (c) on day 14.

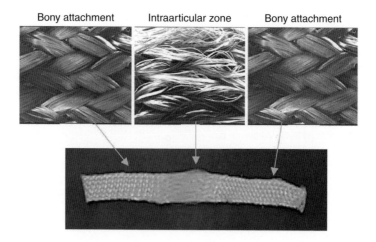

Fig. 2.24   General configuration of ligament design for 3-D rectangular braid.

To engineer functional ACL scaffolds, Cooper *et al.* fabricated 3-D fibrous scaffolds using customized, 3-D circular and rectangular braiding machines [67]. The PGLA fibers were laced to produce yarns with a yarn density of 18 yarns per yarn bundle (Fig. 2.24). The PGLA yarns were then placed in a custom-built circular braiding loom with a 3 × 16 carrier arrangement. The circular braiding machine used the sequential motion of the carriers (alternating tracks) to form 48-yarn, 3-D circular braids with braiding angles that ranged from 26 to 31°. For comparison in architecture and as an alternative design, scaffolds were also fabricated using a 3-D rectangular braiding system in which PGLA fibers were laced to produce yarns with yarn densities of 9, 30, and 60 yarns per bundle to investigate the effects on mechanical and porosity parameters due to fiber number. The resultant microporous scaffold exhibited pore diameters ranging from 175 to 233 µm and initial mechanical properties of the construct approximated those of the native ligament. The 3-D braided scaffold was composed of three regions: femoral tunnel attachment site, ligament region, and tibial tunnel attachment site. The attachment sites had high angle fiber orientation at the bony attachment ends and lower angle fiber orientation in the intra-articular zone. This predesigned heterogeneity in the grafts was aimed to promote the eventual integration of the graft with bone tissue.

To avoid the potential long-term stress shielding by the non-absorbable ACL substitutes and to ensure progressive full load transfer to regenerated tissue, Kobayashi *et al.* developed resorbable LAD consisting of a braided PLLA fibers [68]. When biomechanics and tissue response to the PLLA–LAD were examined by using bone–patella tendon–bone (BTB) for ACL construction in a goat model up to 2 years, the PLLA–LAD was found to induce no deterious, histological effects during this period. Based on this promising results, Ishimura *et al.* performed clinical trials of the PLLA augmentation device for ACL repairs in combination with PT but without any cell seeding [69]. It was found that this PLLA fibers effectively augmented BTB–ACL reconstruction with no evidence of increased laxity as the device lost strength. This bioabsorbable PLLA device will not evoke any stress-shielding effect on the augmented natural tissues in contrast to nondegradable poly-propylene fibers.

2.3.2. Tendons

Tendons are densely packed connective tissues and transmit forces between muscle and bone. They are stiff in tension but flexible enough to conform to their anatomical environment. The material properties of tendon tissue can be attributed to the parallel fibrils of collagen that make up approximately 75% of the dry weight of adult tendons. In the resting state the fibrils display a periodic wavy pattern, defined as the crimp. As a tendon is stretched the crimped collagen fibrils begin to straighten out, and as a result the tendon becomes stiffer with increasing mechanical strain. Tendons have a low cell density, about 20% of the tissue volume, but the fibroblasts are integral in the development and maintenance of the tissue. The distinct spatial orientation of tendon fibroblasts is associated with the organization of collagen fibers into the hierarchical tendon structure.

Tendons are thought to have reparative ability, but when either the tendon is completely destroyed or the defect site is too large to allow for reapposition of the ends, tendon replacement is usually required. Tendon defects are difficult to treat because of the lack of a proper source for tendon autografts. Tissue engineering, using *in vitro*–expanded cells to generate an autologous tissue, offers the possibility of providing autografts for tendon repair. Because of its relatively avascular nature, attempts have been made to create biologically based tendons *in vitro*, but these have met with limited success because of the difficulty in creating a construct that is both mechanically and biologically compatible with the *in vivo* environment. Mechanical difficulties can arise from the reliance on artificial scaffolds when attempting to engineer tendon. Type I collagen is one of the most widely used scaffold materials, since fibroblasts will contract a collagen gel to form a tissue-like structure. Collagen would appear to be the ideal foundation for an artificial tendon, but at present the mechanical properties of *in vitro* fibroblast–collagen constructs are much inferior to those of native tissues because of disorganized fibrils of gelled collagen. Furthermore, native tendons possess an ECM composed of many proteins, GAGs, and proteoglycans that control the assembly of the collagen fibril, the load-bearing unit, and contribute to the formation of the tissue hierarchy. Fibroblasts rely on cell-matrix signaling pathways during development to properly assemble the fibrils and maintain form and function after maturation. To induce the formation of a tendon-like structure with adequate mechanical properties based solely on biological products, the successfully engineered tendon will need to include all the necessary ECM components. The ECM will give the construct mechanical properties similar to those of the replaced tendon so that mobilization can resume as soon as possible, accelerating the process of healing. If tendon fibroblasts can be induced to secrete and assemble their own ECM, the resulting mechanical and morphological properties will resemble those of native tendons.

Chapter 2

Ouyang *et al.* evaluated the morphology and biomechanical function of the Achilles tendons generated with knitted scaffold loaded with allogeneic BMSCs [70]. To manufacture a knitted scaffold possessing sufficient porosity and mechanical strength, they used PGLA (GA:LA = 90:10) fibers of 20 filaments/yarn (diameter of filament: 25 μm). Plain weft-knitted fabrics was manufactured from three yarns with six stitches per centimeter and 1-mm loop size. Knitted PGLA scaffold was sutured to the ends of the tendons in group I (*n* = 19 legs). Ten million BMSCs immobilized in 0.3 ml of fibrin glue were seeded onto the knitted PGLA scaffold. The procedure was repeated for group II contralateral hind legs, using 0.3 ml of fibrin glue and scaffold but without seeding of BMSCs. For group III, hind legs were used as normal controls. Figure 2.25 shows the results of biomechanical testing for scaffold/bMSC-treated tendons 12 weeks after surgery. When mean tensile stiffness was expressed as a percentage of normal tendon values using data from group III, the stiffness of group I and II reached 87 and 56% of normal tendon stiffness at 12 weeks after implantation, respectively. The mean tensile modulus of group I and II calculated on the basis of nominal stress was 63 and 53% of normal tendon, respectively. Even though the regenerated tendons possessed well-organized types I and III collagen, their fiber length was shorter and the ratio of type III collagen to type I was higher when compared with normal tendon. It seems likely that at the 12-week time point the healing may not have been complete.

Using a hen model, Cao *et al.* tested whether tendon tissue could be engineered with a tissue structure similar to that of native tendon in an immunocompetent animal [71]. Autologous tenocytes were first extracted from donor tendons harvested from one foot, and the isolated and suspended tenocytes were mixed with PGA fibers to form a cell–scaffold construct in the shape of a tendon, which was further wrapped

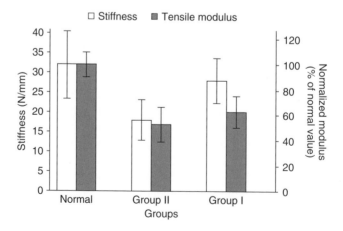

Fig. 2.25   Histogram comparing the mean (±1 SD) tensile stiffness and normalized modulus of scaffold/bMSC-treated (group I) tendons 12 weeks after surgery, scaffold-alone-treated (group II) tendons 12 weeks after surgery, and normal tendons (group III).

with acellular intestinal submucosa membrane. At the contralateral foot, a tendon defect of 3–4 cm in length was created at the second flexor digitorum profundus tendon by resecting a tendon fragment. The defects were bridged either with a cell–scaffold construct in the experimental group or with scaffold material alone in the control group.

Calve *et al.* developed a method that provided appropriate environmental cues *in vitro* for tendon fibroblasts to generate their own ECM and assemble into a 3-D construct [72]. An amenable substrate was created by coating 35-mm culture dishes with silicone elastomer. After curing, mouse LN was applied to the dishes in PBS. The PBS was allowed to evaporate, leaving a layer of LN-coated silicone. The anchors, 6 mm segments of size 0 silk suture, dipped in mouse LN, were pinned 12 mm apart with 0.1-mm-diameter stainless steel minute pins. Sutures were allowed to dry, and the plates were filled with growth medium, enough to cover the top of the sutures. After incubation, the growth medium was aspirated and cells suspended in growth medium were seeded onto each plate and supplemented with L-ascorbic acid 2-phosphate. Fresh AA was added each time the growth medium was changed, every 2–3 days. Cells were grown to confluence in culture and allowed to self-assemble into a cylinder between two anchor points. The resulting scaffold-free tissue was composed of aligned, small-diameter collagen fibrils, a large number of cells, and an excess of noncollagenous ECM; all being characteristics of embryonic tendon. The stress–strain response of the constructs also resembled the nonlinear behavior of immature tendons, and the UTS was approximately equal to that of embryonic chick tendon, roughly 2 MPa.

Because harvesting autologous tendon tissue may not be optimal in clinical repair, Liu *et al.* used dermal fibroblasts to replace tenocytes as cells to seed for tendon engineering [8]. In a porcine model, both tenocytes and dermal fibroblasts were isolated from harvested autologous tissues. A 3 cm-long defect in the flexor digitorum superficialis tendon was created and the defect was then repaired with a construct made from dermal fibroblasts (experimental group) or with a construct made from tenocytes (control 1 group), or with a biomaterial construct alone (control 2 group). At 6 weeks, a neotendon bridging the defect was formed in all groups and abundant cells were aligned with a relatively longitudinal pattern in the experimental group and control 1 group. In the control 2 group, relatively fewer invading fibroblasts were observed in the peripheral area but not in the central area. At 14 weeks, cell numbers in both experimental and control 1 groups had fallen remarkably and the collagen fibers were more longitudinally aligned in the first two groups. Randomly arranged cells and collagen fibers were observed in the third group. Similar tensile strength was found among the three groups. At 26 weeks, tendons of the first two groups appeared to be similar to each other and similar to natural tendon in color and morphology. Tendon engineered by either tenocytes or dermal fibroblasts matched the natural counterpart in cell:collagen ratio and collagen alignment pattern. In contrast, tissue formed in the third group was more similar to fibrotic tissue morphologically, which had a severe adhesion to surrounding tissues. Histology also showed randomly aligned collagen fibers and proliferating fibroblasts along with many vascular

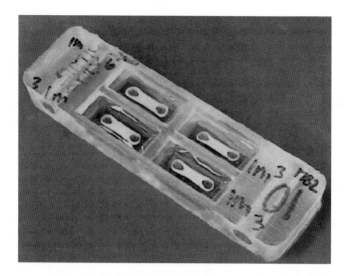

Fig. 2.26   Each silicone dish consists of four wells into which cells and collagen gels are pipetted. The cells contract the gels around two posts located at the base of each well.

structures, suggesting that tissue formed in the control 2 group resembled the histological structure of fibrous/scar tissue. In addition, the tensile strengths of the first two groups were similar to each other but greater than that of the third group.

Juncosa-Melvin *et al.* fabricated autologous tissue-engineered constructs (Fig. 2.26) at four cell-to-collagen ratios (0.08, 0.04, 0.8, and 0.4 M/mg) by seeding MSCs from 16 adult rabbits at one of two seeding densities (0.1×10^6 and 1×10^6 cells/ml) in one of two collagen concentrations (1.3 and 2.6 mg/ml) [73]. The highest two ratios (0.4 and 0.8 M/mg) were damaged by excessive cell contraction and could not be used in subsequent *in vivo* studies. The remaining two sets of constructs were implanted into bilateral full-thickness, full-length defects created in the central one-third of the PT. A second group of rabbits ($n = 6$) received bilateral acellular implants with the same two collagen concentrations. No significant differences were observed in any structural or material properties or in histological appearance among the two cell-seeded and two acellular repair groups. Average maximum force and maximum stress of the repairs were approximately 30% of corresponding values for the central one-third of normal PT and higher than peak *in vivo* forces measured in rabbit PT. However, average repair stiffness and modulus were only 30 and 20% of normal PT values, respectively.

2.4. Rotator Cuff

Rotator cuff injuries are one of the most common causes of upper extremity disability, originating from intrinsic and extrinsic factors. Intrinsic factors include aging, structural changes of the tendon itself, low vascularity, and inflammatory processes.

These lead to degeneration beginning at the supraspinatus tendon and constitute the most frequent cause of rotator cuff failures and low healing response. Extrinsic factors include bone and soft tissue, which act by restricting the free gliding of the rotator cuff tendons under the coracoacromial arch. Surgical therapy addresses mostly the extrinsic etiology. Muscle-derived cells (MDCs) isolated and purified from skeletal muscle tissue display characteristics of the early developmental stage of the myogenic lineage. These cells seem to be heterogeneous, and a small minority of these cells seems to be capable of expressing stem cell markers and differentiating into myogenic and osteogenic lineages. To study the feasibility of applying these MDCs as a gene delivery vehicle to the supraspinatus tendon, Pelinkovic *et al.* isolated MDCs from the supraspinatus tendon of mice and transfected the cell population with the linear pNEO lacZ plasmid carrying the β-galactosidase marker gene and the neomycin resistance gene [74]. The final colonies harvested from the cultivated cells were 98% positive for the galactosidase marker gene. The genetically engineered, highly purified MDCs were injected into the supraspinatus tendons of nude rats and were monitored for 3 weeks. *In vivo*, the engineered MDCs did not express vimentin which is generally thought to be expressed by all fibroblastic cells. β-Galactosidase marker gene expression of the injected cells was detected up to 21 days. From day 7 after injection, the cell nuclei became spindle shaped, cells were integrated into the tendon collagen bundles, and the cells showed differentiation into vimentin expressing fibroblastic cells.

2.5. Skeletal Muscle

Congenital diaphragmatic hernia (CDH) defect is one of the most common causes of neonatal morbidity and mortality. Current surgical treatment of the defect utilizes either a muscular flap or a prosthetic patch. Although both procedures have increased the infant survival rate, disadvantages include high blood loss for the muscular flap and reherniation for the prosthetic patch. If tissue engineering creates *in vitro* a skeletal muscle patch that possesses the mechanical and contractile properties of the diaphragm and is able to grow with the surrounding tissue, it would become an important element in restorative surgery of congenital defects. However, the tissue engineering should address several issues. Among them is orientation of muscle cells parallel to each other to ensure directed force production. Neumann *et al.* challenged this problem using a device which facilitated arrays of polymer microfibers functioning as scaffold of C2C12 mouse skeletal myoblasts [75]. Figure 2.27 illustrates how to fabricate the fiber array. Laminin-coated polypropylene (PP) fibers between 10 and 15 μm in diameter were used in this study. The microfibers were sandwiched between the sticky sides of two Mylar frames to keep the fibers stationary. The optimal fiber spacing needed to achieve cell alignment with the lowest possible content of scaffold material was found to be 50–55 μm. It seems possible to create larger, 3-D structures by arraying fibers in several layers or by stacking cellular sheets.

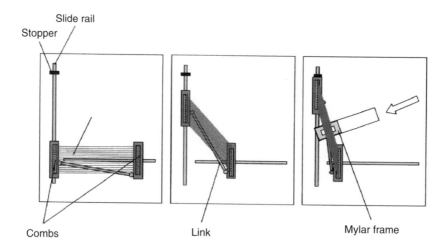

Fig. 2.27 Schematic drawing of fiber deposition onto Mylar frames. Left: Fibers are strung in between the teeth of two lice combs, mounted on slide rails. Middle: The combs are moved, keeping the fibers parallel to each other while the fiber spacing decreases. Right: Fiber spacing is adjusted to the desired distance, and the Mylar frame is raised by a micromanipulator until the fibers stick to the adhesive.

De Coppi *et al.* combined the isolation and genetic engineering of myoblasts with tissue transplantation in an attempt to create well-vascularized muscle tissue [76]. Myoblasts were obtained from a single explant of adult Lewis rat myofibers and transfected with a bicistronic plasmid encoding VEGF and GFP or with a plasmid encoding a nonfunctional VEGF–ALP fusion protein. The VEGF expression and GFP expression *in vitro* assessed by Western blot analysis ELISA and fluorescence microscopy, respectively, showed that the myoblasts were successfully expressing the recombinant proteins. The transfected cells suspended in type I collagen were injected subcutaneously into nude mice. Analysis of the retrieved engineered muscle tissues by RT-PCR immunostaining and fluorescence showed expression of VEGF and GFP proteins. Immunohistochemical analysis of the muscle tissues 1, 3, and 4 weeks after implantation confirmed the muscle phenotype. Neovascularization and muscle tissue mass significantly increased with functional VEGF-transfected cells compared with nonfunctional VEGF-transfected cells.

2.6. Joints

In any joint, uniquely configured apposing articular surfaces are sheathed by synovial cells, which in turn are encased in an osteoarticular skeleton held intact by complex ligamentous and capsular structures. Moreover, the articular surfaces are composed of hyaline cartilage and an underlying subchondral bone. The reconstruction of severely traumatized or congenitally deficient joints in young patients is a challenging problem, since the reconstruction needs multiple complicated procedures. The reconstruction with prosthetic joints is compromised by the limited durability of the non-biological materials, intractable infection, and replacement of the implanted prosthesis when the child grows.

2.6.1. Large Joints

Endoprosthetic reconstruction of the immature skeleton of infants has been performed, but problems such as loosening, failure, and lack of growth are particularly vexing with available methods. Therefore, biological solution would be important for absent or destroyed large joints in the immature skeleton. As the shape of a joint is crucial to its function, development of a prototype model system for engineering a biological joint (apposing articular surfaces allowing some stable motion) is a useful basis for further refinement and application of new principle of regenerative medicine. Zaleske *et al.* assessed the feasibility of using a devitalized knee as a scaffold for a engineered chimeric joint [77]. Developing chick knees (19 days old) were devitalized by lyophilization or multiple freeze–thaw cycles and used as scaffolds for repopulation with bovine articular chondrocytes (bACs). The bACs were seeded into porous 3-D collagen sponges and cultured for 1 day before fabrication of chimeric constructs. A pair of cell-seeded sponges was inserted into the joint space to contact preshaved articular surfaces, as illustrated in Fig. 2.28. In some constructs, a membrane of expanded PTFE (ePTFE) was inserted between the collagen sponges. This study identified several design parameters crucial for successful engineering of a chimeric joint. The collagen lattice provided immediate attachment to the chick scaffold and maintained the chondrocytes in the desired location. The femoral and tibial sponges became confluent at their interface, creating a single mass of neocartilage and transgressing the scaffold provided by the devitalized chick joint. The joint space was

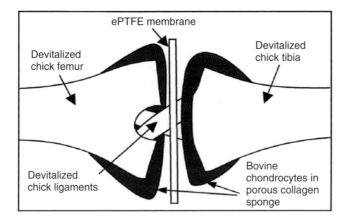

Fig. 2.28 Schematic of a posterior/anterior section, representing the relationship of the components in a chimeric construct with an ePTFE membrane.

maintained in all constructs by placement of an ePTFE membrane between the sponges. Penetration of neocartilage into the chick scaffold tended to increase with time in culture. Lyophilization completely obliterated evidence of chick chondrocyte gene expression. Storage at $-20°C$ and freeze–thaw cycles were not sufficient to completely devitalize the scaffold.

2.6.2. Small Joints: Phalangeal Joints

To generate a phalangeal joint using tissue engineering techniques, Isogai *et al.* attempted an approach in which three different cell types—derived from bovine periosteum, cartilage, and tendon—were used [78]. Figure 2.29 presents three different types of absorbable scaffolds used for the study. One type of scaffold, used for cartilage and tendon support, was simply a non-woven mesh of PGA fibers alone, whereas the second type of scaffold was created by cutting a non-woven PGA mesh into rectangles with dimensions approximate for the formation of phalanges. The PGA mesh was immersed in PLLA solution and, after solvent evaporation, the resulting PGA–PLLA composite was subjected to a temperature of at least 195°C for 90 min. In the case of Group III, chondrocyte–polymer sheets were sutured to periosteum–polymer constructs of the distal and middle phalanges to create apposing articular surfaces, and a joint was formed by wrapping this composite with additional

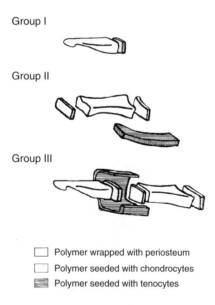

Group I

Group II

Group III

☐ Polymer wrapped with periosteum
☐ Polymer seeded with chondrocytes
▨ Polymer seeded with tenocytes

Fig. 2.29 Schematic drawings of three different types of composite tissue structures in the experimental groups. The structures were constituted *in vitro* by suturing to create models of a distal phalanx (Group I), a middle phalanx (Group II), and a distal interphalangeal joint (Group III). The sutured tissues then were implanted subcutaneously in athymic mice.

PGA sheets seeded with tenocytes. This created a distal interphalangeal joint. All constructs were implanted into the dorsal subcutaneous space of athymic nude mice. The implants from Groups I, II, and III were removed after 20 weeks. The subcutaneous implantation resulted in the formation of new tissue with the shape and dimensions of human phalanges with joints. Histological examination revealed mature articular cartilage and subchondral bone with a tenocapsule that had a structure similar to that of human phalanges and joints.

The articular destruction of metacarpophalangeal (MCP) joints in rheumatoid arthritis (RA) greatly impairs the function of the hand, especially the ability to grasp objects of large size. Current small joint surgeries have been done with biostable joint prostheses, generally Swanson prosthesis which is a one-piece silicone implant with stems and a spacer in the middle. A range of 5–82% breakage of the implants prevalence has been reported in different studies. In 1994 a concept of bioreconstructive joint scaffolds was developed in a Finish group by performing a first prospective study using commercially available Vicryl (PGLA) and Ethisorb fleeces folded into small, rectangular scaffolds [79]. The folded scaffold was intended to behave in a similar way as the tendon. However, the resorption time of both tested materials was so short that the tissue did not have enough time to regenerate and mature, and the joint space collapsed. Therefore, a scaffold consisting of an L,D-lactide copolymer with an L:D monomer ratio of 96:4 (PLA 96) with longer absorption time was used as a temporary support. In animal tests, PLA 96 scaffolds implanted in rat subcutis were filled with tissue in 3 weeks, and were totally absorbed within 3 years. Based on these findings, Honkanen *et al.* evaluated the clinical, radiological, and functional outcomes of PLA 96 joint scaffold arthroplasty in severe arthritic destruction and revision operation after failure of silicone arthroplasties in the MCP joints of RA patients [80]. The four-ply multifilament yarn melt–spun from PLA 96 was knitted to a tubular mesh. The knitted tube was rolled to cylindrical scaffolds. The diameter of single filament varied between 70 and 80 μm. A typical scaffold is shown in Fig. 2.30. Scaffolds had open and highly interconnected porosity of 80% throughout the structure. In the operation, resection of the bone was equal to Swanson arthroplasty. When there was ulnar deviation, the proximal bony attachments of both collateral ligaments were released. Deliberation of the volar capsule under metacarpal bone and release of the volar plate were performed to achieve adequate correction of volar subluxation. Ulnar intrinsic muscle contractures were released when required. The abductor digiti minimi of the fifth finger was always dissected. Intermedullary bone grafting was performed in revision arthroplasties. Balancing and tightening of the collateral ligaments were performed by duplicating or refixing the ligament more proximally through drill holes in the proximal metacarpal bone. At the end, the extensor tendon was centralized. In this prospective study, 23 RA patients (80 joints) were operated on, using PLA 96 implants. Fifteen patients (54 joints) were monitored for at least 1 year. Pain alleviation was well achieved. Range of motion improvement was achieved to extension direction of functional arc. The average ulnar deviation was preoperatively 26°, and in the follow-up it was 6°.

Fig. 2.30   The PLA96 joint scaffold implant.

Volar subluxation was noticeable in 56% of joints postoperatively and in 6% at 1-year follow-up. Remodeling of the bone and thickening of the cortices were clearly seen and there was a space between the metacarpus and the proximal phalange. When the follow-up of these patients to determine the long-term results is completed, it will be proven whether this first formation of a living, functional joint *in situ* by means of a synthetic bioreconstructive joint scaffold really creates a new era in the construction of damaged joints.

3. CARDIOVASCULAR AND THORACIC SYSTEM

Tissue engineering of blood vessels from autologous cells may affect a large portion of the adult population because of the prevalence of vascular disease. Cardiac surgeons further encounter children with congenital disease and adult patients with complex malformations, which require synthetic prostheses for operation. However, a readily usable and durable composite material with growth and remodeling potential is not yet available for congenital heart surgery. Currently used prosthetic or bioprosthetic materials, even allogeneic materials, lack growth potential and inevitably require re-intervention within 10–15 years after the definitive surgery in pediatric patients. Young patients have an increased calcium turnover, and anti-mineralization treatments can decrease this only partially. Since replacement structures created from autologous cells and bioabsorbable polymer scaffolds using tissue engineering techniques contain living cells, these structures have growth potential, will remodel to an appropriate size and shape, cause less calcification, infection, and a reduced incidence of thromboembolic complications, and do not require donor cells.

3.1. Blood Vessels

3.1.1. Large-Calibered Blood Vessels

To create patches for cardiovascular repair, Ozawa *et al.* studied histologic changes of three absorbable biomaterials: gelatin, PGA, and a poly(LLA-co-ε-caprolactone) (50:50) (PCLA) [81]. A gelatin sponge (Gelform™) was tested because the use of fetal cardiocytes and the pliable gelatin scaffold enabled them to create *in vitro* and *in vivo* beating patch. A non-woven PGA mesh was studied because it is one of the most common synthetic absorbable polymers for tissue engineering field. They found that a PCLA spongy matrix reinforced with PLLA-knitted (KN) or PLLA-woven (WV) fabrics would favor *in vitro* and *in vivo* cell growth. A sheet of ePTFE was used as a control for comparison. Each of rectangular pieces was sutured along the margin of the purse-string suture to cover the transmural defect surgically created in the right ventricular outflow tract (RVOT) of adult rat hearts ($n = 5$). At 8 weeks after implantation, the biomaterials were excised. The PTFE patch itself did not change in size except for increasing in thickness because of fibroblast and collagen coverage of both of its surfaces. Host cells did not migrate into the PTFE biomaterial. In contrast, cells migrated into the bioabsorbable gelatin, PGA, and KN– and WV–PCLA scaffolds. Cellular ingrowth per unit patch area was the highest in the KN–PCLA patch, as shown in Fig. 2.31. The KN–PCLA patch increased modestly in size and thickness, whereas the WV–PCLA patch did not change in both size and thickness. Fibroblasts and collagen were the dominant cellular infiltrate and ECM formed in the absorbable scaffolds. All the subendocardial patch surfaces were covered with ECs and no thrombi were observed. The PCLA grafts held promise to become a suitable patch for surgical repair because the spongy matrix structure of the PCLA patch favored cell colonization relative to the other patches, and the strong, durable outer PLLA fabric layers in these patches offered physical

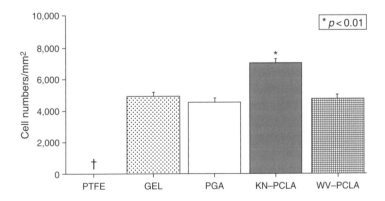

Fig. 2.31 Numbers of host cells in patch scaffolds at 8 weeks after implantation into defect in the right ventricular outflow tract of adult rats. Patches studied ($n = 5$ for each biomaterial) were PTFE, gelatin (GEL), PGA, KN–PCLA, and WV–PCLA.

and bioabsorbable advantages relative to the other bioabsorbable materials studied. Ozawa *et al.* further reported the morphological and histological changes of an implanted cultured autologous smooth muscle cells (SMC)-seeded PCLA patch used to repair a surgically created RVOT defect in an adult rat [82]. The SMCs seeded in the patch were in a synthetic phenotype prior to implantation. At 8 weeks after implantation, the spongy component of the scaffold was absorbed and replaced by cells and ECM. After 22 weeks of implantation, the phenotype of the SMCs in the scaffold changed slowly from a synthetic to a more contractile phenotype. Elastin fibers were colocalized with the SMCs, and the elastin sites were observed throughout the patches.

Chang *et al.* investigated the tissue regeneration patterns in the biological patch to repair a defect in the pulmonary trunk in a canine model (Fig. 2.32) [83]. The canine pulmonary trunk was more extensible than the GA- and genipin-fixed acellular bovine pericardia. The degree of inflammatory reaction observed for the implanted genipin-fixed acellular patch (AGP) was found to be significantly less than the GA-fixed acellular patch (AGA). At 1 month postoperatively, intimal thickening was found on the inner surfaces of both the groups. The intimal thickening observed on the AGA was significantly thicker than AGP. An intact layer of ECs was found on the intimal thickening of the AGP, whereas ECs did not universally and totally cover the entire surface of the AGA. Additionally, fibroblasts with neocollagen fibrils and myofibroblasts were observed in the acellular patches for both the groups.

Toshimitsu *et al.* assessed the applicability of a cell-free bioabsorbable polymer sheet for the reconstruction of the hepatic vein or inferior vena cava (IVC) during living-donor liver transplantation or gastrointestinal surgery [84]. The patch grafted onto the IVC was a sponge-like polymer sheet composed of an L-lactide-ε-caprolactone copolymer, P(CL/LA), and reinforced with PGA fibers. The sheet was cut into pieces of 3×2 cm^2 area and 1 mm in thickness. With a porosity of 95% or more, the patch was designed to be absorbed into the body within 6–8 weeks. Five hybrid pigs weighing 15–30 kg served as the recipients of the patch. Using forceps, a side clamp was placed on the ventral side of the

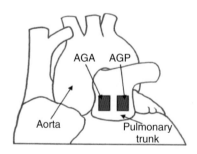

Fig. 2.32 Schematic drawing of the GA-fixed acellular patch (AGA) and the genipin-fixed acellular patch (AGP) implanted on the pulmonary trunk in a canine model.

infrahepatic IVC, followed by the removal of a 3×2 cm^2, elliptical portion of the IVC. The IVC was patched with the polymer sheet using continuous 5–0 proline sutures. Three months after operation, the five pigs grafted with the patch were laparotomized again under general anesthesia. The tissue around the patch was observed for evidence of inflammation, followed by the removal of the patch site and then the animals were sacrificed. All the five pigs survived until they were sacrificed 3 months after operation. On gross examination, the polymer sheet grafted onto the IVC was completely absorbed, and the graft site was morphologically similar to the native IVC. In all the five pigs, the patched IVC was free of stenosis or deformation. Immunohistochemistry revealed that the patch site was lined with ECs and that smooth muscle was present under the epithelium. Like the native IVC, the patch site tested was positive for factor VIII.

Shin'oka *et al.* reported for the first time the usefulness of a tissue-engineered autograft using cultured cells seeded onto a bioabsorbable scaffold in humans [85]. This challenge was based on several animal experiments. A four-year old girl who had a simple right ventricle and pulmonary atresia had undergone pulmonary-artery angioplasty and the Fontan procedure at the age of three years and 3 months. Angiography 7 months later revealed total occlusion of the right intermediate pulmonary artery. An approximately 2 cm segment of peripheral vein was explanted from the girl, and cells from its walls were isolated. The cell count in the culture increased to 12×10^6 cells by eight weeks. A tube that served as a scaffold for these cells was composed of P(CL/LA), reinforced with woven PGA. The absorbable polymer conduit (10 mm in diameter, 20 mm in length, and 1 mm in thickness) was designed to degrade within eight weeks. Ten days after seeding, the graft was transplanted in May 2000. The occluded pulmonary artery was reconstructed with the tissue-engineered vessel graft. No postoperative complications occurred. On follow-up angioplasty, the transplanted vessel was noted to be completely patent. Seven months after implantation, the patient was doing well, with no evidence of graft occlusion or aneurismal changes on chest radiography. Using the same absorbable conduit and cell culture technique, Naito *et al.* treated a 12-year-old boy who had been diagnosed of polysplenia, complete atrioventricular septal defect, small left ventricle, common atrium, bilateral superior vena cava, and hemiazygous connection [86]. A modified Blalock–Taussig shunt and extracardiac Fontan operation (ECFO) with a Hemashield graft were performed at the ages of 9 and 11 years, respectively. Catheterization demonstrated occlusion of the extracardiac graft. The replacement of the occluded conduit was performed by transplantation of the tissue-engineered graft of 20 mm diameter. The conduit was anastomosed to the hepatic venous stump. The main pulmonary artery end was opened along its previous anastomotic orifice into the right pulmonary artery, and the other end of the conduit was anastomosed with the orifice. Intraoperative transesophageal echocardiography showed patent graft flow, and the patient was discharged from the hospital in good clinical condition. Postoperative CT done 4 months after operation revealed a patent graft.

The procedures applied to the two above-mentioned clinical cases have used vessel walls harvested from patients. The process of cell isolation may introduce a risk of infection and possibly culture does not produce sufficient cells for seeding onto scaffolds. Moreover, culturing cells requires substantial time, which is not available for emergency use, and the use of serum from other species in the culture medium reduces the merit of the procedure. To overcome these drawbacks, Hibino *et al.* have developed a method for creating a tissue-engineered vascular graft by using BMCs, which can be obtained easily and used immediately without cell culture [87]. Resorbable scaffolds made from P(CL/LA) were seeded with different types of cells (group V: cultured venous cells; group B: BMCs without culture; and group C: non-cell-seeded graft [as control]), followed by implantation into the inferior venae cavae of dogs. The grafts were explanted at 4 weeks and assessed histologically and biochemically. In the histologic examination, a regular layer of Masson-staining collagen fiber and a layer of factor VII–stained endothelial and anti-α-SMC antigen-immunoreactive cells stained in groups V and B like native vascular tissue, whereas no such stained regular lining was detected in group C. A 4-hydroxyproline assay in group C showed significantly lower levels than groups V and B or native tissue. The DNA content of the tissue-engineered vascular graft tended to be higher in group C than in groups V and B or in native tissue. Based on these findings, Shin'oka *et al.* switched to using BMCs as the cell source in tissue engineering for clinical applications. Noishiki *et al.* reported that BMCs transplanted onto the surface of an artificial graft led to endotheliazation in a dog model, suggesting that seeding BMCs onto a scaffold would result in a valuable tool for constructing tissue-engineered materials [88]. When BMCs are used, it is possible to obtain sufficient cells on the day of surgery, and, in consequence, patients do not need extra-hospitalization for vein harvesting. In addition, using BMCs as the cell source removes consideration of cell culture medium, and there is a low risk of contamination. Less culture medium and other reagents lead to improved cost-effectiveness.

Matsumura *et al.* classify their protocol into three categories with regard to cell preparation [89]. Procedure I: vein harvesting, cell culture, expansion onto a bio-absorbable scaffold, and culture. Procedure II: direct bone-marrow seeding. Procedure III: mononuclear bone-marrow cell seeding. They performed Procedure I in 3 patients, Procedure II in 10 patients, and Procedure III in 7 patients. In Procedures II and III, they began cardiac surgery under general anesthesia and, at the same time, aspirated 1~4 ml/kg of BMCs from the anterior superior iliac spine through puncture. Aspirated BMCs were passed through a nylon cell strainer to remove fat and bone fractions. For Procedure III, bone marrow was centrifuged to obtain mononuclear BMCs, which were then seeded onto a scaffold by pipetting. The seeded scaffold was kept in culture medium for 2~4 h at 37°C in 100% humidity and a 5% CO_2 atmosphere until use. The culture medium for Procedures II and III was RPMI-1640(Sigma) supplemented with other factors including recombinant human VEGF, HGF, and bFGF, but without FCS. A clinical case is described here as an example which was performed using Procedure III. A 3-year-old girl underwent left mBTd in January 2001; surgeons prepared a tissue-engineered sheet using mononuclear BMCs and performed a total cavo-pulmonary connection, and venous flow from the hepatic vein was separated with a tissue-engineered sheet in the common atrium. Postoperative catheterization showed excellent flow through the hepatic vein, and no residual shunt was shown in the atrium. Figure 2.33 shows the cine angiograms.

In the reconstruction of tissue-engineered vascular autografts (TEVA) described above, the surgeons used BMCs because they hypothesized that BMCs would contribute to histogenesis in TEVAs. In another study to test this hypothesis, they seeded labeled BMCs onto the same absorbable scaffold as used in the human clinical trials, and studied the characteristics, functions, and roles of seeded BMCs after implantation into the inferior vena cava of dogs [90]. To trace the seeded BMCs and distinguish between seeded BMCs and host ECs, blood cells, or BMCs that might have migrated in the graft, 5-(and 6)-carboxyfluorescein-diacetate-succinimidyl-ester (CFDA) was added to the BMC aspirated from the iliac bone of dogs. There was

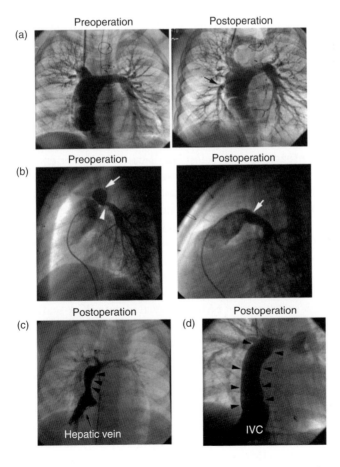

Fig. 2.33 Cine angiograms of cases 1–4: (a) case 1, (b) case 2, (c) case 3, and (d) case 4.

neither stenosis nor obstruction caused by TEVAs implanted into the dogs. Immuno-histochemically, the seeded BMCs expressing EC lineage markers, such as CD34, CD31, Flk-1, and Tie-2, adhered to the scaffold. This was followed by proliferation and differentiation, resulting in expression of ECs markers, such as CD146, factorVIII, and CD31, and smooth muscle cell markers, such as α-smooth muscle cell actin, SMemb, SM1, and SM2. The VEGF and angiopoietin-1 were also produced by cells in TEVAs. Another study revealed that non-seeded TEVAs showed stenosis or atresia *in vivo*, whereas the TEVAs that had been seeded with BMCs did not show these negative characteristics and instead generated what appeared to be normal tissue. These findings provide direct evidence that the seeded BMCs homed in to, adhered to, proliferated in, and differentiated in the scaffold to construct components of a new vessel wall.

3.1.2. Coronary Artery

Despite surgical advances allowing the utilization of internal thoracic artery to bypass occluded coronary arteries, which has a cumulative patency rate of up to 93% after 5 years, the use is limited by the availability of donor arteries of appropriate length and diameter. A number of approaches to create small-diameter vascular prostheses include the use of natural and synthetic polymeric materials, pre-endothelialization of existing polymer grafts *in vitro*, and the creation of bioartificial or tissue-engineered blood vessels. Synthetic grafts that are smaller than 6 mm in diameter display frequent thrombosis. In addition, intimal and anastomotic hyperplasia, with graft wall thickening and luminal occlusion, are a common cause of graft failure after 6 months.

Another approach is to develop small-calibered arteries using absorbable polymers with techniques of tissue engineering. The ideal vascular graft should have differentiated, quiescent small muscle cells, a confluent and quiescent endothelium, and enough mechanical integrity to tolerate physiologic arterial pressures over the long term. Tissue-engineered arteries have been reported to grossly resemble native vessels, but in animal studies in which tissue-engineered vessels generated *in vitro* were evaluated *in vivo*, their performance was inferior to that of autologous veins.

Niklason *et al.* developed techniques for vessel culture using PGA scaffolds. In their technique, a suspension of SMCs is pipetted onto PGA scaffolds supported in bioreactors. The bioreactors are filled with medium and the SMCs are cultured under conditions of pulsatile radial stress for 8 weeks. At the end of this time, the gross appearance of the cultured vessels is indistinguishable from that of native arteries. Histologically, the lab-grown vessels showed an inner polymer region close to the vessel lumen, surrounded by a dense outer cellular region. The engineered vessels cultured in pulsatile bioreactors displayed some contractile function to phar-macologic stimuli, and portions of the vessel walls contained differentiated SMCs. However, some SMCs in engineered arteries had a highly mitotic, dedifferentiated phenotype, and these cells were almost always in very close proximity to residual PGA fragments. To determine if changes in local SMC environment in engineered vessels constitute to increase SMC mitosis and dedifferentiation, Higgins *et al.* carried out a study assessing possible means by which PGA breakdown products (PGA-BP) could lead to changes in SMC phenotype [91]. They collected PGA-BP from a PGA degradation experiment that involved the long-term degradation of 100 mg samples of PGA scaffold in 50 ml volumes of PBS. The breakdown products that were soluble in PBS were isolated using a filtration apparatus. The result of the SMC counting when PGA-BP was added to culture medium is given in Fig. 2.34. As is seen, SMC numbers declined as the percentage of PGA-BP rose, being significant at highest concentrations. Western blotting revealed that SMCs grown in higher concentrations of PGA-PB showed decreased expression of calponin, a marker for SMC differentiation. The same was true for SMCs grown in glycolic acid, which also showed decreased expression of proliferating cell nuclear antigen, a marker for SMC proliferation. In contrast, cells grown in varying amounts of NaCl or HCl showed little change in differentiation. Based on these findings they concluded that, independent of acidity or osmolality, plausible products of PGA degradation induced dedifferentiation of porous SMCs *in vitro*.

In efforts to minimize residual PGA fragments and their dedifferentiating effect in vascular SMCs, the Niklason group further studied several polymer pre-treatments designed to accelerate PGA degradation [92]. The treatments included varying exposure to heat, 1.0 M NaOH, and γ-irradiation. Immersion of the mesh in 1.0 M NaOH for 6 min most markedly increased the rate of mass loss, with <20% of the starting mass remaining after 8 weeks. Treatment with γ-irradiation and heat proved less effective in accelerating mass loss, with degradation profiles not very different from control. The treatment of PGA with NaOH decreased tensile strength at the start of culture. Tissue-engineered arteries cultured for 8 weeks or longer had no significant PGA mechanical properties. The mechanical properties of the final cultured arteries therefore derived entirely from the cellular and ECM components. Characteristics of engineered vessels cultured on control and treated PGA meshes are shown in Table 2.8. For the vessel culture, tubular scaffolds were connected to custom-made bioreactors filled with culture medium, followed by seeding with

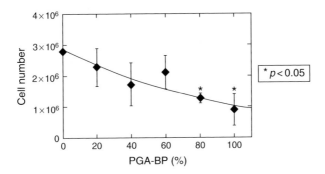

Fig. 2.34 SMC number relative to experimental PGA-BP concentration.

Table 2.8

Characteristics of engineered vessels[a]

Vessel Type	Collagen Content (% Dry Weight)	Suture Retention (g)	SMC Density (10^8 cells/cc)	Wall Thickness (cm, $n = 1$)
Native	49 ± 6 ($n = 3$)	233 ± 46 ($n = 9$)	2.87 ± 0.14 ($n = 2$)	0.032
Pulsed vessels				
Control	44 ± 10 ($n = 6$)	89 ± 6 ($n = 3$)	1.01 ± 0.13 ($n = 5$)	0.037
NaOH	45 ± 14 ($n = 2$)	85 ± 14 ($n = 3$)	0.75 ± 0.019 ($n = 3$)	0.039
Heat	25 ± 9 ($n = 2$)	[b]	1.20 ± 0.22 ($n = 2$)	0.023
γ-Irradiation	52 ($n = 1$)	50 ($n = 1$)	0.99 ($n = 1$)	0.043

[a] Constructs were cultured for 8 weeks. Measurements of native vessels are from porcine arteries stripped of adventitia. Suture retention evaluations were performed in triplicate for each sample.
[b] Insufficient data: heat-treated vessels were too weak to be tested for suture retention.

SMCs. The pulsatile perfusion loop consisted of a pulsatile pump, a flexible tissue culture flask containing Dulbecco's PBS, and tubing connected to the bioreactor side-arms. Constructs were pulsed at 165 beats/min. Weekly medium exchange and supplementation of AA were continued for the 8-week culture period until harvest. Engineered vessels cultured on control PGA contained SMCs with an undifferentiated phenotype, especially in the dense residual polymer resin sequestered around the vessel lumen wall. Vessels grown on NaOH- and γ-modified PGA had a new native-like morphology, with more even distribution of SMCs throughout the entire vessel wall, and with far fewer polymer residual fragments than control constructs. They concluded that decreased initial mass of scaffold and accelerated PGA degradation were important in producing vessels of approximate morphology. Wall thickness was not significantly altered with reduced starting mass of PGA, even in constructs grown with NaOH-treated mesh, which started with only half of the initial mass of control vessels. Furthermore, decreased initial mass coupled with accelerated PGA loss did not substantially affect SMC density. Although the cellular contents of some engineered vessels were approximately the same as those of native vessels, mechanical integrity as assessed by suture retention of the engineered vessels was less than native.

Borschel *et al.* produced a tissue-engineered vascular graft that could withstand arterial pressures [93]. Rat arteries were acellularized with a series of detergent solutions, recellularized by incubation with a primary culture of ECs, and implanted as interposition grafts in the common femoral artery. Acellular grafts that had not been recellularized were implanted in a separate group of control animals. No systemic anticoagulants were administered. When grafts were explanted at 4 weeks, 89% of the recellularized grafts and 29% of the control grafts remained patent. Elastin staining demonstrated the preservation of elastic fibers within the media of the acellular grafts before implantation. Immunohistochemical staining of explanted grafts demonstrated a complete layer of ECs on the lumenal surface in grafts that remained patent. Smooth muscle cells were observed to have repopulated the vessel walls. The mechanical properties of the matrix were comparable to native vessels.

Clerin *et al.* explored an alternative approach—elongating intact arteries *ex vivo* by subjecting them to increased longitudinal stress/strain in contrast to traditional approaches of generating tissue-engineered vessels *in vitro* [94]. They hypothesized that the vessels generated by this approach would more closely resemble the structure and function of native arteries than would arteries constructed from isolated cells. They harvested carotid arteries from neonatal (~5 kg), juvenile (~30 kg), and adolescent (~100 kg) pigs. Arteries were cannulated onto stainless-steel rods in an artery chamber. After installation of the artery, the chamber was filled with culture medium and then connected to a perfusion system. Longitudinal strain was applied daily to the artery by manual displacement of the steel rods. Six carotid arteries from juvenile pigs were successfully lengthened in the perfusion system over 9 days, 48.1% up from the initial physiological loaded length. The corresponding increase from initial to final *ex vivo* unloaded length on removal from the system was 20.5%. Control and elongated arteries displayed native appearance, excellent viability, normal vasoactivity, and similar mechanical properties as compared with freshly harvested arteries. Growth, as opposed to just redistribution of existing mass, contributed to elongation as evidenced by an increase in artery weight.

3.1.3. Angiogenesis

Marui *et al.* investigated whether simultaneous application of bFGF and hepatic growth factor enhances blood vessel formation in murine ischemic hindlimb compared with bFGF or HGF applied alone [95]. Collagen microspheres (CM) were used as a sustained-release carrier for bFGF and HGF. In the back subcutis of the mice, the remaining bFGF radioactivity released from the collagen sheet gradually decreased over 4 weeks, while the radioactivity of the solution form of bFGF applied in a bolus subcutaneous injection disappeared within 3 days after administration (Fig. 2.35). The release profile of bFGF closely correlated with the degradation profile of the collagen sheet. Similarly, the remaining HGF radioactivity also gradually decreased over 4 weeks nearly in accordance with the collagen degradation. Bolus injection of the solution form of HGF also rapidly disappeared within 3 days. Unilateral hindlimb ischemia was created in C57BL/6 mice. A single intramuscular injection of 5 μg or less of bFGF-incorporated CM (bFGF/CM) into the ischemic limb did not significantly increase the laser Doppler perfusion image index (LDPII) compared with the control (no treatment) 4 weeks after the treatment. Similarly, 20 μg or less of HGF/CM did not increase LDPII. Based on these results, they compared the dual release of CM incorporating 5 μg of bFGF and 20 μg of HGF with either the single release of 5 μg of bFGF/CM alone or 20 μg of HGF/CM alone. The results are shown in Fig. 2.36. The LDPII of the dual release was higher than other single releases. Furthermore, the LDPII in the dual release was equivalent to that with 80 μg of bFGFμ alone or 80 μg of HGF/CM alone. The capillary density at 4 weeks in the dual release was higher than that in single releases. The percentage of mature vessels assessed by α-smooth muscle actin staining was also higher in the dual release.

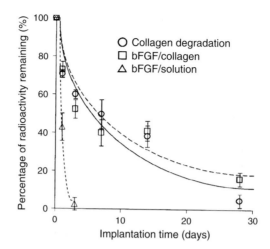

Fig. 2.35 *In vivo* release profile of bFGF from collagen. Sustained release of bFGF (square) from collagen was observed over 4 weeks, and the time profile correlates with that of the collagen degradation (circle). The radioactivity of the bolus injection of bFGF solution (triangle) disappeared within three days.

The BMSCs have characteristics of stem cells for mesenchymal tissues and secrete many angiogenic cytokines, raising the possibility that marrow implantation into ischemic limbs could enhance angiogenesis by supplying EPCs and angiogenic cytokines or factors. Shintani *et al.* showed that bone-mononuclear cell implantation into ischemic limbs or myocardium in animals promoted collateral vessel formation,

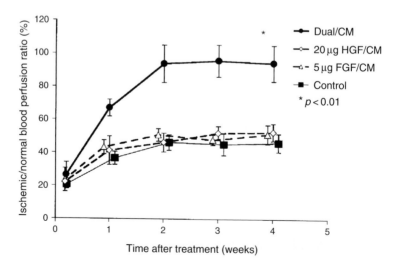

Fig. 2.36 Restoration of the blood perfusion with dual/CM (CM incorporating 5 μg of bFGF and 20 μg of HGF), 20 μg of HGF/CM alone, 5 μg of bFGF/CM alone, and the control.

with incorporation of EPCs into new capillaries, and that local concentrations of angiogenic factors (bFGF, VEGF, and angiopoietin-1) or angiogenic cytokines (interleukin 1β and tumor necrosis factor (TNF)α) were increased in implanted tissues [96]. On the basis of these results in animals, Tateishi-Yuyama *et al.* started a clinical trial to test cellular therapy with autologous bone-marrow-mononuclear cells (BM-MNCs) in patients with ischemic limbs [97]. They first did a pilot study, in which 25 patients (group A) with unilateral ischemia of the leg were injected with BM-MNCs into the gastrocnemius of the ischemic limb and with saline into the less ischemic limb. They then recruited 22 patients (group B) with bilateral leg ischemia, who were randomly injected with bone-marrow-mononuclear cells in one leg and peripheral blood-mononuclear cells in the other as a control. Primary outcomes were safety and feasibility of treatment, based on ankle-brachial index (ABI) and rest pain. At 4 weeks in group B patients, ABI was significantly improved in legs injected with BM-MNCs compared with those injected with peripheral blood-mononuclear cells. Similar improvements were seen for transcutaneous oxygen pressure, rest pain, and pain-free walking time. These improvements were sustained at 24 weeks. Similar improvements were seen in group A patients. Two patients in group A died after myocardial infarction unrelated to treatment.

Although transplantation of bone marrow–derived mononuclear cells has already been applied for the treatment of critical limb ischemia, as described above, little information is available regarding comparison of the angiogenic potency between MSC and MNC. To study the difference, Iwase *et al.* injected equal numbers of MSC or MNC in a rat model of hindlimb ischemia and compared their therapeutic potential [98]. Immediately after creating hindlimb ischemia, rats were randomized to receive MSC transplantation (MSC group), MNC transplantation (MNC group), or vehicle infusion (Control group). Three weeks after transplantation, the laser Doppler perfusion index was significantly higher in the MNC group than in the Control group (0.69 ± 0.1 versus 0.57 ± 0.06). However, there was a much more marked improvement in blood perfusion in the MSC group (0.81 ± 0.08). Capillary density was higher in the MSC group, as shown in Fig. 2.37. The number of transplanted cell-derived ECs was higher in the MSC group than in the MNC group. Transplanted

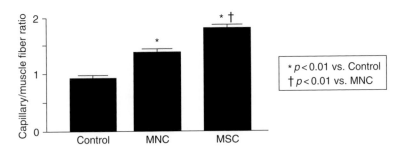

Fig. 2.37 Quantitative analysis of capillary density in ischemic hindlimb muscles. Capillary density is shown as capillary/muscle fiber ratio.

cell-derived VSMCs were detected only in the MSC group. *In vitro*, MSCs were more tolerant to apoptotic stimulus (serum starvation and hypoxia) than MNCs.

3.1.4. Neovascularization

To provide a vascular supply at an intermuscular space, Balamurugan *et al.* induced neovascularization by implanting a synthetic mesh bag with a collagen sponge and gelatin microspheres containing bFGF [99]. In the preliminary experiment, various concentrations of bFGF were tested to standardize the optimal concentration for the induction of angiogenesis. The optimal concentration was achieved by using 50 μg of bFGF impregnated in 2 mg of gelatin hydrogels. In the experimental rat group, many new blood vessels formed around the bag; the extent of the new blood vessels was greater than that in control groups. When the bags were examined under the microscope, numerous fine capillaries were visible. In addition, histologic examination revealed blood vessels in and around the bag, including a few blood vessels that were incorporated inside the collagen sponge from outside the bag. Neovascularization was greater when the bags were examined at 14 days than at 7 days after implantation.

To augment neovascularization and bone regeneration from bone marrow in femoral bone defects of rabbits, Hisatome *et al.* used gelatin microspheres containing bFGF for the sustained release [100]. They implanted fluorescent-labeled autologous BM-MNCs into the defects together with gelatin microspheres containing bFGF on a collagen gel scaffold. The autologous BM-MNCs expressed CD31, an endothelial lineage cell marker, and induced efficient neovascularization at the implanted site 2 weeks after implantation to a significantly greater extent than either BM-MNCs or bFGF on their own, as shown in Fig. 2.38.

Narmoneva *et al.* demonstrated that short self-assembling peptides formed scaffolds that provided an angiogenic environment promoting long-term cell survival and

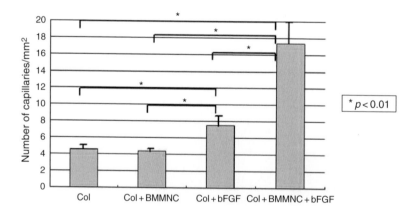

Fig. 2.38 Capillary density at implant sites 2 weeks after implantation ($n = 5$ in each group).

capillary-like network formation in 3-D cultures of human microvascular ECs [101]. In contrast to type I collagen, the peptide scaffold inhibited EC apoptosis in the absence of added angiogenic factors, accompanied by enhanced gene expression of the angiogenic factor VEGF, as shown in Fig. 2.39. In addition, the process of capillary-like network formation and the size and special organization of cell networks could be controlled through manipulation of the scaffold properties, with a more rigid scaffold promoting extended structures with a larger inter-structure distance, as compared with more dense structures of smaller size observed in a more complicated scaffold.

Engineered blood vessels have been found to be immature and unstable. Genes that enhance the survival and/or proliferation of vascular cells—ECs and mural cells—can be introduced to extend the life span of the engineered vessels, but these may prove to be oncogenic. Koike *et al.* created long-lasting blood vessels without such genetic manipulation [102]. To create stable vessels, they first seeded human umbilical vein ECs (HUVECs) and 10T1/2 mesenchymal precursor cells in a 3-D FN-type, type I collagen gel. The 10T1/2 cells differentiated into mural cells through heterotype interaction with ECs. Initially, the HUVECs formed long, interconnected tubes with many branches that showed no evidence of perfusion. Subsequently, they connected to the mouse's circulatory system and became perfused. This led to a rapid increase in the number of perfused vessels in the first two weeks, followed by stabilization, whereas the number of non-perfused vessels gradually decreased and eventually disappeared. By contrast, constructs prepared from HUVECs alone showed minimal perfusion and had generally disappeared after 60 days, even though early morphological changes were similar in the two types of construct. Immunohistochemistry revealed that human cells positive for CD31 lined the lumen of the engineered vessels and that these vessels were fortified by 10T1/2-derived cells that expressed the mural-cell marker smooth-muscle α-actin. Temporal changes in functional density of engineered vessels (total length

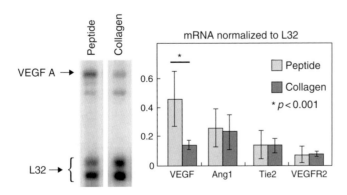

Fig. 2.39 Levels of VEGF mRNA expressed by cells seeded on 1% peptide and collagen gel scaffolds.

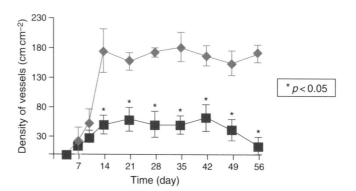

Fig. 2.40 Temporal changes in functional density of engineered vessels.

of perfused vessel structure per unit area) are shown in Fig. 2.40. Asterisks indicate significant difference between the two groups (top curve: HUVEC + 10T1/2; bottom curve: HUVEC-alone construct).

3.2. Heart Valves

Cardiac valvular defects affect both adult and pediatric patients and frequently require surgical interventions such as valve replacement or repair. Prosthetic heart valves even with the most advanced designs are far from ideal. Bioprosthetic and mechanical valve replacements have certain disadvantages that limit their long-term benefits, as listed in Table 2.9 [103]. The criteria for an ideal valve replacement were outlined by Harken and Curtis many years ago, as represented in Table 2.10 [104]. As these prosthetic valves are made of foreign materials, they are potentially thrombogenic and susceptible to infection throughout their life span. In addition, they lack the ability to grow, repair, or remodel. These inherent problems of prosthetic valves have significantly limited durability, especially in the pediatric age group. Bioprostheses and homografts are commercially available and these tissue-derived valves have unique beneficial qualities as well as inherent limitations. For instance, porcine valves have low resistance to flow, larger diameter, lack of anticoagulation requirement compared to mechanical valves, but they suffer a much higher rate of inflammatory deterioration with valve failure compared to mechanical valves. Approximately 90% of all bioprosthetic heart valves fabricated from porcine aortic valves fail with tearing, many failing with little or no calcification. In an attempt to overcome these shortcomings of current valve options, attention has been directed to a tissue engineering approach. Tissue-engineered heart valves (TEHVs) are generally constructed by seeding autologous cells onto anatomically shaped, porous scaffolds. However, tissue-engineered pulmonary valve replacements have developed moderate amounts of central valvular regurgitation, despite the *in vivo* evolution of the engineered valve tissue into a tri-layered structure which resembles

Table 2.9

Limitations of current-state-of-the-art valves

Mechanical valve
  Foreign body response
  Lack of growth
  Mechanical failure
  Need for life-long anticoagulation
  Thrombosis

Tissue valve (xenograft)
  Foreign body response
  Lack of growth
  Short durability
  Calcification

Allogeneic valve (homograft)
  Foreign body response
  Lack of growth
  Donor organ scarcity
  Rejection

Tissue valve (autoxenograft)
  Foreign body response
  Lack of growth
  Rejection

Tissue-engineered valve (autograft)
  Foreign body response
  Growth potential
  No rejection

native valve tissue. Only very limited information is currently available on the extent to which an engineered heart valve duplicates the biomechanical function and structure of the native pulmonary valve cusp, and on the events and mechanisms that guide the remodeling process.

Because of the difficulty in finding synthetic polymers that meet the requirements as scaffolds for heart valves, a number of research groups have fabricated the scaffold matrix for heart valve tissue engineering from allogeneic or xenogeneic

Table 2.10

Characteristics of an ideal valve substitute

No inflammatory and/or foreign body response
No immunologic response → autologous tissue
Viable structure with autorepair potential and life-long durability
Unlimited supply
Antithrombogenic surface → no anticoagulation
Growth potential
Individual custom-made manufacturing

biological sources. Natural materials (*e.g.*, decellularized valve tissue) may possess proper anatomical structure, innate bioactivity, and tissue-like mechanical properties, but their inherent variability and potential immunogenicity represent significant hurdles. The acellularization process for biological matrix scaffolds may not remove all native cells or cell debris. It was reported that after acellularization of porcine tissue, up to 2% of naïve DNA was still detectable with the matrix. These findings might indicate an increased risk for cross-species porcine endogenous retrovirus (PERV) transmission after implantation. Leyh *et al.* investigated whether acellularized porcine heart valve scaffold causes cross-species transmission of PERV in a sheep mode [105]. Porcine pulmonary valve conduits were harvested with a thin ridge of subvalvular muscle tissue proximally and a sheet segment of the truncus pulmonalis distally. For acellularization, the conduits were placed in a bioreactor filled with a 0.05% trypsin and 0.02% EDTA for 48 h, followed by flushing with PBS solution for 48 h to remove all cell debris. Five acellularized valve conduits were seeded first with myofibroblasts and coated with ECs, resulting in a uniform cellular restitution of the pulmonary valve conduit surface. With the heart beating, the pulmonary artery was transected, all 3 naïve leaflets removed, and the repopulated pulmonary valve conduits implanted into the same 5 lambs from which the initial vessels were harvested. In addition, the acellularized porcine heart valve scaffolds were implanted in the remaining 3 sheep. The animals were killed 6 months after operation. The DNA of PERV was detectable in acellularized porcine heart valve tissue, but PCR and RT-PCR results showed no fragments of PERV sequences in TEHVs after 6 months of implantation. Furthermore, examination of peripheral blood monocytes and plasma of sheep by means of PERV-specific PCR revealed negative results up to 6 months after implantation of TEHVs based on acellularized porcine scaffolds. Kallenbach *et al.* also acellularized porcine pulmonary arteries and implanted them into sheep in orthotopic position to examine whether acellularized porcine vascular scaffolds cause cross-species transmission of PERV in a xenogeneic model [106]. Blood samples were collected regularly up to 6 months after the operation, and cellular components were tested for PERV infection by PCR and RT-PCR. The PERV DNA was detectable in acellularized scaffolds of porcine matrices. Acellular porcine pulmonary arteries scaffolds were repopulated *in vivo* by autologous cells of the host, leading to a vessel consisting of all cell components of the vessel wall. The PERV sequences were detectable neither in all tested peripheral blood samples nor in tissue samples of *in vivo* recellularized grafts up to 6 months after implantation.

The porcine cell-specific α-Gal epitope is known to be responsible for hyperacute rejection in xenotransplantation. In tissue engineering, residual α-Gal epitope may induce severe inflammation in humans and may lead to graft failure. Kasimir *et al.* decellularized porcine pulmonary conduits with Triton X-100, sodium deoxycholate, Igepal CA-630, and ribonuclease treatment and compared the decellularized conduits with specimens of the commercially available porcine decellularized SynerGraft regarding cell removal and elimination of the α-Gal epitope [107]. In addition, samples of a porcine bioprosthesis were examined for the presence of the α-Gal epitope. They detected the presence of the α-Gal epitope in clinically used porcine bioprostheses and the first generation of a commercial TEHV. In contrast, complete cell and α-Gal removal was achieved by this decellularization procedure.

Shin'oka *et al.* tested the feasibility of constructing heart valve leaflets in lambs using autologous myofibroblasts and ECs seeded *in vitro* onto a bioabsorbable scaffold [108]. However, the relatively non-pliable nature of Polyglactin™ (Vicryl™) (PGLA) mesh created a logistic problem for the design of a whole leaflet valve, because the non-pliable valves will inevitably result in the development of severe pulmonary stenosis if all 3 leaflets are replaced. Stock *et al.* used a flexible polymer, polyhydroxyoctanoate (PHO), seeded with autologous vascular cells in an initial feasibility study for the creation of a 3-leaflet heart valve in the supravalvular pulmonary valve position [109]. As illustrated in Fig. 2.41, the conduit (20 mm in length and 18 mm in inner diameter) was composed of nonporous PHO film (240 µm thickness) with layers of PGA felt (1 mm thickness) on the inside and outside of the PHO. These constructs were seeded with autologous medial cells on 4 consecutive days and coated once with autologous ECs. Thirty-one days after cell harvesting, 8 seeded and 1 unseeded control constructs were implanted to replace the pulmonary valve and main pulmonary artery on cardiopulmonary bypass. Postoperative echocardiography of the seeded constructs demonstrated no thrombus formation with mild, nonprogressive, valvular regurgitation up to 24 weeks after implantation. The unseeded construct developed thrombus formation on all 3 leaflets after 4 weeks. Although histologic examination after 6 weeks did not show any significant PGA, PHO material was still evident in the conduit and the leaflets at 24 weeks. Analysis of the residual PHO revealed a molecular weight loss of only 26% at 24 weeks, indicating that PHO is virtually a non-absorbable polymer. There was no increase in diameter or length of the constructs over the observed time period.

Dohmen *et al.* started a trial to implant tissue-engineered pulmonary valves during Ross procedures [110]. They reported the first clinical case of a TEHV to reconstruct the RVOT. A viable heart valve was created from decellularized cryopreserved

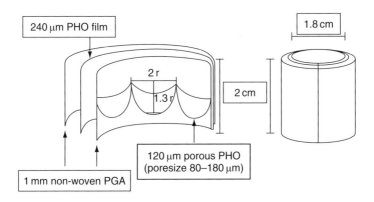

Fig. 2.41 Sketch of the scaffold design for valve.

Table 2.11
Hemodynamic evaluation of the valve function as well as the heart function

	Preoperative	Discharge	3 Months	6 Months	1 Year
LVEF (%)	52	48	66	62	68
LV mass (g)	210	150	140	130	140
LVEDD (cm)	5.7	5.2	5.2	5.1	5.1
Aortic pressure gradient (mmHg)	44	2.5	2.0	1.9	1.3
Aortic flow velocity (m/s)	2.5	1.08	0.99	0.98	0.88
Aortic regurgitation	II	None	None	None	None
Pulmonary pressure gradient (mmHg)	None	1.3	1.4	0.7	1.6
Pulmonary flow velocity (m/s)	NA	1.04	1.05	0.53	0.69
Pulmonary regurgitation	None	I	I	I	I

LV = left ventricle; LVEDD = left ventricular end diastolic diameter; LVEF = left ventricular ejection fraction; NA = not available.

pulmonary allograft that was seeded with viable autologous vascular endothelial cells (AVEC). The matrix used to seed AVEC was commercially available. Four weeks before the operation a piece of forearm vein was harvested to separate, culture, and characterize AVEC. After 4 weeks of culturing, 8.34×10^6 AVEC were available to seed a 27 mm decellularized pulmonary allograft. Trypan blue staining confirmed 96.0% viability. Re-endothelialization rate after seeding was 9.0×10^5 cells/cm^2. Follow-up was completed at discharge, 3, 6, and 12 months postoperatively. Transthoracic echocardiography and magnetic resonance imaging (MRI) revealed the excellent hemodynamic function of the TEHV and the neoaortic valve as well, as shown in Table 2.11. Multislice computed tomography revealed no evidence of valvular calcification. After 1 year of follow-up the patient was in excellent condition without limitation and exhibited normal aortic and pulmonary valve function.

3.3. Myocardial Tissue

Ischemic heart disease and myocardial infarction are major causes for end-stage heart failure. High-risk coronary surgery, left ventricular restoration with or without mitral valve surgery, implantation of artificial assist devices, and biventricular pacemaker implantation are widely used as treatment concepts in ischemic heart failure. Rapid reperfusion of the infarct-related coronary artery is of great importance in salvaging ischemic myocardium and limiting the infarct size in patients with acute myocardial infarction. However, myocardial necrosis starts rapidly after coronary occlusion, usually before reperfusion can be achieved. The loss of viable myocardium initiates a process of adverse left-ventricular remodeling, leading to chamber dilatation and contractile dysfunction in many patients. Percutaneous transluminal coronary angioplasty with stent implantation is the method of choice to re-establish coronary flow.

Sakaguchi *et al.* designed an HGF-incorporating gelatin hydrogel sheet (HGF sheet) to release HGF for more than 2 weeks *in vivo* and investigated whether the HGF sheet could prevent the progression of heart failure in stroke-prone spontaneously hypertensive rats [111]. Rats at the age of 25 weeks received placement of a

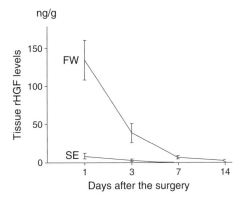

Fig. 2.42 Time course of tissue rHGF levels in the left-ventricular free wall (FW) and the septum (SE).

recombinant HGF (rHGF) sheet on the left-ventricular free wall or sham operation. Tissue rHGF levels in the free wall (FW) and the septum (SE) are shown in Fig. 2.42. The myocardial rHGF levels were higher in the free wall than the septum. These levels were maximal 1 day after surgery and thereafter decreased. The rHGF was not detectable in the septum 7 days after surgery. There were two deaths in the control group and no deaths in the rHGF group during the 4 weeks. Fractional shortening was significantly higher and left-ventricular diastolic dimension was significantly smaller in the rHGF than in the control group. The slope of the peak early diastolic filling velocity and the ratio of that slope to the slope of the peak filling velocity at arterial contraction were significantly lower in the rHGF than the control group. Myocardial fibrosis was lower and capillary density was significantly higher in the rHGF than the control group. Placement of the hydrogel sheet alone did not affect any cardiac function compared with sham operation. The survival rate at 10 weeks after the surgery was much higher in the HGF than the control group.

Other strategies have evolved, aiming at restoring diseased areas of the heart. These approaches include cellular transplantation and tissue engineering of myocardial tissue. Bone marrow–derived MSCs have been used for cardiomyoplasty and to induce neovascularization when they are injured into infarcted myocardium. Cardiac transfer of unfractionated BMCs, or stem cells and progenitor cells derived from bone marrow, can enhance functional recovery after acute myocardial infarction. Based on these data, stem cells and progenitor cells derived from bone marrow have been proposed for use in the repair of cardiac tissue after acute myocardial infarction in patients. Preclinical studies investigating bone marrow–derived cells as treatment for ischemic myocardium have been performed preferentially in the acute ischemia model; data in the chronic ischemia model are lacking. It is unknown whether cell transplantation strategies are able to successfully restore severely injured myocardium. Damaged tissue within infarction areas rapidly undergoes irreversible necrosis, leading to structural alternatives, such as scar tissue development. Some clinical studies suggest improvements of heart function after injection of skeletal myoblasts [112] or BMSCs [113]. However, it is controversial whether these effects are due to the ability of these cell types to create sufficient amounts of new myocardium-like tissue within the infarction area and to participate in synchronized heart contraction. The reported effects might be simply caused by remodeling of connective tissue and ECM.

Christman *et al.* examined the effects of fibrin glue as an injectable scaffold and wall support in ischemic myocardium [114]. The left coronary artery of rats was occluded for 17 min, followed by reperfusion. Echocardiography was performed 8 days after infarction. One to two days later, either 0.5% bovine serum albumin (BSA) in PBS, fibrin glue alone, skeletal myoblasts alone, or skeletal myoblasts in fibrin glue were injected into the ischemic left ventricle. Echocardiography was again performed 5 weeks after injection. The animals were then sacrificed and the hearts were fresh-frozen and sectioned for histology and immunohistochemistry. Fractional shortening (FS) as a measure of systolic function was calculated as FS (%) = $[(LVID_d - LVID_s)/LVID_d] \times 100$, where LVID is the left-ventricular internal dimension, d is diastole, and s is systole. As shown in Table 2.12, both the FS and infarct wall thickness of the BSA group decreased significantly after 5 weeks. In contrast, both the measurements for the fibrin glue group, cells group, and cells in fibrin glue group did not change significantly.

To investigate whether transplanted MSCs induce myogenesis and angiogenesis and improve cardiac function in a rat model of dilated cardiomyopathy, Nagaya *et al.* isolated MSCs from bone-marrow aspirates of isogenic adult rats and expanded *ex vivo* [115]. Cultured MSCs secreted large amounts of angiogenic, antiapoptotic, and mitogenic factors: VEGF, HGF, adrenomedullin, and insulin-like growth factor-1. Five weeks after immunization, MSCs or vehicle was injected into the myocardium. Some engrafted MSCs were positive for cardiac markers—desmin, cardiac troponin T, and connexin-43—whereas others formed vascular structures and were positive for von Willebrand factor or smooth muscle actin. Compared with vehicle injection, MSC transplantation significantly increased capillary density and decreased the collagen volume fraction in the myocardium, resulting in decreased left-ventricular

Table 2.12

Echocardiography data[a]

	Before Injection	5 Weeks Postinjection	*p* Value
Fractional shortening (%)			
Control group	45 ± 8	22 ± 6	0.0005
Fibrin group	26 ± 5	23 ± 8	0.18
Cells group	29 ± 14	28 ± 2	0.89
Cells in fibrin group	42 ± 10	33 ± 6	0.19
Infarct wall thickness (cm)			
Control group	0.29 ± 0.08	0.24 ± 0.04	0.02
Fibrin group	0.26 ± 0.04	0.23 ± 0.06	0.40
Cells group	0.30 ± 0.08	0.26 ± 0.06	0.44
Cells in fibrin group	0.30 ± 0.04	0.32 ± 0.02	0.43

[a] *p* Values are for internal comparisons between measurements 1 week after infarction and 5 weeks after injection.

end-diastolic pressure and increased left-ventricular maximum dP/dt. Silva *et al.* determined whether bone marrow–derived MSC transplantation would improve the morphology and function of the heart in a chronic canine model of myocardial ischemia [116]. Twelve dogs underwent ameroid constrictor placement. Thirty days later, they received intramyocardial injections of either MSCs or saline only. All were euthanized at 60 days. Mean left-ventricular ejection fraction was similar in both groups at baseline but significantly higher in treated dogs at 60 days. White blood cell count and C-reactive protein levels were similar over time in both groups. There was a trend toward reduced fibrosis and greater vascular density in the treated group. The MSCs colocalized with endothelial and SMCs but not with myocytes.

Wollert *et al.* conducted a randomized controlled trial to assess the effect of intracoronary transfer of autologous BMCs on left-ventricular functional recovery in patients after acute myocardial infarction and successful percutaneous coronary intervention (PCI) [117]. After PCI for acute ST-segment elevation myocardial infarction, 60 patients were randomly assigned to either a control group ($n = 30$) that received optimum postinfarction medical treatment, or a bone-marrow-cell group ($n = 30$) that received optimum medical treatment and intracoronary transfer of autologous BMCs 4.8 days after PCI. Primary endpoint was global left-ventricular ejection fraction (LVEF) change from baseline to 6-month follow-up, as determined by cardiac MRI. Image analyses were done by two investigators blinded for treatment assignment. The result is given in Table 2.13. Global LVEF at baseline (determined 3.5 days after PCI) was 51.3 in controls and 50.0 in the BMC group. After 6 months, mean global LVEF had increased by 0.7% points in the control group and 6.7% points in the bone-marrow-cell group. Transfer of BMCs enhanced left-ventricular systolic function primarily in myocardial segments adjacent to the infarction area. Cell transfer did not increase the risk of adverse clinical events, in-stent restenosis, or proarrhythmic effects.

Engineered myocardial tissues might serve as a basis for the development of tissue, which is capable of replacing human myocardium in many disease states of the failing heart [118]. Kofidis *et al.* used the clinically available hemostatic collagen matrix for the construction of bioartificial myocardial tissue (BMT) [119]. Pieces ($20 \times 15 \times 2$ mm^3) of Tissue Fleece™ were placed into wells in cell culture dishes. The collagen matrix, Tissue Fleece (Baxter Deutschland, Germany), is a mesh-like scaffold that consists of type I collagen fibrils of equine origin. It typically serves as a hemostypticum for the management of diffuse surgical bleeding, particularly in the field of thoracic and cardiovascular surgery (Tissue Fleece costs 10-fold less than Matrigel™). One milliliter of cell suspension containing 2×10^6 neonatal rat cardiomyocytes was added to each Tissue Fleece segment. The mixture was allowed to gel at 37°C for 4 h before culture medium was added. Bioartificial myocardial tissue was cultured in MEM plus 10% FCS and 110 μM 5-bromo-2'-deoxyuridine. Gelation of the collagen–cell mixture in culture lasted until day 2. The resulting size of BMTs was approximately $20 \times 15 \times 2$ mm^3. The shape remained stable throughout the total culturing period (Tissue Fleece is designed to disintegrate *in vivo* and *in vitro* at body temperature). When placed in

Table 2.13

Left-ventricular volume and mass indices, global LVEF, and late enhancement as determined by contrast-enhanced MRI at baseline and 6-month follow-up

	Baseline		6-Month Follow-up		Change		BMC Treatment Effect*	p
	Controls	BMC Group	Controls	BMC Group	Controls	BMC Group		
LVEDV index (ml/m²)	81.4 (16.9)	84.2 (17.2)	84.9 (21.9)	91.7 (26.0)	3.4 (11.1)	7.6 (20.0)	4.0 (−4.4 to 12.5)	0.32
LVESV index (ml/m²)	40.6 (16.9)	43.0 (14.7)	42.6 (23.5)	42.4 (23.9)	2.0 (11.1)	−0.6 (14.9)	−3.2 (−9.7 to 3.3)	0.33
Global LVEF (%)	51.3 (9.3)	50.0 (10.0)	52.0 (12.4)	56.7 (12.5)	0.7 (8.1)	6.7 (6.5)	6.0 (2.2 to 9.9)	0.0026
LVM index (g/m²)	78.2 (18.3)	82.7 (18.7)	71.7 (14.2)	71.9 (14.6)	−6.5 (12.8)	−10.8 (10.6)	−2.5 (−7.3 to 2.3)	0.30
LE (ml)	30.3 (17.4)	33.0 (21.1)	19.8 (9.8)	18.9 (12.2)	−10.5 (10.6)	−14.1 (13.0)	−2.2 (−5.4 to 1.0)	0.18

BMC = bone-marrow cell. Data are mean (SD) unless otherwise stated. * Treatment effects expressed as differences in least-squares means (ANCOVA model) with 95% CI. LVM = left-ventricular mass. LE = late contrast enhancement. There were no differences between groups at baseline.

aqueous solution at 37°C, Tissue Fleece disintegrated completely within a few hours. By directly pouring cells, rather than injecting them, onto the surface of the scaffold, and allowing 30 min of hydration with culture medium, cells were able to infiltrate through the porous structure of the scaffold and build junctions to the fibrils. In this way the scaffold obtained greater integrity and remained solid for weeks. Despite the dense fibril mesh and the direct fibril-to-cell junction, the syncycial contractility at common frequencies did not appear to be inhibited. Contractions first became apparent 36 h after casting and reached the maximal frequency and strength on day 4 in 87 of 100 BMTs. Forty percent of the BMTs displayed ongoing contractility for 12 weeks in culture. The frequency of contraction ranged between 40 and 220 "beats" per min, with an average frequency of 125 beats/min. The BMT strips contracted spontaneously and continuously. As shown in Fig. 2.43, force generation in unstimulated BMTs reached 8.6 μN. Maximal force was achieved with 2.5 mm of elongation of the BMTs, leading to a 119% increase in force (18.8 μN) compared with unstretched BMTs. Contractile force increased after topical administration of Ca^{2+} and adrenaline. Stretch led to the highest levels of contractile force. As represented in Fig. 2.44, elasticity of BMT was similar to that of native rat myocardium.

For myocardial restoration, Kofidis *et al.* seeded undifferentiated GFP-labeled mouse ES cells in Matrigel [120]. In a Lewis rat heterotopic heart transplant model an intramural left-ventricular pouch was fashioned after ligation of the left anterior descending coronary artery. The liquid mixture was injected in the resulting infarcted area within the pouch and solidified within a few minutes after transplantation. Five recipient groups were formed: transplanted healthy hearts (group I), infarcted control hearts (group II), matrix recipients alone (group III), the study group that received matrix plus cells (group IV), and a group that received ES cells alone (group V). After echocardiography 2 weeks later, the hearts were harvested and stained for GFP and cardiac muscle markers (connexin 43 and α-sarcomeric

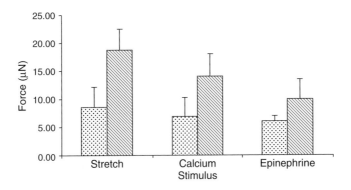

Fig. 2.43 Impact of stretch and pharmacological agents on force development (left columns: baseline; right columns: after stimulus administration).

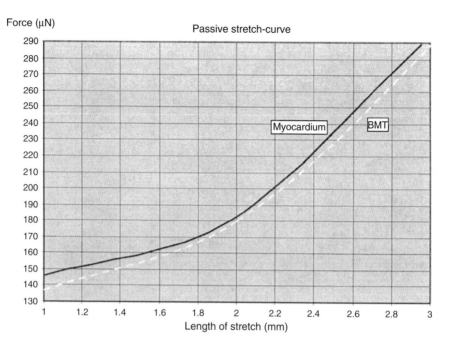

Fig. 2.44   BMT and rat myocardial tissue elasticity.

actin). The graft formed a sustained structure within the injured area and prevented
ventricular wall thinning. The incubated cells remained viable and expressed connexin 43 and α-sarcomeric actin. Fractional shortening and regional contractility
were better in animals that received bioartificial tissue grafts compared with control
animals (group I, 17.0%; group II, 6.6%; group III, 10.3%; group IV, 14.5%; group
V, 7.8%).

Birla *et al.* described a method of engineering contractile 3-D cardiac tissue with
the incorporation of an intrinsic vascular supply [121]. Neonatal cardiac myocytes
were cultured *in vivo* in silicone chambers, in close proximity to an intact vascular
pedicle. Silicone tubes were filled with a suspension of cardiac myocytes in fibrin gel
and surgically placed around the femoral artery and vein of adult rats. At 3 weeks, the
tissues in the chambers were harvested for *in vitro* contractility evaluation and
processed for histologic analysis. By 3 weeks, the chambers had become filled with
living tissue. The H&E staining showed large amounts of muscle tissue situated
around the femoral vessels. Electron micrographs revealed well-organized intracellular contractile machinery and a high degree of intercellular connectivity. Immunostaining for von Willebrand factor demonstrated neovascularization throughout the
constructs. With electrical stimulation, the constructs were able to generate an average
active force of 263 μN with a maximum of 843 μN. Electrical pacing was successful
at frequencies of 1–20 Hz. In addition, the constructs exhibited positive inotropy
in response to ionic calcium and positive chronotropy in response to epinephrine.

Boublik *et al.* fabricated hybrid cardiac constructs from neonatal rat heart cells, fibrin (Fn), and biodegradable knitted fabric (Knit) [122]. Initial (2-h) constructs were compared with native heart tissue, studied *in vitro* with respect to mechanical function and compositional remodeling (collagen, DNA), and implanted *in vivo*. For 2-h constructs, stiffness was determined mainly by the fibrin and was half as high as that of native heart, whereas UTS, failure strain (ε_f), and strain energy density (E) were determined by the Knit and were 8-, 7-, and 30-fold higher than native heart, respectively. Over 1 week of static culture, cell-mediated, serum-dependent remodeling was demonstrated by a 5-fold increase in construct collagen content and maintenance of stiffness not observed in cell-free constructs. Cyclic stretch further increased the construct collagen content in a manner dependent on loading regimen. The presence of cardiac cells in cultured constructs was demonstrated by immunohistochemistry (troponin I) and Western blot (connexin 43). However, *in vitro* culture reduced Knit mechanical properties, decreasing UTS, ε_f, and E of both constructs and cell-free constructs and motivating *in vivo* study of the 2-h constructs. Constructs implanted subcutaneously in nude rats for 3 weeks exhibited the continued presence of cardiomyocytes and blood vessel ingrowth by immunostaining for troponin I, connexin 43, and CD-31.

3.4. Trachea

Some airway-compromising lesions require extensive circumferential tracheal resections. Because a completely satisfactory allograft or prosthesis does not yet exist, tracheal reconstruction after such resections presents a significant surgical concern. The most troubling complication of prosthetic devices is stenosis, caused by mesenchymal granulation tissue, a narrowing that often coincides geographically with poor epithelialization. In the native trachea, epithelial ulceration is often associated with the development of substantial granulation tissue, which sometimes leads to life-threatening obstruction. Absence of EC migration and proliferation often accompanies failed tracheal repairs. Despite many attempts at tracheal replacement for more than 50 years, no predictable and dependable solution has yet been found. The indication for tracheal replacement is indeed rare as the evolution of surgical techniques now allows for resection of about half of the adult and one-third of the pediatric trachea. A variety of materials have been used in the tracheal prostheses, such as various inert materials alone or in combination with autologous tissue, but none of them has been effective. As there is a general agreement that autologous tissues are the first choice in reconstructive surgery, particularly in tracheal surgery, several reports have described experimental tracheal reconstruction with autotransplants, but none of them has proven to be satisfactory for human use. Akl *et al.* [123] and Osada and Kojima [124] reported experimental reconstruction of a tracheal defect of longer than 10 cm using a rotated main bronchus. The major drawback of the method is loss of part of the lung, and the method cannot be used for high mediastinal tracheal defects. Cavadas created a suitable native trachea, which was composed of an autologous microvascular jejunal transfer with a cartilage skeleton for major tracheal reconstruction [125]. This method is ideal but is complicated by the need of a tracheal stent and the need for a 10–15-cm jejunum with its vessels.

The limitations of these tracheal replacements have led to an interest in regeneration of tracheal tissue. Reconstruction of other cartilaginous structures, such as the ear and nose, has intensively been attempted with tissue-engineered techniques, but few studies have focused on reconstruction of cartilage in the trachea. The trachea must maintain flexibility in the longitudinal direction to allow for free movement of the head, while maintaining the rigidity necessary to prevent collapse of the trachea during breathing. This is accomplished in native tissue by using the cartilaginous rings, which are not adequately modeled by a cartilaginous tube. Kojima *et al.* evaluated the ability of autologous, tissue-engineered, helical cartilage to form the structural component of a functional tracheal replacement [126]. Samples of nasal septal cartilage were obtained from six sheep. Chondrocytes and fibroblasts were isolated from the cartilage and the connective tissue separated from the underlying cartilage, respectively. Suspensions of chondrocytes and fibroblasts were placed on $100 \times 10 \times 2$ mm^3 and 50×10 mm^2 meshes of PGA fibers, respectively. The chondrocyte-seeded mesh was placed in the grooves of a 20-mm diameter $\times$ 50-mm long helical template made from Silastic silicone, and the entire template was covered with the fibroblast-seeded mesh. The implants were placed either in a subcutaneous pocket in the nude rat or in the mesh of a sheep just above the sternocleidomastoid muscle. After 8 weeks, the autologous tissue-engineered trachea was harvested from the neck, and the cervical trachea was resected at 5 cm in length. The autologous, tissue-engineered trachea was implanted with end-to-end anastomoses. Sheep receiving the tissue-engineered tracheas breathed spontaneously with no subcutaneous emphysema and were voluntarily ambulatory. Animal survival time ranged from 2 to 7 days as shown in Table 2.14. The gross appearance of the tissue-engineered trachea in both groups was very similar to that of native trachea. There was a distinct structure containing both cartilage and connective tissue that was similar to that seen in native trachea. Safranin-O staining showed that rat tissue-engineered tracheas and sheep tissue-engineered tracheas had similar morphologies to native tracheal cartilage. As shown in Fig. 2.45, GAG contents were similar to those seen in native tracheal tissue in rat

Table 2.14

Postoperative course of the sheep undergoing tracheal reconstruction with tissue-engineered trachea (TET)

No.	Postoperative Survival Period (days)	Trachea Status at Death	Sheep TET Condition
1	7	Malacia	Soft
2	5	Malacia	Soft
3	2	Stenosis	Small
4	2	Stenosis	Small
5	2	Stenosis	Small
6	2	Stenosis	Small

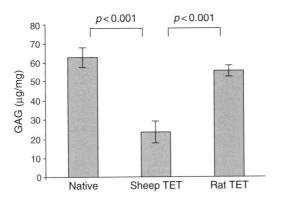

Fig. 2.45   GAG content of native and engineered tissue (data represent $n = 6 \pm$ SD); TET = Tissue-engineered Trachea.

tissue-engineered tracheas. Collagen and cell contents of sheep tissue-engineered trachea were elevated, compared with that of normal tracheas, whereas the proteoglycan content was less than that found in normal tracheas.

Kamil *et al.* evaluated the feasibility of using tissue-engineered cartilage for laryngotracheal reconstruction in a pig model [127]. Auricular cartilage was harvested from three young swine. The cartilage was digested, processed, and suspended and a cell culture was obtained. The cells were then suspended in a solution of Pluronic F-127. This suspension was then implanted subcutaneously into each pig's dorsum. Eight weeks after implantation, the cartilage was harvested with the surrounding perichondrial capsule. All three pigs survived to the 3-month postoperative interval with no evidence of stridor or airway distress. Interval bronchoscopy revealed a normal patent airway with a mucosalized graft. Histopathological analysis revealed incorporation of the tissue-engineered cartilage graft in the cricoid area, which correlated with results of bronchoscopic evaluation. Okamoto *et al.* investigated whether BMP-2, released slowly from a gelatin sponge, could induce cartilage regeneration in a canine model of tracheomalacia, and evaluated the long-term results [128]. A 5-cm gap was created in the anterior cervical trachea by removing 5-cm long strips of 10 sequential cartilagines. In the control group, the gaps were left untreated. In the gelatin sponge group, a gelatin sponge soaked in a buffer solution was implanted in each defect. In the BMP group, a gelatin sponge soaked in a buffer solution containing 12 µg BMP-2 was implanted in each defect. No regenerated cartilage was detected in the control or gelatin sponge groups, even 6 months after surgery. In contrast, regenerated cartilage, which had developed from the host perichondrium, was observed around the stumps of the resected cartilagines in the BMP group. This regenerated cartilage maintained the integrity of the internal lumen for more than 6 months. A compressive fracture test result given in Table 2.15 revealed that the tracheal cartilage in the BMP group was significantly more stable than that in the gelatin sponge and control groups.

Table 2.15

Stress for breaks of anterior tracheal cartilage

	Stress (10^6 N/m^2)
Control group	0.68 ± 0.43*†
Gelatin sponge group	1.68 ± 1.26‡§
BMP group	4.92 ± 1.32**
Normal trachea	3.52 ± 1.03

Data are mean ± SD.
* $p = 0.0001$ vs. gelatin sponge group.
† $p < 0.001$ vs. normal trachea.
‡ $p = 0.0015$ vs. BMP group.
§ $p = 0.002$ vs. normal trachea.
** $p = 0.013$ vs. normal trachea.

Kojima *et al.* evaluated the feasibility of using autologous sheep marrow stromal cells cultured onto PGA mesh to develop helical engineered cartilage equivalents for a functional tracheal replacement [129]. The BMCs obtained by iliac crest aspiration from sheep and cultured in monolayer for 2 weeks were seeded onto non-woven PGA mesh. Cell–polymer constructs were wrapped

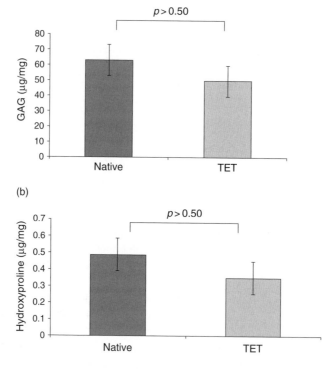

Fig. 2.46 (a) GAG contents of native trachea and tissue-engineered trachea (TET). (b) Hydroxyproline contents of native trachea and TET.

around a silicone helical template and then coated with microspheres incorporating 0.5 μg TGF-β2. The cell–polymer–microsphere structures were then implanted into a nude rat. Cell–polymer constructs with TGF-β2 microspheres formed stiff cartilage *de novo* in the shape of a helix after 6 weeks. Control constructs lacking TGF-β2 microspheres appeared to be much stiffer than typical cartilage, with an apparently mineralized matrix. Tissue-engineered trachea was similar to normal trachea and histologic data showed the presence of mature cartilage. The GAG and OHP contents were also similar to native cartilage levels, as shown in Fig. 2.46.

4. NERVOUS SYSTEM

4.1. Neuron

Brain tissue engineering in the postinjury brain represents a promising option for cellular replacement and rescue, providing a cell scaffold for either transplanted or resident cells. Some investigations of animals and short-term human bone-marrow transplants suggest that bone marrow can repair brain. Cogle *et al.* investigated whether adult human hematopoietic cells can contribute to long-term adult human neuropoiesis without fusing [130]. They examined autopsy brain specimens from three sex-mismatched female bone-marrow-transplantation patients, a female control, and a male control. Hippocampal cells containing a Y chromosome were present up to 6 years post-transplant in all three patients. Transgender neurons accounted for 1% of all neurons; there was no evidence of fusion events since only one X chromosome was present. Moreover, transgender astrocytes and microglia made up 1~2% of all glial cells.

4.2. Spinal Cord

It is recognized that injured mammalian central nervous system (CNS) induces connective tissue invasion and glial scar formation at the surface facing the connective tissue. Scar formation by reactive astrocytes is considered to protect the CNS environment from the surrounding connective. Astrocytes function to close the open lesion, demarcating the CNS tissue from the surrounding connective tissue compartment. To elucidate early cellular events involved in the outgrowth of regenerating axons and astrocyte migration and/or elongation of astrocytic processes into implanted alginate at the borders of transected spinal cord, Kataoka *et al.* resected two segments (Th7-8) and implanted alginate in the lesion [131]. As controls, collagen gel was implanted in place of alginate, or the lesion was left without implantation. Two and four weeks after surgery, many regenerating axons, some of which were accompanied by astrocytic processes, were found to extend from the stump into the alginate-implanted lesion. In the all non-implanted animals, large cystic cavities were formed at both interfaces with no definite axonal outgrowth into the lesion. In collagen-implanted animals, cavity formation was found in some rats, and regenerating axons once formed at the stumps did not extend further into the lesion. Astrocytic processes extending into alginate-implanted lesion had no basal laminae, whereas those found in control experiments were covered with basal laminae.

To improve neuronal cultures, Tian *et al.* modified HAc with poly-D-lysine (PDL) by the covalent bonding of lysine via surface hydroxyl groups to mimic coating PDL on the substrate [132]. A porous composite matrix of PDL and HAc was prepared by a freeze-drying technique to mimic the ECM of the brain. The compatibility of the hydrogel with brain tissue was evaluated in a traumatic brain injury model in the adult rat. The incorporation of PDL peptides into the HAc–PDL hydrogel allowed for the modulation of neuronal cell adhesion and neural network formation. Macrophages and multinucleated foreign body giant cells found at the site of implantation of the hydrogel in the rat brain within the first weeks postimplantation decreased in numbers after 6 weeks, consistent with the host response to inert implants in numerous tissues. The infiltration of the hydrogel by glial fibrillary acidic protein-positive cells (reactive astrocytes) and the contiguity between the hydrogel and the surrounding tissue were demonstrated by SEM.

To examine whether BMCs can improve the neurologic functions of complete spinal cord injury (SCI) in patients, Park *et al.* evaluated the therapeutic effects of autologous bone marrow cell transplantation (BMCT) in conjunction with the administration of granulocyte macrophage-colony stimulating factor (GM-CSF) in six complete SCI patients [133]. The BMCT in the injury site (1.1×10^6 cells/μl in a total of 1.8 ml) and subcutaneous GM-CSF administration were performed on five patients. One patient was treated with GM-CSF only. The follow-up periods were from 6 to 18 months. The results of the clinical follow-up are shown in Fig. 2.47. Sensory improvements were noted immediately after the operation. Sensory recovery in the sacral segment was noted mainly from 3 weeks to 7 months postoperatively. Significant motor improvements were noted 3–7 months postoperatively. Four patients showed neurologic improvements in their American Spiral Injury Association Impairment Scale (AIS) grades (from A to C). One patient improved to AIS grade B from A and the last patient remained in AIS grade A. Side effects of GM-CSF treatment such as a fever ($>38°C$) and myalgia were noted. Serious complications increasing mortality and morbidity were not found. The follow-up study with MRI 4–6 months after injury showed slight enhancement within the zone of BMT. Syrinx formation was not definitely found.

4.3. Peripheral Nerve

As the spinal cord travels through the intervertebral foramen, it sends 31 pairs of spinal nerves emerging from the foramen through intervertebral foramina distributed in the cervical, thoracic, lumber, and sacral regions. For instance, the sciatic nerve runs backwards down to the knee and bifurcates, giving rise to two nerves denominated "tibial" and "peroneal"; the former in turn forms the digital nerves, whereas the latter gives rise to the popliteal and saphenous nerves. As shown in Fig. 2.48 [134], a

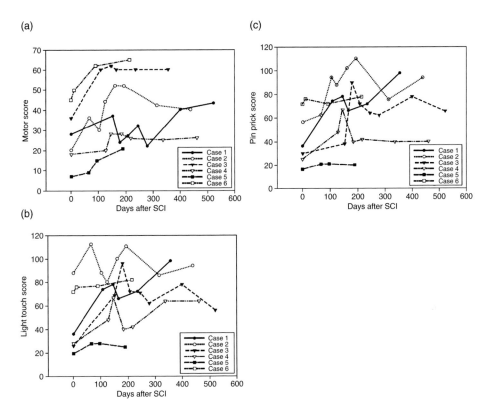

Fig. 2.47 Motor (a), sensory (Light touch) (b), and pin prick (c) score changes after treatment.

peripheral nerve consists of nerve fibers (axons) that may vary in size, being either myelinated or non-myelinated, and transmit impulses either to or from the nervous system. Peripheral nerves are frequently referred to as mixed nerves because they include both sensitive and motor fibers. The nerve's structural organization changes along it as a result of the repeated division and union of various nervous fascicles, giving rise to complex fascicle formations. A peripheral nerve may comprise thousands of axons, but their number varies from one peripheral nerve to another. In mammals, damage of a peripheral nerve induces a complex although reproducible sequence of histopathological events. After the nerve is sectioned, the first response takes place at the nerve fiber's distal end, a phenomenon known as "anterograde degeneration" (Wallerian degeneration); afterwards, the lesion becomes evident in proximal fibers, a phenomenon known as "retrograde neuronal degeneration"; and progressive degenerative changes occur at the cell soma leading to neuronal death. The regeneration capacity of peripheral nerves contrasts with the lack of capacity of the CNS. This regeneration capacity of the peripheral nervous system (PNS) has been attributed to the microenvironment provided by Schwann cells (SC) located at the distal nervous

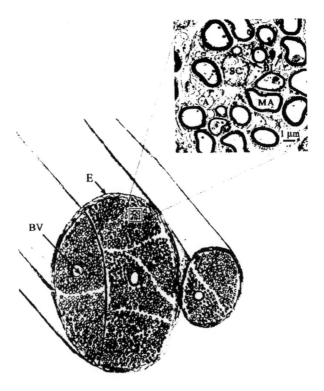

Fig. 2.48   Illustration and TEM image of a cross-section of a typical peripheral nerve. Axons (A) (*i.e.*, nerve cell extensions), support cells (*e.g.*, Schwann cells (SC)), blood vessels (BV), and an ECM are held within the cable by a sheath called the "epineurium" (E). Schwann cells wrap around individual axons to form myelinated axons (MA). The ECM includes the epineurium, basal laminae (BL), and the interstitial endoneurium. Basal laminae surround myelinated axons, and the interstitial endoneurium provides support between the basal laminae.

stump. Despite the regeneration capacity, the functional recovery following peripheral nerve damage is still deceiving, even after the immediate microsurgery to repair the damaged nerve. Clinical experience has demonstrated that functional recovery is particularly poor in relation of damage in big nerves, such as those emerging from the lumbar plexus, given that they travel through long distances to reinnervate again their target cells, previously denervated as a result of the damage. The slow regeneration rate, 1–3 mm/day, translates into months or even years.

Nerve injury is debilitating in both the PNS, where regeneration is limited to small gaps, and the CNS, where regeneration is essentially non-existent due to both physical (*i.e.*, glial scar) and chemical (*i.e.*, myelin proteins) inhibitors. Peripheral injuries can result from mechanical, thermal, chemical, congenital, or pathological etiologies. Failure to restore these damaged nerves can lead to loss of muscle function, impaired sensation, and painful neuropathies. When severed, peripheral nerves attempt to grow toward and re-establish a connection (*i.e.*, innervate) with the muscle or sensory site they previously controlled. The most successful cases of regeneration occur when severed axons can locate and grow through the remnants of the distal nerve cable still attached to the original innervation site. There are currently two clinical approaches for the repair of peripheral nerve gaps. When the nerve gap is small, the two injured ends are approximated and sutured. When the gap is longer than 4 mm, surgeon will be unable to repair by simple suturing, and autografts are commonly used as bridges. Here a segment of nerve obtained from the patient (typically, sural nerve) is grafted between the proximal and the distal stumps of the injured nerve. The transfer of normal donor nerve from an uninjured part of the body, which is the "gold standard" method for nerve restoration, is limited by tissue availability, risk of disease spread, donor site deformities, and potential differences in tissue structure and size. It is reported that less than 3% of patients who undergo median nerve repair directly or through median nerve autograft regain normal sensation, and less than 25% regain motor function at the wrist level. The use of autografts for the repair of peripheral nerve defects results in multiple surgeries and loss of function at the donor site. Over the past decade, end-to-end nerve repair and nerve grafts in adults have yielded an excellent result in 0–67% of the patients reconstructed [135, 136].

There exists, then, an impetus to develop another method of managing short nerve gaps. Weiss introduced the tubulization technique in 1940 as an alternative method for repairing and regenerating sectioned peripheral nerves that have lost some segments [137]. This may act as a bridge, providing support and contact guidance for extending axons and invading cells. Since then, a variety of artificial materials have been used for nerve guide (nerve tube, chamber, guide, or conduit) to bridge the gap between transected nerves and allow the concentration of axoplasmal exudates which contain the neurotrophic and mitogenic factors required to induce nervous growth, besides facilitating the orientation of the axonic bud. Regenerating nerve fibers are allowed to grow toward the distal nerve stump, whereas neuroma formation and ingrowth of fibrous tissue into the nerve gap are restricted. Trophic and tropic factors are necessary for the survival and growth of the damaged regenerating axons. Strategies to promote axonal extension through a site of injury include both the provision of nervous system growth factors and the implantation of substrates to support axon extension, such as cellular grafts. In general, however, the growth of axons is highly random and does not extend past the lesion site and into host tissue. Physically guiding the linear growth of axons across a site of injury, in addition to providing neurotrophic and/or cellular support, would help to retain the native organization of regenerating axons across the lesion site and into distal host tissue, and would potentially increase the probability of achieving functional recovery.

The materials first used for the tubular conduits include silicone, PTFE, and acrylic copolymers. Silicone had been used in the majority of studies, but it was permanent, and long-term tubulization of a nerve produced a localized compression with resultant decreased axonal conduction. Moreover, if these are placed for neural conduits for patients they have to be eventually removed. Despite numerous advances in neuroreparative techniques, functional outcome after surgical repair of injured peripheral nerves remained a significant challenge. Plastic and orthopedic surgeons are continually confronted with the limitations in surgical outcome for repair of nerves injured or resected in the PNS. A possible alternative in peripheral nerve repair is the use of bioabsorbable nerve guides. Because nondegradable nerve guides may even have a negative influence on the regenerated nerves, such as compression of the nerve due to a chronic foreign body reaction, nerve guides should be absorbed after functioning as a temporary scaffold for nerve regeneration. Furthermore, nerve guides should have a thin wall with little swelling during degradation to avoid nerve compression. However, with an internal diameter that is too large, it is more likely that fibrous tissue grows into the lumen of the nerve guide, thereby hampering nerve regeneration and maturation. After bridging the gap, maturation and remodeling of the newly formed nerve tissue need to occur. In this respect, the ECM type III and IV collagen as a part of the nerve tissue are of interest. Collagens constitute an important fraction of ECM involved in peripheral nerve functioning in vertebrates. Type III collagen is a component that is also present in peripheral nerve tissue. Jansen *et al.* showed that poly(DLLA-ε-CL) nerve guides yielded good nerve regeneration and types III/IV collagen distribution patterns as compared with the nonoperated control side, although the delineation of matrix was clearer in the control side [138]. In the numerous entubulation studies reported to date, some of the best results have been achieved in the PNS over short distances with nerve guides consisting of PGA, LA/CL copolymers, or collagen.

The first randomized prospective multicenter evaluation of a PGA conduit for nerve repair was reported by Weber *et al.* in 2000 [139]. The PGA was fashioned into a tight-weave mesh, rolled into a 2-mm diameter tube and crimped. The tight weave served to separate the internal milieu from the external tissues and retard fibroblast ingrowth while allowing the diffusion of oxygen; the crimping was performed to provide crush resistance and prohibit kinking while being bent. The study enrolled 98 subjects with 136 nerve transactions in the hand and prospectively randomized the repair to two groups: standard repair, either end-to-end or with a nerve graft; or repair with a PGA conduit. Two-point discrimination was measured by a blinded observer at 3, 6, 9, and 12 months after repair. There were 56 nerves repaired in the control group and 46 nerves repaired with a conduit available for follow-up. When the two groups as a whole were analyzed with respect to gap length, there was no statistically significant variation between them, as shown in Fig. 2.49. There was a difference, however, when the two groups were broken down into clinically important gap-length groups. For those nerves with a deficit of 4 mm or less than 4 mm, which is generally accepted to be the maximum gap length for digital nerves to be repaired by the end-to-end method with minimum tension, the conduit repair had significantly better results than did the standard end-to-end repair. When the distance between nerve ends is greater than 4 mm, the nerve must be either repaired under tension (with joint positioning providing a possible way to reduce tension across the repair) or reconstructed with a nerve guide. There was no statistically significant difference in the outcome of repaired nerves with 5- to 7-mm

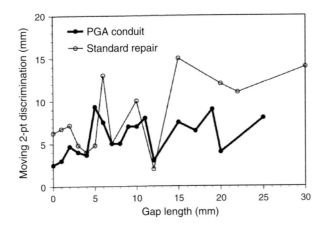

Fig. 2.49 Moving two-point discrimination, by type of repair and gap length.

deficits, more of which were grafted in the standard repair group. The lack of difference may result from the fact that in this subgroup, there were no crush or avulsion injuries in the standard-repair patients. In contrast, nearly one-third of the conduit repairs were performed for crush/avulsion injuries. Repair of crush injuries was reported to have a poor outcome than did all other mechanisms of injury. In cases in which the gap length was 8 mm or greater, all of the standard-repair nerves were reconstructed with a nerve graft. Comparison of the eight nerve-grafted repairs with 17 conduit repairs showed that the conduit group achieved significantly better sensory recovery. Although it was pointed out that the data presented in this study should be interpreted with caution because of the complicated design of the study, the lack of standardization, many incontrollable variables, and the potential evaluation problems [140], it seems that reconstruction of any nerve injury with less than 3-cm gap in a sensory nerve can be reconstructed better with a bioabsorbable conduit than with either a traditional nerve repair or an autologous nerve graft. NeuroGen™ nerve guide, a semipermeable collagen tube developed by Integra NeuroScience, has been used clinically for repairing peripheral nerve injuries.

In efforts to further improve regeneration, numerous studies have been done to fill guidance channel lumens with SCs, ECM molecules, neurotrophic factors, and/or scaffolds that increase surface area and provide both haptotactic and chemotactic cues. In development, axons in the nervous system are guided to their targets by a combination of substrate-mediated cues from ECM or cell adhesion proteins, and diffusible cues that include neurotrophic factors. The ECM proteins affect cell interactions during the development, maintenance, and regeneration of the nervous system, while neurotrophic factors support growth, differentiation, and survival of neurons in the nervous system. Laminin-1(LN-1) and nerve growth factor (NGF) promote both the survival and the growth of axons in the PNS. The SC migration into the lumen seems to be a critical step for successful regeneration. In fact, when SCs are pre-seeded in nerve guidance channels, the injured peripheral nerves regenerate at a faster rate and over longer distances. This may result in part from the production of cell adhesion molecules and neurotrophic factors from SCs. Among those are LN-1 and NGF. The linear character of nerve growth raises a question of which factors orient the outgrowing axons. Mechanisms that have been suggested to orient migratory cells and the growth cone may be determined by the configuration and physicochemical properties of their contact environment and nerve fibers may be guided by specific neurite-promoting macromolecules within the ECM during regeneration.

Because orientation effects on growing axons were reported to be exerted by collagen substrata when prepared for alignment of the collagen fibrils along a common axis [141], Yoshii *et al.* developed a nerve guide made of collagen filaments and assessed its effect in peripheral nerve regeneration [142]. Type I collagen filaments, 20 μm in diameter, were crosslinked with PEG diglycidyl ether (2%, 25°C, and 24 h) and additionally by UV irradiation. And 2000 (Fiber Group A) and 4000 (Fiber Group B) collagen filaments were used to make a 31-mm-long nerve guide. A crosslinked 30-mm-long collagen tube (Tube Group) of 1.5 mm in ID and 140 μm in the thickness was used as a control. The right and left sciatic nerves of Wistar rats were exposed and divided into the tibial and the fibular division. A 20-mm-long proximal portion of the tibial division of the right sciatic nerve was removed to create a 30-mm defect between the proximal stump of the right sciatic nerve and the distal stump of the left sciatic nerve. These stumps were sutured to the opposite ends of a collagen-filament nerve scaffold with three epineural sutures to bridge the proximal right sciatic nerve and distal left sciatic nerve. In the Tube Group, the stumps of the right proximal and left distal tibial divisions of the sciatic nerves were inserted 1 mm into the opposite ends of a collagen tube and were held in place with three epineural sutures. At 12 weeks postoperatively, the mean number of regenerated nerve axons was significantly large in Fiber Group B compared with Fiber Group A, whereas there was no large difference in the mean fiber diameter of regenerated myelinated axons between Fiber Groups A and B. In contrast, in the distal end of the collagen tube, no regenerated axon was found. To evaluate whether a graft made from acellular muscle enriched with cultivated SCs can bridge extra large gaps where conventional conduits usually fail, a well-established rat sciatic nerve model was used with an increased gap length of 50 mm [143]. The conduits consisted of freeze-thawed or chemically extracted homologous acellular rat rectus muscles and implanted SCs. Autologous nerve grafts were used for control purposes. After 12 weeks the control group showed superior results regarding axon counts, histologic appearance, and functional recovery compared with the muscle grafts. The chemically extracted conduits completely failed to support nerve regeneration. They were not stable enough to bridge longer nerve gaps with an expanded regeneration time.

The application of collagen tubes with different inner structures and different diameters was reported by Stang *et al.* for peripheral nerve reconstruction [144]. They bridged a 2-cm gap of rat sciatic nerve. Porcine collagen conduits consisting of types I/III were used to produce three types of conduits, either hollow or filled with an inner collagen skeleton. Two had an inner diameter of 2.6 mm, an outer diameter of 4 and 3.5 mm and a length of 20 mm. The third type of collagen tubes had a reduced lumen with an inner diameter of 1.5 mm. In case of SC/collagen conduits, a suspension of cultivated SCs in DMEM was transferred via a cannula into the lumen of the conduits immediately before implantation. The conduits were placed and sutured to the proximal and distal nerve stump in a sleeve technique. After 8 weeks the regeneration process was monitored clinically, histologically, and morphometrically. The amount of fibers within the collagen tube was clearly influenced by the presence of SCs. Grafts without SCs contained no detectable axons at all, whereas those with SCs showed a slightly better regeneration. Comparison of hollow tubes with SCs and tubes with inner skeleton and SCs demonstrated that SCs were not the only parameter to determine the axonal regeneration. Although both types of collagen tubes were equal in diameter and wall thickness, the tubes with inner skeleton and SCs revealed no regeneration at all. The inner skeleton impaired nerve regeneration independent of whether SCs were added or not.

Gamez *et al.* evaluated nerve tissue regeneration potentials of three tubes (inner diameter, 1.2 mm; outer diameter, 2.4 mm; wall thickness, 0.6 mm; and length, 15 mm) made of photocured gelatin: a plain photocured gelatin tube (model I), a photocured gelatin tube packed with bioactive substances (LN, FN, and NGF) coimmobilized in a photocured gelatin rod (model II), and a photocured gelatin tube packed with bioactive substances coimmobilized in multifilament fibers (model III) [145]. These tubes were implanted between the proximal and the distal stumps 10 mm off the dissected right sciatic nerve of 70 adult rats for up to 1 year. The highest regenerative potentials were found using the model III, followed by the model II. Markedly retarded neural regeneration was observed using the model I.

For the creation of longitudinally oriented channels in poly(2-hydroxyethyl methacrylate) (PHEMA) gels, Flynn *et al.* used a fiber templating technique [146]. First, PCL fibers were extruded and grouped into bundles. The PCL bundles were inserted into 4-mm inner diameter glass tubing, and the tubes were sealed with rubber septa. The appropriate quantities of 2-hydroxyl methacrylate (HEMA), water, ethylene glycol, initiator, and accelerating agent were mixed and injected into the fiber-filled polymerization tube following degassing. After polymerization, the PHEMA/PCL composites were placed in vials filled with acetone. Sonication of the vials resulted in the complete dissolution of the PCL, leaving longitudinally oriented fiber-free channels distributed in the PHEMA. Figure 2.50 demonstrates representative images of two types of scaffolds produced. The small channel scaffolds had 142 channels, with 75% of the channels in the 100–200 μm size range, while the large channel scaffolds had 37 channels, with 77% of the channels in the 300–400 μm range. Ahmed *et al.* described a novel method to create a long nerve gap contact guidance conduit from large cables of fibronectin (LFn) [147]. To form LFn-cables, a 1-ml syringe was used to pick up a concentrated solution of Fn (78 mg/ml) in 6 M urea, pH 7.4, and the solution was squeezed into a small beaker containing a solution of 0.25 M HCl, 2% $CaCl_2$, and pH 0.9. The FN fibers precipitated immediately and were manually drawn together to form LFn-cables. Large cables which were 14 cm long and 1.5 cm in diameter were made. As is seen in Fig. 2.51, SEM of freeze-dried LFn-cables showed a predominant fiber orientation throughout the cable, while the

Fig. 2.50 Representative (a) optical (8×) and (b) electron micrographs of small channel scaffolds and (c) optical (8×) and (d) electron micrographs of large channel scaffolds.

cross-sectional structure of these cables was porous, with pore sizes between 50 and 100 μm in diameter. Dried cables hydrated rapidly to 1.6 and 4.8 times their original length and diameter, respectively. Once hydrated, these cables had pores that ranged from 10 to 100 μm through which SCs and fibroblasts were able to grow *in vitro* and align with the axis of the fibrils by contact guidance. The porosity of the cable was enhanced by the natural dissolution of protein over a 3-week duration in culture with cells, such that 50–200 μm pores were observed.

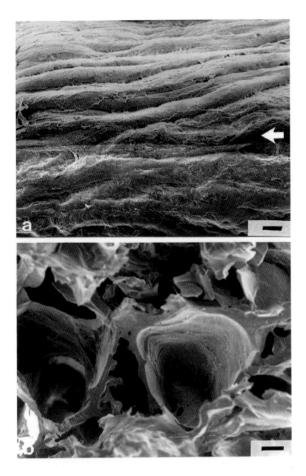

Fig. 2.51 Scanning electron microscopy of cross-sectional surfaces of freeze-dried LFn-cables. The fibrous alignment is indicated by the arrow in the freeze-dried specimen. Scale bar: 100 μm.

To confirm whether exogenous LN-1 and NGF facilitate peripheral nerve regeneration across gaps, Yu and Bellamkonda designed a bridge presenting LN-1 and/or NGF to the nerve gap [148]. The LN-1 was covalently coupled to the backbone of a thermoreversible hydrogel, agarose, through carbonyldiimidazole chemistry. The efficiency of LN-1 coupling to agarose was approximately 20%, and the resulting amount of LN-1 covalently coupled to agarose was 67 μm/ml of 1% agarose gel. For slow release of NGF from agarose gel, a lipid microtubule (LMT)-based system was used. The LMTs had an average length of 40 μm and a luminal diameter of 0.5 μm. The NGF was loaded at a protein:lipid ratio of 1% by hydration of 10 mg of lyophilized LMTs with 1 ml of PBS containing 100 μg of NGF. The NGF-loaded LMTs were embedded into agarose or LN-1-coupled agarose by mixing LMTs with agarose or LN-agarose solution. With the use of this method in which NGF was released by diffusion from the open ends of the LMTs, a sufficient amount of NGF could be loaded into the LMTs to allow for physiological concentrations (>5 ng/ml) of NGF to be released for at least 42 days *in vitro*. As it usually takes 1–2 months for peripheral nerve to regenerate across a 10-mm-long gap, the 42-day release of NGF was assumed to be appropriate for the specific application. They used semipermeable polysulfone tubes with an inner diameter of 1.6 mm, an outer diameter of 3.2 mm, and a length of 12 mm as guidance conduits in which agarose gel was filled. The MW cutoff of the tubes was 50,000 Da. The implants were separated into seven categories as described in Table 2.16. The proximal and distal stumps of sciatic nerve of adult rats were secured 10 mm apart in 12-mm long polysulfone conduits carrying the experimental or control gel formations or autografts, using a nylon suture. Figure 2.52 shows the success rate of regeneration which was determined as the number of conduits with continuous nerve cable containing myelinated axons divided by total number of conduits implanted for 2 months. As is seen, the PBS group had the lowest success rate of cable formation, whereas the AG–LN–NGF (LN-1-coupled agarose gels embedded in NGF–LMTs) and AU (autograft, which was resected from the rat sciatic nerve) groups had the highest success rate of cable formation. The total number of myelinated axons and the density of myelinated axons in AG–LN–NGF scaffolds were significantly higher in positive control and experimental groups (AU, AG–LN, AG–NGF, and AG–LN–NGF) than in the negative control groups (PBS, AG, and AG–PBS). Also, AG–LN–NGF scaffolds performed comparably to AU when functional measures that include the relative gastrocnemius muscle weight and the sciatic functional index were quantified.

A large number of investigators have transected a peripheral nerve and inserted the cut ends (stumps) of the nerve inside a tube fabricated from a large variety of materials, following the fundamental studies of Lundborg and coworkers [149–152]. This has led to the generation of a large body of data obtained by using a variety of biomaterials and devices, employed as either unfilled or filled tubes. To directly compare the data obtained under conditions that have been arbitrarily selected

Table 2.16

Description and notation for experimental and control groups

Group	Notation	Components
Control group I (negative)	PBS	Phosphate-buffered saline
Control group II (negative)	AG	Plain agarose gel in PBS solution
Control group III (negative)	AG–PBS	Agarose gel embedded in PBS–LMTs
Control group IV (positive)	AU	Autograft, which was resected from the rat sciatic nerve
Experimental group I	AG–LN	LN-1-coupled agarose gels
Experimental group II	AG–NGF	Agarose gel embedded in NGF–LMTs
Experimental group III	AG–LN–NGF	LN-1-coupled agarose gels embedded with NGF–LMTs

Abbreviations: AG, Agarose; AU, autograft; LMT, lipid microtubule; LN, laminin; NGF, nerve growth factor.

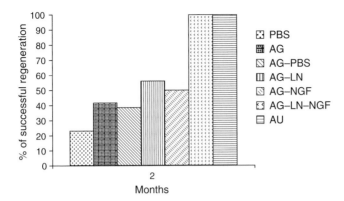

Fig. 2.52 Graph showing the percent success rate of regeneration in control and tissue-engineered scaffolds. The success rate of regeneration was determined as the number of channels with continuous nerve cable containing myelinated axons divided by total number of channels implanted for 2 months.

by independent investigators, Yannas and Hill established procedures by which tabulation data obtained using nonidentical protocols can be reduced to equivalent conditions (normalized) [153]. The experimental parameters that must be controlled to compare the regenerative activity of nerve conduit configurations include the animal species, the gap length, and the identity of the assay employed to evaluate the results. The frequency of reinnervation across a gap, reported as percent of nerves fitted with a nerve conduit that were bridged by myelinated axons, $\% N$, has been used by a large number of investigators as a dimensionless measure of synthesis of a nerve trunk. The discussion of gap length and animal species that will follow is based exclusively on the use of the frequency of reinnervation, $\% N$, as the assay of choice in studies with the rat and mouse models. Use of the frequency of reinnervation as the assay is based on the conclusion that a relatively small increase in the gap length bridged by a silicone tube is followed by a sharp drop in $\% N$, as shown in Fig. 2.53. In this S-shaped characteristic curve, the *critical axon elongation*, L_c, can be measured at the inflexion point; it is the gap length beyond which $\% N$ drops below 50%. The data in Fig. 2.53 show that L_c is 9.7 $\pm$ 1.8 and 5.4 $\pm$ 1.0 mm for the rat and mouse sciatic nerve, respectively. The performance of other nerve conduit devices is compared against the silicone standard by comparing their respective characteristic curves. It is hypothesized that the characteristic curves of various devices are identically S-shaped and differ from the curve for the silicone standard only by a horizontal shift along the gap-length axis. Accordingly, calling the length shift ΔL, we can use the magnitude of ΔL, a difference between the values of L_c of two devices, as a measure of the difference in regenerative activity between two devices. Generalizing along the same direction, the additional gap length (regenerative activity) conferred by a specific parameter X of a device, *e.g.*, the chemical composition of one of the tubes, can be estimated by comparing L_c for the device in the presence and the absence of the device parameter under study.

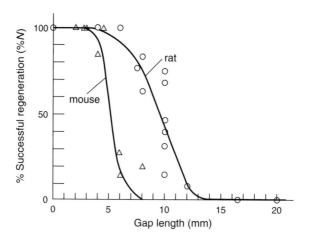

Fig. 2.53 The characteristic curve for an unfilled silicone tube used to bridge gaps of variable length in the sciatic nerve of rat and mouse.

The procedure described above was used to construct Table 2.17. The table lists over 25 entries, based on data from independent investigators of nerve regeneration using a conduit. Mouse data are not shown in Table 2.17; in general, mouse data follow rat data very closely. As can be seen from Table 2.17, insertion of the stumps in a nerve conduit or insertion of the distal stump alone had very significantly higher regenerative activity compared to an experimental configuration in which the stumps were either not inserted in the nerval conduit or the distal stump

Table 2.17

Regenerative activity (incremental axon outgrowth across nerve chamber) of several tubulated configurations compared to silicone tube standard[a]

Experimental Variable X, that affects outgrowth of axons across a gap	L_c in Presence of X (mm)	L_c in Absence of X (mm)	Shift Length, ΔL (mm)[b]
A. Effects of use of nerve chamber, insertion of distal stump and ligation of distal tube end			
Collagen tube vs. no tubulation	≥13.4	≤6.0	>7.4
Distal stump inserted vs. open-ended tube	11.7	≤6.0	>57
Distal stump inserted vs. ligated distal end	11.7	≤6.0	>57
B. Tube wall composition			
Silicone tube (standard)	9.7	(9.7)[c]	0
EVA copolymer[d] vs. silicone standard	≤11.0	(9.7)[c]	≤1.3
PLA, plasticized[e] vs. silicone standard	≥13.4	(9.7)[c]	≥3.7
LA/e-CPL[f] vs. silicone standard	≥13.4	(9.7)[c]	≥3.7
Collagen tube vs. silicone tube	≥13.4	8.0	≥5.4
Collagen tube degrading optimally vs. silicone tube	est. 25.5	(9.7)[c]	est. >15

Table 2.17 (continued)

Experimental Variable X, that affects outgrowth of axons across a gap	L_c in Presence of X (mm)	L_c in Absence of X (mm)	Shift Length, ΔL (mm)[b]
C. Tube wall permeability			
Cell-permeability vs. impermeability	≥19.4	13.4	≥6.0
Cell-permeability vs. protein-permeability	≥11.4	13.4	≥3.9
Protein-permeability vs. impermeability	7.5	8.9	−1.4
D. Schwann cell suspensions			
Schwann cell suspension vs. PBS	21.4	≤14.0	≥7.4
E. Tube filling: solutions of proteins			
bFGF vs. no factor	14.3	≤11.0	>3.3
NGF vs. cyt C[g]	11.1	10.4	0.7
aFGF vs. no factor	15.5	≤11.0	≥4.5
F. Tube filling: gels based on ECM components			
Fibronectin vs. cyt C[g]	19.1	18.6	0.5
Laminin vs. cyt C[g]	18.6	18.6	0
G. Tube filling: insoluble substrates			
Collagen–GAG matrix vs. no matrix	16.1	≤11.0	≥5.1
Oriented fibrin matrix across gap vs. oriented matrix only adjacent to each stump	15.5	≤11.0	≥4.5
Early forming vs. late-forming fibrin matrix	15.5	12.5	3.0
Axially vs. randomly oriented fibrin	11.4	≤6.7	≥4.7
Rapidly degrading CG matrix (NRT[h]) vs. no CG matrix	≥13.4	8.5	≥4.9
Axial vs. radial orientation of pore channels in NRT[h]	≥13.4	10.0	≥3.4
Laminin-coated collagen sponge vs. no laminin coating	11.7	≤6.0	>5.7
Polyamide filaments[i] vs. no filaments	18.4	≤11.0	≥7.4
NRT[h] in collagen tube vs. silicone tube[j]	>25	7.7	>17.3

Refer to the original work for the references.

[a] Data from rat sciatic nerve. Regenerative activity of a configuration is expressed in terms of the critical axon elongation, L_c. (Adapted by permission of author and publisher from Yannas, IV. *Tissue and Organ Regeneration in Adults*. New York: Springer, 2001.)

[b] Regenerative activity increases with the shift length, ΔL, the difference between the L_c value for the test configuration and the internal control (or the unfilled silicone tube standard). Standard: silicone tube, $L_c = 9.7$.

[c] Parentheses indicate use of the value for the silicone standard (used in the absence of internal control data to compile the entry in the table).

[d] EVA, ethylene-vinyl acetate copolymer.

[e] PLA, poly(lactic acid).

[f] LA/ε-CPL copolymer, copolymer of lactic acid and ε-caprolactone.

[g] Cyt C, cytochrome C.

[h] NRT, nerve regeneration template, a graft copolymer of type I collagen and chondroitin 6-sulfate, differing from dermis regeneration template by a higher degradation rate and an axial (rather than random) orientation of pore channel axes along the nerve axis.

[i] Eight polyamide filaments, each 250 μm in diameter, placed inside the silicone tube.

[j] Obtained by implanting the experimental tube in the cross-anastomosis (CA) surgical procedure.

was left outside or was ligated, respectively. Bioabsorbable tube, based on each of two synthetic polymers, plasticized PLA and a copolymer of lactic acid and CL, as well as a natural polymer, type I collagen, induced significantly greater outgrowth of axons than nondegradable tubes based on an ethylene-vinyl acetate copolymer (EVA). The regenerative activity of type I collagen tubes was maximal at an intermediate level of the degradation rate, corresponding to a half-life of about 3 weeks. Also, cell-permeability conferred very significant regenerative activity compared to the impermeable tube, but a protein-permeable tube lacked such relative activity. Highly variable regenerative activity resulted following use of several tube fillings. Suspensions of SCs showed very significant regenerative activity as did also use of solutions of acidic fibroblast growth factor (aFGF) and bFGF but not solutions of NGF. Use of ECM macromolecules, such as collagen, LN, and FN, in the form of solutions or gels had no significant activity; furthermore, when gel concentrations exceeded certain levels, negative activity was noted. Several insoluble substrates showed very significant regenerative activity. Very active substrates included highly oriented fibrin fibers and axially oriented polyamide filaments. Other very active substrates included a family of highly porous ECM analogs characterized by specific surface that exceeded a critical level, axial orientation of pore channel axes, and a sufficiently rapid degradation rate.

   To evaluate techniques for obtaining reproducible data on the recovery of nerve function and to understand the events during recovery, Meek *et al.* studied the recovery of nerve function using a combination of functional tests (walking-track analysis and video analysis) and electrophysiological measurements (withdrawal reflex, nerve conduction velocity, and electromyography) [154]. The results were compared with the opposite, normal side as well as with a control group that was not operated on. They bridged a 15-mm gap in the sciatic nerve of rats with a bioabsorbable p(DLLA-ε-CL) nerve guide. Sensory nerve function was found to recover as measured by electrostimulation. Motor nerve function partly recovered but electromyograms remained abnormal throughout the study. They concluded that functional reinnervation by regenerating axons occurred after bridging a 15-mm nerve gap with the absorbable nerve guide, but the walking patterns remained abnormal.

5. MAXILLOFACIAL SYSTEM

The homeostasis and the maintenance of teeth along with the periodontal structures that anchor them into the alveolar bones of jaw deteriorate progressively both with age and with metabolic maladies such as diabetes and osteoporosis. The oral cavity offers distinct advantages to tissue engineering because of ease of observation, accessibility, and large treatment effects even for small-sized defects, compared with other parts of the body. Indeed, the techniques of guided tissue regeneration (GTR) for periodontal tissues and guided bone regeneration (GBR) for alveolar ridge have been clinically used for patients since 1980s.

5.1. Alveolar Bone and Periodontium

Periodontium comprises connective tissues surrounding and supporting the teeth in the jaws, including cementum, periodontal ligament, alveolar bone, and gingival. For the proper placement of dental implants, sufficient alveolar bone volume should be available. If not available, alveolar bone augmentation should be performed either before or in conjunction with implant placement. As it is often difficult to distinguish the regeneration of lost periodontal tissues from alveolar ridge augmentation, the two tissues will be discussed here in combination.

In the beginning of the 1980s, Swedish researchers found that regeneration of periodontal tissues and alveolar ridge was significantly promoted when space was provided to facilitate new tissue ingrowth, and epidermal down-growth was inhibited by placing a membrane in the defect where tissue regeneration was required. Specifically, the scaffold membrane applied to the tissue defect serves as a substrate for fundamental cellular activities, ideally encouraging migration and matrix formation for a specific cell type while preventing the interference of non-desirable cells. Since then, a large number of clinical trials have been performed to prove the efficacy of the new techniques which are called GTR/GBR. The techniques are designed to facilitate tissue regeneration and healing by directing periodontal precursor cells from adjacent bone edges into the intercalary space. In practice, a membrane barrier is placed around an osseous defect to create a secluded space and retard the immigration of connective tissue cells, thereby ensuring that the intercalary is repopulated with tissue-forming cells. Thus, the design criteria for GTR/GBR devices include biocompatibility, cell/tissue occlusion, space maintenance, tissue integration, and ease of the membrane manipulation. Unless a space providing GTR/GBR device is present, limited or no regeneration of alveolar bone will be observed in supraalveolar periodontal defects and in alveolar ridge defects following gingival flap surgery. Insignificant regeneration of alveolar bone is also seen when the device inadvertently had collapsed or was compressed into the defect, or when the space underneath the membrane during surgery had been filled with slowly resorbing or non-resorbable biomaterial.

In 1984, however, Karaki *et al.* found that osteogenesis in a periodontal environment proceeded in the presence of space provision without strict provisions for gingival tissue occlusion [155]. This concept of space provision without tissue occlusion was supported by more recent studies. In a rat calvaria model, non-resorbable, ePTFE barrier devices with a porosity of 20–25 and 100 μm increased the rate of osteogenesis compared with less porous devices (<8 μm) [156]. Furthermore, a study with a macroporous ePTFE membrane [300 μm pores, 0.8 mm apart] showed that the macroporous ePTFE device was considerably more clinically effective than an occlusive device [157]. None of the sites receiving the macroporous device experienced wound failure in contrast to 50% of the sites receiving the occlusive ePTFE device.

Wikesjo *et al.* evaluated the importance of cell or tissue occlusion for GTR outcomes [158]. They created 5–6-mm, supraalveolar, periodontal defects around the mandibular premolar teeth in dogs. Space providing ePTFE membranes, with (macroporous) or without (occlusive) 300-μm laser-drilled pores, 0.8 mm apart, were implanted to provide for GTR. The gingival flaps were advanced to cover the membranes and sutured. All defect sites of the animals euthanized at 8 weeks post-surgery, irrespective of membrane configuration or history of membrane exposure and removal, exhibited substantial evidence of periodontal regeneration including a functionally oriented periodontal ligament. Alveolar bone regeneration for animals receiving occlusive and macroporous ePTFE membranes was averaged (3.2 ± 1.1 versus 2.0 ± 0.4 mm). Cementum regeneration was enhanced in defect sites receiving the occlusive ePTFE membrane compared to the macroporous membrane (4.7 ± 0.4 versus 2.3 ± 0.2 mm). Based on these results the researchers concluded that tissue occlusion does not appear to be critical determinant for GTR. However, tissue occlusion may be a requirement for optimal GTR therapy. In clinical practice, macroporous devices for GTR/GBR must use bioabsorbable biomaterials because the porous nature of the device results in considerable tissue integration, which may make surgical removal of a non-resorbable device complicated and result in considerable morbidity. To circumvent the need for surgical removal, natural products like collagen-based devices—dura mater, oxidized cellulose, laminar bone—and synthetic bioabsorbable devices based on PLA, PGA, and their copolymers have been developed and marketed for the treatment of patients. While bioabsorbable devices do not require a second surgery, they commonly present limitations including space provision, early/late resorption, and adverse inflammatory reactions in the resorption process, resulting in fragmentation and associated foreign-body reactions. Tatakis et al. reported abscess formation including macrophages in over 50% of a group of patients treated with GTR using a PDLLA-based device [159].

The principle of GBR has been successfully applied to augment resorbed alveolar crests in humans, but bone regeneration is often unsuccessful. This may result from membrane collapse and from the subsequent leak of space maintenance for bone ingrowth. To resolve these problems, titanium-reinforced membranes and membrane-supporting devices such as mini-screws or pins and graft materials have been investigated, but unsuccessful bone regeneration has still been observed. Although the regenerative potential depends on the native osteogenic activity of the sites concerned, GBR membranes cannot enhance osteoblast proliferation, migration, or bone matrix synthesis. An effective means to address the disadvantage of GBR membranes is to use BMP which has a high osteoinductive capacity. Lee et al. studied the effect of rhBMP-2-loaded PLLA/TCP membrane on bone augmentation in rabbit calvaria bone [160]. For fabrication of hemispherically shaped membranes, PLLA/TCP(50/50) solutions were cast on a hemispherical metal mold and dried under vacuum to remove solvent. The dome membrane had a height of 4 mm, a base diameter of 8 mm, and a thickness of 0.5 mm. A horizontal 4-mm rim extended from the base of the bone. The inner concave face of the dome membrane was soaked with a solution of rhBMP-2 so as to contain 5 μg of rhBMP-2 in a membrane. The external cortical plate of New Zealand white rabbits was indented circularly and the demarcated external cortical bone was removed using a round bur.

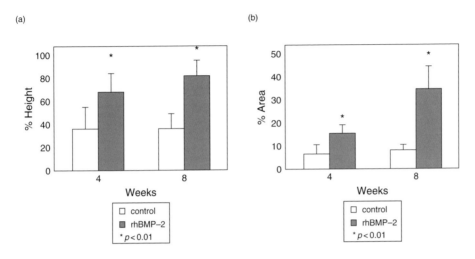

Fig. 2.54 New bone height (a) and area (b) in relation to the total submembranous space available.

The dome-shaped membranes were properly fitted over the circular calvarial defects and then tightly fitted to the defect rim using a membrane fixation pin. Figure 2.54 illustrates the new bone height and area at 4 and 8 weeks in relation to the total submembranous space available. The test group with loaded rhBMP-2 shows significantly increased bone height and area than the control group without rhBMP-2. In the control group, bone height and area did not differ significantly at 4 and 8 weeks, whereas in the test group, bone height and area were significantly greater at 8 weeks. In the 8-week specimens, the original dome shape of the membrane was well conserved, but marginal erosion of the membrane was demonstrated in both groups. In the control group, the new bone remained minimal and localized just above the former calvarial surface, whereas in the test group most of the specimens showed new bone extending to the top of the available space under the membrane.

Wikesjo *et al.* evaluated regeneration of alveolar bone and periodontal attachment, and biomaterials reaction following surgical implantation of a space-providing, macroporous glycolic acid–trimethylene carbonate copolymer (GA–TMC Cop) membrane combined with a rhBMP-2 construct in a discriminating onlay model [157]. They created routine supraalveolar periodontal defects at the mandibular premolar teeth in 9 beagle dogs and used bone-shaped GA–TMC Cop membranes (100–120-μm pores) with rhBMP-2 (0.2 mg/ml) in a bioabsorbable hyaluronan carrier or the membrane with hyaluronan alone. Jaw quadrants receiving the membrane alone experienced exposures at various time points throughout the study, while jaw quadrants receiving the membrane/rhBMP-2 combination remained intact although one site experienced a late minor exposure. Newly formed alveolar bone became incorporated into the macroporous membrane in sites receiving rhBMP-2.

The space-providing membrane was resorbed within 24 weeks and associated with only a minimal inflammatory reaction. Regeneration of alveolar bone at 24 weeks was significantly enhanced in sites receiving the membrane/rhBMP-2 combination compared to control (4.6 ± 1.3 versus 2.1 ± 0.4 mm). Cementum regeneration was significantly increased in sites receiving GA–TMC Cop/rhBMP-2 compared to those receiving the GA–TMC Cop control at 8 weeks (2.1 ± 1.0 versus 0.7 ± 0.4 mm), but similar amounts of cementum regeneration were observed for the GA–TMC Cop/rhBMP-2 and the GA–TMC Cop control treatment at 24 weeks (1.3 ± 0.5 versus 1.7 ± 1.0 mm). Functionally oriented fibers were rare observations in sites receiving GA–TMC Cop/rhBMP-2. In contrast, sites implanted with the macroporous GA–TMC Cop device characteristically showed functionally oriented fibers. Ankylosis was significantly increased in sites receiving GA–TMC Cop/rhBMP-2 compared to sites receiving the GA–TMC Cop control. Root resorption of surface erosion character was observed in all animals. Furthermore, Wikesjo *et al.* evaluated the potential of the space-providing, macroporous ePTFE device to define rhBMP-2-induced alveolar bone formation using an absorbable collagen sponge as a carrier of rhBMP-2 and a discriminating onlay defect model [161]. It was shown that rhBMP-2 significantly enhanced regeneration of alveolar bone in conjunction with the macroporous ePTFE device for GTR.

There are studies in which a bioabsorbable material is filled in a periodontal defect, instead of placing a space-providing membrane in the defect making the inside open. Nakahara *et al.* placed dehydrothermally crosslinked collagen sheet in a periodontal defect [162]. Three-walled alveolar bone defects ($3 \times 4 \times 4$ mm^3) were created bilaterally in edentulous regions made mesially in the canines in both the maxilla and the mandible of 9 beagle dogs. To evaluate the effects of bFGF on the regeneration of periodontal tissues, they incorporated bFGF into the collagen sponge sheet prepared with a freeze-drying technique. To facilitate the sustained release of bFGF from the sheet, bFGF was included in absorbable microspheres made from negatively charged gelatin. The bFGF–gelatin microspheres were physically incorporated in the collagen sponge sheet. A 5-mm-thick collagen sheet, either with or without bFGF was randomly implanted in the bilateral defects and the mucoperiosteal flaps were sutured over the created bone defects. Throughout the observation period of 4 weeks, no ankylosis was seen in either group. However, epithelial down-growth into the bone defects (three of six specimens) and root resorption caused by the newly formed connective tissue (two of six specimens) were recognized in the control group 4 weeks after surgery. Many capillary vessels were observed around the residual gelatin microspheres even 4 weeks after surgery. At 2 and 4 weeks, the number of capillary vessels and the area of the new bone formed in both groups were highly significant statistically, as presented in Fig. 2.55. At 2 weeks, mononuclear cells accumulated at the exposed root surface and formed a dense, new cementum-like matrix. Four weeks after surgery, a new cementum-like structure (2.4 ± 0.9 mm) also was observed on the root surface and functional recovery of the periodontal ligament was indicated in part by the perpendicular orientation of regenerated fibers. A noticeable finding in this study was that, in the control group where crosslinked collagen sponges

(a)

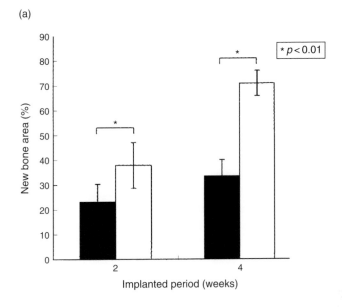

(b)

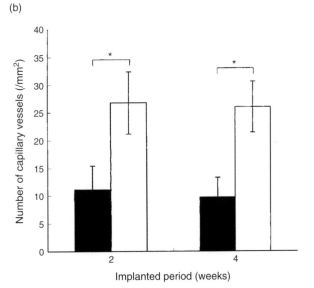

Fig. 2.55 The areas of new bone formation (a) and number of capillary vessels (b) 2 and 4 weeks after surgery. Each experimental group consisted of six defects from three beagle dogs. The control group was implanted with sandwich membrane without bFGF (solid columns); the bFGF-treated group was implanted with the sandwich membrane that contained 100 μg of bFGF (open columns).

alone were implanted without bFGF in the periodontal defects, epithelial down-growth and root resorption occurred, the defects being filled with connective tissue, in marked contrast with implantation of collagen sponges containing bFGF in a sustained-release system.

For maxillary sinus augmentation with the simultaneous placement of dental implants, Rodriguez *et al.* combined the use of PRP and deproteinated bone xenograft [163]. They theorized that PRP, through its angiogenic properties, would lead to an improved revascularization and faster consolidation of the graft when combined with an osteoconductive material. Fifteen patients with less than 5 mm of residual alveolar height in the posterior maxillary alveolus underwent a total of 24 maxillary sinus augmentations. Before surgery, PRP was prepared by obtaining the required amount of autologous blood and then mixed with the bone xenograft. Dental implants were inserted simultaneously in the grafted sinuses. If there were perforations of the sinus mucosal membrane, these were approximated and covered with autologous fibrin glue prepared from the centrifuged serum with bovine thrombin and calcium chloride. Following this, the bovine graft mixed with PRP was inserted in the maxillary sinus surrounding the dental implants. The bone biopsy from the patients showed evidence of viable new bone formation in close approximation to the xenograft. The bone density of the grafted bone was similar to or exceeded the bone density of the surrounding native maxillary bone.

Several studies have reported on PRP-enhanced bone regeneration, but some studies were not designed with matched pairs. Under these conditions, the evaluation of the true effect of PRP on osseous healing is likely to be complicated by variables such as genetics, age, hormones, and function. Especially in the late phase, such variables may ultimately influence treatment outcome regardless of the effect of the PRP component. Therefore, Choi *et al.* undertook a histologic examination of grafted bone at 6 weeks postoperatively to evaluate new bone formation during the early phase, and both sides of each dog's mandible were used to provide matched pairs [164]. The mandibular premolar teeth had been bilaterally extracted previously, and the ridges were allowed to heal for 3 months. After this period, continuity resection was performed on both sides of the mandible. One defect (the PRP group) was reconstructed with the original particulate bone mixed with PRP. As a control, the contralateral defect (non-PRP group) was reconstructed with the original particulate bone alone. Biopsies after 6 weeks showed lower levels of bone formation in the PRP group than in the non-PRP group, and fluorescence microscopy revealed a delay in the remodeling of grafts loaded with PRP. These findings suggest that the addition of PRP does not appear to enhance new bone formation in autologous bone grafts.

Fennis *et al.* performed a study on the use of autologous scaffolds, particulate cortico-cancellous bone grafts, and PRP in goats. Twenty-eight goats underwent a continuity resection of the mandibular angle. In all the goats primary reconstruction was carried out using specially designed pre-shaped osteosynthesis plates and mono-cortical screws [165]. Trays made from the cortices perforated with a round burr with a diameter of 2.3 mm were filled with particulate bone grafts taken from the anterior iliac crest. They were used to bridge the defect in 14 goats. Both groups were divided

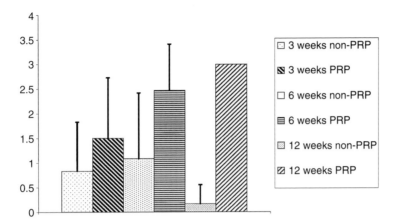

Fig. 2.56 Graphic illustration of bone union.

into three subgroups that were sacrificed after 3, 6, and 12 weeks. Histological and histomorphometric evaluation revealed that the use of PRP enhanced the bone healing, as shown in Fig. 2.56. This effect was statistically significant and particularly visible in the 6- and 12-week groups.

5.2. Temporomandibular Joint

The temporomandibular joint (TMJ), a synovial, bilateral joint, is formed by the condyle of the mandible with the glenoid fossa and articular eminence of the temporal bone, as illustrated in Fig. 2.57 [166]. A TMJ disc (or meniscus) is a specialized fibrocartilaginous tissue located between the mandibular condyle and the glenoid fossa–articular eminence of the temporal bone. The TMJ disc serves to increase congruity between these surfaces and thus distribute load over an increased contact area, absorb shock, assist in joint lubrication, divide the joint into separate compartments for different movements in the joint, and limit how far the condyle is compressed into the temporal bone. The TMJ is surrounded by a capsule, which is attached to the disc around its periphery near its attachment to the condylar neck.

The signs and symptoms associated with temporomandibular disorders (TMDs) include limited mouth opening, pulling of the jaw to the left or right with jaw movement, clicking, and locking, all of which can be the result of a pathology involving the TMJ disc. The TMDs often lead to pain in regions outside of the immediate joint area, such as recurrent headaches and neck pain. The most common TMDs are pain dysfunction syndrome, internal derangement and osteoarthrosis, and traumas. The pathology of particular interest to TMJ disc tissue engineers is the internal derangement which is defined as an abnormal relationship of the articular disc to the mandibular condyle, fossa, and articular eminence (disc displacement) or can also be defined as a localized mechanical interference of smooth joint movement, of which

(a)   (b)

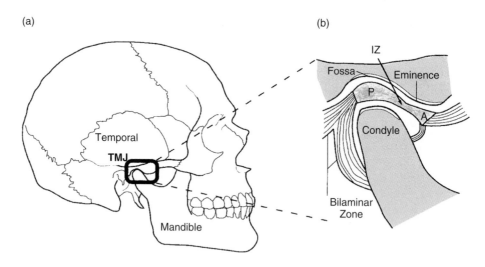

Fig. 2.57   (a) Side view of the human skull, emphasizing the bones of the TMJ. (b) Sagittal section of the TMJ—enlarged highlighted region of (a)—denoting the positioning of the TMJ disc between the condyle and the fossa–eminence. The surfaces of the condyle and the fossa–eminence are lined with articular cartilage. Regions of the TMJ disc are denoted by P (posterior band), IZ (intermediate zone), and A (anterior band).

disc displacement is only one type. The leading course of internal derangement may be macrotrauma to the joint due to impact or hypertension. Patients afflicted with a severe TMD can experience a significantly reduced quality of life, affecting both personal life and work, as everyday activities such as eating, talking, yawning, and laughing can become painful. It is reported that 20–25% of the population have symptoms of TMD, compared with only 3–4% of the population who actually seek treatment for their symptoms [167]. Almost 70% of patients with TMDs suffer from disc displacement, which emphasizes the importance of addressing pathologies of the TMJ disc in the effort to treat patients with TMDs.

When the meniscus is damaged by trauma, disease, or degeneration, reengineering the entire condylar-ramus unit is the best approach. However, because the disc is avascular, it is unable to heal after injury. Discectomy with a Proplast-Teflon disc replacement had been shown to be disastrous, eliminating such an implant as an option. About 85% of discectomies are successful under ideal circumstances, but success is not characterized by asymptomatic function. The TMJ has been shown to adapt to internal derangement in several instances, but the adaptation and recession of symptoms may only be temporary because the underlying disease still exists. Given that discectomy is indicated only when the disc is too damaged to repair, replacing the damaged disc with a healthy native disc equivalent would place the disc below the severity range of discectomy, and traditional repositioning methods could then be used. Because results of traditional repositioning methods have produced better results than discectomy, better results would therefore be expected with such a procedure using a native disc equivalent than with discectomy or disc removal, which is a drastic operation exhibiting a debatable success rate.

Tissue engineering offers a potential solution to replace the irreversibly damaged disc and alleviate the need for discectomy. The goal of TMJ regeneration is to rebuild the subchondral bone, cortical plate, and overlying hyaline articular cartilage to provide a neofunctional unit. Designing a patient-specific osteochondral composite condyle–ramus construct with precisely specified pore size, geometry, and channel orientation may maximize bone and cartilage ingrowth. Although a tissue-engineered disc would be most suitable for reconstruction of the TMJ after full or partial discectomy, tissue engineering of the TMJ disc is still in its infancy, and there exists a gap between the tissue engineering community and the TMJ characterization community. Figure 2.58 shows a schematic of an approach to tissue engineer the TMJ disc, proposed by Detamore and Athanasiou [166]. They emphasize that the studies attempting to tissue engineer the disc should be aware of the wealth of cellular, biochemical, and

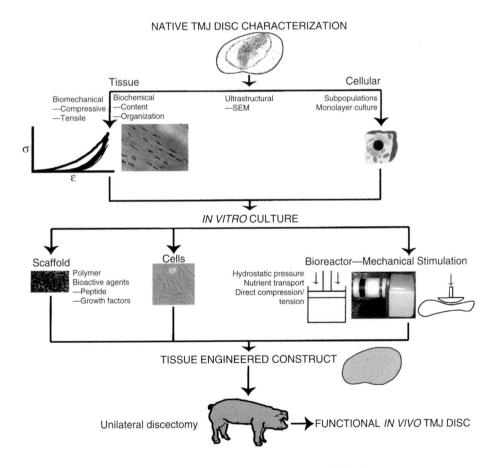

Fig. 2.58 Schematic of an approach to tissue engineer the TMJ disc.

mechanical characterization studies that have been performed for the disc, as they are literally the design and validation criteria for tissue engineering efforts.

5.3. Enamel and Dentin

Regeneration of an entire tooth, including the pulp, remains a very distant, challenging goal. One of the less distant goals of dental tissue engineering may be the regeneration of selected dental tissues such as dentin and cementum. The former could be of value in regeneration of the tooth structure lost due to caries, while the latter can restore the tissue due to periodontal disease. Some of the tissues lost by periodontal disease have already been clinically regenerated by GTR/GBR techniques. However, tissue engineering of enamel and dentin is still in its infancy and most of the associated studies focus on basic research such as cell culture.

The human tooth is protected by enamel of 1–2 mm thickness that is composed of HAp crystals. In early caries lesions, acid-forming bacteria cause microscopic damage to the enamel, creating cavities that are less than 50 μm deep. Such cavities cannot be repaired by simple setting of restorative materials because these do not adhere perfectly to the enamel owing to differences in chemical composition and crystal structure. Yamagishi *et al.* showed that a synthetic material could reconstruct enamel without prior excavation, in a process that not only repairs early caries lesions but can also help to prevent their reoccurrence by strengthening the natural enamel [168]. They prepared a mother solution by mixing 35% H_2O_2 aqueous solution with 85% solution of H_3PO_4 at a volume ratio of 4:1. A paste was made by adding the mother solution to fluorized-apatite powder which was prepared by mixing Ca-deficient HAp powder with NaF solution. The precipitate was filtered and dried. In the treatment, a small amount of mother solution was brushed on the affected part, and the paste was quickly applied before the solution dried without any mechanical removal. The durability of the re-grown layer was tested using a brushing machine. Figure 2.59 shows SEM photograph of the longitudinal section

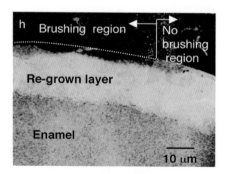

Fig. 2.59 SEM photograph of longitudinal section of paste-treated tooth after 10,000 times of brushing.

of paste-treated tooth after 10,000 times of brushing. The loss of the re-grown layer by brushing was less than 10%.

5.4. Mandible

The loss of the mandibular arch after tumor excisions or traumatic injuries leaves the patient not only with a remarkable deformity of the face but also with a mandible that does not function for chewing, swallowing, or speaking. In such a case, reconstruction has often been performed by bone graft or implantation of prosthesis. Reconstruction by vascular or nonvascular autologous bone grafts has been considered to be reliable, but this treatment has disadvantages including donor site morbidity and insufficient supply. In the late 1990s, distraction osteogenesis was applied to mandibular regeneration, but this procedure is time consuming and it is difficult to reproduce the complicated shape of the mandible by itself. The final target for the technology of mandibular reconstruction is to regenerate physiological bone that makes dental implants or accommodation of dentures possible. In this connection, mandibular reconstruction is an attractive target for tissue engineering.

Fennis *et al.* evaluated a method of mandibular reconstruction using PDLLA scaffold. Six goats underwent a continuity resection of the mandibular angle [169]. The defect was bridged with a pre-shaped PDLLA scaffold, filled with an autologous particulate bone graft from the anterior iliac crest, and fixed with two pre-shaped titanium plates. To accelerate bone healing, autologous PRP was mixed with the particulate bone graft. All goats had an uneventful healing. The osteosynthesis system withstood immediate loading for a period of 6 weeks until sacrifice. The particulate bone grafts within the PDLLA scaffold, which appeared to be narrowed, showed considerable resorption and replacement by fibrous tissue. In all goats, however, callus formation along the reconstructed segment was seen, providing bony continuity and maintaining the original contour of the reconstructed segment.

Autologous particulate cancellous bone and marrow (PCBM) that is rich in osteogenic progenitor cells and bone matrices has excellent properties as bone graft because it has full bone formation ability, and, in addition, spontaneous regeneration of donor sites is possible. However, PCBM does not, by itself, have structural strength and the ability to hold its desired shape and is not able to support the newly formed bone while it acquires enough strength to withstand external force. Furthermore, this framework should be preferably bioabsorbable and disappear from the implanted site after completion of bone repair. To address this issue, Kinoshita *et al.* developed a mesh manufactured from PLLA that can maintain high mechanical strength for a long period of time compared with other bioabsorbable materials [170, 171]. The PLLA monofilaments with diameter of 0.3 and 0.6 mm were woven into mesh sheets, which could be cut with scissors and was easily molded by heating up to 70°C. Figure 2.60 shows the PLLA mesh and tray used for mandible regeneration. After extensive preclinical studies with adult dogs [172], they started clinical studies using the PLLA sheet/tray and autologous PCBM since 1995 to establish a procedure for mandibular reconstruction engineering, as illustrated in Fig. 2.61. In

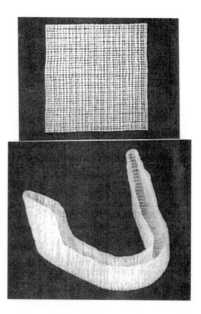

Fig. 2.60 PLLA mesh sheet and mandibular mesh tray.

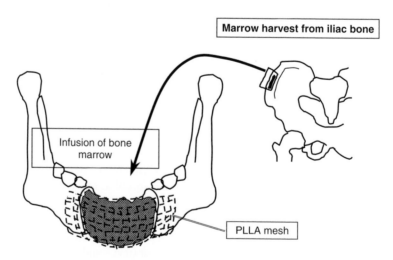

Fig. 2.61 Schematic drawing of mandibular reconstruction by means of PLLA mesh and particulate cancellous bone and marrow (PCBM). The PLLA mesh tray was adjusted to the shape and size of the bone defect with cutting by scissors and warming at about 70°C. The PLLA mesh tray was fixed to the residual bone with stainless steel wires and filled with PCBM taken from the iliac bone.

Table 2.18

Patients' characteristics and medical treatments (age: 15–75 years old [average: 45.2 years old])

Characteristics		Treatments	
Benign tumor	23	Marginal resection	21
Malignant tumor	11	Segmental resection	17
Cyst	4	Hemimandiblectomy	2
Osteomyelitis	2		
Total	40	Total	40

8 hospitals in Japan, 62 cases underwent mandibular reconstruction between 1995 and 2001 [173]. Mesh trays were used in 28 cases and mesh sheets in 6 cases. The PCBM was harvested from the iliac bone of patients and 10–40 g of PCBM was transplanted to individual patients. Table 2.18 shows the characteristics of 40 patients. The clinical results were evaluated as *excellent* when the area of osteogenesis, based on X-ray films 6 months after the surgery, was over two-thirds in comparison to right after operation. Results were evaluated as *good* when osteogenesis was less than two-thirds with no reconstruction required. All other results were graded as *poor*. And 40 cases (56%) were judged as being *excellent*, 17 cases (27%) as being *good* and 10 cases (16%) as being *poor*. As can be seen from the clinical outcomes given in Table 2.19, of the 40 cases falling under *excellent* and *good* evaluations, 32 showed absorption of the regenerated bone at less than 10%, 6 cases at 10–20%, and 2 cases at 20–30% absorption by X-ray findings 6 months after surgery. One case showed latent inflammation due to partial exposure of the PLLA tray, but it was successfully treated by excision of the related part of the tray. Twenty-two cases had dentures and four cases had dental implants in the regenerated bone. In cases having early dentures or implants, there was little bone absorption or advanced ossification. This method seems also to be able to provide a good facial relationship between the upper and the lower jaws, which is important for dental implantation or denture accommodation for long-term use.

Table 2.19

Long-term observation for 40 cases over 1 year for mandibular reconstruction with PLLA mesh and PCBM

Late inflammation resection	1	(2.5%)
Absorption of the reproduced bone (panoramic radiography)		
<10%	32	(80.0%)
10–20%	6	(15.0%)
20–30%	2	(5.0%)
Denture prosthesis	22	
Dental implant	4	

Kinoshita *et al.* further evaluated the effect of bFGF on regeneration of experimental mandibular defect, as this growth factor seemed to promote angiogenesis and bone regeneration which would be required for patients with poor blood circulation at the implanted site or reduced supply of osteogenic progenitor cells [174]. For sustained release of bFGF, this protein was incorporated into gelatin microspheres with various water contents. They were placed in defects created in the mandibular base of adult dogs. Twelve weeks after surgery, the area of bone formed was found to be wider in the groups with bFGF-incorporated microspheres than in the control group without bFGF, and active bone formation was observed around the bFGF-incorporated microspheres. The gelatin microspheres with water contents of 95 and 90% were absorbed 4 and 12 weeks after implantation, respectively, whereas those with water contents of 85 and 75% still remained at the implanted site 12 weeks after implantation.

5.5. Orbital Floor

Orbital floor reconstruction is usually carried out for a defect caused by facial trauma or tumor ablation. On the other hand, orbital floor is often fractured and fragmented into thin, small pieces 10–15 mm in size. These bone fragments descend due to the weight of the glove, making their reduction difficult. By reconstructing the bony wall or floor, herniation of orbital fat or entrapment of ocular muscle can be avoided. The final form, shape, and volume of the bony orbit can also be restored, thus leading to acceptable functional and aesthetic outcome. Reconstruction of the fractured orbital floor has employed various techniques, including grafting of the maxilla or iliac bone, nasoseptal cartilage, and the insertion of prostheses. Autologous bone grafts are biologically excellent, but they create a burden in regard to donor site and can hardly facilitate delicate shaping of the orbital floor. Silicone sheets were in wide use because of convenience, but they often failed in providing sufficient mechanical strength. In contrast, titanium miniplates have high strength, but are too rigid to fit each orbital floor shape. A disadvantage common to these non-absorbable implants is semipermanent, unavoidable foreign-body reactions. Resorbable implants like Gelfilm™ and Polyglactin 910™ are commercially available, but they are suitable only for repair of small defects because of their poor mechanical properties and rapid resorption.

A resorbable poly(*p*-dioxanone) (PDS) plate was once applied for reconstruction of orbital floor, but postoperatively unpredictable globe positions took place [175]. Baumann *et al.* revealed that PDS sheets with which most orbital floor or wall defects had been reconstructed were absorbed too fast to enable bone formation at the level of reconstruction [176]. Kontio *et al.* evaluated therefore the reconstruction capacity of orbital wall when poly-L/D,L-lactide(96/4) (PLDLA96) implants were used for large blow-out defects in 18 sheep, as the degradation time of the PLDLA96 is shorter than that of pure PLA *in vitro*, while strength properties are comparable to those of pure PLA [177]. The polymer was melt and extruded into continuous plates with a flat film die. The plates were reheated to 70°C and biaxially stretched into the shape of a bowl over a hemispherical steel die with a radius of 35 mm. The contralateral side, where the defects healed spontaneously, served as controls. At first, the implants were surrounded by elastic capsules, which gradually ossified. At 36 weeks, 60% were still visible and deformed but surrounded by bone. Light microscopy revealed a low grade inflammatory reaction. Shear strength decreased gradually and was not measurable after 16 weeks. Crystallinity increased steadily from 1.5 to 29.3% and molecular weight decreased from 49,000 to 4,190 Da. The final bony defect was smaller in the reconstructed sides than in the controls. The PLDLA96 implant provoked a local inflammation, which did not prevent bone healing. The deformation of the implant, however, indicated that this PLDLA96 plate was not suitable for orbital floor reconstruction.

Yamazaki *et al.* used PLLA meshes for repair of human orbital floor defects, together with PCBM [178], since the combination of PLLA mesh and PCBM has been shown to be effective in the reconstruction of mandibular defects [173], as described above. They treated 9 patients with orbital floor fractures using PLLA meshes shaped mostly to ellipses ranging from 22×27 to 10×15 mm^2 in size. The largest amount of PCBM (4 g) was needed for the patient to whom a mesh sheet of 22×27 mm^2 was applied. The surgical approach was performed by a skin incision in the eyelid for all patients. Displaced infraorbital borders were reduced and fixed with titanium miniplates. The PLLA mesh was trimmed with a cutter to the size suitable to the fracture site and curved to the shape of the orbital floor defects by heating, with their edges being smoothened by melting. After fitting the mesh to the orbital floor defect, small holes were made in the infraorbital edge, and the mesh sheet was immobilized with stainless steel wire. The necessary amount of PCBM was taken readily using a bone curet after simply making a skin incision of 2-cm length in the waist and making a small hole of 7–8 mm diameter in the cortical bone of the iliac bone. The collected PCBM was filled in the gaps of the PLLA mesh. Postoperative symptoms were extremely mild, and few patients complained of pain. The CT images showed new bone formation at the orbital floor 3 months after operation. In 6 patients, globes recovered in the normal position, as summarized in Table 2.20. In another patient who had the fracture in the furthest place of orbital floor, the low globe position improved from 3 to 1 mm. Diplopia was resolved in all patients. Examination of the visual field using the Hess screen showed clear improvements in all patients and showed almost symmetrical visual fields. No displacement of globe recurred for the longest follow-up of 61 months. It seemed probable that reduced globes were mainly supported by PLLA mesh sheets at an early stage postoperatively, effectively preventing postoperative globe displacement, and replaced by new bone formed on the meshes at a later stage. The PLLA mesh was more maneuverable and stable in the implanted site than a PLLA plate, probably because of penetration of blood vessels and fat into the mesh, resulting in interlocking. Inflammatory reaction was minimal, probably because of a balanced decrease in mechanical strength of PLLA. The surface of PLLA in contact with the tissue was large and circulation of blood was good, while the quantity of implanted PLLA mesh was small.

Table 2.20
Clinical findings during examination of patients with orbital floor fracture treated with PLLA mesh and PCBM

Patient				Preoperative				Postoperative				Follow-Up Duration (months)
Sex	Age	Side	Fracture	Diplopia	Globe Position	Restricted Eye Motility	Hypoestheia N. infraorbital	Diplopia	Globe Position	Restricted Eye Motility	Hypoestheia N. infraorbital	
1 F	23	L	blow-out	unknown	low	yes	yes	no	normal	no	no	51
2 M	13	R	blow-out	yes	low–3 mm	no	no	no	low–1 mm	no	no	49
3 M	56	L	orbital floor medial wall	yes	low–3 mm	yes	yes	no	normal	yes	yes	47
4 M	51	R	orbital floor medial wall	yes	low–3 mm	yes	yes	no	normal	no	yes	45
5 M	36	L	blow-out	yes	low–2 mm	yes	no	no	normal	yes	no	41
6 F	23	L	blow-out	yes	low–2 mm	yes	no	no	normal	no	no	37
7 M	64	R	orbital floor medial wall	yes	low–3 mm	yes	yes	no	normal	no	no	34
8 M	23	L	orbital floor medial wall	yes	low–2 mm	yes	yes	no	normal	no	no	6

6. GASTROINTESTINAL SYSTEM

6.1. Esophagus

Congenital or acquired esophageal disease that requires surgical resection and repair often is associated with high morbidity and a variety of potential postsurgical complications. Traumatic injury accounts for significant morbidity among children especially after the ingestion of caustic substances. Many of these children have esophageal stricture with subsequent need for patch repair or full circumference replacement. Multiple strategies to reconstruct pharyngogastric transit currently exist. Most commonly, the colon or the denervated stomach replaces the esophagus. Free jejunal grafts can also be used. Although these conduits have each served as acceptable substitutes, the choice of conduit still remains open to debate, and none exactly recapitulate esophageal architecture and function. In addition, they require an exchange in which some native tissue and its *in situ* function are lost to replicate a most important function. Options for esophageal replacement might be limited in some patients by previous resection or intestinal disorders.

Grikscheit *et al.* proposed to use a tissue-engineered esophagus for replacement of the abdominal esophagus [179]. Esophagus organoid units were produced by dissecting the abdominal esophagus without mesentery from Lewis rat pups or adults, which was cut into full-thickness 2×2-mm^2 sections after lengthwise opening along the antimesenteric border. Scaffold polymers were constructed from 2-mm-thick nonwoven PGA, which formed into 1-cm tubes and sealed with PLLA solution and coated with type I collagen. Esophagus organoid units were paratopically incorporated in the bioabsorbable polymer tubes, which were implanted in syngeneic hosts. Four weeks later, the tissue-engineered esophagus (TEE) was either harvested or anastomosed as an onlay patch or total interposition graft. And 8 Lewis rats were implanted with neonatal-derived organoid units and 3 were implanted with adult-derived organoid units. Implantation was achieved through a 1.5-cm upper abdominal incision, through which the greater omentum was externalized and wrapped around the TEE construct secured with a suture, and returned to the peritoneum before closing the abdomen in layers. At 4 weeks, 2 groups of 3 neonatal-derived TEE animals each underwent an additional operation. Histology revealed a complete esophageal wall, including mucosa, submucosa, and muscularis propria, which was confirmed by means of immunohistochemical staining for α-actin smooth muscle. The TEE architecture was maintained after interposition or use as a patch, and animals gained weight on a normal diet. The GFP-labeled TEE preserved its fluorescent label, proving the donor origin of the TEE.

Badylak *et al.* used porcine-derived, xenogeneic ECM derived from either the SIS or urinary bladder submucosa (UBS) as a tissue scaffold for esophageal repair in a dog model [180]. Patch defects measuring approximately 5 cm in length and encompassing 40–50% of the circumference of the esophagus or complete circumferential segmental defects measuring 5 cm in length were created by surgical resection in healthy adult female dogs. The defects were repaired with ECM scaffolds derived from either SIS or UBS. The xenogeneic scaffolds used for repair of the patch defects were resorbed completely within 30–40 days and showed replacement by skeletal muscle, which was oriented appropriately and contiguous with adjacent normal esophageal skeletal muscle, organized collagenous connective tissue, and a complete and intact squamous epithelium. No signs of clinical esophageal dysfunction were seen in any of the animals with the patch defect repair. However, the xenogeneic scaffolds configured into tubes for repair of the segmental defects all showed stricture within 45 days of surgery.

6.2. Liver

Since the discovery of cyclosporine as an effective immunosuppressant in 1978, liver transplantation has become an accepted treatment for many types of liver disease. The current standard treatment of acute liver failure (ALF) involves supportive care that focuses on bridging patients to either orthotopic liver transplantation or spontaneous recovery. The treatment for patients that have a congenital alteration of hepatic metabolism is also only transplantation of liver, but several problems in liver transplantation remain, including donor organ shortage. Several groups have attempted to increase the supply of donor organs by living-related transplantation and split-liver transplantation.

Extracorporeal liver support systems have been used to treat patients with ALF attempting to either bridge them to recovery or to transplantation. A bioartificial liver (BAL), developed by the group at Cedars-Sinai Medical Center, USA, using cryopreserved porcine hepatocytes, is the most extensively studied of the extracorporeal cell-based liver support systems. Several preliminary uncontrolled clinical studies examining the effect of this therapy in patients with ALF have shown that BAL treatment improved neurologic function, reduced intracranial pressure, and increased cerebral perfusion pressure. Demetriou *et al.* reported the results of the first Phase II/III prospective, randomized, multicenter, controlled trial examining the effect of BAL treatment on survival in patients with ALF [181]. A total of 171 patients (86 control and 85 BAL) were enrolled. Patients with fulminant/subfulminant hepatic failure and primary nonfunction following liver transplantation were included. Patients in the control group received intensive clinical care according to current standard best practices at each study site. For the entire patient population, survival at 30 days was 71% for BAL versus 62% for control. After exclusion of primary nonfunction patients, survival was 73% for BAL versus 59% for control. When survival was analyzed accounting for confounding factors, in the entire patient population, there was no difference between the two groups. However, survival in fulminant/subfulminant hepatic failure patients was significantly higher in the BAL compared with the control group.

While the extracorporeal liver assist device is promising, this approach would not offer permanent treatment. A potential alternative approach to correct various types of hepatic insufficiency is hepatocyte transplantation on bioabsorbable polymer scaffolds. This technique may correct, in part, single enzyme defects associated with liver. Using the principles of tissue engineering, hepatocytes are seeded onto polymer scaffolds and then implanted into the recipient. The polymer scaffold allows cell attachment, guides cell organization and proliferation, and induces neovascularization. Vacanti's group showed functional engraftment of hepatocytes in biodegradable polymers transplanted into heterotopic positions in rat, mouse, and dog [182, 183, 184]. They used the term "engraftment" literally to mean initial hepatocyte survival and proliferation. Because the optimal location for implantation of hepatocyte-polymer constructs is unknown, they transplanted hepatocytes on biodegradable polymers into the subcutaneous space of the abdominal wall, between mesenteric leaves of the small bowel, or within the omentum of Lewis rats [185]. These three areas were chosen on the basis of the presence of a relatively large surface area. Hepatocytes isolated from the liver of adult, in-bred Lewis rats were seeded onto porous PLLA discs with dimensions of 18 mm (diameter) $\times$ 1 mm (thickness). Cell–polymer constructs were implanted into the omentum mesentery, and subcutaneous space, and secured into place with suture. In one group of recipients, an end-side portacaval shunt was performed 2 weeks before construct implantation, since portacaval shunt was shown to stimulate survival of hepatocytes transplanted into the mesentery. Hepatocyte viability in all isolations was greater than 95% and all constructs retained their discoid shape 2 weeks after implantation. All specimens were firmly adhered to surrounding tissue. Hematoxylin and eosin (H&E) staining revealed survival of clusters of organized hepatocytes within the constructs surrounded by fibrovascular ingrowth and polymer remnants. The clusters of organized hepatocytes were located largely on the periphery of the polymers, usually adjacent to blood vessels. A branching capillary network was seen within the larger clusters of hepatocytes. Three groups of specimens were obtained: constructs 2 weeks after implantation (group 1), constructs 2 weeks after implantation into portacaval shunted animals (group 2), and constructs 4 weeks after implantation (group 3). Table 2.21 summarizes the results of morphometric analysis in each of the groups with the constructs implanted in the subcutaneous space, the mesentery, and the omentum. In each of the groups, engraftment in the omentum was the greatest. Portacaval shunt before implantation significantly increased engraftment in the mesentery and the omentum. Figure 2.62 shows differences in hepatocyte engraftment among the three implantation positions. Overall, the greatest engraftment is seen in the omentum in portacaval shunted animals. In all groups, engraftment in the subcutaneous space was minimal.

One of the main issues in hepatocyte transplantation within porous scaffolds is the limited viability of transplanted cells. Kedem *et al.* addressed this issue by

Table 2.21
Mean engraftment for each group and each site of implantation[a]

	Subcutaneous (μm^2)	Mesentery (μm^2)	Omentum (μm^2)
Group 1 (2 weeks)	810 ($n = 4$)	20,229 ($n = 3$)	31,881 ($n = 4$)
Group 2 (portacaval shunt/2 weeks)	5,543 ($n = 5$)	77,134 ($n = 5$)	211,182 ($n = 5$)
Group 3 (4 weeks)	6,788 ($n = 5$)	28,846 ($n = 5$)	61,683 ($n = 5$)

[a] In each group, engraftment in the omentum is the greatest and engraftment in the subcutaneous space is the least.

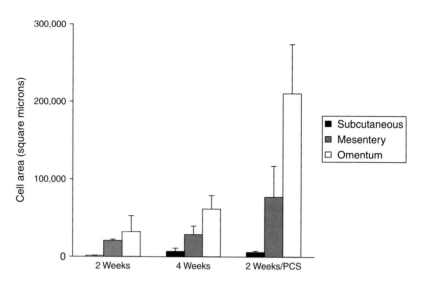

Fig. 2.62 Graph summarizing differences in hepatocyte engraftment (expressed in square micrometers on y axis) in the subcutaneous space, the mesentery, and the omentum in three groups: 2 weeks after implantation, 2 weeks after implantation in portacaval shunted animals, and 4 weeks after implantation.

enhancing scaffold vascularization before cell transplantation via sustained delivery of VEGF, and by examining the liver lobes as a platform for transplanting donor hepatocytes in close proximity to the host liver [186]. The vascularization kinetics of unseeded VEGF-releasing scaffolds on rat liver lobes were evaluated by analyzing the microvascular density and tissue ingrowth in implants harvested on days 3, 7, and 14 postimplantation. Capillary density was greater at all times in VEGF-releasing scaffolds than in the control scaffold without VEGF supplementation. On day 14, it was 220 ± 33 versus 139 ± 23 capillaries/mm^2. Furthermore, 35% of the newly formed capillaries in VEGF-releasing scaffolds were larger than 16 μm in diameter, whereas in control scaffolds only 10% exceeded this size. The VEGF had no effect on tissue ingrowth into the scaffolds. Hepatocyte transplantation onto the implanted VEGF-releasing or control scaffolds was performed after 1 week of prevascularization on the liver lobe in Lewis rats. Fifty implants were harvested on days 1, 3, 7, and 12 and the area of viable hepatocytes was evaluated. The enhanced vascularization improved hepatocyte engraftment. Twelve days after transplantation, the intact hepatocyte area (136,910 μm^2/cross-section) in VEGF-releasing scaffolds was 4.6 higher than in the control group.

When hepatocytes with the ECM components extracted from Engelbreth-Holm-Swarm cells were transplanted, the cells survived for at least 140 days and formed small liver tissues [187]. Liver engineering in hemophilia A mice reconstituted 5–10% of normal clotting activity, enough to reduce the bleeding time and have a therapeutic benefit. Conversely, the subcutaneous space did not support the persistent survival of hepatocytes with Engelbreth-Holm-Swarm gel matrix. Ohashi *et al.* hypothesized that establishing a local vascular network at the transplantation site would reduce graft loss. To test this idea, they provided a potent angiogenic agent before hepatocyte transplantation into the subcutaneous space. With this procedure, persistent survival was achieved for the length of the experiment (120 days).

6.3. Bile Duct

An extrahepatic bile duct affected by cancer or stenosis is currently treated by surgery consisting of removal of the affected portion and anastomosis of the hilar bile duct to the small intestine. However, the postoperative course is frequently complicated by retrograde infection via the intestine or stenosis at the anastomosis that requires re-anastomosis. In the worst case, bile stagnation due to impaired bile flow may require liver transplantation despite the impairment limited to the bile duct. As bile duct stenosis will lead to liver graft failure even though pathology is confined to the bile duct, this may require re-transplantation.

Miyazawa *et al.* investigated tissue engineering of bile duct using BMCs [188]. The cells were harvested from the swine sternum and seeded onto the interior of tubular, porous scaffolds made from P(LA/CL) (50:50) and reinforced with PGA fibers. One hour after seeding, the cell–polymer constructs with a diameter of 5 mm were implanted into each of the pigs from which BMCs had been collected. The common bile duct of pigs was cut around the confluence with the cystic duct, and the duodenal end of the common bile duct was anastomosed to the construct. Then, a hole of 5 mm diameter was made in the descending duodenum, to which the remaining end of the bile duct construct was sutured, as shown in Fig. 2.63. The engineered bile duct was recovered 6 months after implantation. The cell–polymer constructs grew to form a shape similar to the native bile duct while functioning

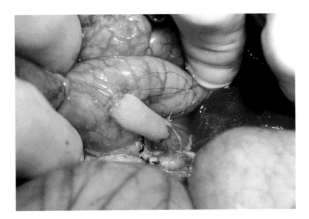

Fig. 2.63 Bile duct organoid units anastomosed to the native common bile duct at the one end and to the descending duodenum at the other to reconstruct the bile duct.

effectively as a bile duct. On macroscopy, the implanted constructs had grayish white surface and were similar to the native common bile duct in morphology. On histology, the tissue-engineered bile duct was almost similar to the native common bile duct, and its portion possibly corresponding to bile duct epithelium cells was positive for cytokeratine 19 just as the native bile duct was.

6.4. Abdominal Wall

Free muscle transfer from local or distant sites is commonly employed for the surgical repair of muscle-tissue defects, but this practice is frequently associated with significant donor site morbidity. Alternatives include the *in vitro* development of a functional 3-D muscle for transplantation or the construction of an implantable biomimetic scaffold with seeded cells. Satellite cells are committed myogenic progenitors that mediate both postnatal growth and regeneration after injury of muscles. The cells reside beneath the basal lamina of adult skeletal muscles, accounting for 2–5% of sublaminal nuclei, and normally mitotically quiescent. Upon activation, satellite cells, now called "myogenic precursor cells", undergo multiple rounds of division before their terminal differentiation. One of the strategies for muscle tissue engineering involves the harvesting of satellite cells, their expansion *in vitro*, and their subsequent autologous implantation *in vivo* into the sites requiring repair or replacement. One of the main obstacles in the formation of new muscle tissue is the lack of an adequate support for expanded satellite cells.

Conconi *et al.* explored the possibility of using a biological matrix seeded with myoblasts for muscle-tissue engineering [189]. Myoblasts were obtained by culturing single muscle fibers isolated by enzymatic digestion from rat flexor digitorum brevis, and their phenotype was confirmed by myogenic differentiation factor, myogenic factor-5, myogenin, and desmin. Cultured myoblasts were harvested and seeded on patches of homologous acellular matrix (ACM), obtained by detergent-enzymatic treatment of abdominal muscle fragments. Myoblast-seeded patches were inserted between obliqui abdominis muscles on the right side of 1-month-old rats, while non-seeded patches were implanted on the left side. Thirty days after surgery, non-seeded patches were completely replaced by fibrous tissue, while the structure of myoblast-seeded patches was well preserved until the second month. Seeded patches displayed abundant blood vessels and myoblasts, and electromyography evidenced in them single motor-unit potentials, sometimes grouped into arithmic discharges. Ninety days after implantation, the thickness of myoblast-seeded patches and their electric activity decreased, suggesting a loss of contractile muscle fibers.

6.5. Small Intestine

Short bowel syndrome is a morbid product of massive small bowel resection. Tissue engineering techniques may offer reconstructed, functional small bowel grown from a small sample of autologous human tissue, but there is a great challenge for intestinal tissue engineers. That is to reprogram autologous adult intestinal cells so that they develop into replacement organs. One possibility for this approach is to utilize "organoid units", which are mesenchymal cores surrounded by polarized epithelium, each 0.25×1.0 mm^2 in size. The epithelial–mesenchymal cell interactions in the units are thought to be important for neointestine wall development. Organoid units can be yielded by mechanical and enzymatic disruption of whole thickness immature bowel. The combination of organoid units with microporous absorbable scaffolds may produce complex cystic structures with organized mucosa and a distinct outer wall. The wall consists of cells that stain with smooth muscle immunohistochemical markers, fibroblasts, and ECM. The neomucosa acquires many of the characteristics of normal mucosa, including sucrose activity and functional transport characteristics. Techniques available to enhance neomucosal maturation include anastomosis to native bowel and the creation of a portacaval anastomosis. The effectiveness of this venous shunt suggests that blood-borne growth factors may be important.

Gardner-Thorpe *et al.* identified and quantified the microvascular elements of the neointestine to characterize the angiogenic cell-to-cell signaling that occurs during intestinal growth, and to compare the tissue growth factor profile in engineered intestine wall with that of growing native bowel [190]. To produce intestinal organoid units, they harvested small bowel from 3-day-old Lewis rats. The purified organoid units were seeded at a density of 1×10^5 onto microporous non-woven PGA cylinders treated with PLLA and collagen solution. The constructs were paratopically transplanted into the omenta of 23 adult Lewis rats and fixed using a suture. At sacrifice 1–8 weeks after implantation, the neointestine was dissected from recipient tissues. Nineteen implants (83%) developed into neointestinal cysts, which progressively increased in mean volume to 12.6 cm^2 at 8 weeks. The mass increased from 1.3 to 9.7 g. Morphometric analysis of the H&E-stained sections showed a progressive thickening of the mucosal layer of the neointestine, but this layer never attained the thickness of juvenile or adult tissue because of the paucity of villi. Villi were first seen at 6 weeks, but were not a constant finding in every neointesine, juvenile, and adult groups. After 8 weeks the thickness of the outer layer increased to beyond the normal limits of adult and juvenile bowel, as shown in Fig. 2.64. The capillary density in the mucosal layer was similar for each group, and the outer layer of the neointestine had a capillary density similar to adult, rather than juvenile bowel, as shown in Table 2.22. In the first few days after organoid unit implantation, the microporous scaffold facilitated diffusion of gases and metabolites. The neointestine grew to resemble the native small bowel and the capillary density remained constant as the tissue expanded. Areas bare of neoepithelium were observed infrequently. The mucosal capillary density was similar in the neointestine, juvenile, and adult intestine, and did not change over time. Clearly, the mechanism driving neointestinal wall organization differed from that controlling the rapid growth of native juvenile intestine.

Grikscheit *et al.* replaced a vital organ with tissue-engineered small intestine (TESI) after massive small intestine resection [191]. Ten rats underwent TESI implantation with GFP-marked cells or sham laparotomy followed by massive small intestine resection. Side-to-side anastomosis of TESI to proximal small intestine was

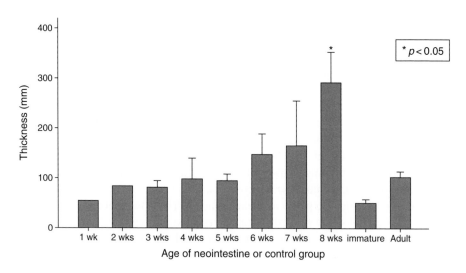

Fig. 2.64   Thickness of the muscular and connective layer.

performed or omitted. All 10 rats initially lost and then regained weight. The initial rate of weight loss was higher in TESI (+) than TESI (−), but the nadir was reached a week earlier with more rapid weight gain subsequently to 98% preoperative weight on day 40 in animals with engineered intestine versus 76%. Serum B12 was higher at 439 pg/ml versus 195 pg/ml. Intestinal ALP mRNA appeared greater in TESI (+) than TESI (−), with constant villi levels. Histology revealed appropriate architecture including nerve and GFP labeling persisted.

Unlike mature skeletal and cardiac myocytes, mature smooth muscle cells (SMCs) are reported to retain the developmental potential to dedifferentiate, both *in vivo* and *in vitro* [192]. To regenerate small intestine with autologous tissues, Nakase *et al.* used collagen scaffolds seeded with SMCs in a canine model [193]. Autologous SMCs were isolated from stomach wall and cultured. Two types of cell–scaffold construct were fabricated: In SMC (+), cultured SMCs were mixed with collagen solution and poured into a collagen sponge; in SMC (−), SMC seeding was omitted. Both loops were isolated together with several feeding and

Table 2.22
Capillary density[a] expressed as the number of capillaries per 1000 nuclei in the microscope field

	Juvenile Bowel (Mean ± SEM)	Mature Bowel (Mean ± SEM)	Neointestine (Mean ± SEM)
Muscle layer	14.49 ± 2.77	75.75 ± 3.07 ($p < 0.001$)[b]	82.95 ± 4.81 ($p < 0.001$)[b]
Mucosal layer	74.54 ± 4.80	82.50 ± 7.30 (NS)	89.54 ± 5.68 (NS)

[a] Capillary density is expressed as the number of capillaries per 1000 nuclei in the microscope field.
[b] $p < 0.001$ compared with the juvenile group. NS, not significantly different from the juvenile group.

draining vessels, while the interrupted bowel was reanastomosed using a two-layer technique to maintain a normally functioning alimentary canal. Defects were created in the small intestine at the middle of the isolated ileal loops, and patched with silicone sheets, which were necessary to protect the scaffolds from infection and digestion. The SMC (+) scaffolds or SMC (−) scaffolds were placed below the silicone sheet. They were fixed in place with absorbable sutures and covered on the serosal aspect with omentum. The ileal loops were used to construct a double ileostomy on both sides of the abdominal incision. The abdominal incision was closed in two layers. Two animals were killed at 4 weeks after implantation, and another two animals at 8 weeks. The last four animals were divided evenly. At 8 weeks, one subgroup underwent reanastomosis of the SMC (+) ileal loop, and the other group underwent reanastomosis of the SMC (−) ileal loop to prevent disuse atrophy. Dogs in both subgroups were killed at 12 weeks after implantation. At 12 weeks the SMC (−) group showed a luminal surface covered with a regenerated EC layer with very short villi, but only a thin smooth muscle layer was observed, representing the muscularis mucosae. In the SMC (+) group, the luminal surface was completely covered with a relatively well-developed epithelial layer with numerous villi. Implanted SMCs were seen in the lamina propria and formed a smooth muscle layer.

7. UROGENITAL SYSTEM

7.1. Bladder

The role of the urinary bladder is to store urine while maintaining significant chemical gradients between the urine and the blood, often for prolonged periods. The bladder permeability barrier therefore plays a role in maintaining normal homeostasis. A wide variety of clinical conditions encountered in urology may cause decreased bladder capacity and/or poor compliance that could lead to recurrent urinary tract infections, incontinence, renal parenchymal damage, and renal impairment. Bladder reconstruction using bowel segment is considered as the "gold standard", but this corrective measure is associated with deleterious side effects secondary to the loss of the urothelial barrier, such as infection, acid–base unbalance, urinary calculi, and carcinogenic transformation. *In vivo* tissue engineering technologies for bladder reconstruction frequently utilize naturally derived biodegradable materials that are placed into the host with or without pre-seeding of cells. From a tissue engineering perspective, collagen-based scaffolds such as SIS and bladder acellular matrix (BAM) have been utilized for bladder replacement. The use of large BAM segments showed that on long-term follow-up, the inflammatory response toward the matrix resulted in contracture or resorption of the graft [194]. It seems probable that when BAM is used as a bladder substitute, urine leakage might occur and induce inflammation and fibrosis. On the other hand, large pore sizes and connectivity are required to accommodate and deliver a cell mass sufficient for tissue repair.

The SIS is obtained from various sources and processed in different ways, thereby making it difficult to determine the single most important factor in obtaining reliable and consistent bladder regeneration using the unseeded technique. In a canine bladder augmentation model, Kropp *et al.* directly compared SIS derived from proximal segments of jejunum with SIS derived from distal segments of ileum [195]. All 6 animals underwent 40% cystectomy and immediate augmentation with 1 of the 2 SIS materials. They assessed the harvested bladders for the amount of regeneration, SIS graft size, and shrinkage 10 weeks after augmentation. Both SIS materials demonstrated some evidence of bladder regeneration but it appeared that only in sow aged greater than 3 years distal ileum SIS produced consistent bladder regeneration without bone formation and severe shrinkage. It appeared that not all material, even if obtained from different segments of 1 sow's bowel, had the same regenerative results.

Farhat *et al.* estimated the porosity/permeability properties of BAM using two porosity indicators: deionized water as the representative bladder fluid and dextran (fluorescein-labeled, and molecular weight: 10,000) as the macromolecules to measure porosity [196]. Bladders harvested from pigs of 20–50 kg were stirred in hypotonic solution (EDTA, Triton X-100, and Pefabloc Plus) for 24 h, followed by a hypertonic solution (1.5 M KCl) for another 24 h. To reduce the thickness of the BAM mimicking an expanded stretched bladder, the acellularization process was performed with the bladder filled to 120 ml with the different solutions used for acellularization. The bladder outlet was tied and then immersed sequentially in the respective solutions. Using morphometric analysis, quantifiable thinner ACM was confirmed (thick ACM range 3–4 mm, and thin ACM range 1.8–2.2 mm). For the thick ACM, the mean abluminal porosity index (0.03813 ml/min·cm^2) was significantly higher than the luminal side (0.0325 ml/min·cm^2). In addition, for the thin ACM, the abluminal mean porosity index (0.032 ml/min·cm^2) was also higher than the luminal side (0.02865 ml/min·cm^2). When porosity was plotted against time, the volume collected in any direction (luminal and abluminal) increased linearly with time, indicating a constant rate of flow. The dextran diffusivity is shown in Fig. 2.65, where the mean values for the porosities were calculated from the absorbance of aliquots sampled at different time intervals. In the thicker ACM, the mean abluminal absorbance, 0.0715, was noted to be higher than the luminal 0.02425. On the other hand, for the thinner ACM, the mean luminal absorbance (0.0485) was greater than the abluminal (0.02625). When absorbance was plotted as a function of time, the diffusivity reached a plateau at 30 min.

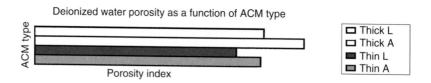

Fig. 2.65   Porosity as depicted by absorbance with different ACM types.

Schultheiss *et al.* generated a vascularized, autologous, reseeded bladder substitute and evaluated immediate vascularization and perfusion of the graft after implantation into the recipient organism in a porcine model [197]. The ACM was processed from porcine small bowel segments by subsequent mechanical, chemical, and enzymatic decellularization, preserving the jejunal arteriovenous pedicles. The matrix was reseeded with primary bladder SMCs and UCs, and its vascular structures were resurfaced with EPCs. To evaluate graft perfusion short-term implantation was performed. The acellular scaffold was successfully repopulated with multilayers of ingrowing SMCs and superficial UCs. After reseeding the jejunal arteriovenous pedicles with EPCs and cultivation for 3 weeks the larger vessels as well as the intramural scaffold capillary network were repopulated with cell monolayers expressing endothelial specific proteins. Perfusion stagnation and implant thrombosis occurred within 30 min after the implantation of acellular scaffolds not reseeded with EPCs. In the EPC seeded group the vascular system revealed intact perfusion, and no relevant thrombus formation was observed after 1 or 3 h.

7.2. Ureter

Osman *et al.* evaluated the effectiveness of ACM used as a tube for replacement of a relatively long segment of the canine ureter [198]. The ACM was obtained by the excision of the whole ureter of donor dogs that were sacrificed and not included in the study group. Retrieved ureters were treated to have complete cell lysis, while maintaining the fiber framework. The study included 10 dogs in which a 3-cm segment was excised from one ureter and replaced by a tube of ACM of the same length and width. The new tube was sutured at proximal and distal ends by watertight interrupted sutures around a Double-J stent that remained for 6 weeks. Excretory urography was done 1 and 2 weeks after stent removal and the dogs were then sacrificed. All dogs survived surgery except one which died one week postoperatively of a malpositioned stent and urinary ascites. There was no clinically apparent postoperative complication during the presence or after the removal of the ureteral stents. One week after stent removal, excretory urography showed ipsilateral mild to moderate hydroureteronephrosis in 3 dogs and no dye excretion in 6 with a normal contralateral kidney. One week later no dye excretion was detected in all except one dog, which showed more radiological deterioration. At the time of sacrifice there was moderate to marked hydroureteronephrosis above the level of the new tube in all dogs. Although the graft was intact in all subjects, marked shrinkage was observed. On ureteral calibration there was significant narrowing of the lumen up to complete occlusion. At 8 weeks histopathological examination showed extensive fibrosis.

7.3. Urethra

Patients with urethral strictures often require additional tissues for repair. Autologous non-urethral tissue grafts or flaps from genital and extragenital skin, bladder, rectal and buccal mucosa, tunica vaginalis and peritoneum have been used. However, use of non-urethral tissues often requires additional procedures for graft retrieval and may be associated with prolonged hospitalization and donor site morbidity. Synthetic non-absorbable materials such as silicone, PTFE, and polyester have been tried previously for urethral reconstruction, but these materials have not been optimal substitutes due to associated problems, such as erosion and dislodgement. El-Kassaby *et al.* explored the feasibility of using a bladder submucosa collagen–based inert matrix as a free graft substitute for urethral stricture repair [199]. The submucosa of cadaveric human bladder tissue was microdissected and isolated from the adjacent muscular and serosal layers. The cyclic washing and chemical treatment were repeated until decellularization was confirmed. The inert collagen matrix was trimmed to size as needed for each patient and the neourethra was created by anastomosing the matrix in an onlay fashion to the urethral plate with continuous absorbable sutures. The size of the created neourethra ranged from 1.5 to 16 mm. After 36- to 48-month follow-up, 24 of the 28 patients had a successful outcome. The remaining 4 patients had a slight caliber decrease at the anastomotic sites on urethrography. A subcoronal fistula developed in 1 patient which closed spontaneously 1 year after repair. Mean maximum urine flow rate increased from the preoperative value of 9 ± 1.29 to 19.7 ± 3.07 ml per second postoperatively. Cystoscopic studies revealed adequate caliber conduits and normal appearing urethral tissues. Histological examination of the biopsy specimens showed the typical urethral stratified epithelium.

7.4. Vaginal Tissue

Various procedures have been used in the past for vaginal reconstruction, and different tissue sources have been employed for reconstructive surgery, but the use of non-vaginal tissue for vaginal reconstruction is associated with limited functionality. Tissue engineering is expected to offer a solution for challenging cases when a shortage of local tissue exists. However, there is a paucity of information regarding the tissue engineering of female reproduction and genital tissues. Filippo *et al.* investigated the feasibility of using vaginal epithelial cells and SMCs for the engineering of vaginal tissues *in vivo* [200]. The cells were isolated from the vaginal tissue of New Zealand white rabbits, grown, expanded in culture, and characterized immunocytochemically. Both the vaginal epithelial and the SMCs were seeded onto opposite sides of PGA scaffold coated with PLGA (LA:GA = 50:50) in a stepwise fashion at a concentration of 10×10^6 and 20×10^6 cells/cm^3, respectively. Seeded and unseeded scaffolds were implanted subcutaneously into athymic mice. One week after implantation, the retrieved polymer scaffolds demonstrated multilayered tissue strips of both cell types, and penetrating native vasculature was also noted. There was no evidence of tissue formation in the controls. It is likely that vaginal epithelial cells and SMCs could survive *in vivo* for prolonged periods and could self-organize toward seemingly normal structural orientation. Longitudinal strips of native and tissue-engineered vagina were compared in organ bath analyzes. The strips were attached by suture to a tissue support hook at one end and an isometric force transducer at the other end. Serial field stimulation was applied with 3-min

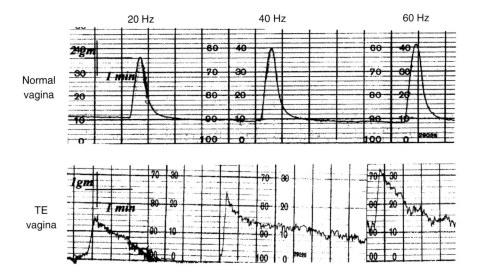

Fig. 2.66 Electrical field stimulation demonstrating evoked potentials at various levels of electrical stimulation for both normal and tissue-engineered (TE) vaginas 6 weeks after implantation.

resting intervals between stimulations, and active tension was measured. As shown in Fig. 2.66, the tissue-engineered vaginal constructs were capable of producing contractile forces similar to those seen with native vaginal tissue when stimulated with a series of electrical impulses.

7.5. Corporal Tissue

Conditions such as ambiguous genitalia, the exstrophy–epispadias complex, and traumatic injury often require surgical intervention. The aim of penile reconstruction is to achieve structurally and functionally normal genitalia. Corporal tissue reconstruction has been a challenge in the treatment of several clinical conditions because of the limited amount of native tissue available. Because of the unique anatomical structure of the corpora cavernosa, its replacement with alternative tissues has not been entirely successful. Autologous tissue flaps or prosthetic devices have been used as corporal substitutes. However, engineered autologous tissue may be preferable for long-term use. Kershen *et al.* showed that human corporal human SMCs seeded on bioabsorbable polymer scaffolds formed vascularized corpus cavernosum muscle *in vivo* [201]. Capillary formation was facilitated by the addition of ECs [202], but 3-D tissue structurally similar to native corpus cavernosum could not be engineered because of the type of polymer matrices used. As a next step, this research group developed a naturally derived scaffold which is structurally similar to native corpora [203]. The new matrix is a 3-D acellular collagen matrix derived from donor rabbit cavernosa because the corporal structures of this species are similar to those of humans. All cellular components were removed via a multistep procedure. Treatment with several cycles of distilled water promoted cell lysis and removed cellular debris. Subsequent treatment with Triton X eliminated the nuclear components, whereas ammonium hydroxide lysed the cell membranes and eliminated the cytoplasmic proteins. Human corpus cavernosal muscle and ECs were seeded on the acellular matrices. A total of 80 matrices, 20 without cells and 60 with cells, were implanted subcutaneously into athymic mice. An additional 36 matrices seeded with cells were maintained in culture for up to 4 weeks. The implanted matrices showed neovascularity into the sinusoidal spaces 1 week after implantation. Increasing organization of SMCs and ECs lining the sinusoidal walls was observed at 2 weeks and continued with time. The matrices were covered with the appropriate cell architecture 4 weeks after implantation. The retrieved tissue was composed of 38% muscle, which represents the lower limit of muscle content present in normal human corpus cavernosum (38.5–52%). The EC content in the retrieved corporal tissue was 12%, whereas the composition of ECs in normal corporal tissue was 2.8%. Organ bath studies showed that the cell-seeded corporal tissue matrices responded to electrical field stimulation, whereas the unseeded implants failed to respond.

8. OTHERS

8.1. Skull Base

In adults and children over 2 years of age, large cranial defects do not reossify successfully, posing a substantial biomedical burden. Cowan *et al.* used ADAS cells seeded onto apatite-coated scaffolds to heal critical-sized skull defects in mice [204]. The ADAS cells were harvested from the subcutaneous anterior abdominal wall of mice. Inguinal fat pads excised, washed, and finely minced in PBS were digested with type II collagenase. Neutralized cells were centrifuged to separate mature adipocytes and the stromal-vascular fraction. Floating adipocytes were removed, and pelleted stromal cells were passed through a 100-μm cell strainer before plating. To fabricate scaffolds from PGLA (85/15), PGLA/chloroform solutions were mixed with sucrose to obtain 92% porosity, and compressed into thin sheets. After freeze-drying overnight, scaffolds were immersed in water to dissolve the sucrose. After scaffold fabrication, scaffolds were coated with apatite. Immediately before the coating process, dried PGLA scaffolds were subjected to argon-plasma etching to improve wetting and coating uniformity. Scaffolds were cut into discs and seeded with 8×10^4 cells per cm^2 of scaffold by pipetting. Twenty-four hours after seeding, cell attachment was checked by identifying nuclei. For implantation, the pericranium of adult mice was removed and defects were made in the parietal bone. After placement of the scaffold into the defect, the skin was sutured. Implanted, apatite-coated PGLA scaffolds seeded with ADAS cells produced significant intramembranous bone by 2 weeks and areas of complete bone bridging by 12 weeks. The contribution of implanted cells to new bone formation was 84–99%.

8.2. Dura Mater

To replace the dura mater, Yamada *et al.* fabricated a bioabsorbable sheet with sandwich structure as illustrated in Fig. 2.67 [205]. A non-woven PGA fabric was used to reinforce two P(LA/CL) sheets. This composite sheet displayed good mechanical properties and was completely absorbed 24 weeks after implantation in the back of rats. And 2 weeks to 26 months after implantation into 31 rabbits with dural defects,

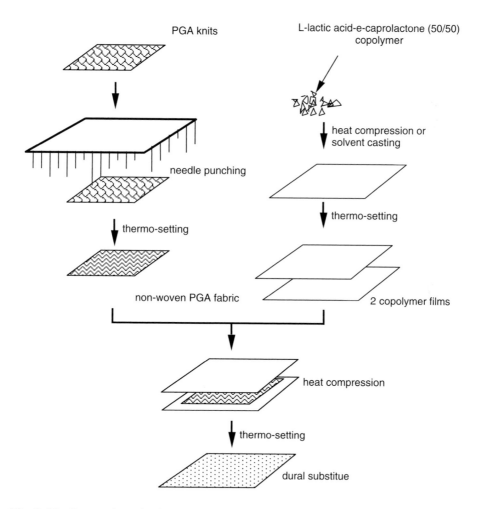

Fig. 2.67 Preparation of a three-layered sheet from P(LA/CL) and PGA for dura mater regeneration.

infection, cerebrospinal fluid (CSF) leakage, evidence of convulsive disorders, significant adhesion to underlying cortex, or calcification was not noticed in any case. The regenerated dura-like tissue had a high pressure–resistant strength 2 weeks after implantation. Yamada *et al.* applied this absorbable sheet to 53 patients during neurosurgical procedures [206]. The average follow-up was 35.5 months. The handling properties and biocompatibility of this sheet were satisfactory without any significant complication. In patients who underwent a second surgery performed more than 18 months after the initial operation, this sheet was found to have been replaced by autologous collagenous tissue, and chronic foreign-body reactions to this material were negligible.

8.3. Cornea

Because cornea stroma represents the major component responsible for corneal transparency, a primary effort was made by Liu *et al.* to test the possibility of stromal engineering [207]. Stromal fibroblasts were isolated from newborn rabbits. After *in vitro* expansion, cells were mixed with non-woven PGA fibers and transplanted into the cornea of the mother rabbit from which the newborn rabbits were derived. The construct, which was opaque at the time of implantation, became a nearly transparent stroma at 8 weeks postimplantation. The newly formed stroma was proven as an engineered stroma by the presence of GFP-labeled stromal cells. The mechanism by which engineered stroma became transparent remains incompletely understood. Compared with stroma, the engineering of corneal epithelium is simple. For instance, corneal ECs are isolated and then grown on an acellular human amnionic membrane until confluent. Histology of the engineered corneal epithelium shows a structure similar to that of its native counterpart. To engineer a composite cornea tissue *in vitro*, Wang *et al.* isolated rabbit or human corneal epithelial, stromal, and ECs and mixed them with type I collagen as a stromal matrix and cultured them *in vitro* [208]. Histology revealed a composite structure including stratified epithelium, stroma, and a single layer of endothelium, suggesting that a similar strategy might be useful for *in vivo* engineering of composite cornea.

To reconstruct corneal surfaces and restore vision in patients with bilateral severe disorders of the ocular surface, Nishida *et al.* studied the use of autologous oral mucosal ECs as a source of cells for the reconstruction of the corneal surface [209]. They harvested 3×3-mm^2 specimens of oral mucosal tissue from four patients with bilateral total corneal stem-cell deficiencies. Tissue-engineered epithelial-cell sheets were fabricated *ex vivo* by culturing harvested cells for two weeks on temperature-responsive cell-culture surfaces with 3T3 feeder cells that had been treated with mitomycin C. After conjunctival fibrovascular tissue had been surgically removed from the ocular surface, sheets of cultured autologous cells that had been harvested with a simple reduced-temperature treatment were transplanted directly to the denuded corneal surfaces (one eye of each patient) without sutures. Complete re-epithelialization of the corneal surfaces occurred within 1 week in all 4 treated eyes. Corneal transparency was restored and postoperative visual acuity improved in all eyes. During a mean follow-up period of 14 months, all corneal surfaces remained transparent.

Yoshizawa *et al.* sought to develop full-thickness *ex vivo*–produced human conjunctiva and mucosa equivalents using a serum-free culture system without a feeder layer and to compare conjunctiva and oral mucosa equivalents to assess their suitability as graft material for eyelid reconstruction [210]. Human conjunctiva and oral mucosal keratinocytes were cultured, expanded, and seeded onto AlloDerm™ to produce *ex vivo*–produced full-thickness mucosa equivalents. Progressive epithelial stratification was observed on day 4, 11, and 18 in conjunctiva and oral mucosa equivalents. A proliferation marker (Ki-67) immunoreactivity progressively increased with cultured time in both types of equivalents, indicating the continued presence of actively proliferating cells. A membrane antigen (GLUTI) immunoreactivity, concentrated in the basal keratinocytes of stratified epithelia of both types of equivalents, mimicked native tissue and indicated a high glycolytic state of the basal cells.

8.4. Prenatal Tissues

Anatomic congenital anomalies occur in approximately 3% of all newborns and account for more than 30% of deaths within the neonatal period. Congenital cardiac defects are the leading noninfectious cause of death in this age group. Although many anomalies can be surgically corrected at birth, the use of non-functional prosthetic material for anatomic reconstruction contributes significantly to short- and long-term morbidity. Advances in fetal imaging have made possible the early gestational diagnosis of most anatomic congenital anomalies. Advances in prenatal intervention and fetal tissue sampling have resulted in consideration of a strategy of perinatal tissue engineering: that is, the prenatal harvest, isolation, and *in vitro* expansion of autologous fetal cells for the purpose of engineering a tissue construct with subsequential surgical reconstruction immediately after birth. To date this concept has been explored as an experimental therapeutic strategy for reconstruction of muscular diaphragmatic defects, bladder exstrophy, as well as superficial skin defects. Consideration of this approach for bladder-based applications is limited by the variety of cells necessary for proper reconstruction of extensive congenital malformations involving several organ systems or anatomic defects requiring a variety of tissues for reconstruction, such as congenital cardiac anomalies. The application of this concept to cardiac tissue engineering is further complicated by the practical concerns of donor site morbidity and the difficulty of expanding differentiated cardiac myocytes *in vitro*. These concerns would be overcome by the ability to isolate and expand a population of multilineage progenitor cells from the fetus.

As the fetal liver is the primary hematopoietic organ *in utero*, Krupnick *et al.* hypothesized that the fetal liver might be a rich source of mesenchymal progenitor cells as well [211]. They evaluated the ovine fetal liver as a potential source of multilineage mesenchymal progenitor cells, with the specific goal of using such a cell population for autologous prenatal tissue engineering. Liver stromal cells were isolated from a portion of the right lateral hepatic lobe of midgestation fetal lambs and expanded *in vitro*. Passage I undifferentiated fetal liver stromal cells, labeled *in vitro* with 5-bromo-2-deoxy-uridine (BrdU), were organized into type I collagen/Matrigel-based 3-D organoid reinforced by porous non-woven PLLA. The cells were suspended within the collagen matrix during its liquid phase at 4°C. This cell-containing mixture was then allowed to polymerize at 37°C *in vitro* within the interstices of the PLLA, trapping the cells within the collagen/Matrigel matrix. A donor heart of an immunocompromised adult male nude rat was then transplanted as a vascularized heterotopic infrarenal abdominal graft into a recipient from the same colony after sublethal 500-cGy irradiation of the recipient from a cesium source. Immediately before reperfusion of the heterotopic heart, the organoid described above, containing BrdU-labeled fetal liver stromal cells, was implanted directly into the left ventricle and sutured to the surrounding myocardium with a running suture, as shown in Fig. 2.68. Implantation of this construct was performed within 6–12 h of organoid assembly, thus limiting *in vitro* incubation. And 2–4 weeks after implantation the animals were killed, and engraftment and differentiation of the transplanted fetal cells were evaluated. Passage 1 cells displayed a uniform fibroblast-like morphology but could be induced to differentiate into skeletal muscle, adipocytes, chondrocytes, and ECs by selective medium supplementation. By manipulating the ECM *in vitro*, spontaneously contracting cardiac myocyte-like cells could be generated as well. Multilineage differentiation was confirmed by morphology, protein expression, and upregulation of lineage-specific mRNA. The potential for engineering myocardial tissue was then investigated by transplanting early-passage progenitor cells, organized on a 3-D matrix, into the ventricular tissue engineering. Survival, incorporation into the host myocardium, and cardiomyocytic differentiation of the transplanted cells were confirmed.

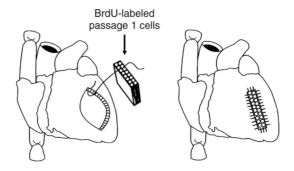

Fig. 2.68 Surgical manipulation for ventricular tissue engineering. After cold cardioplegia and transplantation, but before reperfusion of the heterotopic heart, the left ventricle was incised and a 3-D type I collagen-based progenitor cell–seeded construct was implanted into the ventricle and sutured to the surrounding myocardium.

REFERENCES

1. I.V. Yannas, J.F. Burke, D.P. Orgill *et al.*, Wound tissue can utilize a polymeric template to synthesize a functional extension of skin, *Science*, **215**, 174 (1981).
2. J.G. Rheinwald and H. Green, Formation of a keratinizing epithelium in culture by a cloned cell line derived from a teratoma, *Cell*, **6**, 317 (1975).
3. N.E. O'Connor, J.B. Mulliken, S. Banks-Schlegel *et al.*, Grafting of burns with cultured epithelium prepared from autologous epidermal cells, *Lancet*, **317**, 75 (1981).
4. S.T. Boyce and R.G. Ham, Calcium-regulated differentiation of normal human epidermal keratinocytes in chemically defined clonal culture and serum-free serial culture, *J. Invest. Darmatology*, **81**, 33s (1983).
5. M.S. Higham, R. Dawson, M. Szabo *et al.*, Development of a stable chemically defined surface for the culture of human keratinocytes under serum-free conditions for clinical use, *Tissue Engineering*, **9**, 919 (2003).
6. K. Izumi, S.E. Feinberg, H. Terashi *et al.*, Evaluation of transplanted tissue-engineered oral mucosa equivalents in severe combined immunodeficient mice, *Tissue Engineering*, **9**, 163 (2003).
7. E. Bell, H.P. Ehrlich, D.J. Buttle *et al.*, Living tissue formed *in vitro* and accepted as skin equivalent tissue of full thickness, *Science*, **211**, 1052 (1981).
8. W. Liu, L. Cui, and Y. Cao, A closer view of tissue engineering in China: The experience of tissue construction in immunocompetent animals, *Tissue Engineering*, **9**, Suppl.1, S-17 (2003).
9. N. Morimoto, Y. Saso, K. Tomihata *et al.*, Viability and function of autologous and allogeneic fibroblasts seeded in dermal substitutes after implantation, *J. Surg. Res.*, **125**, 56 (2005).
10. A.F. Black, C. Bouez, E. Perruer *et al.*, Optimization and characterization of an engineered human skin equivalent, *Tissue Engineering*, **11**, 723 (2005).
11. A.J. Beck, J. Phillips, L. Smith-Thomas, R. Short *et al.*, Development of a plasma-polymerized surface suitable for the transplantation of keratinocyte-melanocyte cocultures for patients with vitiligo, *Tissue Engineering*, **9**, 1123 (2003).
12. R.M. France, R.D. Short, R.A. Dawson *et al.*, Attachment of human keratinocytes to plasma co-polymers of acrylic acid/octa-1,7-diene and allyl amine/octa-1,7-diene, *J. Mater. Chem.*, **8**, 37 (1998).
13. W. Deng, Q. Han, L. Liao *et al.*, Engrafted bone marrow-derived Flk-1$^+$ mesenchymal stem cells regenerate skin tissue, *Tissue Engineering*, **11**, 110 (2005).
14. Y. Cao, J.P. Vacanti, K.T. Paige *et al.*, Transplantation of chondrocytes utilizing a polymer-cell construct to produce tissue-engineered cartilage in the shape of a human ear, *Plast. Reconstr. Surg.*, **100**, 297 (1997).
15. J.C. Britt and S.S. Park, Autogenous tissue-engineered cartilage: Evaluation as an implant material, *Arch. Otolaryngol. Head Neck Surg.*, **124**, 671 (1998).
16. A.B. Saim, Y. Cao, Y. Weng *et al.*, Engineering autogenous cartilage in the shape of a helix using an injectable hydrogel scaffold, *Laryngoscope*, **110**, 1694 (2000).
17. A. Haisch, S. Klaring, A. Groger *et al.*, A tissue-engineering model for the manufacture of auricular-shaped cartilage implants, *Eur. Arch. Otorhinolaryngol.*, **259**, 316 (2002).
18. D.W. Hutmacher, K.W. Ng, and C. Kaps, Elastic cartilage engineering using novel scaffold architectures in combination with a biomimetic cell carrier, *Biomaterials*, **24**, 4445 (2003).
19. S.-J. Shieh, S. Terada, and J.P. Vacanti, Tissue engineering auricular reconstruction: *in vitro* and *in vivo* studies, *Biomaterials*, **25**, 1545 (2004).

20. N. Isogai, H. Kusuhara, Y. Ikada *et al.*, Comparison of different chondrocytes for use in tissue engineering of cartilage model structures, *Tissue Engineering*, in press.

21. H. Yanaga, M. Koga, K. Imai *et al.*, Clinical application of biotechnically cultured autologous chondrocytes as novel graft material for nasal augmentation, *Aesthetic. Plast. Surg.*, **28**, 212 (2004).

22. J. Borges, M.C. Mueller, N.T. Padron *et al.*, Engineered adipose tissue supplied by functional microvessels, *Tissue Engineering*, **9**, 1263 (2003).

23. T. Korff and H.G. Augustin, Engineered adipose tissue supplied by functional microvessels, *J. Cell Biol.*, **143**, 1341 (1998).

24. H. Yamashiro, T. Inamoto, M. Yagi *et al.*, Efficient proliferation and adipose differentiation of human adipose tissue-derived vascular stromal cells transfected with basic fibroblast growth factor gene, *Tissue Engineering*, **9**, 881 (2003).

25. S.-W. Cho, S.-S. Kim, J.W. Rhie *et al.*, Engineering of volume-stable adipose tissues, *Biomaterials*, **26**, 3577 (2005).

26. Y.S. Choi, S.-N. Park, and H. Suh, Adipose tissue engineering using mesenchymal stem cells attached to injectable PLGA spheres, *Biomaterials*, **26**, 5855 (2005).

27. A. Alhadlaq, M. Tang, and J.J. Mao, Engineered adipose tissue from human mesenchymal stem cells maintains predefined shape and dimension: Implications in soft tissue augmentation and reconstruction, *Tissue Engineering*, **11**, 556 (2005).

28. R. Dorotka, U. Windberger, K. Macfelda *et al.*, Repair of articular cartilage defects treated by microfracture and a three-dimensional collagen matrix, *Biomaterials*, **26**, 3617 (2005).

29. C. Willers, J. Chen, D. Wood *et al.*, Autologous chondrocyte implantation with collagen bioscaffold for the treatment of osteochondral defects in rabbits, *Tissue Engineering*, **11**, 1065 (2005).

30. X. Guo, C. Wang, Y. Zhang *et al.*, Repair of large articular cartilage defects with implants of autologous mesenchymal stem cells seeded into beta-tricalcium phosphate in a sheep model, *Tissue Engineering*, **10**, 1818 (2004).

31. X. Guo, C. Wang, C. Duan *et al.*, Repair of osteochondral defects with autologous chondrocytes seeded onto bioceramic scaffold in sheep, *Tissue Engineering*, **10**, 1830 (2004).

32. T. Tanaka, H. Komaki, M. Chazono *et al.*, Use of a biphasic graft constructed with chondrocytes overlying a β-tricalcium phosphate block in the treatment of rabbit osteochondral defects, *Tissue Engineering*, **11**, 331 (2005).

33. T.S. Johnson, J. Xu, V.V. Zaporojan *et al.*, Integrative repair of cartilage with articular and nonarticular chondrocytes, *Tissue Engineering*, **10**, 1308 (2004).

34. H. Tanaka, H. Mizokami, E. Shiigi *et al.*, Effects of basic fibroblast growth factor on the repair of large osteochondral defects of articular cartilage in rabbits: Dose-response effects and long-term outcomes, *Tissue Engineering*, **10**, 633 (2004).

35. T. Nishida, S. Kubota, S. Kojima *et al.*, Regeneration of defects in articular cartilage in rat knee joints by CCN2 (connective tissue growth factor). *J. Bone Mineral Res.*, **19**, 1308 (2004).

36. N. Rotter, F. Ung, A.K. Roy *et al.*, Role for interleukin 1α in the inhibition of chondrogenesis in autologous implants using polyglycolic acid–polylactic acid scaffolds, *Tissue Engineering*, **11**, 192 (2005).

37. N.D. Case, A.O. Duty, A. Ratcliffe, R. Mueller *et al.*, Bone formation on tissue-engineered cartilage constructs *in vivo*: Effects of chondrocyte viability and mechanical loading, *Tissue Engineering*, **9**, 587 (2003).

38. M. Brittberg, A. Lindahl, A. Nilsson *et al.*, Treatment of deep cartilage defects in the knee with autologous chondrocyte transplantation, *New Eng. J. Med.*, **331**, 889 (1994).

39. L. Peterson, T. Minas, M. Brittberg *et al.*, Treatment of osteochondritis dissecans of the knee with autologous chondrocyte transplantation: Results at two to ten years, *J. Bone Joint Surg.*, **85-A**, 17 (2003).

40. M. Ochi, Y. Uchio, K. Kawasaki *et al.*, Transplantation of cartilage-like tissue made by tissue engineering in the treatment of cartilage defects of the knee, *J. Bone Joint Surg.*, **84**, 571 (2002).

41. S. Wakitani, K. Imoto, T. Yamamoto *et al.*, Human autologous culture expanded bone marrow mesenchymal cell transplantation for repair of cartilage defects in osteoarthritic knees, *Osteoarthritis and Cartlige*, **10**, 199 (2002).

42. M. Ishihara, M. Sato, S. Sato *et al.*, Usefulness of photoacoustic measurements for evaluation of biomechanical properties of tissue-engineered cartilage, *Tissue Engineering*, **11**, 1234 (2005).

43. S.C. Dickinson, T.J. Sims, L. Pittarello *et al.*, Quantitative outcome measures of cartilage repair in patients treated by tissue engineering, *Tissue Engineering*, **11**, 277 (2005).

44. Y. Miura, J.S. Fitzsimmons, C.N. Commisso *et al.*, Enhancement of periosteal chondrogenesis *in vitro*. Dose-response for transforming growth factor-beta 1 (TGF-beta 1), *Clin. Orthop, Related Res.*, **301**, 271 (1994).

45. S.W. O'Driscoll, A.D. Recklies, and A.R. Poole, Chondrogenesis in periosteal explants. An organ culture model for *in vitro* study, *J. Bone Jt Surg. Am.*, **76**, 1042 (1994).

46. M.M. Stevens, H.F. Qanadilo, R. Langer *et al.*, A rapid-curing alginate gel system: Utility in periosteum-derived cartilage tissue engineering, *Biomaterials*, **25**, 887 (2004).

47. P.J. Emans, D.A.M. Surtel, E.J.J. Frings *et al.*, *In vivo* generation of cartilage from periosteum, *Tissue Engineering*, **11**, 369 (2005).

48. W. Bensaid, K. Oudina, V. Viateau *et al.*, *De novo* reconstruction of functional bone by tissue engineering in the metatarsal sheep model, *Tissue Engineering*, **11**, 814 (2005).

49. M. Nishikawa, A. Myoui, H. Ohgushi *et al.*, Bone tissue engineering using novel interconnected porous hydroxyapatite ceramics combined with marrow mesenchymal cells: Quantitative and three-dimensional image analysis, *Cell Transplant.*, **13**, 367 (2004).

50. T. Yoshikawa, H. Ohgushi, K. Ichijima *et al.*, Bone regeneration by grafting of cultured human bone, *Tissue Engineering*, **10**, 688 (2004).

51. Y. Yoneda, H. Terai, Y. Imai *et al.*, Repair of an intercalated long bone defect with a synthetic biodegradable bone-inducing implant, *Biomaterials*, **26**, 5145 (2005).

52. H.G. Schmoekel, F.E. Weber, J.C. Schense *et al.*, Bone repair with a form of BMP-2 engineered for incorporation into fibrin cell ingrowth matrices, *Biotech. Bioeng.*, **89**, 253 (2005).

53. B. Peterson, J. Zhang, and R. Iglesias, Healing of critically sized femoral defects, using genetically modified mesenchymal stem cells from human adipose tissue, *Tissue Engineering*, **11**, 120 (2005).

54. C.M. Cowan, O.O. Aalami, Y.-Y. Shi *et al.*, Bone morphogenetic protein 2 and retinoic acid accelerate *in vivo* bone formation, osteoclast recruitment, and bone turnover, *Tissue Engineering*, **11**, 645 (2005).

55. A. Hokugo, M. Ozeki, O. Kawakami *et al.*, Augmented bone regeneration activity of platelet-rich plasma by biodegradable gelatin hydrogel, *Tissue Engineering*, **11**, 1224 (2005).

56. M.M. Stevens, R.P. Marini, D. Schaefer *et al.*, *In vivo* engineering of organs: The bone bioreactor, *Proc. Nat. Acad. Sci.*, **102**, 11450 (2005).

57. P. Kasten, J. Vogel, R. Luginbuehl *et al.*, Ectopic bone formation associated with mesenchymal stem cells in a resorbable calcium deficient hydroxyapatite carrier, *Biomaterials*, **26**, 5879 (2005).

58. J.H.P. Hui, L. Li, Y.-H. Teo *et al.*, Comparative study of the ability of mesenchymal stem cells derived from bone marrow, periosteum, and adipose tissue in treatment of partial growth arrest in rabbit, *Tissue Engineering*, **11**, 904 (2005).

59. M.-H. Cheng, E.M. Brey, A. Allori *et al.*, Ovine model for engineering bone segments, *Tissue Engineering*, **11**, 214 (2005).

60. C.A. Vacanti, L.J. Bonassar, M.P. Vacanti *et al.*, Replacement of an avulsed phalanx with tissue-engineered bone, *New Engl. J. Med.*, **344**, 1511 (2001).

61. H. Ohgushi, N. Kotobuki, H. Funaoka *et al.*, Tissue engineered ceramic artificial joint—*ex vivo* osteogenic differentiation of patient mesenchymal cells on total ankle joints for treatment of osteoarthritis, *Biomaterials*, **26**, 4654 (2005).

62. V. Gangji and J.-P. Hauzeur, Treatment of osteonecrosis of the femoral head with implantation of autologous bone-marrow cells. A pilot study, *J. Bone Joint Surg.*, **86-A**, 1153 (2004).

63. S.M. van Gaalen, W.J.A. Dhert, A. van den Muysenberg *et al.*, Bone tissue engineering for spine fusion: An experimental study on ectopic and orthotopic implants in rats, *Tissue Engineering*, **10**, 231 (2004).

64. T. Wang, G. Dang, Z. Guo *et al.*, Evaluation of autologous bone marrow mesenchymal stem cell–calcium phosphate ceramic composite for lumbar fusion in rhesus monkey interbody fusion model, *Tissue Engineering*, **11**, 1159 (2005).

65. J.C. Goh, H. Ouyang, S. Teoh *et al.*, Tissue-engineering approach to the repair and regeneration of tendons and ligaments, *Tissue Engineering*, **9**, Suppl.1, S-31 (2003).

66. F. Van Eijk, D.B.F. Saris, J. Riesle *et al.*, Tissue engineering of ligaments: A comparison of bone marrow stromal cells, anterior cruciate ligament, and skin fibroblasts as cell source, *Tissue Engineering*, **10**, 893 (2004).

67. J.A. Cooper, H.H. Lu, F.K. Ko *et al.*, Fiber-based tissue-engineered scaffold for ligament replacement: Design considerations and *in vitro* evaluation, *Biomaterials*, 26, 1523 (2005).

68. H. Kobayashi, Y. Kawamoto, S. Hara *et al.*, Study on bioresorbable poly(L-lactide) (PLLA) ligament augmentation device (LED), *Proc. Second Far-Eastern Symposium on Biomedical materials*, 1995, p. 175.

69. M. Ishimura, H. Ohgushi, K. Inoue *et al.*, Bioabsorbable PLLA artificial ligament as a ligament augmentation device, *Knee* (in Japanese), **21**, 86 (1995).

70. H.W. Ouyang, J.C.H. Goh, A. Thambyah *et al.*, Knitted poly-lactide-co-glycolide scaffold loaded with bone marrow stromal cells in repair and regeneration of rabbit Achilles tendon, *Tissue Engineering*, **9**, 431 (2003).

71. Y.L. Cao, Y.T. Liu, W. Liu *et al.*, Bridging tendon defects using autologous tenocyte engineered tendon in a hen model, *Plast. Reconstr. Surg.*, **110**, 1280 (2002).

72. S. Calve, R.G. Dennis, P.E. Kosnik II *et al.*, Engineering of functional tendon, *Tissue Engineering*, **10**, 755 (2004).

73. N. Juncosa-Melvin, G.P. Boivin, M.C. Galloway *et al.*, Effects of cell-to-collagen ratio in mesenchymal stem cell-seeded implants on tendon repair biomechanics and histology, *Tissue Engineering*, **11**, 448 (2005).

74. D. Pelinkovic, j.-Y. Lee, M. Eengelhardt *et al.*, Muscle cell-mediated gene delivery to the rotator cuff, *Tissue Engineering*, **9**, 143 (2003).

75. T. Neumann, S.D. Hauschka, and J.E. Sanders, Tissue engineering of skeletal muscle using polymer fiber arrays, *Tissue Engineering*, **9**, 995 (2003).

76. P. De Coppi, D. Delo, L. Farrugia *et al.*, Angiogenic gene-modified muscle cells for enhancement of tissue formation, *Tissue Engineering*, **11**, 1034 (2005).

77. D. Zaleske, G. Peretti, F. Allemann *et al.*, Engineering a joint: A chimeric construct with bovine chondrocytes in a devitalized chick knee, *Tissue Engineering*, **9**, 949 (2003).

78. N. Isogai, W. Landis, T.H. Kim *et al.*, Formation of phalanges and small joints by tissue-engineering, *J. Bone Joint Surg.*, **81-A**, 306 (1999).
79. M. Lehtimaki, S. Paasimaa, S. Lehto *et al.*, Development of the metacarpophalangeal joint arthroplasty with a bioabsorbable prosthesis, *Abstract of Scand. Hand Soc. Meeting*, Oslo, Norway, 1998, No. 2.
80. P.B. Honkanen, M. Kellomaeki, M.Y. Lehtimaeki *et al.*, Bioreconstructive joint scaffold implant arthroplasty in metacarpophalangeal joints: Short-term results of a new treatment concept in rheumatoid arthritis patients, *Tissue Engineering*, **9**, 957 (2003).
81. T. Ozawa, D.A.G. Mickle, and R.D. Weisel, Histologic changes of nonbiodegradable and biodegradable biomaterials used to repair right ventricular heart defects in rats, *J. Thorac. Cardiovasc. Surg.*, **124**, 1157 (2002).
82. T. Ozawa, D.A.G. Mickle, R.D. Weisel *et al.*, Tissue-engineered grafts matured in the right ventricular outflow tract, *Cell Transplant.*, **13**, 169 (2004).
83. Y. Chang, H.-C. Liang, H.-J. Wei *et al.*, Tissue regeneration patterns in acellular bovine pericardia implanted in a canine model as a vascular patch, *J. Biomed. Mater. Res.*, **69A**, 323 (2004).
84. Y. Toshimitsu, M. Miyazawa, T. Torii *et al.*, Tissue engineered patch for the reconstruction of inferior vena cava during living-donor liver transplantation, *J. Gastrointest. Surg.*, **9**, 789 (2005).
85. T. Shin'oka, Y. Imai, and Y. Ikada, Transplantation of a tissue-engineered pulmonary artery, *New Engl. J. Med.*, **344**, 532 (2001).
86. Y. Naito, Y. Imai, T. Shin'oka, J. Kashiwagi *et al.*, Successful clinical application of tissue-engineered graft for extracardiac Fontan operation, *J. Thorac. Cardiovasc. Surg.*, **125**, 419 (2003).
87. N. Hibino, T. Shin'oka, G. Matsumura *et al.*, The tissue-engineered vascular graft using bone marrow without culture, *J. Thorac. Cardiovasc. Surg.*, **129**, 1064 (2005).
88. Y. Noishiki, Y. Tomizawa, Y. Yamane *et al.*, Autocrine angiogenic vascular prosthesis with bone marrow transplantation, *Nat. Med.*, **2**, 90 (1996).
89. G. Matsumura, N. Hibino, Y. Ikada *et al.*, Successful application of tissue engineered vascular autografts: Clinical experience, *Biomaterials*, **24**, 2303 (2003).
90. G. Matsumura, S. Miyagawa-Tomita, T. Shin'oka *et al.*, First evidence that bone marrow cells contribute to the construction of tissue-engineered vascular autografts *in vivo*, *Circulation*, **108**, 1729 (2003).
91. S.P. Higgins, A.K. Solan, and L.E. Niklason, Effects of polyglycolic acid on porcine smooth muscle cell growth and differentiation, *J. Biomed. Mater. Res.*, **67A**, 295 (2003).
92. V. Prabhakar, M.W. Grinstaff, J. Alarcon *et al.*, Engineering porcine arteries: Effects of scaffold modification, *J. Biomed. Mater. Res.*, **67A**, 303 (2003).
93. G.H. Borschel, Y.-C. Huang, S. Calve *et al.*, Tissue engineering of recellularized small-diameter vascular grafts, *Tissue Engineering*, **11**, 778 (2005).
94. V. Clerin, J.W. Nichol, M. Petko *et al.*, Tissue engineering of arteries by directed remodeling of intact arterial segments, *Tissue Engineering*, **9**, 461 (2003).
95. A. Marui, A. Kanematsu, K. Yamahara *et al.*, Simultaneous application of basic fibroblast growth factor and hepatocyte growth factor to enhance the blood vessels formation, *J. Vasc. Surg.*, **41**, 82 (2005).
96. S. Shintani, T. Murohara, H. Ikeda *et al.*, Augmentation of postnatal neovascularization with autologous bone marrow transplantation, *Circulation*, **103**, 897 (2001).
97. E. Tateishi-Yuyama, H. Matsubara, T. Murohara *et al.*, Therapeutic angiogenesis for patients with limb ischaemia by autologous transplantation of bone-marrow cells: A pilot study and a randomised controlled trial, *Lancet*, **360**, 427 (2002).

98. T. Iwase, N. Nagaya, T. Fujii *et al.*, Comparison of angiogenic potency between mesenchymal stem cells and mononuclear cells in a rat model of hindlimb ischemia, *Cardiovasc. Res.*, **66**, 543 (2005).

99. A.N. Balamurugan, Y. Gu, Y. Tabata *et al.*, Bioartificial pancreas transplantation at prevascularized intermuscular space: Effect of angiogenesis induction on islet survival, *Pancreas*, **26**, 279 (2003).

100. T. Hisatome, Y. Yasunaga, S. Yanada *et al.*, Neovascularization and bone regeneration by implantation of autologous bone marrow mononuclear cells, *Biomaterials*, **26**, 4550 (2005).

101. D.A. Narmoneva, O. Oni, A.L. Sieminski *et al.*, Self-assembling short oligopeptides and the promotion of angiogenesis, *Biomaterials*, **26**, 4837 (2005).

102. N. Koike, D. Fukumura, O. Gralla *et al.*, Tissue engineering: Creation of long-lasting blood vessels, *Nature*, **428**, 138 (2004).

103. T. Shin'oka, Tissue engineered heart valves: Autologous cell seeding on biodegradable polymer scaffold, *Artificial Organs*, **26**, 402 (2002).

104. D.W. Harken and L.E. Curtis, Heart surgery—legend and a long look, *Am. J. Cardiol.*, **19**, 393 (1967).

105. R.G. Leyh, M. Wilhelmi, T. Walles *et al.*, Acellularized porcine heart valve scaffolds for heart valve tissue engineering and the risk of cross-species transmission of porcine endogenous retrovirus, *J. Thorac. Cardiovasc. Surg.*, **126**, 1000 (2003).

106. K. Kallenbach, R.G. Leyh, E. Lefik *et al.*, Guided tissue regeneration: Porcine matrix does not transmit PERV, *Biomaterials*, **25**, 3613 (2004).

107. M.-T. Kasimir, E. Rieder, G. Seebacher *et al.*, Presence and elimination of the xenoantigen gal (α1, 3) gal in tissue-engineered heart valves, *Tissue Engineering*, **11**, 1274 (2005).

108. T. Shin'oka, P.X. Ma, D. Shum-Tim *et al.*, Tissue-engineered heart valves. Autologous valve leaflet replacement study in a lamb model, *Circulation*, **94** (suppl. II), II-164 (1996).

109. U.A. Stock, M. Nagashima, and P.N. Khalil, Tissue-engineered valved conduits in the pulmonary circulation, *J. Thorac. Cardiovasc. Surg.*, **119**, 732 (2000).

110. P.M. Dohmen, A. Lembcke, H. Hotz *et al.*, Ross operation with a tissue-engineered heart valve, *Ann. Thorac. Surg.*, **74**, 1438 (2002).

111. G. Sakaguchi, K. Tambara, Y. Sakakibara *et al.*, Control-released hepatocyte growth factor prevents the progression of heart failure in stroke-prone spontaneously hypertensive rats, *Ann. Thorac. Surg.*, **79**, 1627 (2005).

112. B.Z. Atkins, M.T. Hueman, J.M. Meuchel *et al.*, Myogenic cell transplantation improves *in vivo* regional performance in infarcted rabbit myocardium, *J. Heart Lung Transplant.*, **18**, 1173 (1999).

113. S. Tomita, R.K. Lin, R.D. Weisel *et al.*, Autologous transplantation of bone marrow cells improves damaged heart function, *Circulation*, **100** (suppl), II 247 (1999).

114. K.L. Christman, H.H. Fok, R.E. Sievers *et al.*, Fibrin glue alone and skeletal myoblasts in a fibrin scaffold preserve cardiac function after myocardial infarction, *Tissue Engineering*, **10**, 403 (2004).

115. N. Nagaya, K. Kangawa, T. Itoh *et al.*, Transplantation of mesenchymal stem cells improves cardiac function in a rat model of dilated cardiomyopathy, *Circulation*, **23**, 1128 (2005).

116. G.V. Silva, S. Litovsky, J.A.R. Assad *et al.*, Mesenchymal stem cells differentiate into an endothelial phenotype, enhance vascular density, and improve heart function in a canine chronic ischemia model, *Circulation*, **111**, 150 (2005).

117. K. Wollert, G.P. Meyer, J. Lotz *et al.*, Intracoronary autologous bone-marrow cell transfer after myocardial infarction: The BOOST randomised controlled clinical trial, *Lancet*, **364**, 141 (2004).

118. T. Kofidis, P. Akhyari, J. Boublik *et al.*, *In vitro* engineering of heart muscle: Artificial myocardial tissue, *J. Thorac. Cardiovasc. Surg.*, **124**, 63 (2002).

119. T. Kofidis, P. Akhyari, B. Wachsmann *et al.*, Clinically established hemostatic scaffold (tissue fleece) as biomatrix in tissue- and organ-engineering research, *Tissue Engineering*, **9**, 517 (2003).

120. T. Kofidis, J.L. de Bruin, G. Hoyt *et al.*, Injectable bioartificial myocardial tissue for large-scale intramural cell transfer and functional recovery of injured heart muscle, *J. Thorac. Cardiovasc. Surg.*, **128**, 571 (2004).

121. R.K. Birla, G.H. Borschel, and R.G. Dennis, Myocardial engineering *in vivo*: Formation and characterization of contractile, vascularized three-dimensional cardiac tissue, *Tissue Engineering*, **11**, 803 (2005).

122. J. Boublik, H. Park, M. Radisic *et al.*, Mechanical properties and remodeling of hybrid cardiac constructs made from heart cells, fibrin, and biodegradable, elastomeric knitted fabric, *Tissue Engineering*, **11**, 1122 (2005).

123. B.F. Akl, J. Mittelman, D.E. Smith *et al.*, A new method of tracheal reconstruction, *Ann. Thorac. Surg.*, **36**, 265 (1983).

124. H. Osada, and K. Kojima, Experimental tracheal reconstruction with a rotated right stem bronchus, *Ann. Thorac. Surg.*, **70**, 1886 (2000).

125. P.C. Cavadas, The split medial gastrocnemius muscle flap, *Plast. Reconstr. Surg.*, **101**, 937 (1998).

126. K. Kojima, L.J. Bonassar, A.K. Roy *et al.*, Autologous tissue-engineered trachea with sheep nasal chondrocytes, *J. Thorac. Cardiovasc. Surg.*, **123**, 1177 (2002).

127. S.H. Kamil, R.D. Eavey, M.P. Vacanti *et al.*, Tissue-engineered cartilage as a graft source for laryngotracheal reconstruction: A pig model, *Arch. Otolaryngol. Head Neck Surg.*, **130**, 1048 (2004).

128. T. Okamoto, Y. Yamamoto, M. Gotoh *et al.*, Slow release of bone morphogenetic protein 2 from a gelatin sponge to promote regeneration of tracheal cartilage in a canine model, *J. Thorac. Cardiovasc. Surg.*, **127**, 329 (2004).

129. K. Kojima, R.A. Ignotz, T. Kushibiki *et al.*, Tissue-engineered trachea from sheep marrow stromal cells with transforming growth factor beta2 released from biodegradable microspheres in a nude rat recipient, *J. Thorac. Cardiovasc. Surg.*, **128**, 147 (2004).

130. C.R. Cogle, A.T. Yachnis, E.D. Laywell *et al.*, Bone marrow transdifferentiation in brain after transplantation: A retrospective study, *Lancet*, **363**, 1432 (2004).

131. K. Kataoka, Y. Suzuki, M. Kitada *et al.*, Alginate enhances elongation of early regenerating axons in spinal cord of young rats, *Tissue Engineering*, **10**, 493 (2004).

132. W.M. Tian, S.P. Hou, J. Ma *et al.*, Hyaluronic acid–poly-D-lysine-based three-dimensional hydrogel for traumatic injury, *Tissue Engineering*, **11**, 513 (2005).

133. H.C. Park, Y.S. Shim, Y. Ha *et al.*, Treatment of complete spinal cord injury patients by autologous bone marrow cell transplantation and administration of granulocyte-macrophage colony stimulating factor, *Tissue Engineering*, **11**, 913 (2005).

134. T.W. Hudson, S.Y. Liu, and C.E. Schmidt, Engineering an improved acellular nerve graft via optimized chemical processing, *Tissue Engineering*, **10**, 1346 (2004).

135. L.R. Williams, H.C. Powell, G. Lundborg *et al.*, Competence of nerve tissue as distal insert promoting nerve regeneration in a silicone chamber, *Brain Res.*, **293**, 201 (1984).

136. N. Terada, J.M. Bjursten, M. Papaloizos *et al.*, Resorbable filament structures as a scaffold for matrix formation and axonal growth in bioartificial nerve grafts: Long term observations, *Restor. Neurol. Neurosci.*, **11**, 65 (1997).

137. P. Weiss, The technology of nerve regeneration: A review—sutureless tubulation and related methods of nerve repair, *Neurosurgery*, **1**, 400 (1944).

138. K. Jansen, M.F. Meek, J.F.A. van der Werff *et al.*, Long-term regeneration of the rat sciatic nerve through a biodegradable poly(DL-lactide-epsilon-caprolactone) nerve guide: Tissue reactions with focus on collagen III/IV reformation, *J. Biomed. Mater. Res.*, **69A**, 334 (2004).

139. R.A. Weber, W.C. Breidenbach, R.E. Brown *et al.*, A randomized prospective study of polyglycolic acid conduits for digital nerve reconstruction in humans, *Plast. Reconstr. Surg.*, **106**, 1036 (2000).

140. G. Lundborg, Discussion (A randomized prospective study of polyglycolic acid conduits for digital nerve reconstruction in humans by R.A. Weber, W.C. Breidenbach, R.E. Brown *et al.*), *Plast. Reconstr. Surg.*, **106**, 1046 (2000).

141. J. Brandt, L.B. Dahlin, and G. Lundborg, Autologous tendons used as grafts for bridging peripheral nerve defects, *J. Hand Surg.*, **24B**, 284 (1999).

142. S. Yoshii, M. Oka, M. Shima *et al.*, Bridging a 30-mm nerve defect using collagen filaments, *J. Biomed. Mater. Res.*, **67A**, 467 (2003).

143. G. Keilhoff, F. Praesch, G. Wolf *et al.*, Bridging extra large defects of peripheral nerves: Possibilities and limitations of alternative biological grafts from acellular muscle and Schwann cells, *Tissue Engineering*, **11**, 1004 (2005).

144. F. Stang, H. Fansa, G. Wolf *et al.*, Structural parameters of collagen nerve grafts influence peripheral nerve regeneration, *Biomaterials*, **26**, 3083 (2005).

145. E. Gamez, Y. Goto, K. Nagata *et al.*, Photofabricated gelatin-based nerve conduits: Nerve tissue regeneration potentials, *Cell Transplant.*, **13**, 549 (2004).

146. L. Flynn, P.D. Dalton, and M.S. Shoiet, Fiber templating of poly(2-hydroxyethyl methacrylate) for neural tissue engineering, *Biomaterials*, **24**, 4265 (2003).

147. Z. Ahmed, S. Underwood, and R.A. Brown, Nerve guide material made from fibronectin: Assessment of *in vitro* properties, *Tissue Engineering*, **9**, 219 (2003).

148. X. Yu and R.V. Bellamkonda, Tissue-engineered scaffolds are effective alternatives to autografts for bridging peripheral nerve gaps, *Tissue Engineering*, **9**, 421 (2003).

149. G. Lundborg, L.B. Dahlin, N. Danielsen *et al.*, Nerve regeneration in silicone chambers: Influence of gap length and of distal stump components, *Exp. Neurol.*, **76**, 361 (1982).

150. G. Lundborg, L.B. Dahlin, N. Danielsen *et al.*, Nerve regeneration across an extended gap: A neurobiological view of nerve repair and the possible involvement of neuronotrophic factors, *J. Hand Surg.*, **7**, 580 (1982).

151. G. Lundborg, R.H. Gelberman, and F.M. Longo, *In vivo* regeneration of cut nerves encased in silicone tubes: Growth across a six-millimeter gap, *J. Neuropathol. Exp. Neurol.*, **41**, 412 (1982).

152. G. Lundborg, F.M. Longo, and S. Varon, Nerve regeneration model and trophic factors *in vivo*, *Brain Res.*, **232**, 157 (1982).

153. I.V. Yannas and B.J. Hill, Selection of biomaterials for peripheral nerve regeneration using data from the nerve chamber model, *Biomaterials*, **25**, 1593 (2004).

154. M.F. Meek, J.F.A. van der Werff, F. Klok *et al.*, Functional nerve recovery after bridging a 15 mm gap in rat sciatic nerve with a biodegradable nerve guide, *Scand. J. Plast. Reconstr. Surg. Hand Surg.*, **37**, 258 (2003).

155. R. Karaki, K. Kubota, M. Hitaka *et al.*, Effect of gum-expanding-mesh on the osteogenesis of surgical bony defects, *J. Jpn. Assoc. Periodontol.*, **26**, 516 (1984).

156. G. Zellin and A. Linde, Effects of different osteopromotive membrane porosities on experimental bone neogenesis in rats, *Biomaterials*, **17**, 695 (1996).

157. U.M.E. Wikesjo, W.H. Lim, R.C. Thomson *et al.*, Periodontal repair in dogs: Evaluation of a bioresorbable space-providing macro-porous membrane with rhBMP-2, *J. Periodontol.*, **74**, 635 (2003).

158. U.M.E. Wikesjo, W.H. Lim, R.C. Thomson *et al.*, Periodontal repair in dogs: Gingival tissue occlusion, a critical requirement for GTR?, *J. Clin. Periodontol.*, **30**, 655 (2003).

159. D.N. Tatakis and L. Trombelli, Adverse effects associated with a bioabsorbable guided tissue regeneration device in the treatment of human gingival recession defects. A clinicopathologic case report, *J. Periodontol.*, **70**, 542 (1999).

160. Y.-M. Lee, S.-H. Nam, Y.-J. Seol *et al.*, Enhanced bone augmentation by controlled release of recombinant human bone morphogenetic protein-2 from bioabsorbable membranes, *J. Periodontol.*, **74**, 865 (2003).

161. U.M.E. Wikesjo, A.V. Xiropaidis, R.C. Thomson *et al.*, Periodontal repair in dogs: Space-providing ePTFE devices increase rhBMP-2/ACS-induced bone formation, *J. Clin. Periodontol.*, **30**, 715 (2003).

162. T. Nakahara, T. Nakamura, E. Kobayashi *et al.*, Novel approach to regeneration of periodontal tissues based on *in situ* tissue engineering: Effects of controlled release of basic fibroblast growth factor from a sandwich membrane, *Tissue Engineering*, **9**, 153 (2003).

163. A. Rodriguez, G.E. Anastassou, H. Lee *et al.*, Maxillary sinus augmentation with deproteinated bovine bone and platelet rich plasma with simultaneous insertion of endosseous implants, *J. Oral Maxillofac. Surg.*, **61**, 157 (2003).

164. B.-H. Choi, C.-J. Huh, J.-J. Suh *et al.*, Effect of platelet-rich plasma on bone regeneration in autogenous bone graft, *Int. J. Oral Maxillofac. Surg.*, **33**, 56 (2004).

165. J.P.M. Fennis, P.J.W. Stoelinga, and J.A. Jansen, Mandibular reconstruction: A histological and histomorphometric study on the use of autogenous scaffolds, particulate cortico-cancellous bone grafts and platelet rich plasma in goats, *Int. J. Oral Maxillofac. Surg.*, **33**, 48 (2004).

166. M.S. Detamore and K.A. Athanasiou, Motivation, characterization, and strategy for tissue engineering the temporomandibular joint disc, *Tissue Engineering*, **9**, 1065 (2003).

167. W.K. Solberg, M.W. Woo, and J.B. Houston, Prevalence of mandibular dysfunction in young adults, *J. Am. Dent. Assoc.*, **98**, 25 (1979).

168. K. Yamagishi, K. Onuma, T. Suzuki *et al.*, Materials chemistry: A synthetic enamel for rapid tooth repair, *Nature*, **433**, 819 (2005).

169. J.P.M. Fennis, P.J.W. Stoelinga, M.A.W. Merkx *et al.*, Reconstruction of the mandible with a poly(D,L-lactide) scaffold, autogenous corticocancellous bone graft, and autogenous platelet-rich plasma: An animal experiment, *Tissue Engineering*, **11**, 1045 (2005).

170. Y. Kinoshita, T. Amagasa, E. Fujii *et al.*, Reconstruction of the mandible with poly (L-lactide) mesh tray/sheet and transplantation of particulate cancellous bone and marrow, *Int. J. Oral Maxillofac. Surg.*, **28** (Suppl.1), 161 (1999).

171. Y. Kinoshita, M. Kobayashi, S. Fukuoka *et al.*, Functional reconstruction of the jawbones using poly(L-lactide) mesh and autogenic particulate cancellous bone and marrow, *Tissue Engineering*, **2**, 327 (1996).

172. Y. Kinoshita, M. Kirigakubo, M. Kobayashi *et al.*, Study on the efficacy of biodegradable poly(L-lactide) mesh for supporting transplanted particulate cancellous bone and marrow: Experiment involving subcutaneous implantation in dogs, *Biomaterials*, **14**, 729 (1993).

173. Y. Kinoshita, S. Yokoya, T. Amagasa *et al.*, Reconstruction of jawbones using poly(L-lactic acid) mesh and transplantation of particulate cancellous bone and

marrow: Long-term observation of 40 cases, *Int. J. Oral Maxillofac. Surg.*, **32** (Suppl.1), 117 (2003).

174. Y. Kinoshita, S. Yokoya, S. Fukuoka *et al.*, The efficacy of bFGF incorporated gelatin microspheres for bone formation of madibular defect, *Int. J. Oral Maxillofac. Surg.*, **30** (Suppl.), 110 (2001).

175. T. Iizuka, P. Mikkonen, P. Paukku *et al.*, Reconstruction of orbital floor with polydioxanone plate, *Int. J. Oral Maxillofac. Surg.*, **20**, 83 (1991).

176. A. Baumann, G. Burggasser, N. Gauss *et al.*, PDS orbital floor reconstruction with an alloplastic resorbable polydioxanone sheet, *Int. J. Oral Maxillofac. Surg.*, **31**, 367 (2002).

177. R. Kontio, R. Suuronen, Y.T. Konttinen *et al.*, Orbital floor reconstruction with poly-L/D-lactide implants: clinical, radiological and immunohistochemical study in sheep, *Int. J. Oral Maxillofac. Surg.*, **33**, 361 (2004).

178. T. Yamazaki, Y. Kinoshita, and Y. Ikada, Repair of human orbital floor defects using poly(L-lactide) mesh filled with particulate cancellous bone and marrow, *Int. J. Oral Maxillofac. Surg.*, **30** (suppl.), 92 (2001).

179. T. Grikscheit, E.R. Ochoa, A. Srinivasan *et al.*, Tissue-engineered esophagus: Experimental substitution by onlay patch or interposition, *J. Thorac. Cardiovasc. Surg.*, **126**, 537 (2003).

180. S. Badylak, S. Meurling, M. Chen *et al.*, Resorbable bioscaffold for esophageal repair in a dog model, *J. Pediatr. Surg.*, **35**, 1097 (2000).

181. A.A. Demetriou, R.S. Brown, R.W. Busuttil *et al.*, Prospective, randomized, multicenter, controlled trial of a bioartificial liver in treating acute liver failure, *Ann. Surg.*, **239**, 660 (2004).

182. K. Asonuma, J.C. Gilbert, J.E. Stein *et al.*, Quantitation of transplanted hepatic mass necessary to cure the Gunn rat model of hyperbilirubinemia, *J. Pediatr. Surg.*, **27**, 298 (1992).

183. S.S. Kim, H. Utsunomiya, J.A. Koski *et al.*, Survival and function of hepatocytes on a novel three-dimensional synthetic biodegradable polymer scaffold with an intrinsic network of channels, *Ann. Surg.*, **228**, 8 (1998).

184. K. Sano, R.A. Cusick, H. Lee *et al.*, Regenerative signals for heterotopic hepatocyte transplantation, *Transplant. Proc.*, **28**, 1857 (1996).

185. H. Lee, R.A. Cusick, H. Utsunomiya *et al.*, Effect of implantation site on hepatocytes heterotopically transplanted on biodegradable polymer scaffolds, *Tissue Engineering*, **9**, 1227 (2003).

186. A. Kedem, A. Perets, I. Gamlieli-Bonshtein *et al.*, Vascular endothelial growth factor-releasing scaffolds enhance vascularization and engraftment of hepatocytes transplanted on liver lobes, *Tissue Engineering*, **11**, 715 (2005).

187. K. Ohashi, J.M. Waugh, M.D. Dake *et al.*, Liver tissue engineering at extrahepatic sites in mice as a potential new therapy for genetic liver diseases, *Hepatology*, **41**, 132 (2004).

188. M. Miyazawa, T. Torii, Y. Toshimitsu *et al.*, A tissue-engineered artificial bile duct grown to resemble the native bile duct, *Am. J. Transplant.*, **5**, 1541 (2005).

189. M.T. Conconi, P.D. Coppi, S. Bellini *et al.*, Homologous muscle acellular matrix seeded with autologous myoblasts as a tissue-engineering approach to abdominal wall-defect repair, *Biomaterials*, **26**, 2567 (2005).

190. J. Gardner-Thorpe, T.C. Grikscheit, H. Ito *et al.*, Angiogenesis in tissue-engineered small intestine, *Tissue Engineering*, **9**, 1255 (2003).

191. T.C. Grikscheit, A. Sidduque, E.R. Ochoa *et al.*, Tissue-engineered small intestine improves recovery after massive small bowel resection, *Ann. Surg.*, **240**, 748 (2004).

192. J. Brittingham, C. Phiel, W.C. Trzyna *et al.*, Identification of distinct molecular phenotypes in cultured gastrointestinal smooth muscle cells, *Gastroenterology*, **115**, 605 (1998).

193. Y. Nakase, A. Hagiwara, T. Nakamura *et al.*, Tissue engineering of small intestinal tissue using sponge scaffolds seeded with smooth muscle cells, *Tissue Engineering*, in press.

194. A.L. Brown, W. Farhat, P.A. Merguerian *et al.*, 22 week assessment of bladder acellular matrix as a bladder augmentation material in a porcine model, *Biomaterials*, **23**, 2179 (2002).

195. B.P. Kropp, E.Y. Cheng, H.K. Lin *et al.*, Reliable and reproducible bladder regeneration using unseeded distal small intestinal submucosa, *J. Urol.*, **172**, 1710 (2004).

196. W. Farhat, J. Chen, P. Erdeljan *et al.*, Porosity of porcine bladder acellular matrix: Impact of ACM thickness, *J. Biomed. Mater. Res.*, **67A**, 970 (2003).

197. D. Schultheiss, A.I. Gabouev, S. Cebotari *et al.*, Biological vascularized matrix for bladder tissue engineering: Matrix preparation, reseeding technique and short-term implantation in a porcine model, *J. Urol.*, **173**, 276 (2005).

198. Y. Osman, A. Shokeir, M. Gabr *et al.*, Canine ureteral replacement with long acellular matrix tube: is it clinically applicable?, *J. Urol.*, **172**, 1151 (2004).

199. A.W. El-Kassaby, A.B. Retik, J.J. Yoo *et al.*, Urethral stricture repair with an off-the-shelf collagen matrix, *J. Urol.*, **169**, 170 (2003).

200. R.E. de Filippo, J.J. Yoo, and A. Atala, Engineering of vaginal tissue *in vivo*, *Tissue Engineering*, **9**, 301 (2003).

201. R.T. Kershen, J.J. Yoo, R.B. Moreland *et al.*, Reconstitution of human corpus cavernosum smooth muscle *in vitro* and *in vivo*, *Tissue Engineering*, **8**, 515 (2002).

202. H.J. Park, J.J. Yoo, R.T. Kershen *et al.*, Reconstitution of human corporal smooth muscle and endothelial cells *in vivo*, *J. Urol.*, **162**, 1106 (1999).

203. G. Falke, J. J. Yoo, T. G. Kwon *et al.*, Formation of corporal tissue architecture *in vivo* using human cavernosal muscle and endothelial cells seeded on collagen matrices, *Tissue Engineering*, **9**, 871 (2003).

204. C.M. Cowan, Y.-Y. Shi, O.O. Aalami *et al.*, Adipose-derived adult stromal cells heal critical-size mouse calvarial defects, *Nature Biotechnology*, **22**, 560 (2004).

205. K. Yamada, S. Miyamoto, I. Nagata *et al.*, Development of a dural substitute from synthetic bioabsorbable polymers, *J. Neurosurg.*, **86**, 1012 (1997).

206. K. Yamada, S. Miyamoto, M. Takayama *et al.*, Clinical application of a new bioabsorbable artificial dura mater, *J. Neurosurg.*, **96**, 731 (2002).

207. W. Liu, X.J. Hu, L. Cui *et al.*, Tissue engineering of nearly transparent corneal epithelium and corneal stroma, *Plast. Surg. Res. Council 47th Ann. Meet.*, Abstract No.67A.

208. L.Q. Wang, Y.F. Huang, and Q.X. Huang, Three-dimensional reconstruction of rabbit corneal epithelium and stroma, *Acta. Military Med. Educ.*, **22**, 303 (2001).

209. K. Nishida, M. Yamato, Y. Hayashida *et al.*, Corneal reconstruction with tissue-engineered cell sheets composed of autologous oral mucosal epithelium, *New Engl. J. Med.*, **16**, 1170 (2004).

210. M. Yoshizawa, S.E. Feinberg, C.L. Marcelo *et al.*, *Ex vivo* produced human conjunctiva and oral mucosa equivalents grown in a serum-free culture system, *J. Oral Maxillofac. Surg.*, **62**, 980 (2004).

211. A.S. Krupnick, K.R. Balsara, D. Kreisel *et al.*, Fetal liver as a source of autologous progenitor cells for perinatal tissue engineering, *Tissue Engineering*, **10**, 723 (2004).

Chapter 3

Basic Technologies Developed for Tissue Engineering

As to elemental technologies for tissue engineering, a tremendous amount of investigations has been performed to date. Some of the results have been presented here without careful scrutiny. The bibliography is by no means all-inclusive. At this point in time, due to the large number of publications in this field, it is difficult to determine what research will really promote clinical trial applications of tissue engineering. The reader should review and assess the relevant literature.

1. BIOMATERIALS

1.1. Naturally Occurring Polymers

1.1.1 Proteins

Collagen
Collagenous scaffolds are generally fabricated using crosslinked collagen. The GA is most frequently employed as a crosslinker of collagen, although a toxic concern is accompanied by GA. To reduce the GA concentration, Ma *et al.* added 75–85% deacetylated chitin to mixtures of collagen and GA [1]. Deacetylation of chitin yields chitosan which is soluble in acidic aqueous media due to a large number of amino groups that are crosslinkable with collagen molecules through GA, as shown in Fig. 3.1. The hydrolytic degradation was performed in the presence of 100 μm/ml of collagenase for 12 h. By this hydrolysis the pure collagen (col) was thoroughly degraded, while the addition of chitosan (col/chi) slightly increased the biostability of collagen. After crosslinking with GA, the biostability of the pure collagen (col-GA) was greatly enhanced, while the ability to resist collagenase degradation was further enhanced by the chitosan combination (col/chi-GA). In another study, a water-soluble carbodiimide, and NHS were used to crosslink collagen in the presence of amino acids, which function as a crosslinking bridge between collagen molecular chains. The uncrosslinked collagen was fully degraded after incubation in the collagenase solution. After treatment with EDAC/NHS, biodegradation degree decreased from 100 to 23.2%. Upon addition of glycine, a neutral amino acid, the biodegradation degree was reduced from 23.2 to 18.3%. No significant difference could be detected between glycine/EDAC and EDAC collagens. The lowest biodegradation degree, 9.1%, was achieved for the collagen crosslinked in the presence of lysine, a basic amino acid.

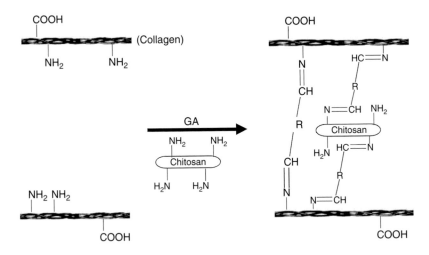

Fig. 3.1 Schematic presentation of collagen crosslinked with GA in the presence of chitosan.

Caruso *et al.* used EDAC as a crosslinking agent of collagen fibers [2]. They hypothesized that EDAC-crosslinked scaffolds could withstand cyclic loading combined with proteolytic enzymes better than UV-crosslinked scaffolds because EDAC is an effective chemical crosslinker, whereas UV irradiation causes partial denaturation of collagen. Collagen fibers were fabricated by extrusion of collagen suspension into a fiber formation buffer. To fabricate collagen fiber scaffolds, 50 collagen fibers were aligned in parallel, divided into 3-cm sections, and tied at each end with nylon thread. Scaffolds made from these fibers were compared to untreated (UNXL) and UV-crosslinked collagen fiber scaffolds. Cyclic loads interacted synergistically with enzymes, rendering UNXL scaffolds unstable and further decreasing the breaking load of UV-crosslinked scaffolds by 35%. In contrast, the breaking load and stiffness of EDAC-crosslinked scaffolds, which were greater than those of UNXL or UV-crosslinked scaffolds, were virtually unaffected by the same load and enzyme treatments. Taguchi *et al.* encapsulated chondrocytes in an alkali-treated collagen (AlCol) gel, using pentaerythritol PEG ether tetrasuccinimidyl glutarate (4S-PEG) as a crosslinker (Fig. 3.2) [3]. AlCol had carboxyl groups generated by hydrolysis of the residual amide groups that exist in asparagine and glutamine of collagen. To encapsulate chondrocytes in gels at pH 7.4 and 37°C, 4S-PEG solution was added to the AlCol–chondrocyte mixture. The mixture was placed in a mold having a silicone rubber spacer. Results of MTT staining showed that cells survived after encapsulation in AlCol gels and biochemical analysis demonstrated that the DNA content in AlCol gels was constant after 3 weeks. The GAG content and mRNA expression of type II collagen and aggrecan increased with culture time.

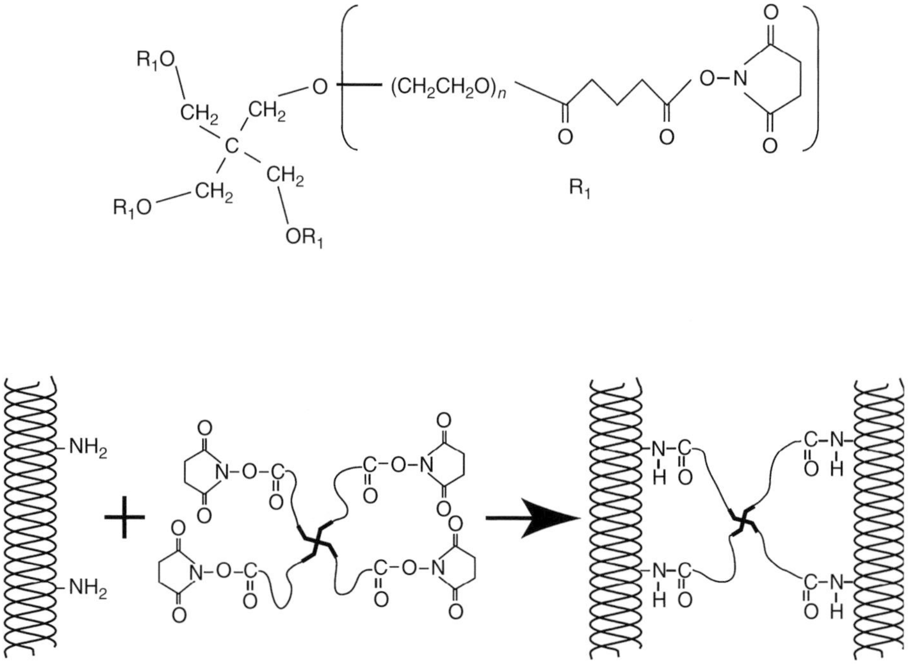

Fig. 3.2 Reaction between alkali-treated collagen and pentaerythritol PEG ether tetra-succinimidyl glutarate.

Transglutaminase (TG), a calcium-dependent enzyme, was reported to success-fully crosslink acid-solubilized bovine type I collagen in the presence of cells [4]. Transglutaminase is distributed intracellularly and extracellularly throughout the body and is responsible for tissue stabilization through the formation of amide crosslinks from carboxyamide and primary amine functionalities. *In vivo*, this transamidation reaction characteristically takes place between the glutamine and the lysine residues of the collagen (Fig. 3.3). In addition to TG, the lysyl oxidase (LO) family of enzymes is also known to catalyze the crosslinking of elastin and collagen fibers through oxida-tion of lysine to α-aminoadipic-δ-semialdehyde, a reactive precursor moiety. Orban *et al.* prepared gels by mixing collagen, Tris-HCl, $CaCl_2$, and TG at 4°C and maintained the mixture in a CO_2 incubator at 37°C in PBS until analysis. Double-layer tubular constructs were fabricated from crosslinked and non-crosslinked BMSC-containing collagen gels for burst strength analysis. The denaturation temperature (T_d) used as a relative measure of tissue strength increased from 38 to 66°C between the control and the foremost crosslinked sample. The average burst pressures between crosslinked (collagen:enzyme = 5,000:1) and non-crosslinked samples were determined to be 71 and 46 mmHg, respectively. Crosslinked and non-crosslinked gels were found to contain a similar number of viable cells at 9 days. A statistical difference existed at 0 and 14 days, with the crosslinked gels having a greater number of viable cells.

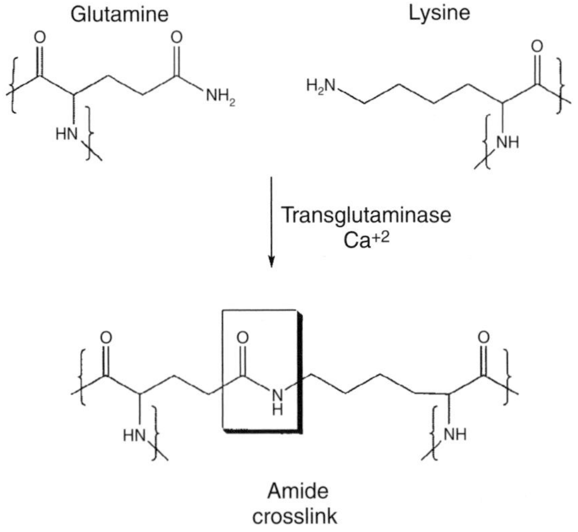

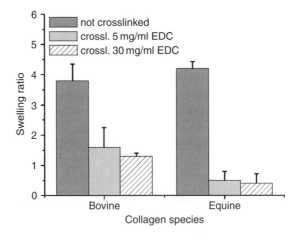

Fig. 3.3   Schematic of the crosslinking reaction of collagen by TG.

Angele *et al.* assessed the difference between the properties of bovine and equine-based collagen [5]. The use of equine collagen would serve as an alternative of bovine collagen, free of the potential risk of infection with exogenous or endogenous virus or prin. The physicochemical differences between equine and bovine collagen were evaluated with and without crosslinking with EDAC and NHS. The NHS was added at an EDAC:NHS ratio of 4:1, followed by crosslinking for 4 h. Figure 3.4 gives the swelling ratio of non-crosslinked and crosslinked collagens. No relevant

Fig. 3.4   Swelling ratio of scaffolds based on collagen of different species and with different degrees of EDAC(EDC)/NHS-crosslinking.

differences could be detected between the swelling ratios of non-crosslinked bovine and equine collagen. However, EDAC/NHS crosslinking resulted in a significant decrease in swelling ratio compared to non-crosslinked collagen. Crosslinking of equine collagen showed a significantly higher reduction of swelling ratios compared with the crosslinked bovine collagen. Without crosslinking, bovine and equine collagens were almost completely degraded after 1.5 h. However, with 5 mg/ml EDAC/NHS-crosslinking, equine collagen revealed only a partial degradation (up to 20%) after 6 days of collagenase treatment, whereas bovine-based collagen showed complete degradation. Equine collagen matrices with 30 mg/ml EDAC/NHS crosslinking were completely resistant to collagenase in contrast to bovine-based collagen matrices, which still had 60% degradation after 6 days of collagenase treatment.

Yost *et al.* developed a counter-rotating cone extrusion device to produce 3-D collagen tubes (Fig. 3.5) [6]. The collagen tubes were produced by loading collagen dispersion (collagen concentration: 25 mg/ml and pH: 2.5–3.0) into a syringe placed into a pump that fed the collagen through the feed port of extruder. As the dispersion was fed between the upper and the lower rotating cones, a collagen tube exited through the final annulus and was collected in a water bath. The collagen tubes were divided into 1.25-cm sections before the addition of neonatal rat cardiac myocytes and exposed to UV. For cell seeding, tubes were placed in culture dishes, and suspension of myocytes was injected into the lumen of each tube, using an intravenous catheter. The tubes were then placed in a rotating-wall bioreactor and the reactor was filled with an additional cell suspension. Figure 3.6 shows an example of hysteresis loops formed during passive stretch of collagen tubes seeded with neonatal cardiac cells (solid circles) and without seeded cells (open circles), along with a hysteresis loop developed with a feline papillary muscle.

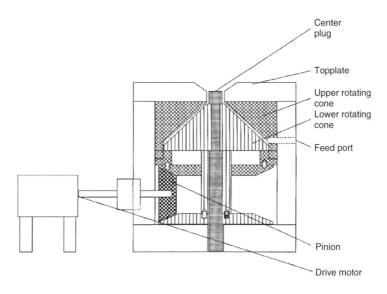

Fig. 3.5 A cutaway view of the counter-rotation cone extruder.

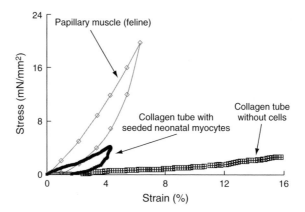

Fig. 3.6 Histeresis loops derived from stretching–unstretching cycles of a feline papillary muscle and collagen tubes with and without seeded myocytes.

Nahmias *et al.* developed a method for patterning cells on arbitrary surfaces including collagen gels with little loss of viability or function [7]. Single-cell suspensions of HUVECs and NIH 3T3 fibroblasts were sprayed with an off-the-shelf airbrush through a mask to create 100-μm scale patterns on collagen gels. Three-dimensional patterns were created by layering a collagen gel on top of the first pattern and patterning the top gel. Co-culture of rat hepatocytes with NIH 3T3 patterns on collagen gels resulted in localized increased activity of cytochrome P-450 along the pattern.

One of the primary limitations of collagen-based hydrogels for use in tissue-engineered grafts is that cells seeded within the gel cause it to contract as much as 70%. By forming a composite gel by adding short collagen fibers, Gentleman *et al.* determined that the contraction due to fibroblasts was decreased and permeability was increased [8]. Lewus *et al.* studied the effect of fiber number and serum concentration on the gel contraction. Short collagen fibers were included in adult rat bone-marrow stem-cell-seeded hydrogels for composite support [9]. The mass of fibers was varied from 1.6 to 31.3 mg per gel, and the effect of serum concentration in the growth medium was examined. Increasing fiber mass and decreasing serum concentration significantly decreased contraction, which plateaued after day 10. Cell number increased throughout the experiment, demonstrating the compatibility of bone-marrow stem cells with the collagen composite gels.

To address the mechanical limitations presented by collagen gels, Sosnik and Sefton produced semi-synthetic collagen-containing networks by photoinitiated free radical polymerization of collagen-containing modified PEO and poly(propylene oxide) block copolymers (PEO–PPO) (70% PEO content and MW = 15,000 Da) with methacrylate groups in aqueous solutions [10]. The resulting matrices showed increased stiffness, while maintaining the advantageous biological features of the collagen, mainly related to the facilitation of cell adhesion. Whereas collagen had a storage modulus (G') around 70 Pa and a loss modulus (G'') of 10 Pa, a crosslinked collagen/PEO–PPO system had G' and G'' values of 7400 and 1000 Pa, respectively.

Gelatin

Combination of gelatin with other polymers has been extensively studied to fabricate scaffolds. Gelatin exhibits good cell attachment presumably because of retention of informational signals such as Arg-Gly-Asp (RGD) sequence. Liu *et al.* covalently co-crosslinked gelatin and low MW HAc into a hydrogel using dithiobis(propanoic dihydrazide) (DTP) [11]. The DTP was first coupled to gelatin and HAc using EDAC followed by reduction of the first formed gels to give soluble thiolated gelatin (gelatin–DTPH) and HAc (HAc–DTPH). The solutions of HAc–DTPH and gelatin–DTPH were mixed in ratios of 75:25, 50:50, and 25:75, and the mixtures were then exposed to air for 12 h until gelation occurred through oxidation. The gels were frozen with liquid nitrogen and lyophilized to give macroporous sponges. As shown in Fig. 3.7, the rate of fibroblast attachment was as low as 3% on the gelatin-free HAc sponge. As the gelatin percentage increased, the cell attachment and proliferation increased, reaching a plateau at 50% gelatin. Cell attachment to the gelatin–HAc sponge could be blocked by preincubation of cell with a soluble FN peptide Gly-Arg-Gly-Asp (GRGD). On the HAc-only sponge, cell number decreased with culture time because the inability to attach led to cell death.

Silk Fibroin

Unger *et al.* revealed the ability of silk fibroin nets to support the growth of endothelial, epithelial, fibroblast, glial, keratinocyte, and osteoblast cell types [12]. The fibroin nets were produced first by a degumming step that removed the sericin of silk followed by a homogenization and drying step that yielded the 3-D, non-woven nets. The nets were approximately 200 μm thick and composed of a mesh of randomly oriented fibers that ranged between 10 and 30 μm in diameter, as shown in Fig. 3.8. Branch points and 3-D open spaces up to 300 μm were distributed throughout the structure and individual fibers generally exhibited a smooth surface as

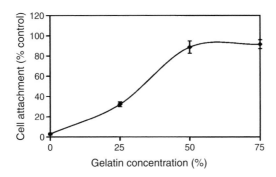

Fig. 3.7 Attachment of HTS fibroblasts on sponges that contained 0, 25, 50, and 75% gelatin–DTPH in HAc–DTPH to form sponges.

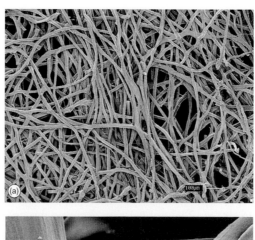

Fig. 3.8   SEM images of non-woven fibroin nets: (a) low-magnification image showing individual fibers and branching and (b) high-magnification image of individual fibroin fibers showing a relatively smooth rounded surface.

revealed by SEM micrographs. Meinel *et al.* compared the *in vitro* response of hMSCs on silk with the responses on tissue culture plastic (β-TCP; negative control), β-TCP with lipopolysaccharide (LPS) in the cell culture medium (positive control), and collagen films [13]. Furthermore, they studied the inflammatory response elicited upon intramuscular implantation of silk and silk–RGD films in comparison to collagen and PLA films. After 9 h of culture, the expression of the pro-inflammatory interleukin 1β (IL-1β) and inflammatory cyclooxygenase 2 (COX-2) in human MSC were comparable for silk, collagen, and β-TCP. After 30 and 96 h, gene expression of IL-1β and COX-2 in MSC returned to the baseline (pre-seeding) levels. The rate of cell proliferation was higher on silk films than on either collagen or β-TCP. *In vivo*, films made of silk, collagen, or PLA were seeded with rat MSCs, implanted intramuscularly in rats and harvested after 6 weeks. Histological and immunohistochemical evaluation of silk explants revealed the presence of circumferentially oriented fibroblasts, few blood vessels, macrophages at the implant–host interface, and the absence of giant cells. Kim *et al.* prepared macroporous 3-D scaffolds from silk fibroin both by an organic solvent process (hexafluoro-2-isopropanol, HFIP) and by an aqueous process [14]. Aqueous-based silk fibroin scaffolds were prepared by adding granular NaCl into aqueous solution of silk fibroin in containers. The containers were covered and left at room temperature, and then immersed in water and the NaCl extracted. The HFIP-derived silk fibroin scaffolds were prepared by adding granular NaCl to HFIP solution of silk fibroin. The solvent was evaporated at room temperature for 3 days. The silk/porogen matrix was then treated in methanol to induce the formation of β-sheet structure.

Peptide
Zhang and coworkers have developed a class of nanofibrillar gels with very high water contents (>99%) crosslinked by the self-assembling of self-complementary amphiphilic peptides in physiological medium. This is called "PuraMatrix™" because of its extreme purity of a single peptide component [15]. PuraMatrix™ is a 16-amino-acid synthetic peptide that is resuspended in water to generate a range of solution concentrations. Upon introduction of millimolar amounts of monovalent cations, Pura-Matrix™ underwent self-assembly into nanofibers with diameters around 10 nm. The physical size relative to cells and proteins, the amphiphilic peptides' charge density, and water-structuring abilities mimicked the *in vivo* ECM. The concentrations of peptide solution to be used for the material production, correlated with the average pore size of 5–200 nm, could be varied from 0.1 to 3% in water. Under appropriate culture conditions, these matrices maintained the functions of differentiated neuronal cells and chondrocytes, and promoted the differentiation of liver progenitor cells.

Genove *et al.* functionalized a peptide scaffold RAD 16-I (AcN-RADARADARADARADA-CONH$_2$) through direct solid phase synthesis extension at the amino terminal with three short-sequence motifs from two major proteins of the basement membrane, laminin 1 (YIGSR, RYVVLPR) and type IV collagen (TAGSCLRKFSTM) [16]. These tailor-made scaffolds increased the formation of confluent cell monolayers with cobblestone phenotype of human aortic endothelial cells (HAEC) in culture. In parallel experiments consisting of culture HAEC on other substrates as controls including type I collagen gels and Matrigel™, similar monolayer formation was observed. Additional assays designed to evaluate EC function indicated that HAEC monolayers obtained on these scaffolds not only maintained LDL uptake activity but also enhanced nitric oxide release and elevated laminin 1 and type IV collagen deposition.

1.1.2. Polysaccharides

Barbucci *et al.* crosslinked polysaccharides to endow good resistance to the physiological environment while maintaining the excellent biological properties [17]. Their approach involved crosslinking of polysaccharides using the carboxylate groups as the point of attack of the reaction while maintaining an incomplete utilization of all the carboxylate groups which are necessary for the expression of the

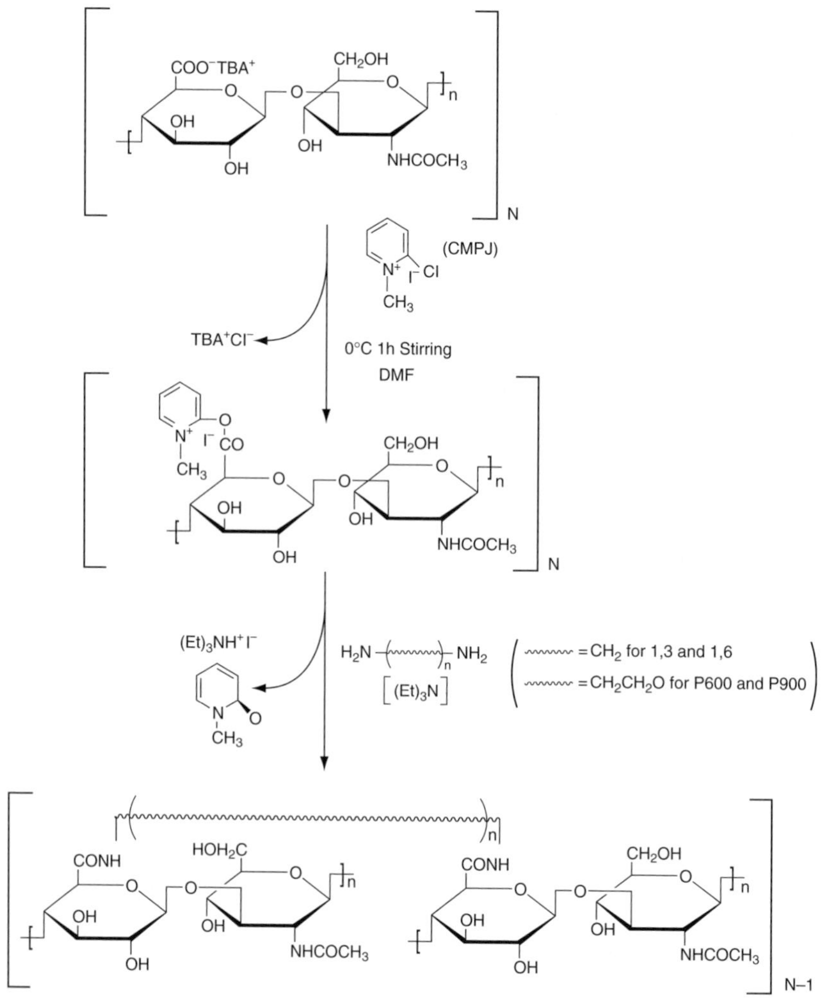

Fig. 3.9   Crosslinking reaction of HAc using CMPJ as the activating agent.

biological characteristics of polysaccharides. Figure 3.9 shows the scheme of the crosslinking reaction using an activating agent, 2-chloro-1-methylpyridinium iodide (CMPJ), and crosslinking agents, diamines including 1,3-diaminopropane (1,3), 1,6-diaminohexane (1,6), *O,O′*-bis(2-aminopropyl) poly(ethylene glycol) 500 (P600), and *O,O′*-bis(2-aminopropyl) poly(ethylene glycol) (P900) [18]. Briefly, a solution of sodium salt of polysaccharide (1%) was subjected to a sodium–hydrogen ionic exchange with the use of Dowex 50WX8 resin and then was added to a 5% tetrabutylammonium (TBA) hydroxide solution until pH 8–9 was reached. The solution was lyophilized to obtain the TBA salt of polysaccharide. The TBA salt was then dissolved in *N–N′* dimethyl formamide under stirring and nitrogen flow. The solution was kept at 0°C before addition of CMPJ. Activation of the HAc carboxylate groups proceeded rapidly by a nucleophilic attack on the carboxylate ion by CMPJ and the liberation of TBA^+Cl^-. Then, the intermediate nucleophilic attack by the diamine on the carbonyl group underwent with elimination of the heterocyclic in the form of 1-methyl-2-pyridone. The reaction was catalyzed by a small amount of triethylamine as hydrogen iodide capture. The reaction mixture was then left at room temperature for 3–4 h to form a hydrogel. The reaction product had characteristics of a gel, even in organic solvent. They applied this crosslinking reaction to alginate and carboxylmethyl cellulose in addition to hyaluronate.

HAc

Shu *et al.* incorporated a latent crosslinking agent into water-soluble HAc. First, they synthesized dithiobis(propanoic dihydrazide) (DTP) and dithiobis(butyric dihydrazide) (DTB) and then coupled them to HAc with CDI chemistry [19]. Next, disulfide bonds of the initially formed gel were reduced using dithiothreitol (DTT) to give the corresponding thiol-modified macromolecular derivatives, HAc–DTPH and HAc–DTBH. The degree of substitution of HAc–DTPH and HAc–DTBH could be controlled from 20 to 70% of available glucuronate carboxylic acid groups. The pKa values of the HAc-thiol derivatives were 8.87 (HAc–DTPH) and 9.0 (HAc–DTBH). The thio groups could be oxidized in air to reform disulfide linkages, which resulted in HAc–DTPH and HAc–DTBH hydrogel films. Further oxidation of these hydrogels with dilute H_2O_2 created additional crosslinks and afforded poorly swellable films. The disulfide crosslinking was reversible, and films could be again reduced to sols with DTT. To demonstrate the capacity for *in vitro* cell encapsulation, Shu *et al.* dissolved HAc–DTPH in a culture medium to give 3% solution under N_2 protection and adjusted the solution pH to 7.4. Murine fibroblasts were mixed with this HAc–DTPH solution to a final concentration of 2×10^6/ml. Next, 0.5 ml of the HAc–DTPH solution was added into a 12-well plate. The cell-loaded plates were incubated until a solid hydrogel formed. Cells could proliferate in the hydrogel after culture for 2 and 3 days and the cell number increased by 15% at day 3.

Leach *et al.* modified HAc with glycidyl methacrylate (GM), yielding a photocrosslinkable conjugate, GMHA [20]. The GMHA was prepared by reacting an aqueous HAc solution with 20-fold molar excess of GM in the presence of excess triethylamine and tetrabutyl ammonium bromide. For creation of cell-adhesive GMHA constructs, they further conjugated GMHA with acrylated forms of PEG and PEG-peptides according to the scheme in Fig. 3.10. Varying the reactant concentrations produced stable hydrogels with peptide conjugation efficiencies up to 80% and peptide concentrations in the range of 1–6 μmol peptide per ml of hydrogel.

Alginates

Stevens *et al.* prepared ionically crosslinked alginate hydrogels from four alginates differing in their plant origin,% α-L-guluronic acid (%G) units, intrinsic viscosity, MW of the sodium salts, as shown in Table 3.1 [21]. Alginates with relatively high percentage of guluronic acid (65–75%) were referred to as G1 and G2, while the

a. EDC-mediated GMHA–peptide conjugates:

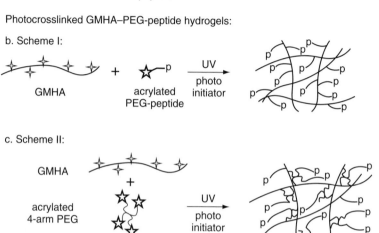

GMHA   peptide

Photocrosslinked GMHA–PEG-peptide hydrogels:

b. Scheme I:

GMHA   acrylated
                        PEG-peptide

c. Scheme II:

GMHA

acrylated
4-arm PEG

acrylated
PEG-peptide

Fig. 3.10 Schematic of peptide conjugation chemistries. Three approaches were evaluated as means of covalently binding peptides: (a) to crosslinked GMHA and (b, c) during the hydrogel crosslinking reaction.

Table 3.1

The plant origin, fraction of the polymer composed of G monomers, MW, and intrinsic viscosity (determined for a 1% solution) for the alginates (sodium salts) employed in the study

Alginate	Plant Origin	%G units	MW (Da)	Intrinsic Viscosity (mPa s^{-1})
M1	*Macrocystic pyrifera*	40	12,000–80,000	20–25
M2	*Macrocystic pyrifera*	40	80,000–120,000	80–200
G1	*Laminaria hyperborea*	65–75	120,000–150,000	20–70
G2	*Laminaria hyperborea*	65–75	300,000–350,000	200–400

ones with a relatively high mannuronic acid (M) content (40%) were referred to as M1 and M2. The influence of alginate and calcium ion concentration on the gel formulations was investigated utilizing $CaSO_4$ and $CaCl_2$ as a source of divalent ions. Gelation could be induced by addition of 200 or 300 mM $CaSO_4$ to 1 and 2% solution of the sodium alginates, as shown in Table 3.2. However, in all cases,

Table 3.2
Exclusion criteria for the various gel compositions

	CaSO$_4$ 200 mM	CaSO$_4$ 300 mM	CaCl$_2$ 50 mM	CaCl$_2$ 75 mM	CaCl$_2$ 100 mM	CaCl$_2$ 150 mM	CaCl$_2$ 300 mM
M1, 1% (w/v)	F, S, P	F, S, P	F	F	F		
M1, 2% (w/v)	S, P	S, P	F	✓	✓		
M2, 1% (w/v)	F, S, P	F, S, P	F	F	F		
M2, 2% (w/v)	S, P	S, P	F	✓	✓		
G1, 1% (w/v)	F, S, P	F, S, P	F	F	F		
G1, 2% (w/v)	S, P	S, P	F	✓	✓		
G2, 1% (w/v)	F, S, P	F, S, P	F	F	F	F	F
G2, 2% (w/v)	S, P	S, P	F	✓	✓	✓	✓
G2, 3% (w/v)				✓	NI	NI	NI
G2, 4% (w/v)				✓	NI	NI	NI

S, slow gelation (>10 min to reach completion); P, presence of precipitates; F, fragmentation of the gel upon handling; NI, unsuitability for injectable delivery. Gels that were not excluded utilizing this set of criteria are indicated with ✓.

a period of several hours was required for gelation to reach completion, and CaSO$_4$ precipitates were apparent throughout the gel, particularly when 300 mM CaSO$_4$ was utilized. By contrast, rapid gelling (<1 min) could be achieved in 1 and 2% solution of the alginates using 50–100 mM CaCl$_2$ solutions. Gels formed with 1% sodium alginate solutions and 50 mM CaCl$_2$ exhibited poor mechanical properties (easily fragmented), whereas gels formed utilizing 2% alginate and 75–100 mM CaCl$_2$ were mechanically more stable and remained cohesive when handled. Gels formed from G1 and G2 which had a mean G-block length ($N_{G>1}$) equal to 16 were favored over the gels formed utilizing M1 and M2 which had a considerably lower guluronic acid content and a mean G-block length ($N_{G>1}$) of 6.3. The G2 consistently produced more homogeneous gels than G1 as determined visually, presumably due to the higher MW and intrinsic viscosity of G2 relative to G1. Figure 3.11 shows Young's modulus as a function of alginate and calcium ion concentration. While Young's moduli showed a significant increase in a manner proportional to increasing alginate concentration, changes as a function of calcium ion concentration were <0.03 MPa. Drury *et al.* studied tensile properties of alginate hydrogels as a function of alginate type, formulation, gelling conditions, incubation, and strain rate [22]. They found that incubation of hydrogels for at least 7 days diminished many of the initial tensile property differences associated with formulation and gelling conditions.

Boontheekul *et al.* investigated whether alginate gel degradation could be controlled by combining partial oxidation of polymer chains prior to gel formation and the utilization of a bimodal MW distribution in gel formation [23]. Myoblast culture studies were performed on these degradable alginate gels to ensure that the changes did not diminish their compatibility with cells. As mammalian cells cannot readily interact with alginates due to the absence of cell surface receptors binding to

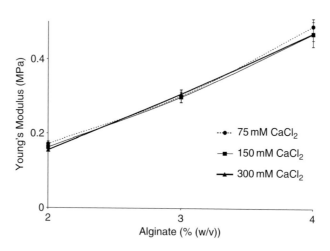

Fig. 3.11   Young's modulus versus % G2 alginate, for gelation induced with 75, 150, and 300 mM CaCl₂ solutions.

alginates and lack of protein adsorption onto alginates, the alginate chains were chemically modified to present cell adhesion ligands. Alginates were partially oxidized to a theoretical extent of 1% with sodium periodate, which created acetal groups susceptible to hydrolysis. The ratio of low MW to high MW alginates used to form gels was also varied, while maintaining the gel forming ability of the polymer. The rate of degradation was found to be controlled by both the oxidation and the ratio of high to low MW alginates, as monitored by the reduction of mechanical properties and corresponding number of crosslinks, dry weight loss, and MW decrease. Myoblasts adhered, proliferated, and differentiated on the modified gels at a comparable rate to those cultured on the unmodified gels.

Chondroitin Sulfate

In order to enhance the bioactivity of a scaffold and potentially improve tissue regeneration, Li *et al.* synthesized photopolymerizable hydrogels derived from chondroitin sulfate (CS) [24]. They added methacrylate groups to CS using glycidyl methacrylate (GM), as shown in Fig. 3.12. They chose GM from methacrylate options such as methacrylic anhydride and methacryloyl chloride because of the efficiency of reaction and lack of cytotoxic by-products compared with other reagents. The CS was dissolved in PBS, followed by addition of GM. And 150 µl of 20% solution of the macromer, methacrylated CS (CS-MA), was placed in tissue culture inserts and polymerized using a UV photoinitiator (Irgacure 2959, 0.05%) and 365-nm light. The CS-MA was also allowed to copolymerize with (PEO)–diacrylate (PEODA) macromers. The amount of water that a hydrogel is able to absorb was related to the crosslinking density and the pore structure which may influence the survival of encapsulated cells. Water absorption of CS-MA hydrogels decreased in inverse proportion to the PEODA

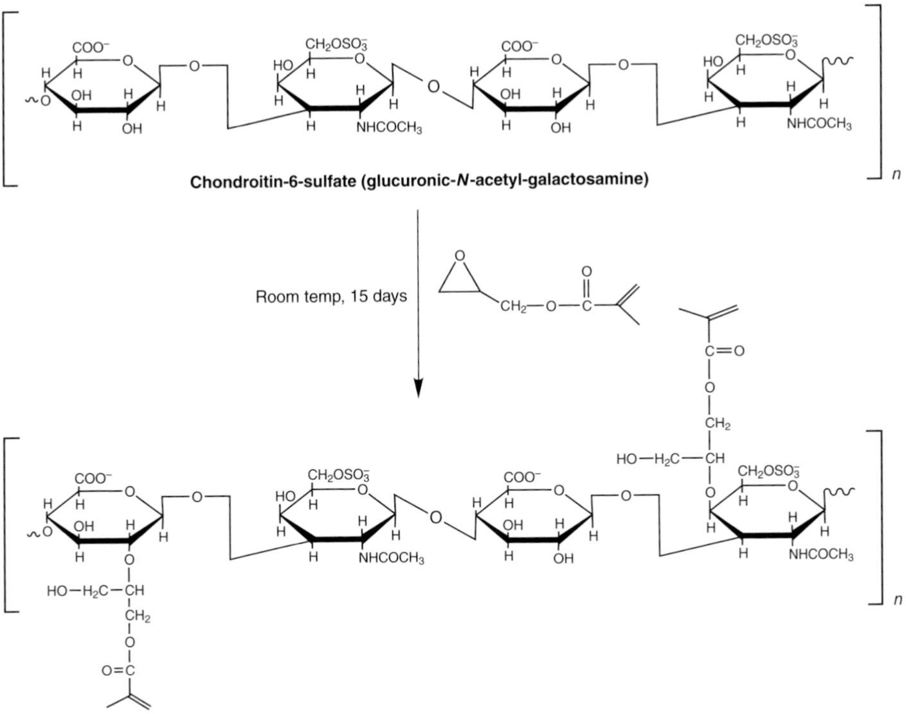

Fig. 3.12 Synthetic pathway of methacrylated chondroitin sulfate (CS-MA). The CS was reacted with GM to form the CS-MA macromer.

content. The CS hydrogels degraded in the presence of chondroitinase and chondrocytes remained viable after photoencapsulation and incubation in the hydrogels.

Chitin and Chitosan

Transparent chitin hydrogel tubes were prepared by Freier *et al.* from chitosan solutions using acylation chemistry and mold casting techniques [25]. Alkaline hydrolysis of chitin tubes resulted in chitosan tubes, with the extent of hydrolysis controlling the resulting amine content. This, in turn, impacted compressive strength and cell adhesion. Chitosan tubes were mechanically stronger than their chitin origins, as measured by the transverse compressive test, where tubes having degrees of acetylation of 1, 3, 18% (*i.e.*, chitosan), and 94% (*i.e.*, chitin) supported loads at 30% displacement of 41, 25, 11, and 9 g, respectively. The chitin-processing methodology could be optimized for compressive strength, by either incorporation of reinforcing coils in the tube wall, or air-drying the hydrogel tubes. Chitin and chitosan supported adhesion and differentiation of primary chick dorsal root ganglion neurons *in vitro*. Chitosan films showed significantly enhanced neurite outgrowth relative to chitin films, reflecting the dependence of nerve cell affinity on the amine content in the polysaccharide.

Other Natural Polymers

In order to mimic the natural cartilage matrix for use as a scaffold for cartilage tissue engineering, Chang *et al.* prepared an artificial ECM from gelatin (0.5 g), CS (0.1 g), and HAc (6 mg), as the percentage dry weight of each component in hyaline cartilage is 15–20% type II collagen, 5–10% CS, and 0.05–0.25% HAc [26]. The three powders were mixed with water and crosslinked using EDAC. The solution was frozen and then lyophilized. The resulting scaffold had a uniform pore size of 180 μm and porosity of 75%. Chondrocytes were uniformly distributed in the scaffold in spinner flask cultures, retained their phenotype for at least 5 weeks, and synthesized type II collagen.

Ferreira *et al.* synthesized absorbable dextran-based hydrogels using a biocatalytic transesterification reaction between dextran and divinyladipate in dimethylsulfoxide via a scheme given in Fig. 3.13 [27]. The hydrogels showed a higher elastic modulus for a given swelling ratio than chemically synthesized dextran-based hydrogels. *In vivo* studies in rats showed that the hydrogel network was bioabsorbable over a range of time scales from 5 to over 40 days, when the degree of substitution was lower than 30%.

Stevens *et al.* fabricated agarose stamps for direct patterning of cells on porous scaffolds [28]. As illustrated in Fig. 3.14, hot solutions of 2–3% agarose were cast against silicone molds to a height of ~3 cm. Once the agarose solution had gelled at room temperature and ambient pressure (~2 h), the agarose stamp was peeled away from the silicone master, and used for cell patterning.

1.2. Synthetic Polymers

1.2.1. *Poly(α-hydroxyacid)s*

Homopolymers

Moran *et al.* determined the effects of polymer composition on the physical and biological properties of PLA/PGA blends with fractional PLA contents [29]. The blends were fabricated by coating non-woven meshes of PGA with the solution of PLA in methylene chloride. The addition of 12% PLA significantly decreased cell adhesion, probably because the PLA coat was less cell adhesive than PGA. Deposition of increasing amounts of PLA increased the compressive equilibrium modulus of the blends, resulting in a more than 20-fold increase in modulus when comparing

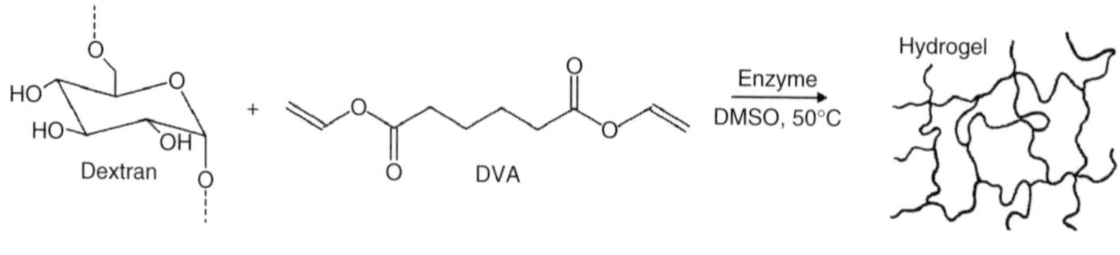

Fig. 3.13 Proleather-catalyzed acylation of dextran with DVA.

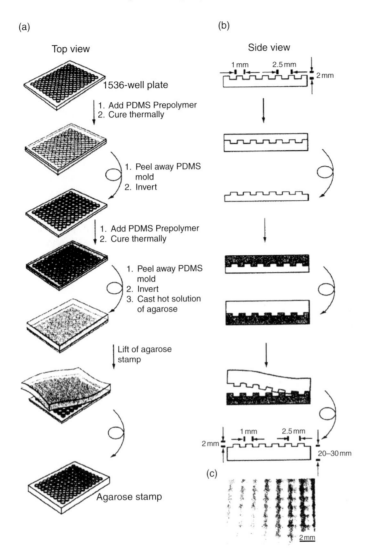

Fig. 3.14 Schematic depicting the fabrication of micropatterned agarose stamps from a 1536-well plate: top view (a) and side view (b); Image of an agarose stamp (c).

68% PLA blend with 0% PLA. A semi-logarithmic plot of mass fraction remaining against time followed pseudo first-order kinetics.

Prokop *et al*. examined soft tissue reactions resulting from LLA–DLLA(70:30) copolymer pins with (C-pin) and without (Polypin) addition of β-TCP. The pins were implanted in the medial femoral condyle of 36 sheep [30]. The pins with 10% TCP showed a decreased initial strength of 128 MPa compared to those without TCP (152 MPa). The combination of PLA and TCP showed a large degradation period than PLA alone. There was no significant difference in the synovial tissues and lymph nodes. In all cases in both groups, the histologic examinations revealed that bone or bony substitute filled the channels of degraded implants. On the examination of the harvested lymph nodes, changes were observed in the inguinal nodes. A foreign-body reaction was observed in bilateral inguinal lymph nodes at 18 months, but there was no indication of such a reaction in subsequent lymph node stations. The natural ageing process, injuries, or previous operations can lead to reactive changes, particularly in the inguinal lymph nodes. Shea *et al.* described up to grade IV reactive changes in the inguinal lymph nodes in 8 out of 18 patients (44%) before implantation of knee or hip prostheses [31]. In 21 of 27 patients (78%) undergoing revision total knee replacement operations, foreign-body reactions up to grade IV were discovered.

Poly(ε-caprolactone) has been used as a biomaterial for scaffold fabrication, but its resorption rate is too low to function as a scaffold. For instance, Pitt *et al.* studied the degradation of PCL in rabbits, rats, and water [32]. Table 3.3 shows the *in vivo* changes in the weight, intrinsic viscosity ($[\eta]$), MW (M_n and M_w),

Table 3.3
In vivo degradation of PCL in rabbit[a]

Time (weeks)	Wt Loss (%)	% Crystallinity (SD)	$[\eta]$[b] (dL/g)	GPC $[\eta]$	GPC $M_n \times 10^{-3}$	GPC M_w/M_n	Elastic Modulus (kg/cm^2)
0	0	45.3 ± 0.3[c]	0.883	0.944	50.9	1.66	2700 ± 100
4	0.3	52.4 ± 0.4	0.830	—	—	—	3100
4	0.4	49.2 ± 1.9	0.820	—	—	—	3020
8	0.8	54.7 ± 1.1	0.810	—	—	—	3160
8	0.1	53.6 ± 0.2	0.793	—	—	—	3240
16	0.7	57.8 ± 1.0	0.733	0.736	35.9	1.75	3010
16	3.6	56.0 ± 0.8	0.663	0.723	34.4	1.78	3240
32	6.4[d]	59.3 ± 1.1	0.610	0.603	26.7	1.79	4040
32	4.5	56.4 ± 2.8	0.607	0.644	24.5	2.21	3370
48	4.6	63.4	0.490	0.471	16.5	2.11	2960
48	5.7[d]	64.2 ± 0.3	0.472	0.467	17.5	1.94	2850
64	—	66.3 ± 0.0	0.356	0.391	12.9	2.14	[e]
64	2.6	64.6 ± 0.1	0.364	0.398	13.1	2.15	
80	15.9	69.0 ± 2.0	0.310	0.291	8.8	2.14	
96	12.2	73.8 ± 1.3	0.150	0.222	6.3	2.15	
112	18.1	—	—	0.191	5.3	2.10	
120	7.4	79.5 ± 1.6	0.184	0.174	4.5	2.20	
120	—	—	0.174	0.166	4.2	2.18	
120	7.6	—	0.176	0.181	4.6	2.24	
129	—	—	0.180	0.164	4.5	2.04	
129	27.3	—	0.208	0.163	4.6	1.98	
144							

[a] Capsules melt extruded; dimensions 5 cm × 2.49 mm (od) × 1.96 mm (id); sterilized by γ-irradiation (2.5 Mrads).
[b] Measured in toluene.
[c] Standard deviation, $n = 2$.
[d] Small part of capsule lost during recovery.
[e] Capsules fragmented after this time.

crystallinity, and elastic modulus as a function of time. It can be seen that this poly-mer undergoes far slower degradation than PLLA.

Likewise, poly(hydroxyalkanoates) (PHAs) have been studied as a candidate for scaffold. However, PHAs also degrade too slowly to use as absorbable materials [33]. Holland *et al.* studied hydrolysis of films prepared from a series of hydroxybuty-late–hydroxyvalerate (HB–HV) copolymers in a pH 7.4 buffer at 70°C over a period of 2500 h [34]. Figure 3.15 shows the weight loss profiles of these films as a function of degradation time. Only the polymer with MW of 3.6×10^4 Da and 20% HV exhibited significant degradation, but this sample was unrepresentative of the general trend of degradation rates of PHAs since the physical strength associated with this low MW material was not adequate for applications where mechanical properties are important. In this study, to accelerate hydrolysis the temperature during hydrolysis was raised to 70°C, much different from the physiological temperature.

Copolymers

Zhu *et al.* first prepared A–B type diblock copolymers from PDLLA and PEG for medical use in 1986 [35]. Lieb *et al.* synthesized diblock copolymers by a ring-opening reaction from D,L-lactide and PEG-monomethyl ether (Me.PEG) using stannous 2-ethylhexanoate (SnOct$_2$) as a catalyst to control cell adhesion and differ-entiation [36]. The MWs of the synthesized diblock copolymers determined by ^{1}H NMR and GPC are listed in Table 3.4. Figure 3.16 shows attachment of rat marrow stromal cells (rMSCs) to four different diblock copolymer films, PLA, PLGA, and tissue culture polystyrene (TCPS). There was a high percentage of the seeding cells attached to the three hydrophobic materials, PLA, PLGA, and TCPS, whereas cell attachment was decreased on films from the diblock copolymers. Com-parison of the four diblock copolymers revealed that cell attachment was inversely related to the Me.PEG content of the copolymers. On the contrary, with respect to cell differentiation, higher activities were found on block copolymer films after 16

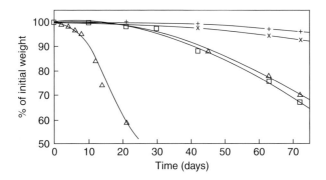

Fig. 3.15 Progressive weight loss of films (discs) of various HB–HV copolymers maintained in an aqueous environment at 70°C and pH 7.4. Initial disc dimensions: 2 cm diam.; 0.15 mm thick. (+) 10% HV, 750k; (×) 20%, 300k; (▵) 12%, 170k; (□) 12%, 100k; (▲) 20%, 36k.

Table 3.4

M_w, M_n, and polydispersity indices of the polymers as determined by gel permeation chromatography

Polymer	Mw[b]	Mn[b]	PI[b]	Mn(PEG)[c]	Mn(PLA)[c]	PEG:PLA[a]
Me.PEG2-PLA40	51,900	28,400	1.8	2,000	38,200	5%:95%
Me.PEG2-PLA20	47,600	23,000	2.1	2,000	20,900	9%:91%
Me.PEG5-PLA45	62,900	45,300	1.4	5,100	47,500	10%:90%
Me.PEG5-PLA20	49,500	35,900	1.4	5,000	19,500	20%:80%
PLGA	84,100	48,300	1.7	—	—	—
PLA	83,300	55,400	1.6	—	—	—

[a] The number-average molecular weight of the PEG and PLA blocks of each polymer was calculated from ^{1}H NMR data. All molecular weight data were rounded to the nearest 100. The last column shows the actual PEG/PLA ratio of the polymers determined from ^{1}H NMR data.
[b] Data obtained by GPC.
[c] Data obtained by ^{1}H NMR.

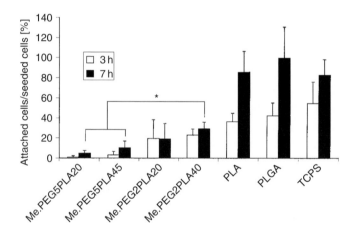

Fig. 3.16 Percentage of plated marrow stromal cells that attached to the different polymers with a seeding density of 5000 cells/cm[2] after 3 and 7 h.

days than on PLA after 10 days. Cells on Me.PEG2–PLA40 produced a significantly larger area of mineralization than did cells on any other materials. About 40% of the film surface was covered with mineralized ECM.

Saito *et al.* synthesized triblock copolymers consisting of two lactide-*p*-dipoxanone(DX) polymer segments and a PEG segment [37]. Ring-opening polymerization of D,L-lactide, DX, and PEG with MW of 3,000 Da was carried out in the presence of stannic chloride as a catalyst at 165°C for 6 h in vacuum. The molar ratio of the LA, DX, and PEG in the resultant copolymers was determined by means of ^{1}H NMR peak area. Figure 3.17 shows the hydrolytic degradation of 9.6K-PLA–PEG (MW = 9.5 × 10[3] Da; PLA:PEG = 68:32) and 9.5K-PLA–DX–PEG (MW = 9.5 × 10[3] Da; PLA:DX:PEG = 45:17:38) in PBS

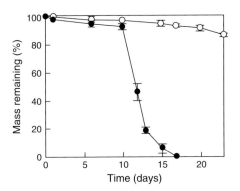

Fig. 3.17 Loss of mass as a function of degradation time for the 9.5K-PLA–DX–PEG (filled circles) and the 9.5K-PLA–PEG (empty circles) polymers *in vitro*.

at 37°C. It is seen that the random insertion of DX in the PLA segment promoted the polymer degradation.

Kim *et al.* fabricated fine fibers using the electrospinning technique [38]. They used four miscible compositions: PLA(PDLLA), PLGA(LA/GA = 50/50), PLA-b-PEG-b-PLA triblock copolymer, and ʟ-lactide used as a catalyst. The role of high MW (HMW) PLA was to provide the overall mechanical strength, the role of lactide was to grossly tune the degradation rate, and the role of PLA-b-PEG-b-PLA copolymer was to control the hydrophilicity. All elecrospun non-woven webs exhibited sub-micron-sized fiber diameters, ranging from 500 to 800 nm, and a density of 0.3–0.4 g/cm^3 or a porosity of 70–75%. Figure 3.18 shows the degradation profile of a blend consisting of 40% of HMW PLA, 15% of lactide, 20% of PLA-b-PEG-b-PLA triblock copolymer, and 25% of low MW PLGA which they thought to exhibit

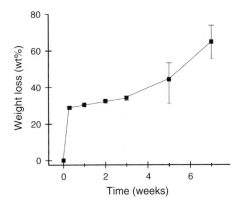

Fig. 3.18 *In vitro* degradation of electrospun scaffolds containing 90K PDLLA 40/LA 15/BC20/10K PLGA 25.

an ideal biodegradation profile, good hydrophilicity, and stable mechanical properties in aqueous conditions. After the initial weight loss due to the escape of lactide
and bridging materials (PLGA and PLA-b-PEG-b-PLA), a distinct change in the
weight loss profile was seen between 3 and 5 weeks, which might be attributed to
the autocatalytic effect of the lactide retained by the bridging materials. The contact
angles among the samples with purified water and the samples with cell-containing
solutions were almost the same. The initial contact angle of the blend based on pure
PLGA (LA/GA = 75/25) was 105°, indicating that the cell solution did not spread
well on the hydrophobic surface. In contrast, the PLGA- or PLA-based blends,
which contained 25–30% of hydrophilic components such as lactide and triblock
copolymer, were hydrophilic, resulting in initial contact angles of less than 60°.

Huang *et al.* prepared (PCL)/PEG diblock and triblock copolymers by
ring-opening polymerization of ε-CL in the presence of PEG, using zinc metal as
catalyst [39]. To tailor the length of each block in block copolymers, Deng *et al.*
synthesized poly(γ-benzyl-L-glutamic acid)-b-PEO-b-PCL (PBLG–PEO–PCL)
based on the reaction schemes described in Fig. 3.19 [40]. First, cyano-terminated
diblock copolymer CN–PEO–PCL was synthesized by the sequential polymerization of ethylene oxide (EO) and CL in acetonitrile and a potassium naphthalene
initiator system and subsequently the cyano-group was converted into amino
group to obtain NH_2–PEO–PCL by catalytic hydrogenation. The NH_2–PEO–PCL
was used as macro-initiator of the ring-opening polymerization of γ-benzyl-L-
glutamate-*N*-carboxyanhydrides (Bz-L-GluNCA) to prepare the triblock copolymer
PBLG–PEO–PCL.

Meek *et al.* synthesized copolymers of CL, LLA, and DLLA instead of
pure LLA to obtain an amorphous biomaterial with faster degradation than semi-
crystalline copolymer of LLA–ε-CL(50/50) [41]. They prepared the copolymers at
a LA:CL molar ratio of 65:35 and an LLA:DLA molar ratio of 85:15. A transparent
tube prepared from this poly(65/35(85:15 L/D) LA–ε-CL) exhibited *in vitro* degradation in buffer solution as shown in Fig. 3.20 and Table 3.5. The tube became
opaque after 5 weeks, and after a period of 12 weeks surface erosion started to
occur. After 22 weeks, the remaining mass was 20%, indicating that 80% of the tube
had been dissolved in the buffer solution as a result of hydrolysis. Initially, T_g of the
completely amorphous material was 18°C and, after 6 weeks, a small but broad
melting peak started to occur between 50 and 100°C, which increased in size and
shifted to a higher temperature region after 22 weeks (60–105°C). Because the
melting temperature of PCL and PLLA are 60 and 188°C, respectively, the observed
melting peak was attributed to LLA-rich regions.

Jeong *et al.* fabricated porous tubular scaffolds from an LLA–ε-CL copolymer,
P(LA/CL), with MWs ranging between 1×10^5 and 3×10^5 Da, seeded with
SMCs, and implanted into nude mice to investigate the tissue compatibility and
in vivo degradation behavior [42]. Histological examination of all the implants with
H&E staining, masson trichrome staining, SM α-actin antibody, and CM-DiI labeling confirmed the regular morphology and biofunction of the SMCs seeded and the
expression of the vascular smooth muscle (VSM) matrices in P(LA/CL) scaffolds.

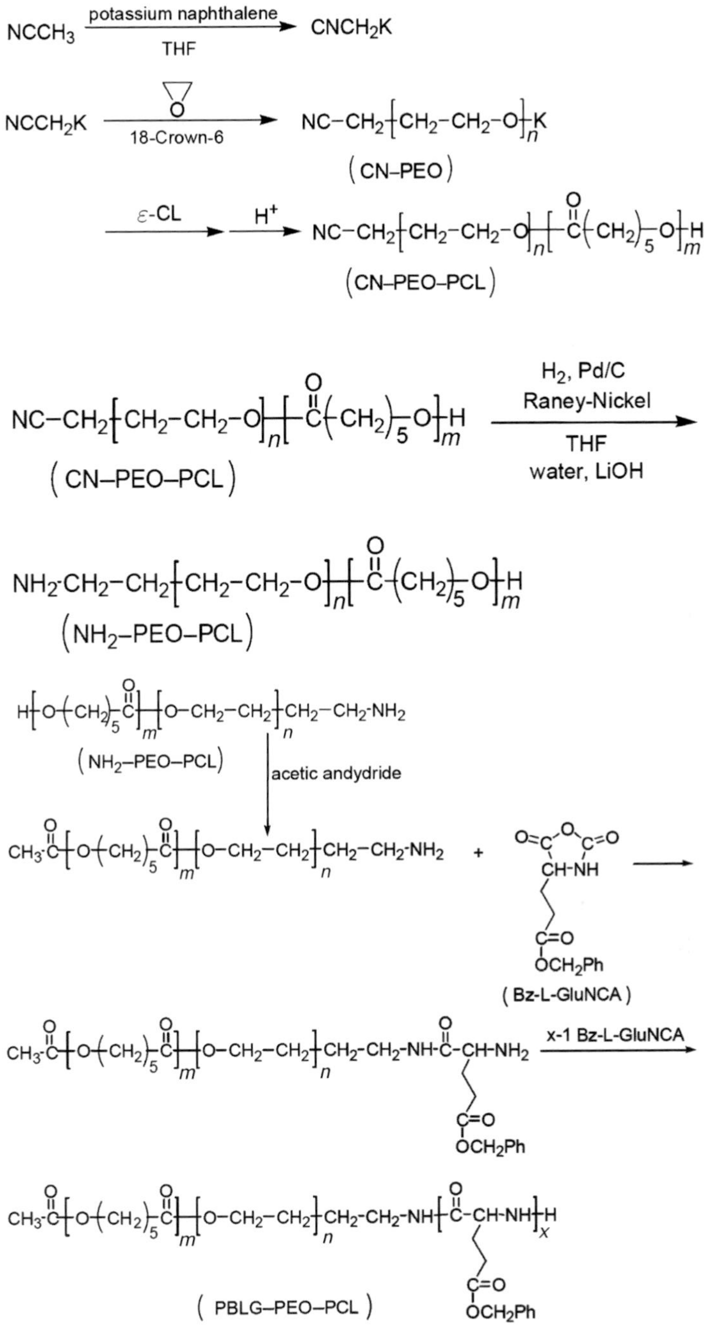

$$NCCH_3 \xrightarrow[\text{THF}]{\text{potassium naphthalene}} CNCH_2K$$

$$NCCH_2K \xrightarrow[\text{18-Crown-6}]{\bigtriangledown O} NC-CH_2\left[CH_2-CH_2-O\right]_n K$$

$$\left(CN\text{-}PEO\right)$$

$$\xrightarrow{\varepsilon\text{-CL}} \xrightarrow{H^+} NC-CH_2\left[CH_2-CH_2-O\right]_n\left[\overset{O}{\overset{\|}{C}}(CH_2)_5-O\right]_m H$$

$$\left(CN\text{-}PEO\text{-}PCL\right)$$

$$NC-CH_2\left[CH_2-CH_2-O\right]_n\left[\overset{O}{\overset{\|}{C}}(CH_2)_5-O\right]_m H \xrightarrow[\substack{\text{THF} \\ \text{water, LiOH}}]{\substack{\text{H}_2,\text{ Pd/C} \\ \text{Raney-Nickel}}}$$

$$\left(CN\text{-}PEO\text{-}PCL\right)$$

$$NH_2\text{-}CH_2-CH_2\left[CH_2-CH_2-O\right]_n\left[\overset{O}{\overset{\|}{C}}(CH_2)_5-O\right]_m H$$

$$\left(NH_2\text{-}PEO\text{-}PCL\right)$$

$$H\left[O-(CH_2)_5-\overset{O}{\overset{\|}{C}}\right]_m\left[O-CH_2-CH_2\right]_n CH_2-CH_2\text{-}NH_2$$

$$\left(NH_2\text{-}PEO\text{-}PCL\right) \quad \Big\downarrow \text{acetic andydride}$$

$$CH_3\text{-}\overset{O}{\overset{\|}{C}}\left[O-(CH_2)_5-\overset{O}{\overset{\|}{C}}\right]_m\left[O-CH_2-CH_2\right]_n CH_2-CH_2\text{-}NH_2 \quad + \quad \underset{\substack{CH-NH \\ | \\ C=O \\ | \\ OCH_2Ph}}{O=\underset{}{C}\diagdown\!\!\!O\!\!\!\diagup\underset{}{C}=O}$$

$$\left(Bz\text{-}L\text{-}GluNCA\right)$$

$$CH_3\text{-}\overset{O}{\overset{\|}{C}}\left[O-(CH_2)_5-\overset{O}{\overset{\|}{C}}\right]_m\left[O-CH_2-CH_2\right]_n CH_2-CH_2\text{-}NH-\overset{O}{\overset{\|}{C}}-\underset{\substack{| \\ C=O \\ | \\ OCH_2Ph}}{CH}\text{-}NH_2 \xrightarrow{x\text{-}1\ Bz\text{-}L\text{-}GluNCA}$$

$$CH_3\text{-}\overset{O}{\overset{\|}{C}}\left[O-(CH_2)_5-\overset{O}{\overset{\|}{C}}\right]_m\left[O-CH_2-CH_2\right]_n CH_2-CH_2\text{-}NH-\left[\overset{O}{\overset{\|}{C}}-\underset{\substack{| \\ C=O \\ | \\ OCH_2Ph}}{CH}\text{-}NH\right]_x H$$

$$\left(PBLG\text{-}PEO\text{-}PCL\right)$$

Fig. 3.19 Schemes for synthesis of a triblock copolymer (PBLG–PEO–PCL).

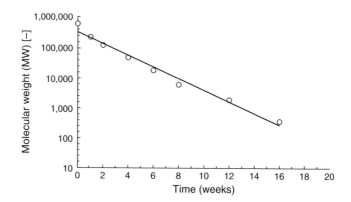

Fig. 3.20 Semilogarithmic plot of M_w of poly(DLLA–ε-CL) nerve guides as a function of degradation time (*in vitro* experiments, PBS, pH 7.5, 37°C, n = 3).

Table 3.5
Physical properties of poly(DLLA–ε-CL) nerve guides as a function of degradation time

Time (Weeks)	Modulus (MPa)	Maximum Strength (MPa)	T_g (°C)
1	3.4 (±0.5)	13.0 (±5.5)	17.7 (±0.5)
2	2.6 (±0.2)	9.9 (±2.8)	16.4 (±1.3)
4	2.2 (±0.3)	3.0 (±1.8)	13.5 (±2.8)
6	1.6 (±0.7)	2.6 (±1.0)	16.5 (±0.9)
8	0.3 (±0.1)	0.1 (±0.0)	11.3 (±0.2)
12	—	—	3.4 (±2.5)
16	—	—	−0.4 (±0.9)
22	—	—	−19.9 (±0.0)

The slow degradation of the implanted P(LA/CL) was explained by the fact that amorphous regions composed of mainly CL moieties degraded earlier than hard domains where most of the LLA units were located.

Pego *et al.* synthesized elastomeric copolymers of 1,3-TMC and ε-CL by ring-opening polymerization in evacuated ampoules using SnOct$_2$ as catalyst [43]. The copolymerization was carried out for 3 days at 130°C. The TMC–CL copolymers were semi-crystalline, depending on the composition, and their T_g varied between −17°C for poly(TMC) and −63°C for PCL. In PBS, TMC–CL copolymers retained suitable mechanical properties for more than a year. Figure 3.21 shows the MW change of the copolymers.

Grijpma *et al.* described the network preparation via UV radical polymerization of linear and star-shaped oligomeric precursors functionalized with fumaric acid monoethyl ester (FAME) [44]. First, star-shaped poly(TMC-co-DLLA) and poly(TMC-co-CL) with hydroxyl groups at the termini were synthesized by ring-opening polymerization (oligomerization) of TMC and DLLA or CL mixtures in the presence of glycerol and SnOct$_2$ at 130°C for 40 h, as shown in Fig. 3.22.

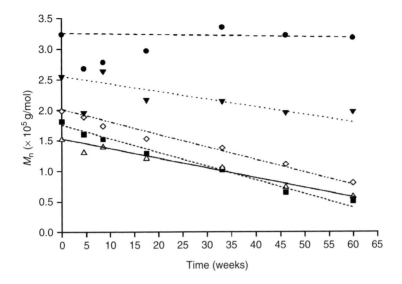

Fig. 3.21 M_n as a function of time for TMC-CL copolymers during degradation at 37°C and pH 7.4. (●) 100% TMC; (▼) 82% TMC; (◊) 31% TMC; (△) 10% TMC; and (■) 100% CL.

A chloroform solution containing *N,N*-dicycrohexylcarbodiimide and 2,2-dimethoxy-2-phenylacetophenone (DMPA) was added to the oligemeric triol solution. The coupling (esterification) reaction was continued for 48 h and functionalization yields were up to 96%. Chloroform solutions containing FAME-functionalized oligomer and DMPA photoinitiator were cast on a glass plate to obtain transparent films. Photocrosslinking of linear and 3-armed macromers by exposing to UV light resulted in the formation of networks with gel percentages up to 96%.

Pego *et al.* assessed the *in vivo* biodegradation and tissue response evoked by poly(TMC) and copolymers of TMC with either 52 mol% DLLA[poly(TMC-DLLA) (52 mol%)] or 89 mol% ε-CL[poly(TMC–CL) (82 mol%)] by subcutaneous implantation of polymer films in rats for period up to one year [45]. The MW and thermal properties of the polymers are compiled in Table 3.6. Initially, the amorphous poly(TMC–DLLA) (52 mol%) and poly(TMC) were transparent, and poly(TMC–CL) (89 mol%) was opaque. As a result of water uptake poly(TMC–DLLA) (52 mol%) samples became opaque after implantation. The shapeless samples that were recovered in 12 weeks were significantly smaller than those of earlier time points. The M_n of poly(TMC–DLLA) (52 mol%) implants decreased from the start of the implantation time, decreasing to 1% of its initial value after 52 weeks. At implantation periods after 12 weeks a rapid decrease in remaining mass occurred and at 52 weeks resorption was nearly complete. For poly(TMC–CL) (82 mol%) a linear and continuous reduction of the M_n was also observed in time but at a much lower rate. At 1 year the mass loss was <7%. In contrast to very slow hydrolytic degradation of poly(TMC) *in vitro*, this polymer

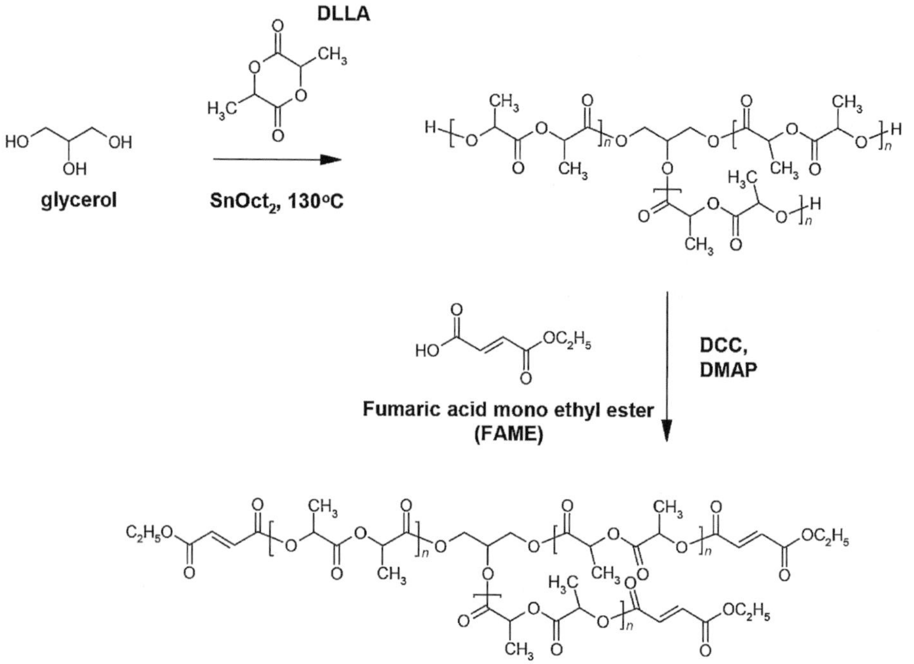

Fig. 3.22 Reaction scheme illustrating the ring-opening polymerization (oligomerization) of DLLA to yield 3-armed hydroxyl-terminated PDLLA oligomers (3-OH precursors) which in second step are functionalized with FAME. The TMC and ε-CL (co)polymers were synthesized and functionalized in an analogous way.

was macroscopically degraded after 3 weeks of implantation. Poly(TMC) implants showed a large decrease in mass from the start without a change in MW. Furthermore, 3 weeks after implantation the MW distribution of the polymer was dimodal, which might be explained by differences in MW between the surface and the bulk of the specimens, in other words, by a surface erosion process. At day 5 after the implantation of poly(TMC) a layer of macrophages and some fibroblasts

Table 3.6
Characterization of the TMC (co)polymers used

Polymer	$M_n \times 10^{-5}$	PDI[a]	$[\eta]$[b] (dL/g)	T_g[c] (°C)	T_m[c] (°C)	Crystallinity[c] (%)
Poly(TMC–DLLA) (48:52 mol %)	2.32	2.33	3.38	17	—	—
Poly(TMC–CL) (11:89 mol %)	1.54	2.06	4.02	−65	43	36
Poly(TMC)	3.16	2.13	4.28	−17	—	—

[a] PDI, polydispersity index.
[b] In CHCl$_3$, at 25°C.
[c] First heating scan (DSC).

surrounded the implants in a comparable way with that of the copolymers. At 12 weeks most of the poly(TMC) had been phagocytosed.

Goraltchouk *et al*. synthesized poly(LLA-co-ethylene glycol) [46]. First, LLA and PEG were reacted together to produce LLA–PEG oligomers and then the LLA–PEG oligomers were reacted with anhydrous methacryoyl chloride to yield methacrylate-terminated LLA–PEG (MA–LLA–PEG). Finally, the MA–LLA–PEG was allowed to polymerize using ammonium persulfate.

Mizutani and Matsuda prepared photocurable biodegradable poly(CL–TMC) copolymers using $SnOct_2$ as a catalyst and by two-step reactions [47]. The first step was ring-opening copolymerization of CL and TMC in the presence of a di-, tri-, or tetra-functional low MW alcohol (trimethylene glycol, trimethylolpropane, or pentaerythritol), PEG, or a four-branched PEG derivative (b-PEG) as an initiator, and the second step was esterification of the hydroxyl terminal groups using acry-loyl chloride. Table 3.7 lists the reaction conditions and compositions and MWs of the resultant copolymers. All the resulting copolymers were viscous liquids. Pho-topolymerization was induced by visible light in the presence of camphorquinone as an initiator. The photocured copolymers were all insoluble in any organic solvent. The photocuring yield (defined as the weight percentage of the insoluble part against the initial weight of liquid) increased with both photoirradiation time and camphorquinone concentration. The percentage of weight loss of the films of (2a) and (4a) showed a few percents of weight loss, whereas PEG-based copolymers, (2b) and (4b), exhibited higher degradation rates than the former ones (Fig. 3.23).

Younes *et al*. synthesized biodegradable elastomers in two steps as shown in Fig. 3.24 [48]. First, a star copolymer (SCP) was manufactured via ring-opening polymerization of CL with DLLA using glycerol as initiator and $SnOct_2$ as cata-lyst. This star copolymer was further reacted with different ratios of a crosslinking monomer, 2,2-bis(ε-CL-4-yl)-propane(BCP), in the presence of ε-CL as a solvent and co-monomer. The elastomers had a low glass transition temperature ($-32°C$) with sol contents ranging from 17 to 37%, and were soft and weak. The physical properties decreased in a logarithmic fashion with time when degraded in PBS. The elastomers degraded relatively slowly, with degradation being incomplete after 12 weeks.

To combine the biological and mechanical properties of chitosan and PLLA, Ding *et al*. immobilized chitosan onto PLLA film surface by plasma graft polymer-ization [49]. After 30 s of argon plasma treatment and subsequent exposure to air, the intensity of the oxygen component increased significantly. The argon plasma may cause the breakage of C–H bonds at the surface of PLLA. The abstraction of hydrogen atoms from the surface will result in the formation of free radicals on the polymer chains. The subsequent exposure of the activated surface to air causes oxy-gen to be incorporated onto the polymer surface, leading to surface oxidation and the formation of peroxides or hydroperoxides species. The peroxides species formed will subsequently initiate the surface free radical coupling reaction with active chitosan species that are activated by the second plasma treatment. When the plasma-pretreated PLLA surface coated with a thin layer of chitosan was subjected

Table 3.7
Acrylated liquid copolymers

Polymer Code Name	Initiator			Poly(CL/TMC)		Acrylated Copolymer		DW^f
	$R(OH)_n$	n^a	Fraction[b]	M_n^c	CL:TMC	Content[d] (mol/g)	M_w^e	
2a	$CH_2(CH_2OH)_2$	2	6.6	2.2×10^3	0.50:0.50	8.1×10^{-4}	2.5×10^3	0.01
2b	$CH_2(CH_2OH)_2$	2	13.2	4.3×10^3	0.49:0.51	4.0×10^{-4}	5.0×10^3	0.01
2c	PEG1000	2	6.6	3.3×10^3	0.52:0.48	5.7×10^{-4}	3.5×10^3	0.32
3a	$CH_3CH_2C(CH_2OH)_3$	3	6.6	3.5×10^3	0.51:0.49	8.1×10^{-4}	3.7×10^3	0.01
4a	$C(CH_2OH)_4$	4	6.6	5.3×10^3	0.49:0.51	8.1×10^{-4}	5.0×10^3	0.01
4b	b-PEG[g]	4	6.6	7.4×10^3	0.49:0.51	5.7×10^{-4}	7.0×10^3	0.32

[a] Multifunctionality of initiator.
[b] Molar fraction of monomer per total OH groups.
[c] Number-average molecular weight of nonmodified copolymer determined by GPC (PEG Standard).
[d] Acrylate content determined by using the corresponding reference copolymers.
[e] Molecular weight determined based on acrylate content.
[f] Degree of water adsorption of photocured film (relative weight of water uptake to polymer).
[g] Diglycerol polyoxyethylene glycol ether (MW 2040 Da) purchased from Shearwater, Inc.

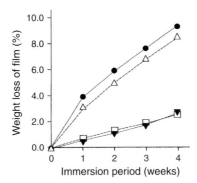

Fig. 3.23 The weight loss of photocured films (liquid film of 15 mm diameter; thickness, 0.4 mm) immersed in PBS of pH 7.4 at 37°C. (2a)(□) and (2c)(△) are difunctional and (4a)(▼) and (4b)(●) are tetra-branched copolymers, (2c) and (4b) are PEG-based copolymers.

to further plasma treatment, the N*is* core-level signal became prominent and the oxygen signal diminished, suggesting that the chitosan was graft-polymerized onto the PLLA surface.

1.2.2. Hydrogels

Depending on the composition and MW, Pluronics, block copolymers from PEG, and poly(propylene glycol) (PPG) can show reversible thermal gelation in aqueous solution under the physiological temperature and pH. Cellesi *et al.* developed a process using thermal gelation and chemical crosslinking ("tandem process") to synthesize hydrogels [50]. The reaction scheme is given in Fig. 3.25. With the presence of an appropriate number of reactive functional groups at the chain terminal, thiols, and electron-poor olefins, Pluronic polymers could also undergo a second stage of irreversible gelation through the occurrence of a pH-sensitive chemical reaction, the Michael-type addition of thiols onto acrylic esters. Subsequently exploiting both gelation mechanisms, hydrogels could be rapidly obtained (via covalent bond formation from the Michael-type addition reaction) when both the temperature and pH of a cold, slightly acidic medium (5°C and pH 6.8) were increased (37°C and pH 7.4). The quick kinetics of the reverse thermal gelation and the harmless character of the Michael-type addition between two sets of terminal groups, acrylates on one set and thiols on the other, allowed irreversibly crosslinked hydrogels to be obtained in a rapid fashion, even when gelation was conducted in direct contact with cells.

Masters *et al.* investigated photo-polymerized HAc hydrogels as a scaffold for cardiac valvular interstitial cells (VICs) that are the most prevalent cell type in native valves [51]. The VICs closely resemble myofibroblasts, which are found in many tissues and play important roles in tissue remodeling. The sodium salt of HAc was methacrylated using fivefold excess methacrylic anhydride relative to free HAc

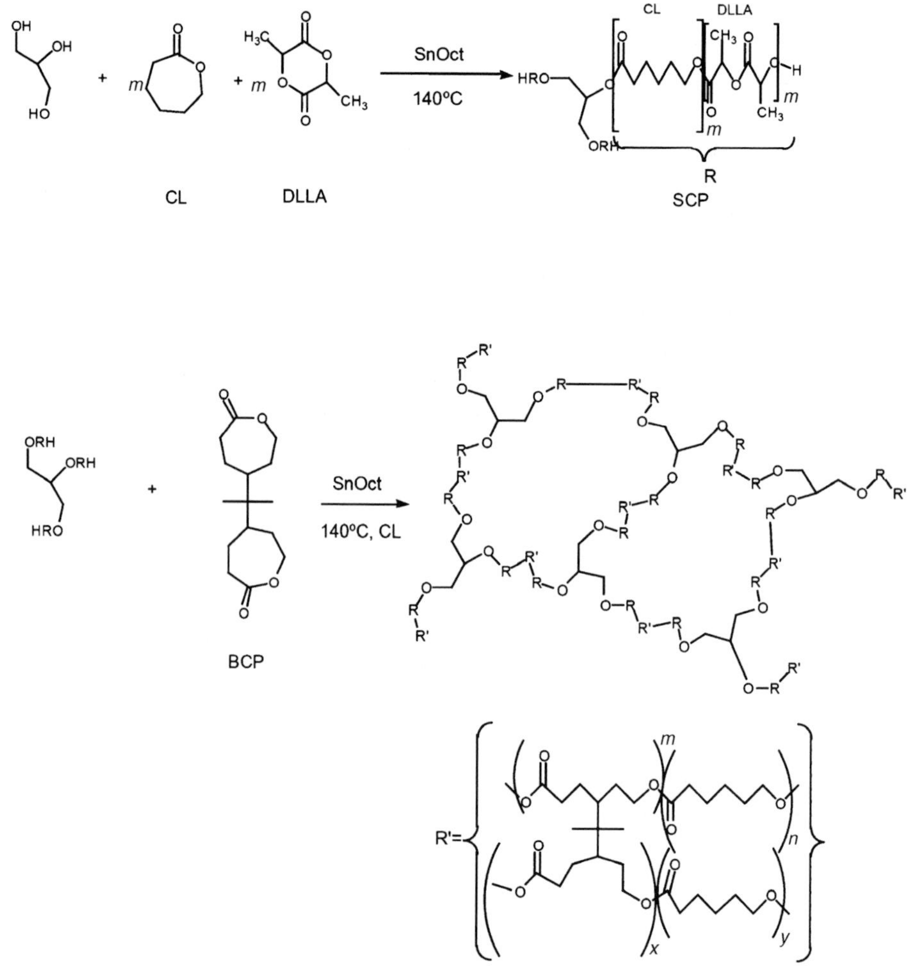

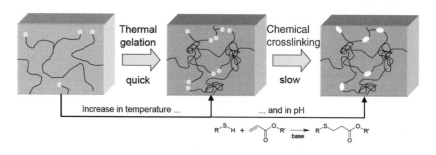

Fig. 3.24   Schematic of the preparation of the SCP and a representation of the formation of a network by reacting the SCP with the BCP.

Fig. 3.25   Reaction scheme.

hydroxyl groups. The methacrylated HAc (HAc–MA) gels were formed by making a solution of the polymer in PBS and exposing to UV light with Irgacure 2959 as a photoinitiator. Degradation products of HAc gels and the starting macromers significantly increased VIC proliferation when added to cell cultures. Low MW HAc degradation products added to VIC cultures also resulted in a fourfold increase in total matrix production and a twofold increase in elastin production over untreated controls. The VIC internalization of HAc, as shown by cellular uptake of fluorescently labeled HAc, likely activated signaling cascades resulting in the biological responses. The VICs encapsulated within HAc hydrogels remained viable, and significant elastin production was observed after 6 weeks of culture.

To entrap cells within a PEO hydrogel scaffold formed by a stereolithography process, Dhariwala *et al.* first prepared an aqueous solution of PEO and PEG dimethacrylate (PEGDM) and a solution of a photoinitiator, 2-methyl-1-[4-(hydroxyethoxy)phenyl]-2-methyl-1-propanone [52]. The initiator solution was added to the polymer solution. Chinese hamster ovary cells were resuspended in the photopolymer solution. The cell-photopolymer solution was placed in a well-plate. This was then placed in the stereolithography machine and exposed to UV laser. The resulting gels with the desired shape were suspended in culture medium and incubated under standard conditions. Mechanical characterization showed the constructs to be compatible with soft tissues in terms of elasticity. High cell viability was achieved and high-density constructs fabricated.

Mironov *et al.* investigated the possibility of using an *in situ* crosslinkable HAc-based synthetic ECM to develop a centrifugal casting technology for tissue engineering [53]. Centrifugal casting combines cell embedding or molding technologies with a rotational device generating centrifugal force. To employ centrifugal casting for cell-seeded materials for tissue engineering, an *in situ* crosslinkable polymer that forms an absorbable hydrogel in the presence of living cells is a prerequisite. Living cells were suspended in a viscous solution of thiol-modified HAc and thiol-modified gelatin, a PEG diacrylate crosslinker was added and a hydrogel was formed during rotation. Tubular tissue constructs consisting of a densely packed cell layer were fabricated with a rotation device operating at 2000 rpm for 10 min.

1.2.3. Polyurethanes

Guan *et al.* synthesized a series of biodegradable poly(ether ester urethane) urea elastomers (PEEUUs) based on poly(ether ester) triblock copolymers [54]. Figure 3.26 shows the synthetic scheme. The copolymers were synthesized by ring-opening polymerization of ε-CL with PEG. The PEEUUs were synthesized from these triblock copolymers and butyl diisocyanate, with putrescine (butylene diamine) as a chain extender. Butyl diisocyanate and putrescine were selected as reactants in the PEEUU synthetic scheme with the expectation that their ultimate degradation products (putrescine) would be non-toxic. The synthetic results and mechanical properties of PCL–PEG–PCL triblock copolymers and PEEUUs are given in Table 3.8.

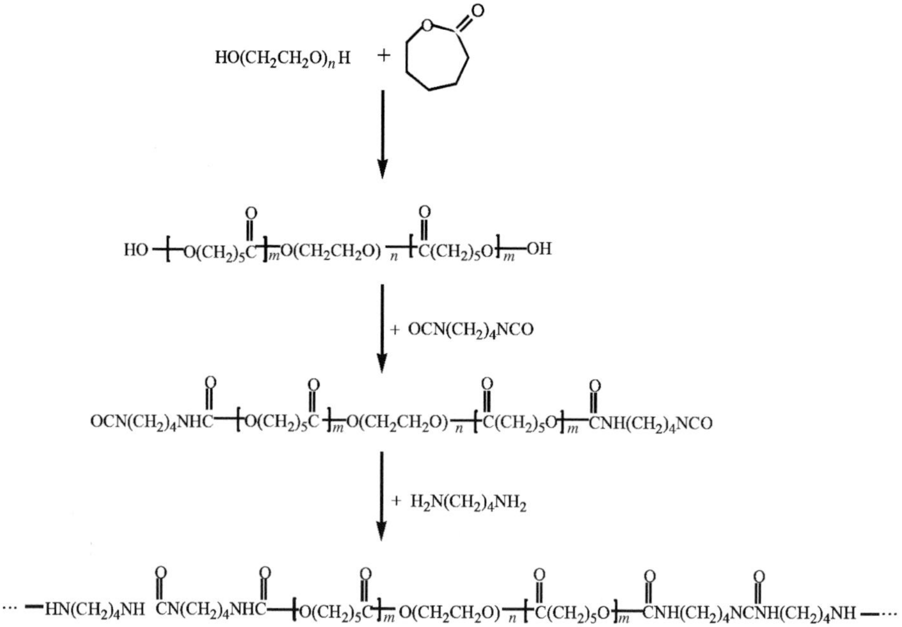

$$HO(CH_2CH_2O)_n H \quad +$$

$$HO + O(CH_2)_5C \underset{\overline{m}}{\overset{O}{\|}} O(CH_2CH_2O) \underset{\overline{n}}{} C(CH_2)_5O \underset{\overline{m}}{\overset{O}{\|}} OH$$

$$+ \; OCN(CH_2)_4NCO$$

$$OCN(CH_2)_4NHC \overset{O}{\|} + O(CH_2)_5C \underset{\overline{m}}{\overset{O}{\|}} O(CH_2CH_2O) \underset{\overline{n}}{} C(CH_2)_5O \underset{\overline{m}}{\overset{O}{\|}} CNH(CH_2)_4NCO$$

$$+ \; H_2N(CH_2)_4NH_2$$

$$\cdots - HN(CH_2)_4NH \; \overset{O}{\underset{\|}{C}}N(CH_2)_4NHC \overset{O}{\|} + O(CH_2)_5C \underset{\overline{m}}{\overset{O}{\|}} O(CH_2CH_2O) \underset{\overline{n}}{} C(CH_2)_5O \underset{\overline{m}}{\overset{O}{\|}} CNH(CH_2)_4NCNH(CH_2)_4NH - \cdots$$

Fig. 3.26   PEEUU synthesis scheme.

Table 3.8
MW and mechanical properties of PEEUU

PEEUU	M_w	M_n	M_wM_n (PDI)	T_g (°C)	T_m (°C)	ΔH_m (J/g)
PEG10/COP25	112,400	50,000	2.25	−51.3	40.5	7.5
PEG10/COP30	100,800	67,800	1.49	−53.4	41.2	10.8
PEG10/COP40	78,500	55,600	1.41	−56.5	52.8	27.0
PEG6/COP21	78,000	49,000	1.59	−46.6	—	—
PEG6/COP26	93,700	44,600	2.10	−48.6	41.2	8.5
PEG6/COP36	87,200	51,900	1.68	−50.0	50.7	13.6

The PEEUUs were designed using the abbreviation of PEGX/COPY, where X and Y represented the MW (divided by 100) of PEG and triblock poly(ether ester), respectively. The synthesized polymers were highly distensible with breaking strains from 325 to 560% and tensile strengths from 8.0 to 20 MPa. Increasing the PCL block length had the effect of increasing both the initial modulus and the tensile strength. A decrease in soft segment PEG block length was associated with a lower initial modulus and higher tensile strength. The *in vitro* degradation of PEEUU films in PBS at 37°C is presented in Fig. 3.27. The PEEUUs with longer PEG blocks degraded to a greater extent than PEEUUs with shorter PEG segments. For PEEUUs with a given PEG block length, the degradation extent increased when the PCL block length decreased. The PEEUU degradation rates corresponded to the water absorbing ratio. These polymers

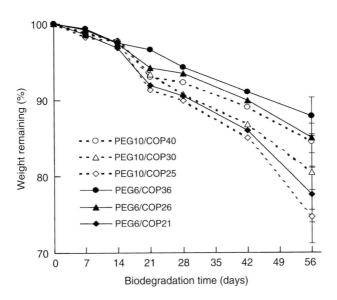

Fig. 3.27 PEEUU mass loss during incubation in an aqueous buffer.

were the thermoplastic elastomer that is amenable to solvent-based processing techniques. The degradation products collected over 4 weeks of degradation from two representative PEEUUs did not appear to be cytotoxic to ECs. However, concerns have been raised with respect to the carcinogenicity and cytotoxicity of the degradation products from the diisocyanate hard segment.

Saad *et al.* synthesized biodegradable PUs using hydroxybutyrate-co-valerate polyol [55], while Grad *et al.* reported synthesis of PUs using hexamethylene diisocyanate, PCL diol with a MW of 530, and isosorbide diol (1,4:3,6-dianhydro-D-sorbitol) as their extenders [56]. Zhang *et al.* synthesized a peptide-based urethane polymer from lysine-diisocyanate (LDI) [57]. They further incorporated ascorbic acid (AA) into the biodegradable PU to promote osteoblast differentiation *in vitro* [58]. The AA-containing urethane polymer (LDI–glycerol–AA) was synthesized as follows: 12 mg of AA and 2.862 g of glycerol (-OH 93.33 mmol) were mixed in a dry flask, flushed with nitrogen, and then the flask was fitted with a rubber septum and sealed. Subsequently, 18 ml of ethyl ester of LDI (-NCO 185.8 mmol) was added to the flask and the reaction mixture was stirred in the flask at room temperature for 5 days. The reaction was stopped when Fourier transform infrared (FTIR) spectra exhibited 50% of isocyanate group left in the reaction mixture. Two milliliters of water was added and stirred for 30 min to produce foam. The polymer degradation was tested *in vitro* by placing known amounts of polymer in PBS at 37°C for 1–60 days. The degradation test indicated that the degradation products were lysine, glycerol, AA, and ethanol. Examination of the degradation products of LDI–glycerol–AA polymer showed the presence of glycerol at one of three concentrations to lysine. The AA release stimulated proliferation of osteoblastic precursor cells, type I collagen, and ALP synthesis.

To prepare flexible PU scaffolds that could be employed in soft tissue engineering or as temporary mechanical scaffolds for application without combining cells, Guan *et al.* synthesized PUs from PCL and 1,4-diisocyanatobutane (BDI) with putrescine used as a chain extender [59]. And BDI was selected as the diisocyanate upon which the hard segment was built since it would be expected to yield a polyamine that is essential for cell growth and differentiation, following complete degradation. In order to increase the hydrophilicity of the scaffold, a PEG containing triblock copolymer was synthesized and used as a soft segment. The thermally induced phase separation was employed to prepare scaffolds. The scaffolds obtained had open and interconnected pores of sizes ranging from several μm to more than 150 μm and porosities of 80–97%. By changing the polymer solution concentration or quenching temperature, scaffolds with random or oriented tubular pores could be obtained. The poly(ester urethane)urea (PEUU) scaffolds were flexible with breaking strains of 214% and higher, and tensile strengths of 1 MPa, whereas the PEEUU scaffolds generally had lower strengths and breaking strains. Scaffold degradation in aqueous buffer was related to the porosity and polymer hydrophilicity. Smooth muscle cells were seeded in the scaffolds and both scaffolds supported cell adhesion and growth, with SMCs growing more extensively in the PEEUU scaffold.

1.2.4. Others

Poly(Propylene Fumarate)

Dean *et al.* synthesized an unsaturated linear polyester, poly(propylene fumarate) (PPF), by a two-step reaction process [60]. Fumaryl chloride was added to a solution of propylene glycol in methylene chloride at 0°C in the presence of K_2CO_3. After addition of fumaryl chloride, the reaction mixture was stirred for an additional 2 h at 0°C. Water was then added to dissolve the inorganic salt. The organic layer was separated and dried with Na_2SO_4. After filtration and evaporation of the solvent, the formed di-(2-hydroxypropyl) fumarate underwent a transesterification reaction at 160°C and 0.5 mmHg to produce PPF. The M_n of the resulting PPF was, for instance, 1875. A solid PPF/TCP disc was prepared by crosslinking PPF with *N*-vinyl pyrrolidone (NVP). After mixing PPF with NVP at room temperature, β-TCP was added to the PPF/NVP mixture. Benzoyl peroxide was dissolved in NVP and the solution was added to the PPF/NVP/β-TCP mixture. The resulting PPF consisted of repeating units that contained one unsaturated double bond that permitted covalent crosslinking and two ester groups that allowed for hydrolysis of the polymer into fumaric acid and propylene glycol [61]. For use as an injectable scaffold, PPF was crosslinked with the crosslinking agent PPF-diacrylate (PPF-DA) through free radical polymerization after addition of a thermal initiation system consisting of benzoyl peroxide and *N,N*-dimethyl-p-toluidine.

Poly(DTE Carbonate)

To develop resorbable fiber-based scaffolds for reconstruction of ACL, Bourke *et al.* fabricated fibers from poly(desamino tyrosyl-tyrosine ethyl ester carbonate) [poly(DTE carbonate)] [62]. The study was performed in three phases. First-generation fibers used in phase I were made by melt extrusion at 60–90°C above the glass transition temperature (91°C). Second-generation fibers were fabricated by melt spinning. Solid polymer pellets were transferred to a pack, sealed with a piston, and compressed with a hydraulic press. Subsequently, the pack was placed in the heating jacket. The temperature was brought up slowly to a spinning temperature of 181–183°C. The fibers had diameters of 79 μm, UTS of 230 MPa, modulus of 3.1 GPa, and MW of 65,000 Da. After 30 weeks of incubation in PBS, the fibers had 87% strength retention. In phase III, a prototype ACL reconstruction device was fabricated from poly(DTE carbonate) fibers with strength values comparable to those of the normal ACL (57 MPa). After 30 weeks of incubation in PBS, poly(DTE carbonate) and PLLA fibers had 87 and 7% strength retention, respectively. Similar trends were observed for MW loss. Fibroblasts attached and proliferated equally well on both scaffold types *in vitro*. Finally, in phase III, a prototype ACL reconstruction device was fabricated from poly(DTE carbonate) fibers with strength values comparable to those of the normal ACL.

Polyphosphazenes
To balance the mechanical properties and the degradation rate of polyphosphazenes, Cui *et al.* synthesized micro-crosslinked polyphosphazenes [63]. Here 2-hydroxyl methacrylate (HEMA) was first attached to the side chain along with glycine ethyl ester to form a precursor with unsaturated substitutes. The co-substituted poly-organophosphazene was mixed with HEMA or acrylic acid, followed by free radical polymerization to prepare micro-crosslinked polyorganophosphazenes. The synthesis route is shown in Fig. 3.28. Polymer 1 was obtained from thermal ring-opening polymerization of hexachlorocyclotriphosphazene (HCCP). The single substituted polyphosphazenes (2 and 3) and the co-substituted polyphosphazenes 4 and 5 were obtained from polymer 1 in different ways. The HEMA-crosslinked polymers 6 and 7 were obtained from 4 and 5, respectively. Ascorbic acid–crosslinked polymers 8 and 9 were obtained from 4 and 5, respectively. Figure 3.29 shows *in vitro* degradation of polymer films in deionized water at 37°C. Polymer 2, investigated as a control, underwent rapid weight decline. Polymer 6 lost half weight in 7 weeks while Polymer 3 in 4 weeks. Polymers 8 and 9 possessed a similar degradation rate to Polymers 6 and 7 during the first 4 weeks while the degradation pace showed a remarkable increase during the following weeks. As a consequence, Polymer 9 lost more than 70% of its weight at the end of degradation study. The marked self-acceleration effect could be ascribed to the introduction of ascorbic acid, which played the role of a potential catalyst to the degradation reaction.

PEOT/PBT
Deschamps *et al.* synthesized segmented poly(ether esters) based on PEG and poly(butylene terephthalate) (PBT) (PEOT/PBT) [64]. The chemical structure is shown in Fig. 3.30. And PBT was selected because this polymer exhibits excellent thermal and mechanical properties. The PEOT/PBT multiblock copolymers were

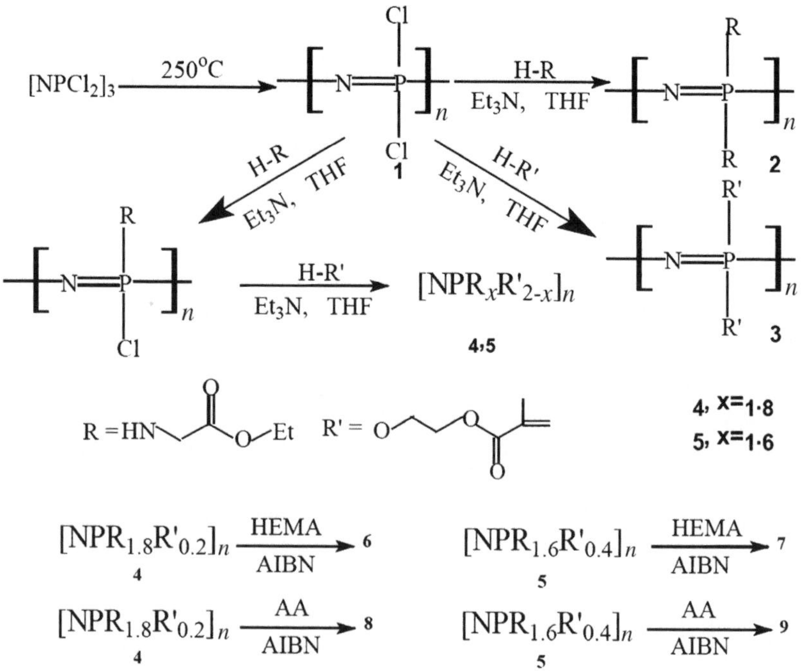

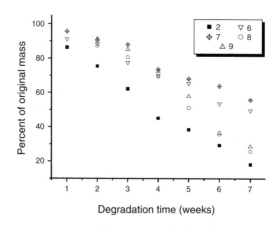

Fig. 3.28 Synthetic passway of poly(organophophazenes).

prepared by a two-step polycondensation of PEG, 1,4-butanediol, and dimethyl terephthalate in the presence of titanium tetrabutoxide as catalyst and Irganox 1330 as antioxidant. The antioxidant was removed by dissolution of copolymers in chloroform and precipitation into an excess of ethanol. The composition of the block

Fig. 3.29 Hydrolysis degradation rate of Polymers 2 and 6–9 in water at 37°C.

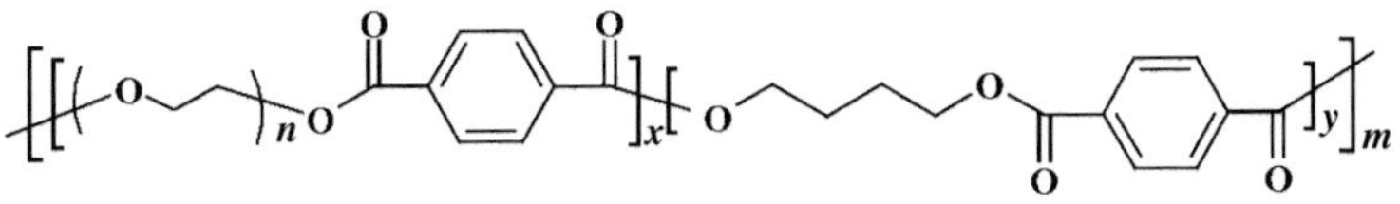

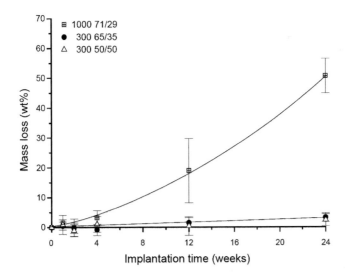

PEOT "soft segment"   PBT "hard segment"

Fig. 3.30   Chemical structure of PEOT/PBT-segmented block copolymers.

copolymers was abbreviated as *a* PEOT *b* PBT *c* in which *a* was the MW of the
PEG used, *b* the weight percentage of PEOT soft segments, and *c* the weight percentage of PBT hard segments. *In vitro* degradation of melt-pressed PEOT/PBT
discs was studied by subcutaneous implantation in rats. Changes in intrinsic viscosity and mass loss during degradation are represented in Fig. 3.31. The copolymers
prepared with PEG 300 (300 PEOT 65 PBT 35 and 300 PEOT 50 PBT 50) showed
only little degradation after 6 months of implantation and kept their mechanical
integrity. Severe fragmentation of 1000 PEOT 71 PBT 29 was observed after 24
weeks of implantation.

Poly(Glycerol Sebacate)
A bioabsorbable elastomer, poly(glycerol sebacate) (PGS), was synthesized by a
polycondensation reaction of glycerol and sebacic acid, generating a viscous prepolymer that could be melt-processed or solvent-processed into the desired shape
[65]. This prepolymer could be further reacted to covalently crosslink the polymer,
resulting in a 3-D network of random coils with hydroxyl groups attached to its

Fig. 3.31   Mass loss as a function of implantation time for: (⊞) 1000 PEOT71PBT29;
(●) 300PEOT65PBT35 and (△) 300 PEOT50PBT50.

backbone. Both the crosslinking and the hydrogen-bonding interactions between the hydroxyl groups contributed to its elastomeric properties. The material had a Young's modulus of 0.28 MPa, a UTS of greater than 0.5 MPa, and a strain to failure greater than 330%. When implanted subcutaneously, the degradation half-life was 21 days, and complete resorption was observed by day 60.

1.3. Calcium Phosphate

Xu *et al.* developed fast-setting and resorbable calcium phosphate cement (CPC) scaffolds with high strength and tailored macropore formation rates [66]. To this end, chitosan, sodium phosphate, and hydroxypropyl methylcellulose (HPMC) were used to render CPC fast-setting and resistant to washout. Absorbable Vicryl™ fibers and mannitol porogen were incorporated into CPC for strength and macropores for bone ingrowth. Hardening time was 70 min for CPC-control, 9 min for CPC–HPMC–mannitol, 8 min for CPC–chitosan–mannitol, and 7 min for CPC–chitosan–mannitol–fiber. The latter three compositions were resistant to washout, whereas the CPC-control paste showed washout in a physiological solution. Immersion for 1 day dissolved mannitol and created macropores in CPC. The CPC–chitosan–mannitol–fiber scaffold had strength of 5 MPa, significantly higher than 1 MPa of CPC–chitosan–mannitol scaffold and 0.3 MPa of CPC–HPMC–mannitol scaffold. The strength of CPC–chitosan–mannitol–fiber scaffold was maintained up to 42 days and then decreased because of fiber degradation.

1.4. Composites

Blaker *et al.* prepared PDLLA/Bioglass™ composite foams using melt-derived Bioglass™ powder with a mean particle size <5 μm. Stiff Bioglass™ particles as filler were used to enhance the elastic constants, mechanical strength, and structural integrity of porous polymer constructs [67]. The PDLLA was dissolved in dimethyl-carbonate to give a polymer weight to solvent volume ratio of 5%. For the composite foams, Bioglass™ powder was added to the solution to result in either 5 or 40% Bioglass™ concentration. The PDLLA/Bioglass™ mixture was transferred to a lyophilization flask and sonicated for 15 min to improve the dispersion of the Bioglass™ into the polymer solution. The flask was immersed in liquid nitrogen and maintained at $-196°C$ for 2 h. The frozen mixture was then transferred to an ethylene glycol bath at $-10°C$ and connected to a vacuum pump. The solvent was sublimed at $-10°C$ for 48 h and then at $0°C$ for 48 h. The foams exhibited high porosities and possessed two distinct pore size macropores of 100 μm average diameter, and interconnected micropores of 10–50 μm diameter. The tubular macropores were highly oriented as a result of the unidirectional cooling process. This porous structure is typical of foams prepared by the phase separation process. The composite foam structures with Bioglass™ particles showed a similarly well-defined tubular and interconnected porous structure, although the 40% Bioglass™-filled foam had more irregular pore morphology.

Table 3.9

Monomer ratio and M_w in ternary (PLGC) and binary (PLCL) copolymers, and calcined temperature of TCP, type of polymer, and mixing ratio of the composites prepared

Sample Abbreviation	L-Lactic Acid (mol %)	Glycolic Acid (mol %)	ε-Caprolactone (mol %)	M_w* (kDa)
PLGC$_{1009}$	75	11	14	250
PLCL$_{1001}$	92	0	8	110
PLCL$_{2004}$	78	0	22	200
PLGC$_{3005}$	48	14	38	190

To create bioabsorbable and thermoplastic materials for GBR/GTR membranes, Kikuchi *et al.* prepared composites of β-TCP and poly(L-LA-co-GA-ε-CL) (PLGC) and poly(L-LA-co-ε-CL) (PLCL) [68]. Table 3.9 gives the chemical compositions of the copolymers used and the blending ratios of the composites. Composites were prepared by a heat-kneading method and maintained mechanical strength for longer times than the copolymers alone, as shown in Fig. 3.32. In addition, the composites had a pH auto-regulation property. Animal tests for GBR indicated that the composite membrane successfully regenerated dog's mandible defects of $10 \times 10 \times 10$ mm^3 in size.

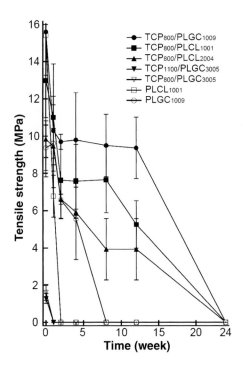

Fig. 3.32 Tensile strength changes of the composites and the polymer as a function of soaking time in physiological saline.

Zou *et al.* prepared porous β-TCP particles/collagen composites by acid-treating the collagen, dispersing fine β-TCP particles and lyophilizing [69]. The acid treatment could disassemble collagen fibrils. In the resulting porous composites, β-TCP particles homogenously existed on the skeleton of the collagen fibril network. The tight bonding between β-TCP particles and collagen fibrils in the composites demonstrated an integrated structure. Kim *et al.* synthesized gelatin/HAp nanocomposites, and *in vitro* osteoblastic cellular responses to the nanocomposites were assessed in comparison with those conventionally mixed gelatin-HAp composites [70]. The osteoblastic MG63 cells attached to the nanocomposites to a significantly higher degree and subsequently proliferated more. The ALP activity and OCN produced by the cells were significantly higher on the nanocomposite scaffolds than on the conventional composite scaffolds. These improved cellular responses on the nanocomposites were ascribed to the increased ionic release and serum protein adsorption on the nanocomposites.

2. FABRICATION OF POROUS SCAFFOLDS

2.1. Freeze Drying

O'Brien *et al.* primarily manufactured porous, type I collagen-GAG (CG) scaffolds using a freeze-drying or lyophilization technique [71]. Using this technique, a suspension of collagen and GAGs was solidified (frozen); the CG co-precipitate was localized between the growing ice crystals, forming a continuous, interpenetrating network of ice and the co-precipitate. Sublimation of the ice crystals led to the formation of a highly porous sponge. The final pore structure depended on the underlying freezing processes during fabrication. The rapid, uncontrolled quench freezing process typically used in fabricating porous scaffolds via freezing-drying resulted in space- and time-variable heat transfer through the suspension, leading to non-uniform nucreation and growth of ice crystals and, ultimately, scaffold heterogeneity. In localized regions of poor contact between the pan in which the suspension was frozen and the freeze-dryer shelf, there was a lower rate of ice-crystal nucreation than in neighboring regions, giving increased variation in pore size; due to poor heat conduction and the increased temperature of the suspension at these points, these areas had been termed "hot spots". To increase scaffold homogeneity, the size of the pan used during the freezing process had to be reduced to increase the pan stiffness and reduce the effects of warping, and the rate of CG suspension freezing had to be slowed and controlled to reduce the heterogeneous freezing processes observed during conventional scaffold fabrication processes.

The structure of the collagen scaffold was controlled by the final temperature of freezing and the heat transfer processes associated with freezing. Furthermore, the formation of ice crystals within the CG suspension was influenced by both the rate of nucreation of ice crystals and the rate of heat and protein diffusion. The nucreation rate defined the number of ice crystals that formed and the rate of heat and protein (collagen-GAG) diffusion away from the point of nucleation defined the size of ice crystals. Both the rate of nucreation and diffusion were mediated by the difference between the temperature of freezing and the actual temperature of the material during the freezing process, termed "undercooling". A larger undercooling led to an increased rate of nucleation of ice crystals with a decreased heat and protein diffusion rate away from the point of nucreation. Therefore, with larger undercooling, smaller ice crystals were formed, leading to a CG scaffold with a smaller average pore diameter following sublimation. Additionally, the direction of heat transfer and the speed of heat transfer influenced the shape of ice crystals; the existence of a predominant direction of heat transfer led to the formation of columnar ice crystals with the major axis aligned in the predominant direction of heat transfer. Creation of a scaffold with an equiaxed pore structure required removing predominant direction of heat transfer from the freezing process. Scaffolds produced using a constant cooling rate of 0.9°C/min were found to have the most homogeneous structure with the lowest coefficient of variance in pore size and the smallest difference in pore size between the longitudinal and the transverse planes in the scaffold. There was reduced variation in pore size throughout the scaffold along with an equiaxed pore structure when the scaffold was manufactured using the constant cooling rate technique (0.9°C/min). In the scaffold produced using the constant cooling rate technique, the collagen fibers (struts) were randomly arranged around approximately equiaxed pores while in the scaffold produced using the original quenching technique, there were roughly parallel planes of collagen membranes separated by thin collagen struts.

Ho *et al.* prepared porous PLLA and PLGA (50:50) scaffolds by using the freeze-extraction method [72]. After freezing the PLLA/dioxane solutions at −20°C, the frozen solutions were immersed in ethanol aqueous solution at −20°C. The ethanol aqueous solution was then removed by drying at room temperature, leading to the formation of porous scaffolds without skin structure. For chitosan, Ho *et al.* used a freeze-gelation method instead of the freeze-extraction method, as it was not easy for this polymer to extract out the solvent (aqueous solution of acetic acid) with a non-solvent. The idea of the freeze-gelation method is not to remove the solvent but to adjust the solvent property to allow for the gelation of chitosan. Chitosan gels in an alkaline environment due to the deprotonation. Based on this fact the freeze-gelation method was developed, in which the frozen chitosan solution was immersed in NaOH solution to adjust the solution pH to bring about gelation. After the gelation of chitosan, the polymer did not redissolve at room temperature and was rigid enough to prevent pore collapse during drying. Similar to freeze extraction, the gelation process was performed at a temperature lower than the freezing point of the chitosan solution to prevent the redissolution of chitosan. Figure 3.33a shows the morphology of the chitosan scaffold fabricated by the freeze-gelation method using the alkaline solution containing NaOH, water, and ethanol. The addition of ethanol was to lower the freezing point. The SEMs of porous alginate scaffold by the freeze-gelation are shown in Fig. 3.33b. The procedures for preparation of alginate scaffolds were similar to those for chitosan scaffolds except for the gelation method. The gelation of alginate was not introduced by

(a)

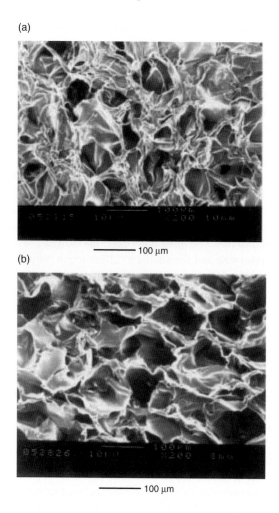

———— 100 µm

(b)

———— 100 µm

Fig. 3.33 Morphology of the chitosan scaffold prepared by freeze-gelation method. (a) Surface and (b) cross-section.

adjustment of pH, but by addition of calcium ion. The solution used to gel alginate consisted of calcium chloride, water, and ethanol. The aqueous solution of calcium chloride provided the calcium ion needed for alginate gelation, and ethanol was added to lower the freezing point so that freeze-gelation could be carried out.

Cao *et al.* fabricated 3-D PLGA scaffolds with the use of the thermally induced phase-separation (TIPS) technique [73]. A dioxane solution of PLGA was cast in a cylindrical aluminum mold and solid–liquid phase separation was induced by decreasing the solution temperature from 45 to −18°C. Two controllable quenching routes, termed "quick quench" and "slow quench", were applied by connecting a temperature-controlled bath to a computer with software developed in-house. They established a series of numerical heat-transfer models in an effort to describe the cooling process. Among them, a 2-D solidification model proved to be the most successful in describing the quenching of the polymer solution and had the potential to be used to infer the various 3-D macroporous architecture created from different quenching conditions.

Stokols and Tuszynski developed a procedure using freeze-drying process to create nerve guidance scaffolds made from agarose, with uniaxial linear pores [74]. Agarose was dissolved in water, heated to 100°C, and injected into glass tubes. The tubes were centrifuged to remove air bubbles, and the agarose was allowed to cool to room temperature. The glass tubes containing agarose were placed in an insulating container such that only the bottom surface of the glass tube was exposed. The glass tube was then placed on the surface of a block of dry ice which in turn rested in a pool of liquid nitrogen, which created a uniaxial thermal gradient. Liquid nitrogen vapor was removed by vacuum. The samples were allowed to freeze for 45 min, at which point linear ice crystals could be seen extending through the polymer, and were lyophilized overnight. As shown in Fig. 3.34, the linear uniaxial freezing gradient produced final scaffolds with linear pores in a honeycomb arrangement. The pores had a mean cross-sectional diameter of 119 μm. Pores extended linearly through the full extent of scaffolds, evidenced by two separate measures. First, egress of fluid that was injected into one end of the channel was examined at the opposite end. In each case, fluid was discharged only from the opposite end of the channel and did not emerge from the side walls. In a second assay of channel continuity, photomicrographs of proximal and distal ends of 2- and 4-mm length scaffolds were compared for persistence of each channel at both ends of the scaffold. While, as to be expected, corresponding proximal and distal channel openings were easier to detect with the shorter length scaffolds, 100% persistence of individual channel

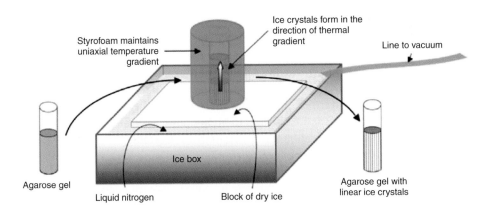

Fig. 3.34 A uniaxial temperature gradient was created by exposing the bottom surface of glass tubes containing polymerized agarose to a block of dry ice cooled in liquid nitrogen. The glass tubes were insulated in Styrofoam, and liquid nitrogen vapor was removed by vacuum to ensure a uniaxial temperature gradient. Arrow indicates the direction of temperature gradient.

patency was found in both 2- and 4-mm-long scaffolds. The hydrated scaffolds were soft, flexible, and stable under physiological conditions without chemical crosslinking, and could be loaded with diffusible growth-stimulating proteins.

2.2. Porogen Leaching

Ma *et al.* fabricated paraffin spheres with different sizes and size distributions by changing the concentration of gelatin which is an effective agent to disperse paraffin liquid to form spherical droplets [75]. They heated 40 g of paraffin and 200 ml of gelatin solution to 80°C under stirring to form a homogeneous suspension, which was then immediately poured into ice water to solidify the paraffin liquid drops. The size of the paraffin spheres increased inversely with the gelatin concentration. For preparation of porous PLLA scaffolds the paraffin spheres and PLLA/1,4-dioxane solution were mixed together in a mold, then frozen at −25°C for 1 h. After the mixture solidified, it was removed from the mold and freeze-dried to remove 1,4-dioxane, then treated several times in boiling hexane solution under vigorous stirring to leach out paraffin spheres. Larger paraffin spheres generated larger pores, with the spherical shapes of the paraffin preserved. Thermally induced phase separation occurred simultaneously in PLLA/1,4-dioxane solution when it was frozen at −25°C, resulting in the formation of micropores (10–30 μm) in the walls of the macroporous scaffolds.

Hacker *et al.* fabricated scaffolds made from amine-reactive diblock copolymers, *N*-succinimidyl tartrate monoamine poly(EG-block-DLLA), which are able to suppress unspecific protein adsorption and to covalently bind proteins or peptides [76]. An appropriate technique for their processing had to be both anhydrous, to avoid hydrolysis of the active ester, and suitable for the generation of interconnected porous structures. Attempts to fabricate scaffolds utilizing hard paraffin microparticles as hexane-extractable porogens failed. Consequently, a technique illustrated in Fig. 3.35 was developed involving lipid microparticles, which served as biocompatible porogens on which the scaffold-forming polymer was precipitated in the porogen extraction media (*n*-hexane). Porogen melting during the extraction and polymer precipitation step led to an interconnected network of pores. Suitable lipid mixtures and their melting points, extraction conditions (temperature and time) and a low-toxic polymer solvent system were determined for their use in processing diblock copolymers of different MWs (22 and 42 kDa) into highly porous off-the-shelf cell carriers ready for easy surface modification toward biomimetic scaffolds.

Tadic *et al.* presented a method to produce ceramics with interconnecting porosity without sintering [77]. They prepared porous objects of carbonated apatite by mixing PVA fibers and sodium chloride as porogens with nanocrystalline carbonated apatite powder. Calcium phosphate powder was ball-milled and sieved to 250–400 μm and was thoroughly mixed with PVA fibers and NaCl crystals of about 250–400 μm diameters. This mixing was cold-isostatically pressed at 4000 bar with a press at 25°C. After pressing, both NaCl and PVA were removed by extraction in warm water for 12 h each. The resulting bioceramic showed an interconnecting porosity with pore diameters in the range of 250–400 μm.

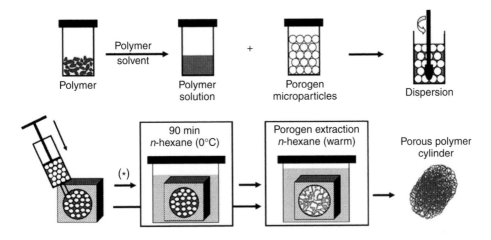

Fig. 3.35 Schematic illustration of the polymer processing procedure. (*) Pre-extraction treatment of molds in *n*-hexane at 0°C applied when lipids were used as porogen material.

2.3. Gas Foaming

A technique was developed by Barbucci and Leone to obtain polysaccharide-based hydrogels with a defined porous morphology [18]. The technique consists of stratifying a crosslinked hydrogel on a cell-culture strainer (filter) with a defined and controlled porosity (40, 70, and 100 μm). The filter was placed on a beaker with the same diameter as the filter and containing a porogen salt ($NaHCO_3$) at the bottom and filled with chloroform, as illustrated in Fig. 3.36. A 0.1 M HCl solution was added with a syringe drop by drop, producing violet effervescence. The formation of CO_2 bubbles and their passage through the filter, first, and then the matrix, second, induced the hydrogel to assume a porous morphology. The pore diameter, density, and the thickness of the walls are listed for hydrogels from alginate, hyaluronate (Hyal), and carboxymethyl cellulose (CMC) in Table 3.10. A correspondence was found between the porosity of the filter and the pore diameter in the hydrogels.

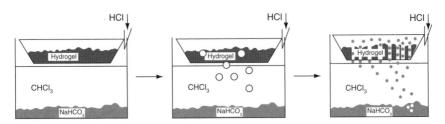

Fig. 3.36 Disposal for the fabrication of porous hydrogels. It is constituted by a beaker having, on the bottom, a stabilized amount of porogen salt ($NaHCO_3$) and filled with 50 ml of chloroform. At the top a cell strainer, on which hydrogel is placed, is collocated. And HCl solution is added with a syringe to facilitate the formation of CO_2 bubbles.

Table 3.10

Pore diameter, density, and thickness of the walls between the pores for polysaccharide-based hydrogels

Native Hydrogel	Materials								
	Alginate No pores			CMC Distance between Laminae: 76 ± 13 µm			Hyal No pores		
	Diameter (µm)	Density (pores/mm²)	Thickness (µm)	Diameter (µm)	Density (pores/mm²)	Thickness (µm)	Diameter (µm)	Density (pores/mm²)	Thickness (µm)
Cell strainer (ϕ 40 µm)	13 ± 4	5.0 ± 1.5	2.2 ± 1.1	14 ± 4	2.0 ± 0.5	2.7 ± 0.9	15 ± 3	3.0 ± 0.5	1.8 ± 0.9
Cell strainer (ϕ 70 µm)	30 ± 2	1.5 ± 0.5	2.4 ± 0.9	30 ± 4	1.50 ± 0.03	2.3 ± 1.4	35 ± 2	1.00 ± 0.01	2.7 ± 0.8
Cell strainer (ϕ 100 µm)	40 ± 4	1.5 ± 0.3	2 ± 1	40 ± 9	1.00 ± 0.01	3.7 ± 2.1	40 ± 4	0.50 ± 0.02	0.70 ± 0.05

Zhang *et al.* created porous PUs using urethane prepolymers that possess remaining isocyanate groups at the chain terminal [78]. When water was added to the prepolymers, PU urea foam was formed via the following reaction:

$$—NCO + H_2O + —OCN \rightarrow —HN–CO–NH— + CO_2 \uparrow$$

Figure 3.37 shows the effect of the amount of water added to a viscous iso-cyanate-functional prepolymer synthesized using the ethyl ester of lysine diiso-cyanate, glycerol, and PEG in a ratio of 4:1:2 (-NCO/-OH = 1.05) in addition to ascorbic acid (AA). The addition of 1 ml of water to 10 g of polymer solution pro-duced a polymer foam with 65% porosity, while a foam of 90% porosity was

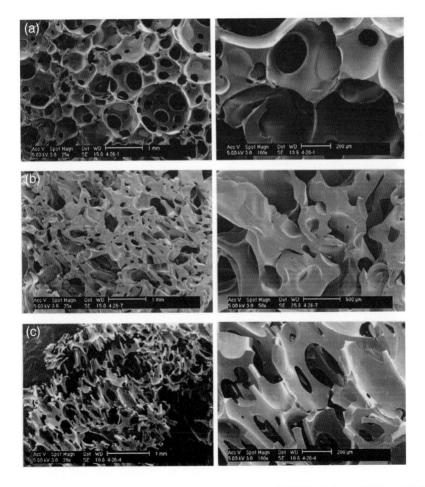

Fig. 3.37 Scanning electron micrograph of LDI–glycerol–PEG–AA scaffolds exhibits spongy internal structure with pores interconnected to support fluid flow: (a) scaffold was obtained by adding 1 ml of water to 10 g of polymer; (b) scaffold was formed by adding 1.25 ml of water to 10 g of polymer; and (c) scaffold was made by adding 1.5 ml of water to 10 g of polymer.

Table 3.11

Composition of poly(propylene fumarate) foam coat
formulation

Chemical	Amount (% w/w)
Poly(propylene fumarate)	50.9
Hydroxylapatite	14.7
1-Vinyl-2-pyrrolidone	13.7
Sodium bicarbonate	1.1
Benzoyl peroxide	2.3
Citric acid	0.9

obtained by adding 1.5 ml of water to 10 g of polymer solution. The cross-sectional view showed sponge-like cavities formed by the liberation of CO_2 during the foaming process. The pore size distribution was typically in the range of 100–500 μm.

Another method for production of spongy scaffolds is to use effervescent foaming agents. Among them is a combination of citric acid and sodium bicarbonate. Lewandrowski *et al.* prepared a porous PPF scaffold as a bone graft extender. The PPF was synthesized by the direct esterification of fumaric acid with propylene glycol [79]. The PPF-based graft extender was prepared by mixing an aqueous solution of NVP and *N,N*-dimethyl-*p*-toluidine (polymerization accelerator, 400 ppm) with a dry powdered mixture of PPF and HAp to form a viscous putty-like paste. Sodium bicarbonate, benzoyl peroxide (initiator), and citric acid were also added to the dry powder formulation. The composition of the formulation is listed in Table 3.11. On mixing of the NVP solution and PPF powder, the reaction of the effervescent agents (citric acid and sodium bicarbonate) resulted in the expansion of the material with pore sizes of 50–500 μm. The PPF foam was characterized by a few large interconnecting pores measuring approximately 200–1000 μm in diameter and a large number of small pores ranging from 50 to 100 μm in diameter.

2.4. Rapid Prototyping

To design and fabricate 3-D scaffolds by printing, Cao *et al.* utilized a rapid prototyping technology called "fused deposition modeling" (FDM) [80]. The pore volume and structure and the porosity of scaffolds are defined mainly by the setting of the computer-controlled FDM machine parameters. They fabricated an osteochondral construct by using FDM-fabricated porous scaffolds with osteogenic and chondrogenic cells. As illustrated in Fig. 3.38, both chondrocytes and osteoblasts were integrated in a scaffold so as to produce a construct that will provide anchorage to the native bone, solving the issue of poor mechanical stability encountered in attaching tissue-engineered cartilage onto bone. The FDM technique allowed designing and fabrication of scaffolds of various lay-down patterns, pore sizes, and porosity. Cao *et al.* also fabricated rectangularly shaped honeycomb-like scaffolds with a three-angle lay-down pattern (0/60/120°) using PCL. Figure 3.39 shows the

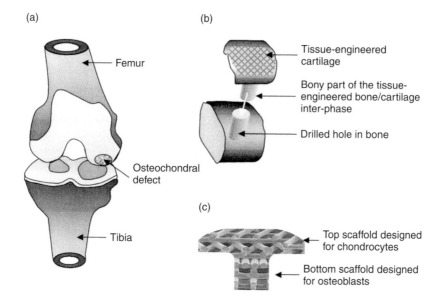

Fig. 3.38 Schematic illustration of osteochondral defect (a) and new concept of scaffold design (b) in which both chondrocytes and osteoblasts can be integrated so as to produce a construct (c) that will provide anchorage to the native bone, solving the issue of poor mechanical stability encountered in attaching tissue-engineered cartilage onto bone.

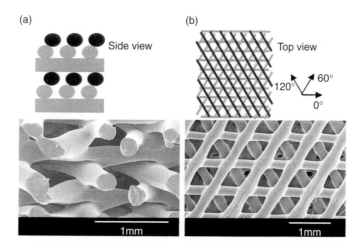

Fig. 3.39 Schematic views of the scaffold design and corresponding SEM photographs of 3-D PCL scaffold structure produced by FDM and having a 0/60/120° lay-down pattern. Freeze-fractured cross-section reveals the side view (a) showing interconnected pore structures; the top view (b) shows the typical array of equilateral triangles.

schematic architecture of the scaffolds. The porosity ranged from 60 to 65% and the pore size ranged from 300 to 580 μm. The scaffolds were cut to measure $10 \times 10 \times 3.2$ mm^3 and were partitioned vertically into two halves with a gap between them. One half (bone compartment) of the partitioned scaffold was designated for BMSC seeding and the other half (cartilage compartment) was designated for chondrocyte seeding. The scaffolds were surface-treated with 5 M NaOH.

To produce artificial constructs that would demonstrate properties of native tissue (microenvironment, 3-D organization, and intercellular contact), Smith *et al.* merged techniques of digital printing and tissue engineering [81]. This direct-write bioassembly system that is capable of using pneumatic or positive displacement pens to deposit material in a controlled 3-D pattern was designed to permit layer-by-layer placement of cells and ECM on a variety of material substrates. Human fibroblasts suspended in PEO/PPO were coextruded through a positive displacement pen delivery onto a polystyrene slide. After deposition, approximately 60% of the fibroblasts remained viable. The Bovine aortic endothelial cells (BAECs) suspended in soluble type I collagen were coextruded via microdispense pen delivery onto the hydrophilic side of flat sheets of PET. After deposition with a 25-gauge tip, 86% of the BAECs were viable. When maintained in culture for up to 35 days, the constructs remained viable and maintained their original spatial organization.

Yang *et al.* developed a biologically benign CO_2 assisted microfabrication process for the production of 3-D tissue scaffolds from premolded 2-D skeletons generated from microembossing [82]. The process involved three steps: (1) photolithography to generate the mold with the planar skeletal structure; (2) microembossing to fabricate 2-D scaffold skeletons with the planar skeletal structure transferred from the mold; and (3) subcritical CO_2 bonding to assemble the laminated multiple 2-D skeletons into 3-D tissue scaffolds. A PLGA film was placed on a hot plate at 220°C. Allowing about 1 min for the PLGA film to melt, the inverse silicone mold was embossed into the molton PLGA film by applying a constant pressure. In bilayer embossing (BLE), two layers of PLGA skeletons were generated simultaneously. A PLGA film was placed on a silicone mold, which was placed on the hot plate. When the PLGA film had melted, another silicone mold in an orthogonal orientation to the bottom mold was pressed down into the molten PLGA film. The resulting PLGA bilayer skeleton was removed from the molds. For CO_2 bonding, four PLGA bilayer skeletons were orthogonally aligned and stacked between two glass slides. A contact pressure was applied by placing weights on the top glass slide. The assembly was placed in a pressure vessel that was immersed in a water bath at 35°C, and a high-pressure syringe pump was used to deliver and control the CO_2 pressure. After saturation with CO_2 for 1 h, the pressure was quickly released and the bonded scaffold removed from the vessel.

For a direct 3-D printing (3-DP) method the final scaffold materials are utilized during the actual 3-DP process. To overcome the limitations of the direct technique, Lee described an indirect 3-DP protocol, where molds are printed and the final materials are cast into the mold cavity [83]. To evaluate the resolution available in this technique, scaffolds with villi features (500 μm diameter and 1 mm height)

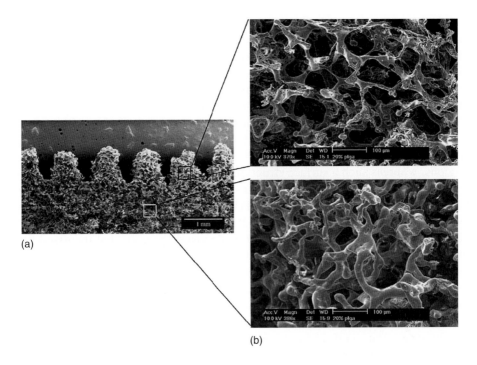

Fig. 3.40 SEM images of the cross-section of villi-shaped PLGA scaffold showing uniform pore distribution (a) and high interconnectivity of pores ranging from 100 to 150 μm (b). Each 1-mm tall, columnar-shaped villus tapers in diameter from ~700 μm at the base to ~500 μm at the villus tip.

were produced by solvent casting into plaster molds, followed by particulate leaching. The SEM shown in Fig. 3.40 reveals open, interconnected pore architecture.

Porous PCL scaffolds were computationally fabricated via selective laser sintering (SLS), a rapid prototyping technique [84]. Cylindrical porous scaffolds with 3-D orthogonal periodic porous architectures were designed using 3-D solid modeling software. The designs were then exported to a Siterstation 2000™ machine in STL file format, and were used to construct scaffolds by SLS processing of PCL powder with a particle size distribution in the 10–100 μm range. The SLS processing of the PCL powder was conducted by preheating the powder to 50°C and scanning the laser at 4.5 W power. Scaffolds were built layer-by-layer using a powder layer thickness of 100 μm. After SLS processing was completed, the scaffolds were allowed to cool inside the machine process chamber.

2.5. Electrospinning

As a series of investigations on electrospinning technology in biomedical application, Kwon *et al.* studied the following: (1) the copolymer composition dependence on the physical state and mechanical properties of electrospun fibers,

(2) structural features including mechanical integrity of electrospun fabrics made of the equimolar copolymer, which were prepared using different solvents under different operation conditions, and (3) the dependence of cell adhesion and proliferation potentials on surface fiber density [85]. To this end, they prepared nano- to micro-structured P(LA/CL) fabrics. Electrospun microfiber fabrics with different compositions of P(LA/CL) (mol % in feed: 70/30, 50/50, and 30/70), PLLA, and PCL were obtained by electrospinning using methylene chloride as a solvent. As shown in Fig. 3.41, the PLLA microfiber exhibited a nanoscale-pore structure with a pore diameter of 200–800 nm at the surface and subsurface regions, whereas such a surface structure was hardly observed in other polymers containing CL. The microfiber fabric made of P(LA/CL) (50/50) was elastomeric. Nanoscale-fiber fabrics with P(LA/CL) (50/50) (0.3 or 1.2 μm in diameter) were electrospun using 1,1,1,3,3,3-hexafluoro-2-propanol as a solvent. The decrease in the fiber diameter of the fabric decreased porosity, but increased fiber density and mechanical strength. Human umbilical vein ECs were adhered well and proliferated on the small-diameter-fiber fabrics (0.3 and 1.2 μm in diameter), both of which were dense fabrics, whereas markedly reduced cell adhesion, restricted cell spreading, and no signs of proliferation were observed on the large-diameter-fiber fabric (7.0 μm in diameter). Mo *et al.* also fabricated an elecrospun nanofiber scaffold from P(LA/CL) [86]. A 3–9 wt% polymer solution in acetone was placed in a syringe fitted with a needle with tip diameters from 0.4 to 1.2 mm. Electrospinning was performed at applied voltages ranging from 9 to 15 kV. The ground collection plate of aluminum foil was located at a distance of

(a)   (b)   (c)   (d)   (e)

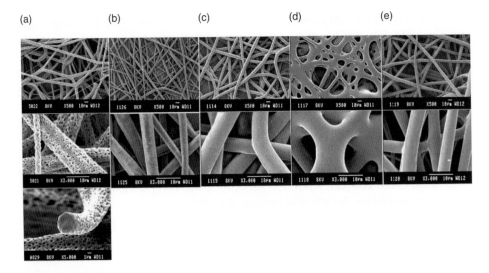

Fig. 3.41   SEM images of electrospun microfiber fabrics made of PLCL (co)polymers: (a) PLLA, (b) PLCL 70/30, (c) PLCL 50/50, (d) PLCL 30/70, and (e) PCL.

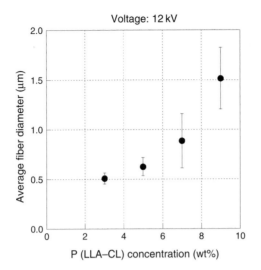

Voltage: 12 kV

Fig. 3.42 Relation between fiber diameter and P(LA/CL) concentration in the electrospinning at a constant applied voltage of 12 kV.

13 cm from the needle tip. The feeding rate of syringe pump was fixed at 2 ml/hr. The polymer solution formed Taylor cone at the tip of the needle by the combined force of gravity and electrostatic charge. A positively charged jet formed from the grounded aluminum foil target. Figure 3.42 shows that the fiber diameter significantly decreased with the decreasing polymer concentration, although there was a limitation to obtain uniform nanofibers without beads. A fiber diameter tended to decrease with the increasing electrospinning voltage, although the influence was not as great as that of polymer concentration.

To prepare PCL/CaCO$_3$ composite nanofibers, Fujihara *et al.* dispersed CaCO$_3$ nanoparticles in a solvent and then dissolved PCL pellet [87]. The prepared solution was placed in a syringe whose needle inner diameter was 0.21 mm. The feeding rate of solution was fixed at 1 ml/h by syringe pump and 20 kV was applied to the needle tip. Electrically charged polymer solution formed Taylor cone from the tip of the needle to the ground collector plate with a fixed distance of 130 mm. During this traveling, solvent was evaporated in the air and randomly oriented nanofibrous meshes were fabricated on the collected plate.

To study the structural and functional effects of fine-textured matrices with sub-micron features on the growth of cardiac myocytes, Zong *et al.* fabricated non-woven PLGA scaffolds by electrospinning [88]. Post-processing was applied to achieve macroscale fiber orientation (anisotropy). Electrospun PLGA membranes were uniaxially stretched to achieve anisotropic fiber architecture with a modified Instron tensile stretching apparatus. Each membrane was first drawn to the desired extension ratio of 200% at a constant rate of 4 mm/min at 60°C, and then cooled down to room temperature under the tension. Typical SEM images of

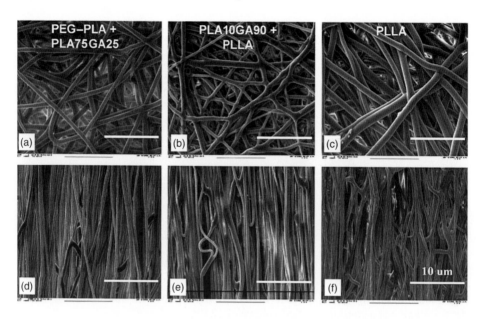

Fig. 3.43 SEM images of as-prepared electrospun membranes: (a) PLA75GA25 + PEG-PLA; (b) PLA10GA90 + PLLA; (c) PLLA. Images of the corresponding oriented electrospun membranes are shown in (d), (e), and (f), respectively.

oriented electrospun scaffolds with different chemical compositions by uniaxial stretching are shown in Fig. 3.43. Apparently, the overall orientation of the non-woven scaffold increased significantly and the degree of orientation was similar to the level reported in the study employing a spinning disc in combination with electrospinning, but the porosity of the membranes decreased by about 20%. Ma *et al.* also showed the ability of electrospinning to produce aligned, core shell–structured, or surface-functionalized polymer nanofibers [89]. Figure 3.44 shows randomly oriented and aligned nanofiber matrices.

A chitosan nanofibrous mat of average fiber diameter of 130 nm was electro-spun from aqueous chitosan solution using concentrated acetic acid solution [90]. The aqueous acetic acid concentration higher than 30% was a prerequisite for chitosan nanofiber formation, because more concentrated acetic acid in water progressively decreased surface tension of the chitosan solution and concomitantly increased charge density of jet without any significant effect on solution viscosity. In a study on electrospinning of protein fibers, Li *et al.* found that the average diameter of gelatin and collagen fibers could be scaled down to 200–500 nm without bead formation, while elecrospun alpha-elastin and tropoelastin fibers were several microns in width [91]. Fibers composed of alpha-elastin, especially tropoelastin, exhibited "quasi-elastic" wave-like patterns at increased solution delivery rates. Figure 3.45 shows the stress–strain curves of the electrospun protein fibers.

(a)

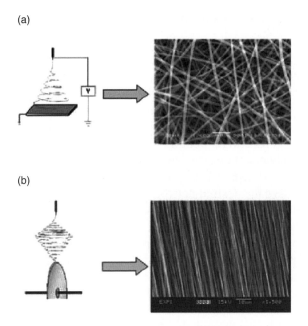

(b)

Fig. 3.44 A randomly oriented nanofiber matrix deposited on a static collector (a) an aligned nanofiber matrix deposited on a rotating collector (b).

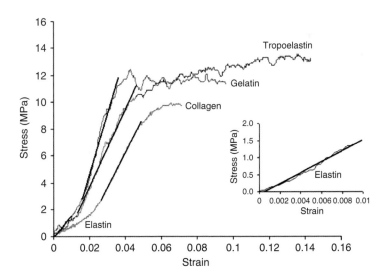

Fig. 3.45 Results of the microtensile test for electrospun natural fiber sheets prepared at concentrations of 8 (gelatin), 10 (collagen), 20 (tropoelastin), and 20% (elastin).

2.6. UV and Laser Irradiation

Leclerc *et al.* tested three fabrication methods using photosensitive biodegradable polymers to produce various kinds of microchannels at a characteristic scale around 100 μm [92]. The first fabrication method was a molding process, presented in Fig. 3.46a. For the molding process, an SU-8 negative master was first made by photolithography (i–iii). Then, a fluorocarbon layer was deposited by a reactive ion-etching machine (iv). The biodegradable polymer was on the mold (v), and then was exposed to UV light for several minutes (vi). After the layer was solidified, it could be peeled off from the mold master (vii). The second fabrication process was very rapid and simple as shown in Fig. 3.46b. It was a direct UV exposure through a mask pattern to have the desired network. The thickness of the exposed polymer could be controlled by spin coating the UV-sensitive polymer on a glass substrate. The third fabrication process (Fig. 3.46c), which combined the UV exposure with the stamping process was used to microfabricate single and multistepwise structures in one layer of photosensitive polymer. After coating the polymer on a flat glass substrate, a negative stamp of the desired microstructures made in silicone was applied on the polymer. This silicone stamp was fabricated also by soft lithography. Because the silicone is transparent to UV light, it was then possible to solidify the polymer by UV irradiation through the stamp.

Woodfield *et al.* constructed porous 3-D scaffolds from poly(ethylene glycol)-terephthalate-poly(butylene terephthalate) (PEGT/PBT) block copolymer using a custom-designed fiber-deposition device [93] (see Fig. 1.12). It consisted of five main components; (1) a thermostatically controlled heating jacket; (2) a molten copolymer dispensing unit consisting of a syringe and nozzle; (3) a force-controlled plunger to regulate flow of molten copolymer; (4) a stepper motor driven x–y–z table; and (5) a positional control unit consisting of stepper-motor drivers linked to a personal computer containing software for generating fiber-deposition paths. Copolymer granules were placed in the syringe purged with nitrogen gas and allowed to melt. A stainless steel plunger was used to apply pressure to the molten polymer. Pressure was regulated by placing the entire deposition device beneath the cross-head of a standard tension-compression test machine in order to control the displacement of the plunger and monitor the resultant force. Stepper motors coordi-nated speed and translation of the x–y–z table, and were controlled by a custom deposition program via the printer port (LPT1). The program required inputs of the overall scaffold dimensions, the spacing between deposited fibers, the number of fiber layers, and the speed at which the x–y–z table translated. By lowering the x–y table one-layer-step in the z-direction, successive layers of rapidly solidifying fibers were laminated to previous layers in a 0–90° pattern creating a consistent pore size and 100% interconnecting pore volume. Fiber layers could be continuously deposited, resulting in scaffolds up to 10 mm in thickness.

To create channels in 3-D hydrogel matrices for guided axonal growth, Luo and Shoichet activated bulk agarose with CDI, followed by reaction with 2-nitrobenzyl-protected cysteine, as shown in Fig. 3.47 [94]. On exposure to UV light, the

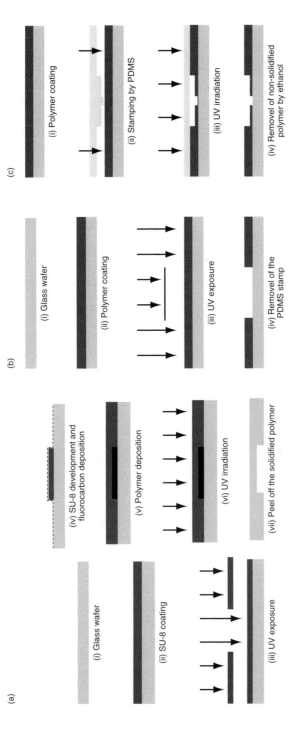

Fig. 3.46 Fabrication process of the microstructures in biodegradable polymer by molding (a), by UV exposure using a mask pattern (b), and by UV exposure and hybrid process with stamping using a poly (dimethyl siloxane) stamp (c).

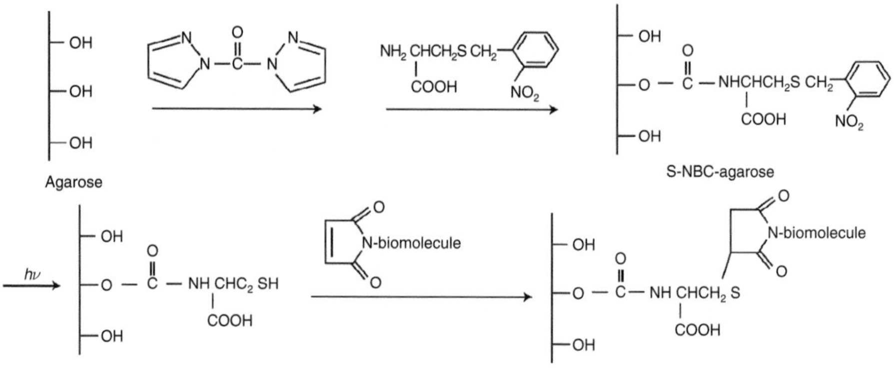

Fig. 3.47 The general strategy used to create adhesive biochemical channels in agarose hydrogel matrices relies on modifying agarose with photolabile groups, focused laser light sources, and biomolecule coupling.

2-nitrobenzyl group was cleaved leaving sulphydryl groups that reacted with maleimido-terminated Gly-Arg-Gly-Asp-Ser (GRGDS). The distribution of the peptide in the gel after laser fabrication was analyzed by confocal microscopy, and the peptide channels were observed penetrating the 1.5-mm-thick sample. The GRGDS peptides were conjugated to the gel in isolated circular channels defined by the shape of the laser beam. These circular peptide domains were observed continuously at all depths in the gel. The diameter of the channels was between 150 and 170 μm over a 1 mm depth, corresponding to the focal spot size of the focused laser beam of 160 μm.

3. NOVEL SCAFFOLDS

3.1. Naturally Derived Scaffolds

3.1.1. ECM-like Scaffolds

For a rational design of tissue engineering scaffolds that should assist cells to form the desired tissue, it will be beneficial to study the biological effect of individual components of the ECM, which is a native scaffold: collagen, elastin, and GAGs. To do so, Daamen *et al.* prepared collagen–elastin–GAG bioscaffolds [95]. They used four different ratios of collagen and elastin (1:0, 9:1, 1:1, and 1:9), with and without chemical crosslinking, and with and without CS. For the preparation of non-EDAC-crosslinked type I collagen scaffolds, a 2% collagen suspension was prepared in 0.5 M acetic acid (pH 2.5), diluted to 1% with ice-cold water, and then deaerated to remove entrapped air bubbles. The collagen suspension was poured into a mold, frozen at -80°C and lyophilized, resulting in porous collagen scaffold formation. For crosslinking, scaffolds were incubated in 2-morpholinoethane sulphonic acid (MES) (pH 5.5) in the presence of 40% ethanol. Subsequently, the scaffolds were crosslinked by immersion in 50 mM MES (pH 5.5) containing 33 mM EDAC and 6 mM NHS.

The EDAC-crosslinking of the scaffolds was also performed in the presence of 2.75% CS which resulted in the covalent attachment of CS to scaffolds. Table 3.12 gives the composition and properties of 12 scaffolds obtained. Collagen–elastin crosslinking using EDAC/NHS resulted in the formation of crosslinks between carboxyl and amine groups with a reduction in the number of amine groups. Collagen could be crosslinked to a high extent than collagen–elastin scaffolds because more amine groups were present. It was not possible to construct a stable scaffold made of only elastin, even not after crosslinking. Coupling of CS to scaffolds generally had no effect on crosslinking efficiency, but increased the water-binding capacity. The tensile strength of the non-crosslinked pure collagen scaffold was the highest (about 100 kPa), and introduction of elastin to the scaffold lowered the tensile strength as well as the E-modulus. Crosslinking of the scaffolds increased both the tensile strength and the E-modulus. Attachment of CS to a scaffold did not result in a change of tensile strength. The collagen and collagen–elastin 9:1 scaffolds could not be extended to 125% while the collagen–elastin 1:9 scaffold to about 150%. All scaffolds showed a porous structure with pores ranging from 20 to 100 μm, as represented in Fig. 3.48. The SEM shows thick elastin fibers and thin collagen fibrils, interacting with each other. Collagen connected separate elastin fibers, like a glue, incorporating elastin in the scaffolds, and elastin fibers were ensheathed by a layer of collagen fibrils. Fibroblasts on the scaffolds after 14 days of culturing formed normal spindle-shaped cells. Myoblasts adsorbed, aligned, and fused to form multinucleated myotubes of several

Table 3.12
Biochemical and biomechanical characteristics of collagen–elastin–chondroitin sulfate scaffolds

Scaffold	Crosslinked with EDC/NHS	Amine Group Content (nmol/mg)	CS Content (ng/mg)	Water-Binding Capacity (# times dry weight)	Tensile Strength (kPa)	E-modulus (MPa)
COL	−	281 ± 7	0	20 ± 1	103 ± 15	0.39 ± 0.07
COL	+	185 ± 3	0	20 ± 1	677 ± 191	1.03 ± 0.08
COL–CS[a]	+	186 ± 8	100 ± 4	33 ± 3	520 ± 105	0.97 ± 0.07
COL–EL 9:1	−	256 ± 8	0	19 ± 2	67 ± 6	0.42 ± 0.10
COL–EL 9:1	+	159 ± 4	0	19 ± 1	420 ± 35	0.78 ± 0.07
COL–EL–CS 9:1[b]	+	164 ± 1	85 ± 5	29 ± 4	394 ± 42	0.76 ± 0.13
COL–EL 1:1	−	147 ± 12	0	16 ± 1	63 ± 23	0.24 ± 0.06
COL–EL 1:1	+	87 ± 5	0	16 ± 3	142 ± 8	0.42 ± 0.04
COL–EL–CS 1:1[c]	+	83 ± 2	58 ± 5	21 ± 3	128 ± 10	0.54 ± 0.11
COL–EL 1:9	−	67 ± 7	0	11 ± 1	ND	ND
COL–EL 1:9	+	57 ± 5	0	11 ± 1	ND	ND
COL–EL–CS 1:9[d]	+	35 ± 4	24 ± 2	12 ± 1	ND	ND

COL = collagen; EL = elastin; CS = chondroitin sulfate; ND = not done; scaffolds were too weak to measure tensile strength and E-modulus. Results are mean ± SD of three independent experiments.
[a] 1 g of COL–CS scaffold contains 900 mg collagen and 100 mg CS.
[b] 1 g of COL–EL–CS 9:1 scaffold contains 823 mg collagen, 92 mg elastin, and 85 mg CS.
[c] 1 g of COL–EL–CS 1:1 scaffold contains 471 mg collagen, 471 mg elastin, and 58 mg CS.
[d] 1 g of COL–EL–CS 1:9 scaffold contains 98 mg collagen, 878 mg elastin, and 24 mg CS.

(a)   (b)

(c)   (d)

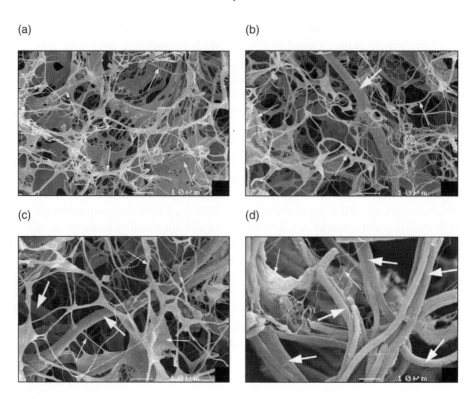

Fig. 3.48   Scanning electron micrographs of the air side of a collagen scaffold (a), a 9:1 collagen–elastin scaffold (b), a 1:1 collagen–elastin scaffold (c), and a 1:9 collagen–elastin scaffold (d). Collagen is present as fibrils and sheets. Note the interactions between (thick) elastin fibers (large arrows) and collagen fibrils (small arrows). Bar is 10 μm.

hundreds of μm in length on all scaffolds, except on the collagen–elastin 1:9 scaffolds. On these scaffolds, cells did not have an elongated form, but a rounder form with small sprout-like structures. This could be caused by diminished proliferation of myoblasts on the collagen-elastin 1:9 scaffolds because myoblasts only start to differentiate when cells contact one another.

Elbjeirami *et al.* hypothesized that lysyl oxidase (LO) would form crosslinks within newly deposited ECM of engineered tissue, resulting in a structure with enhanced mechanical properties [96]. They chose gene therapy as the method of introducing LO because of the high cost of protein purification and issues with long-term delivery. The vascular smooth muscle cells (VSMCs) were liposomally transfected with the LO gene. Both Northern and Western analyzes confirmed increased LO expression. As a control, mock wells received the plasmid DNA vector with no LO cDNA. For the preparation of collagen gels, a solution of dermal type I collagen was dissolved in 0.001 N HCl and seeded with VSMCs. The cell/collagen mixture was allowed to gel at 37°C for 1 h. Then DMEM was added to each gel in wells and the culture medium was supplemented with AA and $CuSO_4$. The gels were cultured for

1 week in the presence or absence of β-aminopropionitrile (BAPN), an LO inhibitor. The LO-seeded gels appeared to be less contracted and compacted as compared to mock-seeded gels. As shown in Fig. 3.49, collagen gels seeded with LO-transfected VSMCs had an elastic modulus of 75 kPa. Controls and BAPN-treated samples were at or below 40 kPa. The UTS increased, with LO transfectants having approximately double the strength of mock controls. Compositional analysis of the ECM deposited by the transformed cells showed similar collagen and elastic levels, and cell proliferation rates, thus attributing increased mechanical properties to ECM crosslinking.

3.1.2. Tissue-Derived Scaffolds

To develop a naturally derived biomaterial that may potentially provide an alternative to the nerve autograft, Hudson *et al.* tested various detergents for their ability to remove axons and SCs from nerve tissue while preserving the ECM [97]. The detergents tested include (1) nonionic detergents Tween 20, Tween 80, and Triton X-100; (2) amphoteric detergents Empigen BB, sulfobetaine-10, sulfobetaine-16, and 3-[(3-cholamidopropyl)dimethylammonio]-1-propanesulfonate; (3) anionic detergents Triton X-200, dodecylbenzene sulfonate, sodium caprylate, and sodium deoxycholate; (4) a cationic detergent hexadecyltrimethylammonium bromide. Using histochemistry and Western analysis, they found that an optimized protocol was created with detergents Triton X-200, sulfobetaine-16, and sulfobetaine-10.

Cartmell and Dunn developed methods to remove cells from rat tail tendons without compromising their mechanical properties of collagen structure, using an ionic detergent sodium dodecyl sulfate (SDS) or an organic solvent tri(*n*-butyl)phosphate (TBP) [98]. Fresh-frozen hindlimbs from New Zealand white rabbits were thawed and

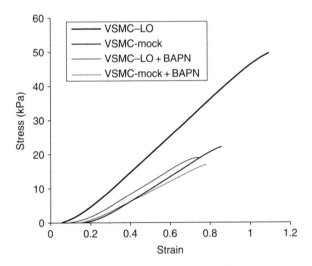

Fig. 3.49   Elastic modulus of collagen gels seeded with transfected VSMCs and cultured in the presence or absence of BAPN for 1 week.

the PT was exposed and freed from its patellar and tibial bone attachments by dissection. The PTs were immersed in an extraction solution. The five experimental groups tested were (1) nontreated controls, (2) SDS for 24 h, (3) SDS for 48 h, (4) TBP for 48 h, and (5) TBP for 72 h. After treatment, they measured tendon cellularity, crimp structure, and mechanical properties. Treatment with either SDS or TBP removed 70–90% of the intrinsic PT cells. Mechanical properties of treated PTs were similar to those of controls, despite changes in appearance. TBP- and SDS-treated PTs were then seeded with fibroblasts and cultured for up to 2 weeks *in vitro*. Fibroblast proliferation was retarded on SDS-treated PTs; in contrast, TBP-treated PTs supported cell proliferation similar to that of untreated controls. Extrinsic fibroblasts were successfully cultured on the TBP-treated PTs *in vitro*, creating viable tissue-engineered grafts.

Brown *et al.* derived bladder acellular matrix (BAM) from full-thickness porcine bladder as a scaffold for *in vivo* bladder tissue engineering [99]. The SDS residual surfactant was removed from the matrix via a final 24-h treatment with a pH-9 tris solution. Samples of freshly processed matrix were evaluated using histology and immunohistochemistry to confirm the effectiveness of the extraction process. The resulting material consisted primarily of collagen and elastin and retained the basement membrane, which was important for UC support. Lu *et al.* incorporated muscle-derived cells (MDCs) into SIS scaffolds. Thirty sheets of SIS were cut into the appropriate size to fit in the specific designed cell culture inserts [100]. The MDCs were harvested from mice hindleg muscle, transfected with a plasmid encoding for β-galactosidase, and placed into single-layer SIS cell culture inserts. Twenty-five MDC and/or SIS specimens were incubated for either 10 or 20 days. Histological results indicated that MDC migrated throughout the SIS after 20 days. The mean areal strain of the 0-day control group was 0.182. After 10-day incubation, the mean areal strain in MDC/SIS was 0.247 compared to 10-day control SIS 0.200. After 20-day incubation, the mean areal strain of MDC/SIS was 0.255 compared to 0.170 of control SIS. Both 10- and 20-days seeded groups were significantly different from that of incubated SIS alone.

As opposed to elastic fibers, elastin fibers do not contain any microfibrils or associated molecules. Purification protocols for elastin generally result in greatly damaged elastin fibers and this likely influences the biological response. Daamen *et al.* described a protocol for the isolation of elastin whereby the elastic fibers stay intact [101]. Elastin fibers were isolated from equine elastic ligaments by several extraction steps and trypsin digestion. Elastin fibers were free of contaminants and had a smooth, regular appearance. Berglund *et al.* focused on methods capable of incorporating exogenous elastins into the vascular grafts rather than relying on *de novo* synthesis and assembly of elastic fibers [102]. Carotid arteries procured from adult pigs were digested by a series of enzymatic, chemical, and thermal treatments aimed at removing cells and nonelastin matrix components. After washes in PBS and removal of excess adventitia, arterial segments were immersed in water and subjected to five autoclave cycles followed by incubations in cyanogen bromide-70% formic acid and a 2-mercaptoethanol-urea solution. Segments were incubated in saline before being incorporated into tissue-engineering vascular grafts. Either human dermal fibroblasts (HDFs) or rat aortic smooth muscle cells (RASMs) were suspended in collagen and poured into tubular molds above the isolated elastin scaffolds. The hybrid constructs exhibited increased tensile strength (11-fold in HDFs and 7.5-fold in RASMs) and linear stiffness moduli (4-fold in HDFs and 1.8-fold in RASMs) compared with collagen control constructs with no exogenous elastin scaffold.

Walles *et al.* sought to grow *in vitro* functional SMCs, chondrocytes, and respiratory epithelium on a biological, directly vascularized matrix as a scaffold for tracheal tissue engineering [103]. Free jejunal segments 10- to 15-cm long with their own vascular pedicle were harvested and decellularized from donor pigs and used as a vascular matrix. Autologous costal chondrocytes, SMCs, and respiratory epithelium and EPCs were first cultured *in vitro* and then disseminated on the previously decellularized vascular matrix. The EPCs re-endothelialized the matrix to such an extent that EC viability was uniformly documented through positron emission tomography. This vascularized scaffold was seeded with functional SMCs and reseeded with viable ciliated respiratory epithelium. Chondrocyte growth and production of extracellular cartilaginous matrix was observed as soon as 2 weeks after their culture.

Mauney *et al.* evaluated the potential of both mineralized and partially or fully demineralized biomaterials derived from bovine bone matrix to be used in tissue engineering strategies [104]. They assessed the ability to support human BMSC osteogenic differentiation *in vitro* and bone-forming capacity *in vivo*. The scaffolds were divided into three separate groups: fully mineralized (FM), partially mineralized (PM), and fully demineralized (DM). The PM and DM scaffolds were demineralized in 0.6 N HCl for 15 min and 12 h, respectively, and then rinsed with water. The BMSCs from some donors expressed significantly higher levels of all osteoblast-related markers following cocultivation in direct cell-to-scaffold contact with FM scaffolds in comparison to DM preparations, while BMSCs from other donors displayed no significant differences in response to various scaffold preparations. The *ex vivo* incorporation of BMSCs into all bone-derived scaffold preparations substantially increased the mean extent and frequency of samples containing *de novo* bone formation over similar non-seeded controls. No statistically significant differences were observed in the extent or frequency of bone formation between various scaffold preparations seeded with BMSCs from different donors.

Freytes *et al.* measured mechanical properties of canine stomach submucosa (SS), porcine urinary bladder submucosa (UBS), urinary bladder matrix (UBM), and porcine SIS multilaminate devices [105]. Representative stress versus strain curves for each device are shown in Fig. 3.50. The failure stress of UBS 2-layer devices was lower than that of UBS +UBM 2-layer devices. The SS 2-layer devices had the highest failure stress and maximum stress tangent modulus (MSTM) values followed by SIS 4-layer devices. The maximum stretches of SS 2-layer devices and USB + UBM 2-layer devices were similar.

3.1.3. Fibrin Gel

Mol *et al.* tested the feasibility of using fibrin gel as a cell carrier combined with a bioabsorbable fiber mesh for engineering cardiovascular tissue, and compared

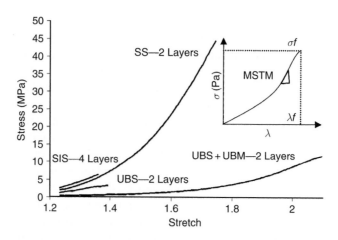

Fig. 3.50   Stress versus stretch curves for ECM devices.

with the conventional cardiovascular tissue engineering procedure by static seeding of cells onto a fiber mesh [106]. Cells were resuspended in a bovine thrombin solution and then added to a bovine fibrinogen solution. After carefully mixing, the fibrin solution containing the cells was dropped onto the scaffold. Seeding with fibrin resulted in less loss of soluble collagen into the medium and a more mature ECM formation, although it did not hamper cell proliferation. Zisch *et al.* sought to develop the natural hydrogel matrix fibrin as platform for extensive interactions and continuous signaling by the vascular morphogen ephrin-B2 that normally resides in the plasma membrane and requires multivalent presentation for ligation and activation of ephrin receptors on apposing EC surfaces [107].

3.1.4. Natural Sponge

Most marine sponges secrete a mineral skeleton composed of microscopic silica or calcite spicules that are joined by, or embedded within, spongin fibers. Spongin is analogous to type XIII collagen. The amino acid composition of spongin is similar to vertebrate collagen with a high percentage of glycosylated hydroxylysine, aspartic acid, and glutamic acid. Glycoproteins and carbohydrates-rich substances are also associated with the collagen component and the adhesion proteins, FN and tenascin, have also been identified in these marine sponges. Green *et al.* examined the capability of a natural marine sponge fiber skeleton to act as a scaffold for osteoprogenitor cells and a delivery vehicle for osteogenic proteins [108]. The marine sponge skeleton selected was an undetermined species of *Spongia* from San Jose, Pinto, West Indies. The overall skeleton of *Spongia* sp. is a fine homogeneous regular network and, when hydrated, is extremely soft, flexible, and elastic. The *Spongia* skeleton had been commercially bleached and prepared for sale by standard commercial methods before purchase, but additional washing in 100% ethanol, storage in 70% ethanol, and UV irradiation ensured that the material was sterile and free of cellular debris. The spaces between collagen fibers ranged from 1500 to 2000 μm. These pore sizes were three to four times greater than those found in human compact bone (150–250 μm) and human trabecular bone (400–500 μm). Four-week culture showed no degradation of this marine sponge skeleton. Histochemical staining for ALP and type I collagen indicated formation of bone matrix. However, a number of caveats exist for their immediate application: (1) marine sponge skeletons must be made biodegradable by chemical or enzymatic pretreatment to weaken collagen crosslinking; (2) the possibility of immunogenicity remains unclear; (3) marine sponges not only promote cell adhesion, but are equally adhesive to bacteria; (4) many natural marine sponges contain silica and sand debris in a central core within the fiber, thus causing potential problems postbiodegradation.

3.2. Injectable Scaffolds

Burg *et al.* developed a concept of injectable composites, wherein cells were seeded onto discrete particles or beads, denoted as "cell carriers" [109]. The cellular carriers were embedded in a gel delivery matrix, thus forming a composite that might be loaded in a syringe and injected into the patient. Cells performed poorly when suspended within a gel lacking a cell carrier. The delivery matrix facilitated delivery of the cellular carrier in a minimally invasive manner, either as a single injection or as several injections, and it maintained distance between individual cellular carriers. McGlohorn *et al.* assessed low-temperature casting of PLLA beads as a possible method to produce synthetic cell-carrier beads [110]. The PLLA pellets were dissolved in chloroform. Glucose and NaCl particles were sieved, and 250–300-μm particles were isolated to serve as porogens. Each porogen was suspended in the PLLA chloroform solution. To form the beads, the suspension was loaded into a syringe and dispensed at 30 ml/h via syringe pump into an ethanol/liquid nitrogen slurry to form beads. Formed beads were isolated and leached with deionized water.

3.3. Elastic Scaffolds

Lee *et al.* developed an elastic scaffold which could be used to engineer tissues under mechanically dynamic conditions [111]. Porous scaffolds were fabricated from GA–ε-CL copolymers using the solvent-casting and particle-leaching technique. NaCl particles (200–300 μm) were added to a copolymer solution in chloroform (5% w/v). The polymer–salt solution was cast onto glass plates coated with poly(vinyl pyrrolidone) and NaCl particles. The resulting copolymer/salt composite membranes were immersed in water and were shaked for 3 days to leach out the salt. The scaffolds had interconnected and uniform pore structures without any nonporous skin layers. The average pore size and porosity of the scaffolds were 250 ± 50 μm and

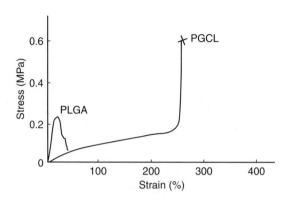

Fig. 3.51   Representative tensile stress–strain curves of PLGA and PGCL scaffolds.

93%, respectively. Figure 3.51 shows representative tensile stress–strain curves of scaffolds from GA–CL(50:50) copolymer(PGCL) and GA–LA(30:70) copolymer (PLGA). The scaffolds were subjected to cyclic loading at various amplitudes between 5 and 20% and at frequencies of 0.1 and 1 Hz for 6 days. The PGCL scaffold was elastic under cyclic loading conditions in air at 37°C for up to 6 days. The permanent deformation was less than 4% of the applied strain magnitude.

3.4. Inorganic Scaffolds

Dong *et al.* prepared porous β-TCP scaffolds [112]. First, β-TCP powder was synthesized by the wet milling method. A slurry consisting of a mixture of $CaCO_3$ and $CaHPO_4 \cdot 2H_2O$ powder at a molar ratio of 1:2 and pure water was prepared in a pot mill. Upon drying this slurry at 80°C, a calcium-deficient HAp powder was obtained. By calcining this powder at 750–900°C, the crystal phase was converted to β-TCP. For preparation of β-TCP, a foaming slurry was prepared by mixing the β-TCP powder, pure water, and a foaming agent. The foaming slurry was dried at room temperature for 1 day and at 40°C for another day to obtain a green body. On sintering the green body at 1050°C for 1 h, a porous β-TCP scaffold was obtained. The average pore size was 200–400 μm in diameter and almost all pores were interconnected via a 100–200-μm path.

Three-D periodic HAp scaffolds were constructed by Michna *et al.* via directwrite assembly of a concentrated HAp ink using a robotic deposition apparatus [113]. The ink was prepared first by making an ammonium polyacrylate (PAA) of pH 9. Then HAp powder was added to this aqueous solution in three aliquots. The HAp suspension was ultra-sonicated using an ultra-sonicating horn. Finally, the suspension was gelled by adding poly(ethyleneimine) (PEI). The HAp ink was then placed on the paint shaker for 10 min between each addition of PEI and 30 min after the final addition. They employed an ink delivery system mounted on a z-axis motion-controlled stage for agile printing onto a moving x–y stage. Three-D periodic HAp scaffolds (6 × 6 × 6 mm³) were assembled, which consisted of a linear

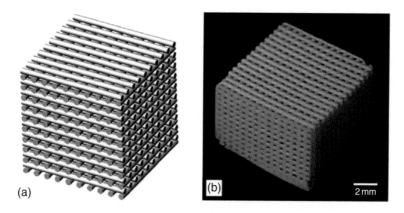

Fig. 3.52 (a) Schematic illustrations of the 3-D periodic scaffold design and (b) corresponding optical image of a HAp scaffold created by direct-write assembly with center-to-center rod distance of 500 μm.

array of parallel rods aligned with the x- or y-axis such that their orientation was orthogonal to the previous layer (Fig. 3.52). The HAp scaffolds were sintered at 1200–1300°C to achieve the desired degree of densification. By controlling their lattice constant and sintering conditions, 3-D periodic HAp scaffolds were produced with a bimodal pore size distribution.

3.5. Composite Scaffolds

Maquet *et al.* fabricated composite scaffolds made of poly(α-hydroxyacid)s and bioactive glass by freeze-drying which involved a solid–liquid phase separation of polymer solutions in organic solvents and subsequent solvent sublimation [114]. This method provided simultaneously versatility because of a variety of scaffold designs and control of the pore size and structure. To simultaneously generate bone and cartilage in discrete regions and provide for the development of a stable interface between cartilage and subchondral bone, Schek *et al.* used materials and biological factors in an integrated approach to regenerate a multi-tissue interface [115]. Figure 3.53 illustrates biphasic composite scaffolds manufactured by image-based design (IBD). Composite scaffolds were composed of two bonded cylinders, one HAp and one PLLA, each 5 mm in diameter and 3 mm in height. The HAp phase contained 800- or 300-μm-diameter orthogonal pores with a porosity of 50%. These were manufactured by indirect solid free-form fabrication. An HAp and acrylic slurry was cast into wax molds obtained from the 3-D printer and sintered. The polymer sponges were created by the salt-leaching technique. A thin PGA film was placed between the two phases before assembly to serve as a barrier to cell infiltration during seeding and cell migration during growth. The two phases were assembled by spreading methylene chloride solution of PLLA on the exposed ceramic face and pressing the polymer

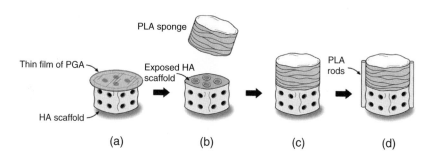

Fig. 3.53 The PLA was used to join the polymer and ceramic phases of the composite scaffold. One face of the ceramic was coated with a thin film of PGA (a). The film was removed from the circumference and PLA was applied to the surface (b). The polymer sponge was pressed onto the ceramic scaffold, allowing the solubilized PLA to serve as adhesive (c). The PLA struts were extruded on two opposite sides of the scaffold to further stabilize the composite (d).

sponge onto the ceramic face. Rods that traversed the ceramic and polymer phases were used to ensure a mechanically stable attachment between the phases. The composite scaffolds were differentially seeded with fibroblasts transduced with an adenovirus expressing BMP 7 in the ceramic phase and fully differentiated chondrocytes in the polymeric phase. After subcutaneous implantation into mice, the biphasic scaffolds promoted the simultaneous growth of bone, cartilage, and a mineralized interface tissue. Within the ceramic phase, the pockets of tissue generated included blood vessels, marrow stroma, and adipose tissue.

Kim *et al.* fabricated HAp and gelatin composites in a foam type. The composite solutions were prepared with addition of different amounts of HAp powder with sizes of 2–5 μm in gelatin solution and followed by freeze-drying [116]. The foams were crosslinked with a WSC at 4°C. The pure gelatin foam had a pore configuration with the porosity and pore size of 90% and 400–500 μm, respectively. The HAp addition made the foam much stronger and stiffer. The osteoblast-like human osteosarcoma cells spread and grew actively on all the foams. Li *et al.* developed a porous scaffold made from chitosan and alginate with improved properties as compared to its chitosan counterpart [117]. The chitosan and alginate scaffold was prepared from solutions of physiological pH to induce coacervation of chitosan and alginate combined with liquid–solid separation. Enhanced mechanical properties were attributed to the complex formation from cationic chitosan and anionic alginate. Bone-forming osteoblasts attached to the composite scaffold, proliferated, and deposited calcified matrix. Calcium deposition occurred as early as the fourth week after implantation.

To develop scaffolds that mimic the function of the native ECM, Stankus *et al.* combined a bioabsorbable, elastomeric PEUU with type I collagen at various ratios [118]. Randomly oriented fibers in the electrospun matrices ranged in diameter from 100 to 900 nm. Matrices were strong and distensible, possessing strengths of 2–13 MPa with breaking strains of 160–280% even with low PEUU contents. Collagen incorporation significantly enhanced SMC adhesion onto electrospun scaffolds.

4. SURFACE MODIFICATION OF BIOMATERIALS AND CELL INTERACTIONS

Park *et al.* treated PLGA scaffolds with 1 N NaOH for 10 min in an attempt to create an optimal surface pertinent for more effective cartilage tissue engineering applications [119]. They speculated that underlying material properties that may have enhanced chondrocyte functions include a more hydrophilic surface (due to hydrolytic degradation of PLGA by NaOH), increased surface area, altered porosity (both percent and diameter of individual pores), and a greater degree of nanometer roughness. To modify PLLA scaffolds with collagen, Ma *et al.* immersed PLLA scaffolds into collagen solution and subjected to suction under reduced pressure [75]. The trapped air bubbles in the internal scaffold appeared as bubbles and evolved out. When no air bubbles were observed, pressure was returned to the ambient value. The collagen solution was pressed into the interior of the scaffold through the open pores. Finally the collagen-containing scaffolds were either dried under vacuum or freeze-dried. The distribution of the collagen in the scaffolds was assessed by confocal laser scanning microscopy (CLSM) to obtain images not of the scaffold surface but of the cross-sections (internal part). The CLSM images of the rhodamine isothiocyanate (RITC)–labeled collagen in the PLLA scaffolds are shown in Fig. 3.54. Only the profiles of collagen can be observed because the PLLA scaffold had not been prestained. The collagen covered closely the internal surfaces of the scaffold when the collagen solution was dried under vacuum at ambient temperature. On the other

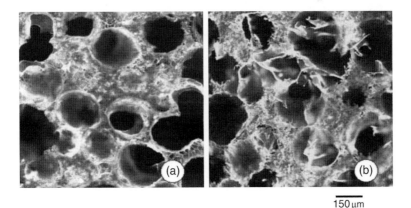

150 µm

Fig. 3.54 CLSM images of the collagen-modified PLLA scaffolds. The RITC-labeled collagen was used. The collagen solution in the scaffolds was either dried under vacuum at ambient temperature (a) or freeze-dried (b).

hand, collagen fibers were formed in the pores of the scaffold due to the phase separation in the collagen solution when it was freeze-dried.

Fahmy *et al.* incorporated an avidin-palmitic acid conjugate in the surface of PLGA by a simple deposition procedure [120]. A drop of the composite was regionally placed on the top of dry porous PLGA scaffolds and allowed to soak in for 15 min at room temperature, followed by washing. They reasoned that this conjugate would naturally position itself at the material surface: the palmitic acid would preferentially partition into the hydrophobic PLGA matrix, whereas the hydrophilic avidin head group would be in display in the hydrophilic external environment, facilitating the attachment of biotinylated ligands to the surface of the scaffolds. Zhu *et al.* also introduced amino groups onto PLLA surface simply by immersing a PLLA film in 1,6-hexanediamine-propanol solution at 50°C, followed by rinsing with water for 24 h at room temperature to remove free 1,6-hexanediamine [121]. These amino groups on the PLLA surface were used for immobilization of bioactive macromolecules onto the surface. Ma *et al.* covalently fixed collagen on the surface of PLLA scaffolds via the reactions illustrated in Fig. 3.55 [122]. First, a PLLA scaffold was immersed in H_2O_2 solution and irradiated under UV light to introduce –OOH groups onto the surface. The photo-oxidized scaffold was then immersed in methyl methacrylic acid (MMA) solution in a glass tube to which Fe^{2+} was injected, and graft polymerization was carried out at 37°C for 1 h. For collagen grafting, the poly(methyl methacrylic acid) (PMMA)-grafted PLLA scaffold was immersed in EDAC solution and allowed to react for 4 h at 4°C, and then to react with collagen solution. To introduce bFGF into PLLA scaffold, the scaffold was allowed to react with collagen solution containing bFGF. Chondrocyte culturing on the collagen-immobilized surfaces showed significantly improved cell spreading and growth. Incorporation of FGFs in the collagen layer further enhanced the cell growth.

Yoon *et al.* immobilized a cell-adhesive peptide moiety, Gly-Arg-Gly-Asp-Tyr (GRGDY), onto the surface of porous PLGA scaffolds for enhancing cell adhesion and function [123]. Figure 3.56 shows the schematic representation for GRGDY surface modification on PLGA films and scaffolds. A carboxylic acid existing as an end group of PLGA was first activated with dicyclohexylcarbodiimide (DCC) and NHS to conjugate hexaethyleneglycol diamine to form an aminated PLGA having a free primary amine in the PLGA chain end. An excess amount of hexaethyleneglycol diamine relative to that of the activated PLGA was used to suppress the coupling reaction of PLGA–PLGA that could possibly occur by a homo-functional crosslinker, hexaethyleneglycol diamine. The synthesized PLGA–NH_2 was blended, in varying weight ratios with PLGA, to produce films and scaffolds. The PLGA scaffolds were fabricated by a gas-foaming/salt-leaching method with the use of ammonium bicarbonate salt particulates. The surface-exposed amine groups were activated with ethyleneglycol-bis-succinimidylsuccinate (EGS) to conjugate N-terminal α-amino group of GRGDY or GRGEY (E:Glu) on the surface of films and scaffolds. Figure 3.57 shows the attachment behaviors of bone-marrow stem cells isolated from rat on the surfaces of various PLGA blend films. It can be seen that the blend films modified with GRGDY exhibited more cell attachment than

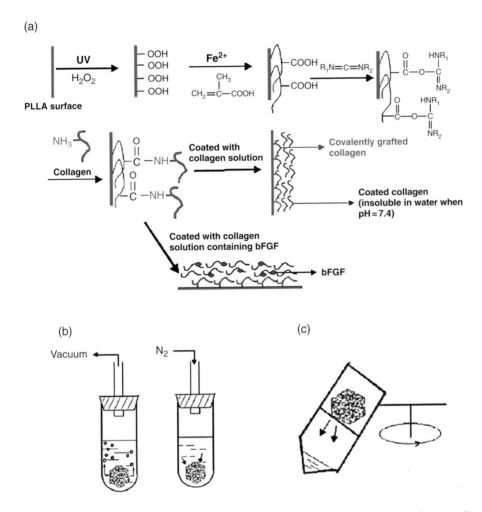

Fig. 3.55 Schematic representation of (a) the reaction scheme in the surface modification of PLLA, (b) the pressing of liquid reagent into PLLA scaffold through vacuum treatment and (c) the removal of liquid reagent out of the PLLA scaffold through centrifugation.

those with GRGEY with the concomitant increase of the blend ratio of PLGA–NH₂. Zhu *et al.* first aminolyzed the surface of a random copolymer of LLA and ε-CL with 1,6-hexanediamine to introduce free amino groups [124]. Using these amino groups as active sites, FN or collagen was separately bonded with GA as a coupling agent. The cytocompatibility to SMCs, fibroblasts, and ECs was evaluated for esophagus tissue engineering scaffold.

To evaluate the effect of RGD-mediated cell adhesion on osteogenesis of bone-marrow-derived stromal cells, Yang *et al.* synthesized a photopolymerizable PEGDA hydrogel [125]. The YRGDS was reacted with acryloyl–PEG–NHS in a 1:2.6 molar

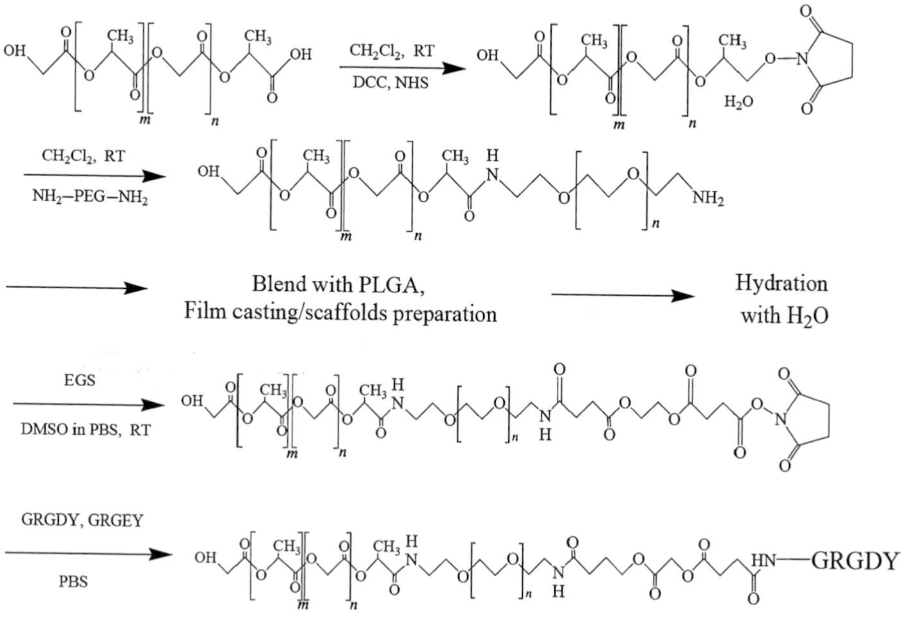

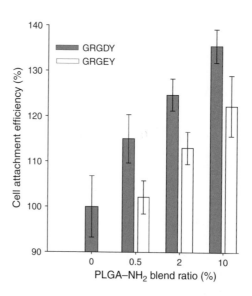

Fig. 3.56 A synthetic scheme of GRGDY surface modification on PLGA film and scaffold.

Fig. 3.57 Cell attachment on the surface of PLGA blend films with different blend ratios. Attached cell number was estimated at 6 h after seeding. The cell attachment value was normalized to PLGA film at 100%.

ratio in buffer. The YRGDS–PEG acrylate was mixed with PEGDA in PBS. Photoinitiator Irgacure D2959 was added to the polymer solution. Passage 3 MSCs were homogenously suspended in the solution to make a concentration of 15 million cells/ml. Seventy-five microliters of the cell–polymer mixture were loaded into molds and exposed to UV light to achieve gelation. The hydrogel samples were incubated in osteogenic medium consisting of DMEM, dexamethasone, ascorbic acid 2-phosphate, β-glycerophosphate, and 10% FCS. Expression of OCN and ALP increased significantly as the RGD concentration increased. Compared with no RGD, 2.5 mM RGD group showed a 1340% increase in ALP production and a 280% increase in OCN accumulation in the medium. The RGD helped MSCs maintain cbfa-1 expression when shifted from a 2-D environment to a 3-D environment. Soluble RGD was found to completely block the mineralization of MSCs.

Shin *et al.* tested the hypothesis that signaling peptides incorporated into synthetic oligoPEG fumarate (OPF) hydrogels induced differentiation and mineralization of marrow stromal cells (MSCs) cultured in the absence of dexamethasone and β-glycerol phosphate [126]. In order to address this hypothesis, Shin *et al.* examined the change in the cell number, expression of typical *in vitro* osteogenic differentiation markers (ALP activity and OP), and calcium deposition from MSCs cultured for 16 days on OPF hydrogels modified with two different peptides. To prepare peptide-modified OPF hydrogels, acrylated peptide was added to the aqueous mixture of OPF and PEGDA with ammonium persulfate and AA. The mixture was cast between two glass plates and allowed to crosslink at 45°C. When MSCs were cultured in media without soluble osteogenic supplements (dexamethasone and β-glycerol phosphate) for 16 days on OPF hydrogels modified with RGD-containing peptides, the normalized cell number was dependent on the peptide concentration between days 0 and 5 and reached comparable values at day 10 regardless of the concentration. The ALP activity of MSCs on the peptide-modified OPF hydrogels was also concentration dependent: ALP activity showed peaks on day 10 or day 13 on the OPF hydrogels modified with 2 and 1 μmol peptides/g, which was significantly greater than those on the OPF hydrogels modified with 0.1 μmol peptides/g or no peptide.

Steffens *et al.* investigated the covalent incorporation of heparin into 3-D collagen matrices and the effects exerted by loading VEGF into these heparinized collagen matrices [127]. The immobilization of heparin was performed with EDAC and NHS. Carboxyl groups on the heparin were activated to succinimidyl esters, which react with amino functions on the collagen to zero length crosslinks. As a first approach to testing angiogenic capabilities, ECs were exposed to nonmodified and heparinized collagen matrices. This exposure led to an increase in EC proliferation. The increase could be further enhanced by loading the heparinized collagen matrices with VEGF. Masuko *et al.* prepared a chitosan–peptide complex based on the selective reaction of chitosan with 2-iminothiolane [128]. A synthetic peptide, RGDSGGC, was readily coupled by disulfide bonds formation with sulfhydryl groups of SH-chitosan in the presence of dimethyl sulfoxide. This polysaccharide–peptide conjugate exhibited excellent capacities for both cell adhesion and cell proliferation of chondrocytes and fibroblasts.

To optimize *in vitro* the 3-D circular braided scaffold designed for ligament tissue engineering, Lu *et al.* focused on material selection and the identification of an appropriate polymer composition based on cellular response, construct degradation, and the associated mechanical properties [129]. They evaluated the effects of polymer surface modification via FN adsorption on scaffold material properties and cell proliferation. The attachment and growth of ACL cells on three types of poly-α-hydroxyester (PGA, PLGA, and PLLA) fibers with varying rates of degradation were examined. While PGA scaffolds measured the highest tensile strength followed by PLLA and PLGA, its rapid degradation *in vitro* resulted in matrix disruption and cell death over time. The PLLA-based scaffolds maintained their structural integrity and exhibited superior mechanical properties over time. As shown in Fig. 3.58, the response of ACL cells was found to be dependent on polymer composition, with the highest cell number measured on PLLA–FN scaffolds. Surface modification of polymer scaffolds with FN improved cell attachment efficiently and effected the long-term matrix production by ACL cells on PLLA and PLGA scaffolds.

Klebe *et al.* paid attention to the alignment of fibroblasts when these were suspended in collagen solution and then subjected to incubation in a mold at 37°C [130]. When the cell–collagen mixture began to form a gel, the boundaries of the gel were left fully constrained, but cutting along selected edges of the gel with a blade allowed the boundaries of the gel to be partially unconstrained. When the gels were further incubated under the fully or partially constrained condition, the embedded fibroblasts adhered to the collagen and contracted the gel through a process that involved interactions between the cells and the surrounding matrix via specific cell surface receptors. If cells exerted the traction force uniformly in all directions, tension at locations farther from the constrained boundaries would be summed over

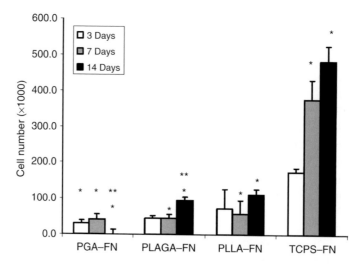

Fig. 3.58 Proliferation of primary rabbit ACL cells on braided scaffolds pre-coated with fibronectin (FN).

more cells and therefore would be greatest along lines paralleled to the long axis of the rectangular gel. On rectangular gels constrained along both short and long axes, cells aligned with the long axis, parallel to the presumed direction of the greatest tension, in the absence of free lateral edges.

Klapperich and Bertozzi used oligonucleotide microarrays to interrogate gene expression profiles associated with cell–scaffold interactions [131]. They seeded collagen/GAG meshes with human IMR-90 fibroblasts and compared transcript levels with control cells grown on tissue culture polystyrene (TCPS). The global gene expression profile of human fibroblasts was found to differ greatly between cells grown on collagen/GAG tissue engineering scaffolds and 2-D TCPS plates. Statistical analyses indicated that 1018 genes were differentially expressed as a function of scaffold (collagen/GAG mesh or TCPS plate). Close examination of the differentially expressed genes revealed several molecular pathways relating to inflammation and wound healing. Mesh-grown cells expressed many pro-angiogenic genes. The upregulation of several genes indicated that the mesh-grown cells were experiencing hypoxia. Some questions regarding this system remain to be answered. For example, the TCPS-grown cells were proliferating during the time course of the study, whereas histological analysis suggested that the mesh-grown cells were not. Since the cells in the study were not synchronized, it remains to be determined whether some transcript changes reflect differences in cell cycling.

Anderson *et al.* described the creation of cell-compatible, biomaterial microarrays that allow rapid, microscale testing of biomaterial interactions with cells [132]. A microarray (also known as a "biochip") is a slide-sized array of many discrete sites to which only one DNA segment, RNA segment, protein, etc., can bind. As shown in Table 3.13, an array of blends of bioabsorbable polymers was tested to examine the ability of this system to efficiently screen biomaterials with an NSC line and primary articular chondrocytes.

5. GROWTH FACTORS AND CARRIERS

5.1. Growth Factor–like Polymers

Mitogenic activities of sulfonated poly(γ-glutamic acid) (γ-PGA-S) were investigated with chlorate-treated L929 fibroblast culture tests [133]. When 72% of the carboxyl groups in γ-PGA were sulfonated (γ-PGA-S72), cell numbers reached a maximum. The activity of γ-PGA-S72 was higher than that of γ-PGA and synthetic heparinoids and was almost comparable to that of heparin, as shown in Fig. 3.59. Ehrbar *et al.* studied three variant forms of VEGF121 (VEGF$_{121}$), each with differential susceptibility to local cellular proteolytic activity, formulated within fibrin matrices [134]. The proteolytic variant α_2PI$_{1-8}$–VEGF$_{121}$ remained immobilized in fibrin matrices until its liberation by cell-associated enzymes, such as plasmin, that degrade the fibrin network; the α_2PI$_{1-8}$ domain served as a site for covalent attachment to fibrin during coagulation. They created a new variant, α_2PI$_{1-8}$–Pla–VEGF$_{121}$ that coupled to fibrin via a plasmin-sensitive sequence (Pla).

Table 3.13
Biomaterial microarray design

1 Poly(1,4-butylene adipate)
2 Poly(ethylene adipate)
3 Poly(1,3-propylene succinate)
4 Poly(1,3-propylene glutarate)
5 Poly(1,3-propylene adipate)
6 Poly(D,L-lactide-*co*-caprolactone) lactide:caprolactone 40:60
7 Poly(D,L-lactide-*co*-caprolactone) lactide:caprolactone 84:16
8 Poly(1,4-butylene adipate), diol-end-capped
9 Poly(ethylene adipate), dihydroxy-terminated
10 Poly(lactide-*co*-glycolide) lactide:glycolide 50:50, MW~60,000
11 Poly(lactide-*co*-glycolide) lactide:glycolide 50:50, MW~18,000, acid-terminated
12 Poly(lactide) D:L 50:50, MW~25,000
13 Poly(lactide-*co*-glycolide) lactide:glycolide 50:50, MW~65,000
14 Poly(lactide-*co*-glycolide) lactide:glycolide 85:15, MW~60,000
15 Poly(lactide-*co*-glycolide) lactide:glycolide 65:35, MW~58,000
16 Poly(lactide-*co*-glycolide) lactide:glycolide 75:25, MW~100,000
17 Poly(lactide-*co*-glycolide) L-lactide, lactide:glycolide 70:30, MW~22,000
18 Poly(lactide-*co*-glycolide) lactide:glycolide 50:50, MW~100,000
19 Poly(lactide) L:DL 60:40, MW~120,000
20 Poly(lactide) D:L 50:50, MW~8,000, acid-terminated
21 Poly(lactide-*co*-glycolide-*co*-glycol) lactide:glycolide:glycol 53:21:26, MW~80,000
22 Poly(ethlyene glycol) MW~300
23 Poly(lactide-*co*-glycolide) lactide:glycolide 65:35, MW~14,000
24 Poly(azelaic anhydride)

Polymers used for microarray synthesis. Polymers were mixed at a 70:30 and 90:10 ratio pairwise in all possible combinations. Three blocks of 1152 polymer blends were printed on each slide, 6 layers deep for each spot, with a center-to-center spacing of 500 μm.

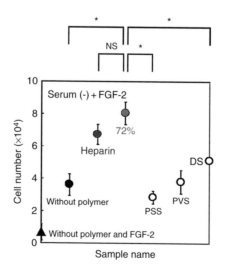

Fig. 3.59   Comparison of FGF-2 activity of γ-PGA-S72 to that of various heparinoids, evaluated by chlorate-treated L929 fibroblasts proliferation in serum-free culture medium with FGF-2 (10 ng/ml) and polymers (10 μg/ml) for 96 h.

Cleavage of this target site by plasmin enabled direct release of $\alpha_2\text{PI}_{1-8}$–Pla–VEGF$_{121}$ from bulk matrix degradation. $\alpha_2\text{PI}_{1-8}$–Pla–VEGF$_{121}$ was released fourfold more quickly than $\alpha_2\text{PI}_{1-8}$– VEGF$_{121}$, both being retained compared to native VEGF$_{121}$; the differences in release could be accounted for based on knowledge of the plasmin sensitivity of the bound growth factor and the structure of the fibrin network. As shown in Fig. 3.60, the bound factors were competent in inducing EC proliferation, the matrix-bound forms being more effective than native VEGF$_{121}$; as well as competent in inducing EPC maturation into ECs.

5.2. Carriers

Schmidmaier *et al.* reported the delivery of insulin-like growth factor-I and TGF-β1 for bone morphogenesis from a PDLLA-coated implant [135]. They prepared solutions of protein and PDLLA directly in chloroform and then formed films on the implants by dip coating. However, this simple method using PLA as a carrier has potentials of deactivation and burst release of proteins. Cho *et al.* focused on the issues of delivering active growth factors from resorbable polymer formulations and simultaneously tailoring the protein release kinetics [136]. They prepared a series of resorbable polymer-based formulations using keratinocyte growth factor (KGF), as shown in Table 3.14. Each formulation containing 15 ppm KGF was prepared using chloroform, based on PLA (PLLA), PGA, a surfactant sodium bis(ethylhexyl) sulfosuccinate(Aerosol-OT, AOT), and different water loadings (R = [water]/[AOT]). They used AOT in an effort to form an environment into which the KGF would reside and remain stable within the resorbable formulation. There is a body of literature on the use of AOT-based reverse micelles to stabilize a wide variety of proteins in organic solvents and on the use of surfactants to form drug/protein-loaded biodegradable formulations and/or to control formulation of biodegradation kinetics.

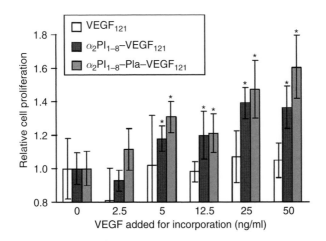

Fig. 3.60 Promotion of endothelial cell growth by VEGFs bound in fibrin.

Table 3.14

Composition range of KGF-doped biodegradable polymer formulations that were prepared and screened[a]

PLA[b]	PGA[c]	AOT (mM)	R[d]	Buffer pH[e]
50–100/2	0–50	3–200	0–30	6–8
50–100/50				
50–100/100				
50–100/200				
50–100/300				

[a] Each formulation contained 15 ppm KGF, and each was prepared using chloroform.
[b] Weight percent PLA/PLA average molecular weight (kDa).
[c] Weight percent PGA.
[d] [water]/[AOT].
[e] Phosphate or TRIS buffer (12 mM).

They defined "active" KGF as the species that is bound by anti-KGF antibodies. To determine the percentage of active KGF release from the formulations [(active/total) $\times$ 100%], they used a steady-state fluorescence anisotropy immunoassay. Anti-KGF antibodies were found to readily recognize native KGF and selectively bind only to the native KGF proteins in a binary mixture that contained active and denatured KGF. Follow-up experiments using PDLLA and chloroform also yielded a large fraction (>95%) of inactive KGF. Figure 3.61 summarizes the effects of composition, PLA/PGA weight ratio, AOT concentration, molar ratio of water-to-AOT(R), and water pH on the percentage of active KGF released from 125 randomly selected formulations. The KGF release kinetics could be tuned by varying the MW of PLA, but KGF was delivered only over several hours in a sustained manner.

Lee *et al.* evaluated the controlled release of rhBMP-2 from bioabsorbable membranes [137]. They loaded 5 µg of rhBMP-2 into stiff membranes made of PLLA and TCP, by soaking the BMP solution into the membranes. To study the release kinetics of rhBMP-2 from membranes, each membrane was immersed in a vial containing PBS at pH 7.4 as a releasing medium. Sealed vials were placed in a shaking water bath at 37°C and shaken at 15 rpm. At predetermined time intervals, samples were withdrawn from the vials, which were replenished with fresh medium. The concentration of rhBMP-2 released into the samples was assayed using a human BMP-2 immunoassay kit. The cumulative release kinetics of rhBMP-2 from the membranes is shown in Fig. 3.62. As can be seen, rhBMP-2 release from the membrane occurred in two phases: an initial burst phase for the first day; and a second, slow-release phase thereafter. This is a typical burst pattern. During the initial phase, 2.2 and 3.5 µg of rhBMP-2 were released into the medium at 12 and 24 h, respectively. Thus, approximately 70% of the rhBMP-2 was released during the first day, following this initial burst. The rhBMP-2 was consistently released at a rate of 7–10 ng/day for up to 4 weeks. They studied also the effect of this rhBMP-2-loaded membrane on bone augmentation in a rabbit calvarial model

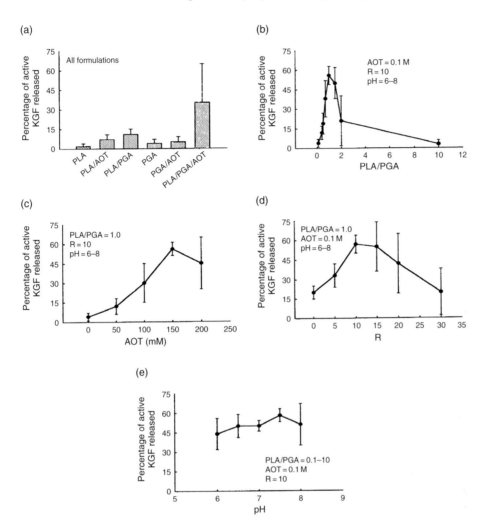

Fig. 3.61 Effects of composition (a), PLA/PGA weight ratio (b), AOT concentration (c), molar ratio of water to AOT (d) and water pH (e) on the percentage of active KGF released from a given formulation.

and found significantly enhanced bone augmentation, compared with the control membranes without rhBMP-2. This suggests that such a low rhBMP-2 concentration as 7–10 ng/day is effective for the enhancement of bone augmentation, as the free rhBMP-2 bursted on the first day in large amounts might have quickly diffused out from the site concerned.

The *in vivo* storage mechanism of FGF-2 (bFGF) has prompted researchers to use heparin or acidic polymers as a carrier of FGF-2 for its sustained release. Such carriers studied include heparin-carrying polystyrene-bound collagen substrates, acidic gelatin hydrogels, alginate gels containing heparin, photocrosslinkable chitosan

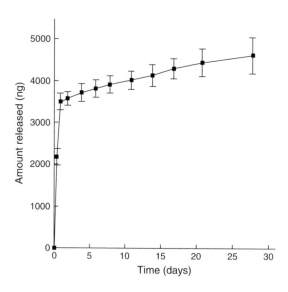

Fig. 3.62   Cumulative release of rhBMP-2 from PLLA/TCP membranes *in vitro.*

hydrogels, and chitosan/heparinoid hydrogels. Tabata *et al.* showed that much longer delivery of growth factors could be achieved using hydrogels prepared from gelatin with an isoelectric point of about 5 as a carrier [138–140]. The representative release kinetics is shown for FGF-2 and TGF-β entrapped in gelatin microspheres following subcutaneous implantation in rats along with those of bolus injection of growth factor solutions. The equilibrated water content of the gelatin hydrogels markedly affected the release kinetics. An example is shown in Fig. 3.63 for the release of BMP-2 from the gelatin hydrogels with water contents ranging from 93.8 to 99.7% [141]. Following subcutaneous implantation of the gelatin hydrogels incorporating ^{125}I-labeled BMP-2 into the back of mice, the skin on the back of mice around the implanted site of microspheres was cut and the corresponding fascia was thoroughly wiped off with a filter paper to absorb ^{125}I-labeled BMP-2. The radioactivity of remaining gelatin hydrogels, excised skin, and filter paper was measured on a gamma counter to assess the time profile of *in vivo* BMP-2 retention. Yamamoto *et al.* evaluated BMP-2-induced ectopic bone formation [141]. Gelatin hydrogels incorporating BMP-2 were implanted subcutaneously into ddY mice, while aqueous BMP-2 solutions were implanted into the mice back as well. The skin tissue including the hydrogel-implanted or injected site was taken out 1, 2, and 4 weeks later. The homogenate of the tissues obtained was centrifuged and the ALP activity of the supernatant was determined by a *p*-nitro phenyl-phosphate method. Figure 3.64 shows the ALP activity of the subcutaneous tissues. Apparently, the application of gelatin hydrogels incorporating 1 and 5 μg of BMP-2 enhanced the ALP activity of subcutaneous tissues to a significantly higher extent than that of free BMP-2. The gelatin hydrogel with a water content of 97.8% and containing 5 μg of BMP-2 and 4 weeks after implantation showed the highest ALP activity.

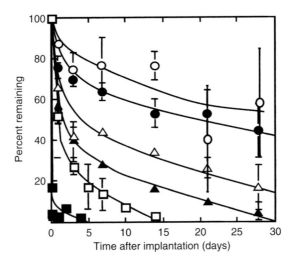

Fig. 3.63 *In vivo* time profiles of the radioactivity remaining after the subcutaneous implantation of gelatin hydrogels incorporating ^{125}I-labeled BMP-2 into the back of mice. The hydrogel water contents are (○) 93.8%, (●) 96.9%, (△) 97.8%, (▲) 99.1%, and (□) 99.7%. The symbol (■) indicates the remaining radioactivity after injection of ^{125}I-labeled BMP-2 solution.

Further, Isogai *et al.* evaluated the effectiveness of FGF-2 impregnated in gelatin microspheres to achieve slow release of growth factor for augmenting the *in vivo* chondrogenic response [142]. The experimental procedure is depicted in Fig. 3.65. Canine chondrocytes were grown *in vitro* onto ear-shaped P(LA/CL)s for one week, and then implanted into the dorsal subcutaneous tissue of nude mice; implants contained FGF-2 either in free solution (PBS) or in gelatin microspheres. A third group underwent

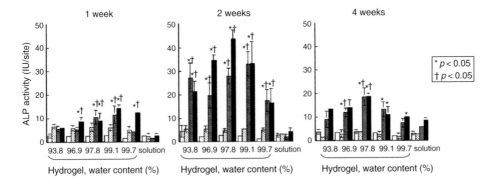

Fig. 3.64 ALP activity of tissues around the implanted site of gelatin hydrogels incorporating BMP-2 and the injected site of BMP-2 solution 1, 2, and 4 weeks later. The BMP-2 doses are (□) 0, (▥) 0.5, (▨) 1, and (■) 5 μg.

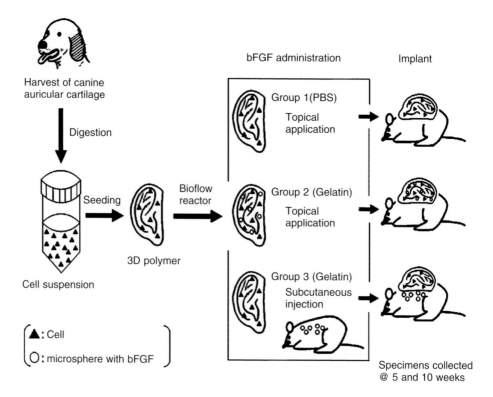

Fig. 3.65 Experimental procedure for the evaluation of FGF-2 release effects on chondrogenesis.

pre-injection of FGF-2 in gelatin microspheres 4 days before chondrocyte–copolymer implantation. The implants with FGF-incorporated microspheres showed the greatest chondrogenic characteristics at 5 and 10 weeks postoperatively: good shape and bio-mechanical trait retention, strong metachromasia, rich vascularization of surrounding tissues, and increased gene expression for type II collagen (cartilage marker) and factor III–related antigen (vascular marker). In the case of implant site pre-administrated with FGF-impregnated microspheres, the implant architecture was not maintained, and reduced vascularization and metachromasia were apparent. The mechanical properties of the samples given in Table 3.15 indicated a more elastic character in PBS group and increased stiffness in gelatin group.

Fujita *et al.* chose chitosan/heparinoid hydrogels for release of FGF-2 [143]. To make chitosan soluble in water at the neutral pH, lactose (lactobionic acid) moieties were introduced to chitosan through the condensation reaction with amino groups of chitosan. The chitosan used had MW of 800–1000 kDa and a deacetylation ratio of 0.8. Native heparin was oxidized with IO_4^- to minimize its strong intrinsic anticoagulant activity. The researchers assumed that basic chitosan molecules complexed with acidic molecules (IO_4-heparin and other heparinoids) formed a hydrogel through ionic interactions and, in addition, that polypeptides such as FGF-2, once ionically

Table 3.15
Biomechanical properties of tissue-engineered auricular implants

	bFGF Delivery	Maximum Tensile Strength (mPa)	Elongation (mm)	Stiffness (mPa/mm)
5 wk	PBS	0.55 ± 0.09	9.71 ± 0.87*	0.06 ± 0.01
	Gelatin	0.46 ± 0.10	4.19 ± 0.87	0.12 ± 0.03
10 wk	PBS	0.78 ± 0.07	4.62 ± 0.72	0.17 ± 0.01
	Gelatin	0.52 ± 0.08	3.17 ± 0.24	0.16 ± 0.01

* $p < 0.01$.

complexed with chitosan or acidic molecules, were released from the hydrogel. The PBS containing FGF-2 and IO_4-heparin was mixed into aqueous solution of water-soluble chitosan (CH-LA). *In vivo* degradation of the resulting hydrogel was examined by incorporating trypan blue to the hydrogel. When the trypan blue–incorporated hydrogel was injected into the back of mice, the amount of dye released decreased with time postimplantation. Approximately 80, 60, and 30% of the dye molecules were retained in the implanted chitosan hydrogel on days 2, 5, and 14, respectively. However, many blue gel fragments near the implanted sites were observed on days 9, 14, and 24, indicating that the injected hydrogel was partially biodegraded *in vivo* in about 21 days after implantation into the back of mice. To evaluate the release of FGF-2 from the hydrogel *in vitro*, hydrogel was spotted on the center of a well in a culture plate. Figure 3.66 shows the release profiles of FGF-2 and IO_4-heparin from the chitosan hydrogels at room temperature. Only a minor amount of IO_4-heparin was released from the hydrogel, and 20% of FGF-2 released within the first day, followed by no further substantial release. To study the effect of FGF-2-incorporated chitosan/IO_4-heparin hydrogels on cell growth, CH-LA solution (50 μL) containing FGF-2 (1.25 μg) and IO_4-heparin (25 μg) was spotted in the center of well and

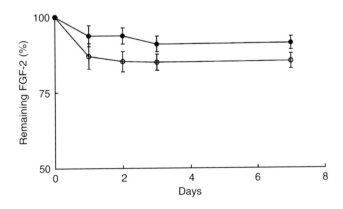

Fig. 3.66 Time course of remaining incorporated FGF-2 and IO_4-heparin in the hydrogel. IO_4-heparin (●) and FGF-2 (○).

HUVECs were plated into the well. The FGF-2-incorporated chitosan hydrogel was able to stimulate HUVEC growth, but washing of hydrogels with the medium for more than 5 days resulted in a loss of stimulating activity. To confirm the effect of FGF-2-incorporated chitosan/IO_4-heparin hydrogel on the vascularization, tissue hemoglobin in subcutaneous tissue in mice around the injected sites of FGF-2-incorporated hydrogels was measured. Figure 3.67 shows the concentration effect of FGF-2 incorporated in the hydrogel on vascularization. More than 1 μg of FGF-2 in the hydrogel (100 μL) had a positive effect to induce neovascularization *in vivo* in a concentration-dependent manner.

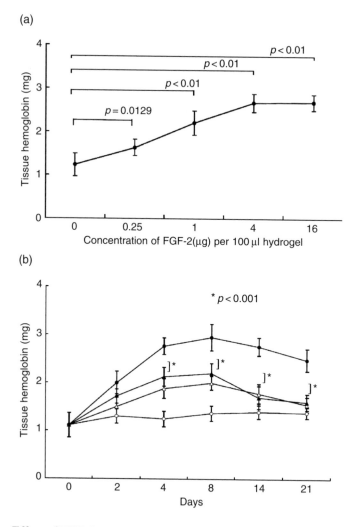

Fig. 3.67 Effect of FGF-2 concentration incorporated in chitosan/IO_4-heparin hydrogels on neovascularization.

Peatti *et al.* described the use of a modified HAc hydrogel for localized delivery of VEGF and FGF-2 [144]. To synthesize HAc–adipic dihydrazide (ADH) and HAc hydrogels, 5 mg/ml solution of HAc in water was treated with ADH and EDAC, and the pH was maintained at 4.75. The reaction was stopped by raising the solution pH to 7.0. The GAG hydrogels were prepared by dissolving the purified HAc–ADH or chondroitin sulfate (CS)–ADH in water. Separately, PEG-propiondialdehyde (diald) was also dissolved in water. Each solution was added to a small polystyrene dish in 1:2 ratio of aldehyde to hydrazide functionalities. A hydrogel formed within 3–5 min, with PEG-diald as a crosslinking agent for the HAc–ADH chains (Fig. 3.68). Hydrogels thus produced were subsequently kept in open dishes at 37°, which allowed them to dehydrate to flexible, dry, durable films. Either VEGF or FGF-2 was incorporated by adding aliquots of a stock solution to the HAc–ADH solution prior to addition of PEG-diald. Crosslinking then proceeded with no other modifications. To test the effect of the growth factor alone, without HAc, a 0.5 ml aqueous liquid aliquot containing 25 ng was delivered to mice. A blunt probe was inserted into the right ear pinna posterior surface through this incision and a 5-mm diameter pocket opened immediately subcutaneously. For those animals receiving an implant, a GAG film disc was positioned within the pocket and the incision was closed without sutures. To distinguish the tissue response to different implants from the response to surgical intervention, vessel growth was re-expressed through the neovascularization index, NI, defined as

$$NI = (treatment - CL) - (sham - CL)/mean\ CL$$

where (treatment $-$ CL) refers to the vessel count from the implanted ear of a particular animal minus that of its contralateral ear, averaged over all the animals, while (sham $-$ CL) represents the same quantity for a sham surgery. Treatment with either growth factor increased microvascular growth over that produced by HAc alone. A synergistic *in vivo* tissue response was seen in the case of co-addition of HAc and VEGF, which produced NI = 6.7 at day 14. This was nearly twice the

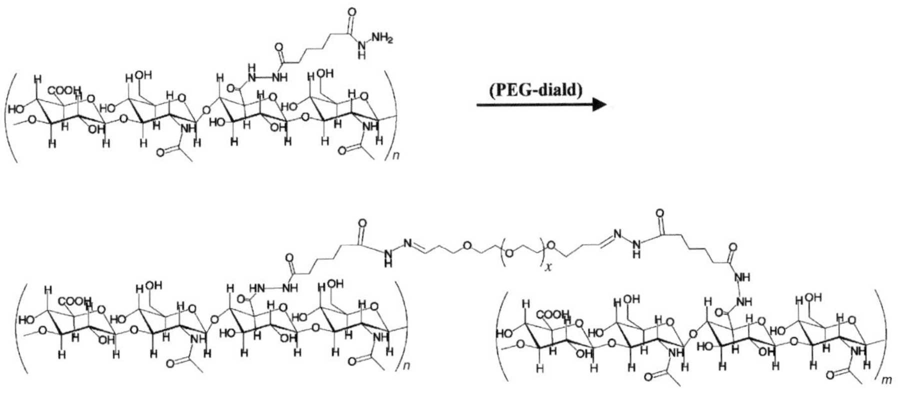

Fig. 3.68 Chemical structure of crosslinked HAc–ADH.

angiogenic effect of any other treatment. Moreover, the response to HAc + VEGF exceeded the sum of the tissue response to HAc alone (NI = 1.8) plus the response to VEGF alone (NI = 1.3) by a factor of 2.2. In contrast, at both time points, the sum of the angiogenic response to HAc alone plus that of FGF-2 alone (NI = 2.9 at day 14) was greater than the effect of co-addition of HAc + bFGF (NI = 3.4 at day 14). In contrast to the potent synergistic angiogenic response elicited by co-addition of HAc + VEGF, co-addition of HAc + bFGF produced no such synergy. The possibility of a synergistic interaction between specific growth factors and o-HAc *in vitro* was raised also by Montesano *et al.* [145]. Using a 3-D collagen gel, they showed that a combination of VEGF and o-HAc could induce an *in vitro* angiogenic response greater than the sum of that provoked by each agent individually.

To load TGF-β1, Lee *et al.* prepared chitosan microspheres using an emulsion-ionic crosslinking method [146]. Chitosan was dissolved in acetic acid solution and TGF-β1 was then dissolved in 4 mM HCl and added to the chitosan solution. The chitosan/TGF-β1 solution was then mixed with an n-octanol solution containing a surfactant and was emulsified in a homogenizer. Tripolyphosphate (TPP) solution was dropped into the emulsion and agitated in the homogenizer. The precipitated microspheres were washed and then lyophilized. The TPP carried five negative charges which allowed the electrostatic interaction with protonated chitosan in aqueous acidic solution; the loading efficiency of TGF-β1 was 23 ng/mg. Particle diameter distributions ranged between 0.3 and 1.5 μm. The release kinetics of TGF-β1 from chitosan microspheres monitored over 7 days is shown in Fig. 3.69. After 3 days the cumulative mass of TGF-β1 released reached a plateau, and the growth factor released over the remaining period was negligible.

An absorbable microparticulate cation exchanger was synthesized by Shalaby *et al.* as a sustained-release carrier for biologically active proteins [147]. They hypothesized that an absorbable microparticulate cation exchanger could be manipulated to

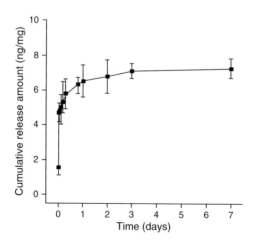

Fig. 3.69   Cumulative TGF-β1 release from chitosan microspheres.

incorporate either relatively high MW proteins such as immunoglobulins through irreversible adsorption or smaller proteins such as cytokines for sustained release through reversible adsorption. To prepare acid-terminated PGA microparticulates (PG-MP), glycolic acid, glycolide, and stannous octoate catalyst were mixed and slowly heated to 100°C under agitation for 20 min. When the reaction mixture became opaque, the temperature was increased to 160°C to achieve high conversion. After cooling, the reaction product was isolated and ground with the use of a mill. The PG-MP with a diameter of 7 μm was treated with the human granulocyte macrophage-colony stimulating factor (GM-CSF) in the absence or presence of poly-L-lysine hydrochloride. The PG-MP was first washed in PBS and incubated with GM-CSF at 100 μg/ml for 4 h at 4°C. After GM-CSF loading, the PG-GM-CSF was washed in PBS again. PG/Lys-GM-CSF was prepared by first incubating PG-MP in PBS containing 5 mg/ml of poly-L-lysine hydrochloride for 4 h at 37°C. The intent of pretreating with Lys was to shift the zeta potential to a more positive value through charge neutralization with the PG-MP surface. Figure 3.70 depicts the release of GM-CSF from both the PG/Lys and the PG-MP. The GM-CSF released from PG-MP surface was nearly complete by day 6, whereas release from the PG/Lys surface was observed for up to 26 days. The ratio of released GM-CSF/microparticulate weight was 52 ng/mg for PG-MP and 137 ng/mg for PG/Lys, respectively.

Controlled material degradation and TGF-β1 release were attempted by encapsulation of TGF-β1-loaded gelatin microparticles within a bioabsorbable polymer OPF [148]. Release studies performed with non-encapsulated microparticles confirmed that at normal physiological pH, TGF-β1 complexed with acidic gelatin resulted in slow release. At pH 4.0, this complexation no longer persisted, and TGF-β1 release was enhanced. However, by encapsulating TGF-β1-loaded microparticles in a network of OPF, release at both pH could be diffusionally controlled.

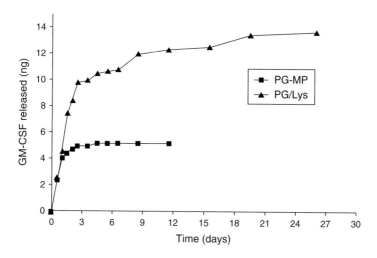

Fig. 3.70 Release of GM-CSF from the surfaces of PG-MP (squares) and PG/Lys (triangles).

However, when microparticle formation was encapsulated in an OPF hydrogel, these values were reduced. Release studies, under conditions that modeled the expected collagenase concentration of injured cartilage, demonstrated that by altering the microparticle crosslinking extent and loading within OPF hydrogels, TGF-β1 release, composite swelling, and polymer loss could be altered. Takahashi *et al.* fabricated gelatin sponges at different contents of β-TCP to allow BMP-2 to incorporate into them [149]. The sponges had an interconnected pore structure with an average pore size of 200 μm, irrespective of the β-TCP content. The *in vivo* release test revealed that BMP-2 was released *in vivo* at a similar time profile, irrespective of the β-TCP content. Figure 3.71 shows the *in vivo* decrement patterns of the remaining radioactivity at the implanted site of gelatin and gelatin-β-TCP sponges incorporating ^{125}I-labeled BMP-2. The period of BMP-2 retention was not influenced by the β-TCP content. When the osteoinduction activity of gelatin or gelatin–β-TCP sponges incorporating BMP-2 was studied following the subcutaneous implantation in rats, homogenous bone formation was histologically observed throughout the sponges, and the extent of bone formation was higher in the sponges with the lower contents of β-TCP. Irrespective of the β-TCP incorporation, the gelatin and gelatin–β-TCP sponges incorporating BMP-2 enhanced the ALP activity to a significantly higher extent than that of the BMP-2-free sponges (Fig. 3.72). The highest ALP activity was observed for the gelatin sponge without β-TCP. The BMP-2-incorporated gelatin sponge without β-TCP exhibited the highest OCN content

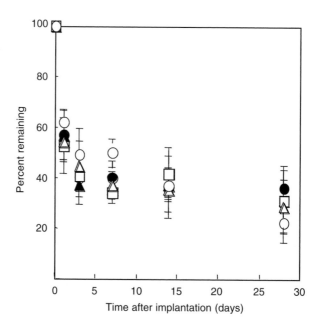

Fig. 3.71 *In vivo* time profiles of the radioactivity remaining after the subcutaneous implantation of gelatin–β-TCP sponges incorporating ^{125}I-labeled BMP-2 into the back of mice. The β-TCP contents of gelatin sponges are (○) 0, (△) 25, (□) 50, (●) 75, and (▲) 90%.

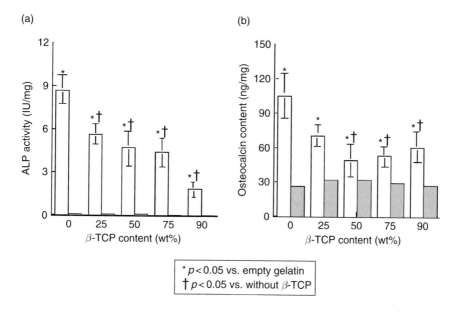

(b)

Fig. 3.72 ALP activity (a) and osteocalcin contents (b) of tissues around the implanted site of gelatin–β-TCP sponges incorporating BMP-2 2 and 4 weeks after implantation, respectively. The BMP-2 doses are (■) 0 and (□) 5 μg.

4 weeks after implantation, while the OCN content decreased with the increasing content of β-TCP.

Smith *et al.* addressed the problem of delay of vascularization in tissue engineering by sustained delivery of VEGF from a porous polymer matrix utilized simultaneously for cell delivery [150]. They combined 3 mg of PLGA microspheres with 10 μg of VEGF dissolved in 0.1% alginate solution. This mixture was lyophilized to form a powder, combined with 50 mg of NaCl and pressed in a die to create discs. The PLGA was gas-foamed by subjecting the disks to CO_2 for 24 h, followed by a rapid reduction of pressure to ambient. The NaCl was leached from matrices by incubation in 0.1 M $CaCl_2$. Figure 3.73 shows an *in vitro* release profile in which 25% of the incorporated growth factor was released within 3 days.

5.3. Combined and Sequential Release of Growth Factors

By analyzing fracture callus, Bourque *et al.* demonstrated that different growth factors predominated in sequence: PDGF, FGF, insulin-like growth factor (IGF), and TGF-β1 [151]. Bostrom *et al.* [152] and Yu *et al.* [153] observed temporal differences in the appearance of various numbers of the BMP family during fracture healing. These observations, combined with the observed influx of various cell types as well as changes in the phenotypes during wound healing, indicate that bone formation and repair occur by a complex cascade involving numerous growth

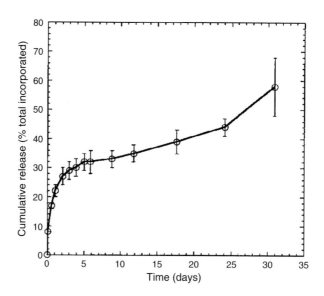

Fig. 3.73   VEGF release from PLGA.

factors and cytokines. On the other hand, inhibition of desired wound healing also has resulted from delivery of growth factors alone or in combination. For instance, delivery combinations of BMP-2 and FGF-2, OP-1 and BMP-2, or BMP-3 and PDGF-BB inhibited bone formation [154–156].

Raiche and Puleo investigated effects of combined and sequential release of two osteotropic growth factors, BMP-2 and IGF-I, in cultures of two types of pluripotent mesenchymal C3H10T1/2(C3H) cells [157]. By altering the extent of crosslinking and the amount of growth factor loaded in coatings, the timing and duration of release could be altered. They designed two-layer heterogeneously loaded and crosslinked gelatin coatings. Top and bottom layers of the coatings were unloaded, loaded with BMP-2 or IGF-I, or loaded with BMP-2 and TGF-I. Crosslinking was controlled to create peaks of release roughly corresponding to expression of growth factors observed in early normal wound healing. The mass of loaded growth factor was altered to produce peaks of equal bioactivity from the top and bottom layers. Whereas treatment with IGF-I alone had no appreciable effect, C3H cells initially exposed to soluble BMP-2 showed a significantly increased specific ALP activity. Treatment on day 2 with BMP-2 followed by combined BMP-2 and IGF-I on day 4 produced the greatest increase in ALP activity. Experimental release of BMP-2 and IGF-I from gelatin coatings compared well with predicted delivery. The shape of the peaks closely matched designed release profiles. Equivalence of model and experimental delivery was similar for IGF-I delivery, with 50–60% of the loaded growth factor released in an immunodetectable form. Different combinations of BMP-2 and IGF-I loading in the top and bottom layers of two-layer coatings resulted in significantly different specific ALP activity. The ALP activity was significantly increased at 7 days for treatments with BMP-2 in the top layer regardless of IGF-I incorporation. Sequential release of BMP-2 from the top layer followed by release of IGF-I from the bottom layer retained the greatest significant increase in ALP activity after 10 days compared to all other treatments.

Simmons *et al.* hypothesized that delivery of bone progenitor cells along with appropriate combinations of growth factors and scaffold characteristics would allow physiological doses of proteins to be used for therapeutic bone regeneration [158]. They tested this hypothesis by measuring bone formation by rat BMSCs transplanted ectopically in SCID mice using alginate hydrogels. The alginate was gamma-irradiated to vary the degradation rate and then covalently modified with RGD-containing peptides to control cell behavior. In the same delivery vehicle, they incorporated BMP-2 and TGF-β3, either individually or in combination. Individual delivery of BMP-2 or TGF-β3 resulted in negligible bone tissue formation up to 22 weeks, regardless of the implant degradation rate. In contrast, when growth factors were delivered together from readily degradable hydrogels, there was significant bone formation by the transplanted BMSCs as early as 6 weeks after implantation. Furthermore, bone formation, which appeared to occur by endochondral ossification, was achieved with the dual growth factor condition at protein concentrations that were more than an order of magnitude less than those reported previously to be necessary for bone formation.

5.4. Gene Transfer

Gene transfer may achieve greater bioavailability of growth factors within defects because of more sustained delivery than direct administration of growth factors. Anusaksathien *et al.* attempted to test the effect of gene transfer by adenovirus encoding PDGF-A and PDGF-B on human gingival fibroblast repopulation and resultant wound fill in a 3-D *ex vivo* wound-healing model [159]. Fibroblast-populated collagen (FPC) gel was prepared on ice-mixing rat tail tendon type I collagen, 1 M NaOH, DMEM, antibiotics, and 3.5×10^4 human gingival fibroblasts/ml. Three-dimensional FPC lattices (FPCLs) were fabricated by plating FPC gel into individual cell culture plates followed by incubation at 37°C. To fabricate a wound bilayered model, each FPCL was "wounded", using a 6-mm biopsy punch and overlaid onto an acellular collagen lattice, as depicted in Fig. 3.74. And 40 μl of collagen gel alone (NT), or collagen gel containing 3.5×10^6 plaque-forming units (PFU) of adenovirus (Ad)/GFP, Ad/PDGF-A, or Ad/PDGF-B per ml was applied to each of the tissue defects. Ten days after gene transfer, each FPCL in the defect area was separated and dissected from the defect periphery. The FPCLs were digested with collagenase and the number of human gingival fibroblast cells yielded from the wound defect and wound periphery areas were measured separately. Significantly greater cell density was detected in the Ad/PDGF-B-treated defects compared with the other treatments at all time points. In addition, cell counting on day 10 was determined in two separate areas including the defect area and the defect periphery. Compared with all treatment groups, Ad/PDGF-B stimulated the greatest increases in cell repopulation in the defect area and defect periphery.

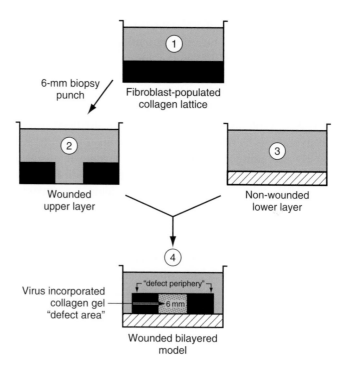

Fig. 3.74 Fabrication of the bilayered wound defect model.

Gene therapy approaches with localized delivery of plasmid DNA encoding for tissue generative factors are attractive for promoting tissue regeneration, but low transfection efficiency inherent in plasmid delivery limits this approach. To overcome this limitation, Huang *et al.* condensed plasmid DNA with nonviral vectors such as poly(ethyleneimine) (PEI), and delivered the plasmid DNA in a sustained and localized manner from PLGA scaffolds [160]. The scaffolds delivering plasmid DNA encoding for BMP-4 were implanted into a cranial critical-sized defect for time periods up to 15 weeks. Histological and microcomputed tomography analysis of the defect sites over time demonstrated that bone regeneration was significant at the defect edges and within the defect site when scaffolds encapsulating PEI-condensed DNA were placed in the defect. In contrast, bone formation was mainly confined to the defect edges within scaffolds encapsulating plasmid DNA (non-condensed), and when blank scaffolds (no delivered DNA) were used to fill the defect. Histomorphometric analysis revealed a significant increase in total bone formation (at least 4.5-fold) within scaffolds incorporating condensed DNA, relative to blank scaffolds and scaffolds incorporating uncondensed DNA at each time point. In addition, there was a significant increase both in osteoid and in mineralized tissue density within scaffolds incorporating condensed DNA, when compared with blank scaffolds and scaffolds incorporating uncondensed DNA.

Winn and associates demonstrated prolonged NGF delivery (up to 13 months), using human NGF transfection of baby hamster kidney cells [161]. These cells were then polymer encapsulated and implanted into the lateral ventricles of healthy adult rats. The potential for collagen deposition from the fibroblast cell line leading to scar formation raised concerns regarding the use of human dermal fibroblasts for NGF delivery and promoted the search for alternative sources of growth factor delivery. Jimenez *et al.* demonstrated that human embryonic kidney cells (HEK-293) could be genetically engineered to deliver bioactive NGF *in vitro* over an extended period of time under tightly controlled inducible expression [162]. Furthermore, through the use of induction agent, NGF protein expression could be extended. McConnell *et al.* assessed the *in vivo* functionality of the NGF expression system [163]. The HEK-293 cells were transfected with human NGF cDNA, and Ponasterone A (PonA) was used as the inducing agent. The NGF secretion from transfected HEK-293 cells was quantitatively determined *in vivo* within collection chambers. The NGF collection chambers were implanted subcutaneously in nude rats. Sealed chambers were filled with one of the following: (1) DMEM, (2) untransfected 293 cells (EcR-293) plus PonA, (3) untransfected EcR-293 without PonA, (4) transfected 293 cells (hNGF-EcR-293) plus PonA, or (5) transfected hNGF-EcR-293 without PonA. As shown in Fig. 3.75, at 24, 48, and 120 h postimplantation, hNGF-EcR-293/PonA($+$) cells possessed significantly higher NGF levels (2.2, 3.1, and 12.9, respectively) when compared with negative controls. There was no statistical difference between the negative control groups at any of the time points, in keeping with the hypothesis that they should yield wound fluid levels of NGF.

To obtain sustained release of TGF-β_1 that is one of the most important factors in the production of cartilage, a cell-mediated gene therapy technique was introduced

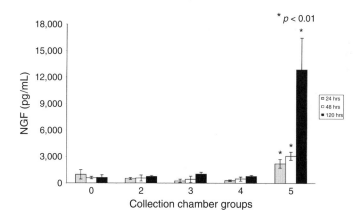

Fig. 3.75 *In vivo* NGF release assayed from collection chambers filled with DMEM (group 1), untransfected cells with ponasterone A {EcR-293/PonA($+$); group 2}, untransfected cells without ponasterone A {EcR-293/PonA($-$); group 3}, transfected cells without ponasterone A {hNGF-EcR-293/PonA($-$); group 4}, and transfected cells with ponasterone A {hNGF-EcR-293/PonA ($+$); group 5}.

by Lee *et al.* [164]. They infected chondrocytes with a retroviral vector carrying the TGF-β_1 gene. The single clone derivative showed sustained TGF-β_1 secretion. It also showed constitutive type II collagen expression. Whereas the TGF-β_1 protein itself was unable to induce formation of cartilage *in vivo*, human chondrocytes engineered to express a retroviral vector encoding TGF-β_1 showed cartilage formation *in vivo* when cells were injected into nude mice intradermally. To prolong gene expression in skin using safe, nonviral gene delivery techniques, an injectable, agarose-based delivery system was tested by Meilander *et al.* The DNA was compacted with polylysine to improve DNA stability in the presence of nucleases [165]. Up to 25 μg of compacted luciferase plasmid with or without agarose hydrogel was injected intradermally in rodents. Bioluminescence imaging was used for longitudinal, noninvasive monitoring of gene expression *in vivo* for 35 days. Injections of DNA in solution produced gene expression for only 5–7 days, whereas the sustained release of compacted DNA from the agarose system prolonged expression, with more than 500 pg (20% of day 1 levels) of luciferase per site for at least 35 days. Southern blotting confirmed that the agarose system extended DNA retention, with significant plasmid present through day 7, as compared with DNA in solution, which had detectable DNA only on day 1. Histology revealed that agarose invoked a wound-healing response through day 14.

6. CELL CULTURE

Kuriwaka *et al.* combined two different cell cultures to find the optimum combination of monolayer and 3-D cultures for cartilage engineering [166]. Figure 3.76 shows the cell culture scheme. In group A chondrocytes were cultured for 3 weeks as a monolayer, while in group B, after 2 weeks of monolayer culture, the cells were isolated, collected, embedded in a collagen gel for 3-D culture, and cultured for 1 week. In group C, chondrocytes were embedded in a collagen gel after 1 week of monolayer culture and cultured for further 2 weeks. In group D, chondrocytes were embedded in a collagen gel without monolayer culturing and cultured for 3 weeks.

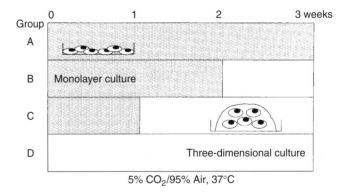

Fig. 3.76 Culture schedule for the four experimental groups.

To assay the CS content in the chondrocyte–collagen composites, they were washed with PBS at the end of 3 weeks of culture, digested with collagenase, and centrifuged. After digestion of the supernatants with chondroitinase and hyaluronidase, the digested mixtures were subjected to HPLC to analyze unsaturated disaccharides derived from isomers of chondroitin 4-sulfate (Δdi-4S) and chondroitin 6-sulfate (Δdi-6S). Table 3.16 gives the assay results. Combination of 2 weeks of monolayer culture followed by 1 week of 3-D culture (group B) resulted in a chondrocyte–collagen composite that had biological, biochemical, and biomechanical characteristics most comparable to articular cartilage. Although group A of monolayer culture only showed the largest increase in cell number, only low amounts of GAG were synthesized, and detectable amounts of type II collagen were not found. However, by 3 weeks, expression of type II collagen and aggrecan mRNA was observed in all groups by RT-PCR.

6.1. Cell Seeding

To study the various stages of cell–material interaction in detail, Lim *et al.* reported on combination of methods applied to osteoblastic cells (hFOB 1.19) using attachment/proliferation rate assays that are sensitive to early stages of bioadhesion (minutes to days, respectively) and measurement of ALP activity that is a marker for hFOB 1.19 differentiation over longer culture intervals (days to weeks) [167]. The substratum materials used included GA-LA copolymers. The maximum number of cells attached ($\%I_{max}$) was estimated from a pseudo-steady-state adhesion plateau observed to occur for hFOB on all surfaces studied herein within a half-maximal attachment time $0 < t^1/_2 < 3$ h. After a significant dwelling time, usually >3 h, cells divided at a rate measured by an exponential rate constant k and doubling

Table 3.16

Characteristics of chondrocytes grown under various conditions[a]

	Group[b]			
	A	B	C	D
Cell count ($\times 10^6$)	24.13 ± 0.36^c	3.00 ± 0.17	1.20 ± 0.11	1.05 ± 2.34
Δdi-6,S (nmol/ml)	2.25 ± 0.43	54.95 ± 3.27	41.48 ± 5.05	28.01 ± 5.48
Δdi-4,S (nmol/ml)	3.02 ± 0.24	27.73 ± 2.39	19.00 ± 3.52	14.30 ± 2.93
Total CS (nmol/ml)	6.92 ± 0.87^c	94.35 ± 6.34^c	69.70 ± 9.72	48.93 ± 9.66
Δdi-6,S/Δdi-4,S	0.74 ± 0.08^c	1.99 ± 0.06	2.20 ± 0.19	1.96 ± 0.11
CS/cell per week (nmol/10^6 cells/week)	—	31.60 ± 3.14	29.02 ± 1.79	16.04 ± 3.72
Stiffness (g/mm)	0.98 ± 0.03^c	3.07 ± 0.02	3.30 ± 0.02	3.38 ± 0.02

Abbreviations: C6S, chondroitin 6-sulfate; C4S, chondroitin 4-sulfate.
[a] Shown are cell count, concentrations of C6S and C4S, the C6S-to-C4S ratio, total CS content per cell per week, and stiffness in four experimental groups.
[b] Values represent mean $\pm$ SD.
[c] $p < 0.05$.

time t_d. These parameters are characteristic of the substratum used in the attachment assay. Table 3.17 lists materials used in the study together with fitted parameters mentioned above. Osteoblast interaction with broad composition range of GA/LA/CL biodegradable polymers prepared as thin films indicated that short-term hFOB interaction with biodegradable polymers broadly correlated with surface energy (water wettability).

Schneider *et al.* performed a study to determine if fibroblasts and osteoblasts would differentially attach to HEMA and PEG hydrogels copolymerized with positive, negative, or neutral charge densities, and if the integrin receptor adhesion ligand GRGDS grafted onto the hydrogels would promote cell attachment [168]. The neutral component of the hydrogels was prepared using a free radical polymerization of PEG dimethacrylate (PEGDMA) and HEMA. To produce negatively charged hydrogels sodium 2-sulfoethyl methacrylate (SEMA) was added and to produce positively charged hydrogels 2-methacryloxy ethyltrimethyl ammonium chloride (MAETAC) was added. The negative SEMA monomer and positive MAETAC monomer remained dissociated over a wide range of pH values and ion concentrations and therefore provided a stable charge. To prepare hydrogels for the addition of adhesion ligands, methacrylic acid (MAAc) monomer was added. The MAAc monomer provided covalent binding sites for the adhesion peptide sequence GRGDS. Specific GRGDS adhesion sites were covalently bound to hydrogel surfaces using a covalent peptide grafting technique based on a tresyl-activation of the hydrogel surface. Osteoblasts or fibroblasts (50,000 cells/10 μL) were plated onto HEMA or PEG-hydrogels containing incorporated charge densities or grafted RGD ligands for 3 h at 37°C in the presence or absence of 10% FCS with or without 25 μg/ml cycloheximide to prevent endogenous protein synthesis and secretion. Cultures were allowed to attach for 1 h before flooding with 1 ml of media. After two additional hours, the media were removed from the wells and unattached cells were quantitated and triplicate cultures were averaged. Osteoblasts and fibroblasts adhered in a similar charge-dependent manner with the greatest attachment on the positively charged surfaces followed by conjugated RGD ligand, negative, and then neutral charged hydrogels (Fig. 3.77). To assess the effect of hydrogels invoked on cell spreading and cytoskeleton development, actin stress fibers were fluorescently labeled in osteoblastic cells and fibroblasts cultured on different hydrogels. Both the cells attached and spread more readily on the positively charged HEMA and PEG hydrogel surfaces, followed by grafted RGD, negative and then neutral conditions.

Lee *et al.* evaluated a mechanism that plays as a predominant role in MSC and chondrocyte adhesion to biodegradable polymers (PLGA, PLA, and PCL) in serum-free conditions, and the effect of L-arginine on early cell adhesion [169]. L-arginine-coated surfaces were prepared by coating with 5% L-arginine solution on PLA-coated glass plates by the dip coating method. As shown in Table 3.18, both MSCs and chondrocytes attached to these polymers at similar rates ($<8\%$), which suggests that polymer characteristics such as roughness and wettability may not predominate during early cell adhesion to synthetic polymers in serum-free conditions. However, L-arginine-coated PLA rapidly increased cell adhesion to 30% within 1 h of culture.

Table 3.17

Adhesion and proliferation of hTOB to materials with varying composition and surface energy

Surface	Contact Angle (degrees)		Adhesion		Proliferation	
	Advancing	Receding	$\%I_{\max}$ (%)	$t_{1/2}$ (min)	k ($h^{-1} \times 10^2$)	t_d (h)
TCPS	55.0 ± 1.5	45.0 ± 0.8	55.3 ± 2.5	31.0 ± 2.8	2.23 ± 0.02	34.0 ± 0.4
Plasma-treated quartz (PTQ)	0 ± 0.0	0 ± 0.0	72.4 ± 4.8	38.0 ± 4.0	2.62 ± 0.02	29.4 ± 0.3
OTS-treated quartz (STQ)	112.7 ± 1.7	98.1 ± 0.8	10.6 ± 2.7	78.5 ± 6.5	1.14 ± 0.06	63.9 ± 3.2
Plasma-treated glass (PTG)	0 ± 0.0	0 ± 0.0	38.6 ± 4.2	61.8 ± 2.6	2.12 ± 0.01	35.6 ± 0.2
OTS-treated glass (STG)	105.9 ± 0.4	97.1 ± 2.7	18.5 ± 2.9	72.7 ± 4.1	1.41 ± 0.01	52.3 ± 0.4
PLGA 5/5 (M_n = 80 k)	71.0 ± 1.8	56.9 ± 2.4	32.6 ± 2.5	64.6 ± 4.1	1.91 ± 0.02	39.4 ± 0.5
PLGA 7/3 (M_n = 96 k)	73.5 ± 2.2	64.4 ± 1.8	23.2 ± 2.5	69.3 ± 5.0	1.70 ± 0.02	43.7 ± 0.4
PLA (M_n = 160 k)	78.0 ± 0.8	66.3 ± 0.6	13.8 ± 2.3	72.6 ± 4.1	1.33 ± 0.02	55.3 ± 1.0
PCL (M_n = 80 k)	76.6 ± 1.2	68.8 ± 4.9	8.2 ± 1.7	74.8 ± 2.8	1.08 ± 0.02	67.2 ± 1.1
PLCL 7/3 (M_n = 82 k)	77.0 ± 0.7	68.3 ± 0.5	11.8 ± 2.6	71.1 ± 3.1	1.25 ± 0.04	58.5 ± 1.8
PLGCL 2.5/2.5/5 (M_n = 60 k)	77.9 ± 1.4	68.3 ± 2.1	19.6 ± 3.1	69.3 ± 3.0	1.49 ± 0.03	49.6 ± 0.9
PLGCL 3.5/3.5/3 (M_n = 54 k)	66.3 ± 6.2	58.4 ± 3.8	23.1 ± 2.1	68.5 ± 3.6	1.64 ± 0.02	45.3 ± 0.5

The unit of variability for contact angle is standard deviation, whereas those for the adhesion and proliferation kinetics are standard errors of the curve fitting. M_n, number-average molecular weight by GPC; PLGA, poly(lactide-co-glycolide); PLCL, poly(lactide-co-caprolactone); PLGCL, poly(lactide-co-glycolide-co-caprolactone).

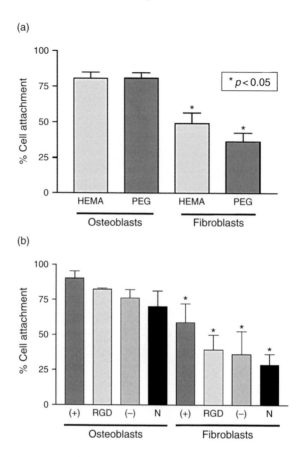

Fig. 3.77 Osteoblasts attach better to HEMA and PEG hydrogels as compared to fibroblasts. Positive (+), RGD ligand, negative (−), then neutral (N, HEMA and PEG) supported osteoblastic cell attachment better than fibroblasts attachment, irrespective of the condition tested.

Moreover, this effect was significantly inhibited by the integrin $\beta 1$ antibody, suggesting that cell-to-polymer interactions were mediated by integrin $\beta 1$.

Godbey *et al.* developed a method for cell seeding that requires less time and fewer cells [170]. The method involved the application of a centrifugal force that facilitates the transfer of cells into porous scaffolds in a homogeneous fashion. A special rotor was constructed to attach to the spindle of a common high-speed centrifuge such that the axis of rotation ran through the central lumen of cylindrical constructs along their entire lengths. The rotor was manufactured with tolerances allowing for up to 6000 rpm. Within the 0–6000 rpm range, the higher rotation speeds often resulted in cell lysis, based upon observations that scaffolds seeded at these speeds contained lower concentrations of cellular protein, and the resultant supernatants contained a greater amount of debris similar to that seen following

Table 3.18

Cell attachment in human MSCs and porcine chondrocytes to synthetic polymers

	Polymers	Non-blocking	Blocking	p Value
A	PLGA	2.8% ± 1.3	2.3% ± 1.4	0.7
B	PLA	3.3% ± 1.7	3.5% ± 2.1	0.8
C	PCL	3.1% ± 1.0	3.3% ± 1.2	0.1
D	PLA + Heparin	6.4% ± 3.5	2.3% ± 1.6	0.2
E	PLA + Arginine	30.2% ± 2.0[a]	5.6% ± 1.3	0.001
	Polymers	Non-blocking	Blocking	p Value
A	PLGA	7.7% ± 2.2	6.7% ± 2.6	0.6
B	PLA	8.0% ± 3.0	7.5% ± 2.8	0.8
C	PCL	6.8% ± 2.0	6.5% ± 1.3	0.9
D	PLA + Heparin	4.2% ± 0.5	5.0% ± 0.9	0.3
E	PLA + Arginine	25.9% ± 2.0[a]	7.4% ± 0.9	0.002

[a] Significant between E and the others (A–D) ($p < 0.05$).

cell lysis. Unbalanced or vibrating rotation resulted in cell shearing and was therefore undesirable. Thus, rotational speed was set constant at 2500 rpm for these experiments. This equates to centrifugal forces of 35.0 and 52.5 g at the inner and outer surfaces of the constructs. The centrifugation experiments utilized PLGA-coated PGA scaffolds and two different cell types, bladder SMCs and human foreskin fibroblasts (HFF) cells. The HFF cells had a proliferation rate that allowed for quicker expansion *in vitro* and a smaller diameter per cell. Different cell diameters directly affected migration rates under centrifugation as well as the ratio of cell to pore diameters. Increased seeding times yielded a greater number of cells being seeded within the PGA scaffolds, up to 10 min of centrifugation (Fig. 3.78). Increasing the centrifugation time to 20 min produced poor results. In addition, breaking up the centrifugation intervals into 1-min segments produced improved seeding results. Probably, some cells were pushed completely through the scaffold after 1 min of centrifugation. Increasing the centrifugation time would not increase the likelihood that one of these cells would re-enter the scaffold interior. However, if centrifugation was halted and the cell suspension was allowed to resettle into the spin tubes, these cells would be available to migrate through the scaffold again on a subsequent centrifugation cycle. Over time, given enough cycles of centrifugation and rest, the number of cells entering the scaffold per spin would theoretically equal the number exiting from the exterior surface of the construct. However, there exists the possibility that some cells are damaged by the centrifugation process, so this theoretical homogeneity would approach zero cells entering and exiting as the number of repeats is increased to a very large value. Because an analysis of protein does not discriminate between living cells, dead cells or cellular debris, MTT assays were performed to assess the mitochondrial activity of cells within the seeded scaffolds. Scaffolds seeded by centrifugation showed a greater amount of cellular activity than scaffolds seeded either by the static method or via spinner

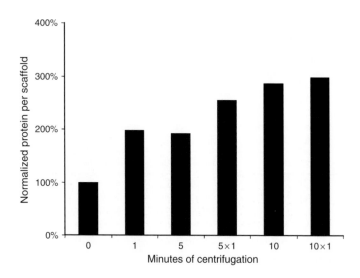

Fig. 3.78 Typical results from scaffold optimization experiments. The data reflect the amount of cellular protein within the scaffolds following various centrifugation duration. "5 × 1" and "10 × 1" denote scaffolds spun for repeated 1-min intervals, respectively.

flask. The spinner flask seeding method was efficient at depositing cells on scaffold exteriors, while the centrifugation method yielded a fairly even distribution of cells throughout the entire thickness of the construct.

Chennazhy and Krishnan evaluated the effect of passage progression on the cell behavior using two different matrices [171]. To prepare gelatin-coated polystyrene (GCPS), culture plates were added with 2% gelatin, incubated for 2 h, and excess solution was aspirated out just before cell seeding. To make fibrin-coated polystyrene (FCPS), culture plates were incubated with bovine thrombin and after removing the excess solution, the surface was added with cryoprecipitate consisting of fibrinogen, FN, Factor XIII, and VEGF. For composite-coated polystyrene (CCPS), it was similar to the process for FCPS but 0.02% gelatin-premixed cryoprecipitate-VEGF was added. The effects of GCPS and CCPS on ECs in serial passages until about 15th passage are compared in Fig. 3.79. Time taken to form 15 passages of confluent monolayer on GCPS was ~85 days, whereas on CCPS in ~50 days 15 passages of monolayer were formed. The ability of ECs to regenerate was found to be deteriorating passage by passage when cells were grown on GCPS. The ECs easily lost their proliferation potential when they were cultured repeatedly on gelatin, turned apoptotic, and overexpressed the prothrombotic protein – von Willebrand factor (vWF). On the contrary, when ECs were grown on a matrix composed of fibrin, FN, gelatin, and VEGF, the cells retained their ability to proliferate, remained viable, and were relatively less thrombogenic, even when passage number progressed.

To develop specialized preservation solutions for cells, Mathew *et al.* investigated several cell systems including two cardiac vasculature cell strains (coronary

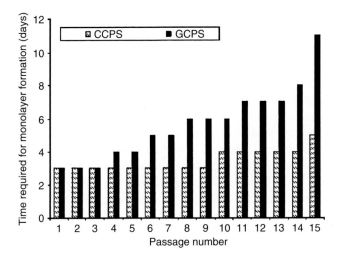

Fig. 3.79 Graphical representation of proliferation rates of HUVEC on GCPS and CCPS/Time taken for each passage to form confluent monolayer on two matrices is shown on y-axis against the passage number on x-axis.

artery smooth muscle cells [CASMCs] and coronary artery endothelial cells [CAECs]), a liver cell line (C3A cells), and a skeletal muscle cell strain (skeletal muscle cells [SKMCs]) [172]. These cells were hypothermically preserved for 2–7 days at 4°C in either cell culture medium, University of Winsconsin Solution (UW or ViaSpan), or HypoThermosol (HTS) variants. The cells were then assayed for viability, using the alamar Blue assay as well as calcein-AM, subsequent to their return to normothermic temperatures for up to 5 days. The CASMC viability was best maintained when preserved in HTS plus Trolox/EDTA; CAEC viability was highest when preserved in HTS plus Trolox; SKMCs stored in HTS plus Trolox/RGD demonstrated enhanced viability; and C3A cells were best preserved in HTS plus FK041. Kisiday *et al.* investigated insulin–transferrin–selenium (ITS) as a complete or partial replacement for FCS during *in vitro* culture of bovine calf chondrocytes in hydrogel scaffolds [173]. Chondrocyte-seeded agarose and self-assembling peptide hydrogels were maintained in DMEM plus 10% FCS, 1% ITS plus 0.2% FCS, or 1% ITS and evaluated for biosynthesis, cell division, and surface outgrowth of fibroblastic-like cells and fibrous capsule formation over several weeks of culture. In peptide hydrogels, cells cultured in ITS plus 0.2% FCS medium exhibited high rates of biosynthesis and showed similar cell division trends as seen in 10% FCS cultures. The ITS medium alone did not support GAG accumulation beyond 5 days of culture, and cell division was less than that in both serum-containing cultures. Extensive cellular outgrowth and fibrous capsule formation were observed in 10% FCS medium, whereas little outgrowth was observed in ITS plus 0.2% FCS and none was seen in ITS medium alone. In agarose hydrogels, chondrocyte biosynthesis and cell division in

ITS medium were similar to that in 10% serum culture over 5 weeks, and cellular outgrowth was eliminated.

A relatively large volume of blood such as 100 ml is needed to prepare patient own serum. In an attempt to decrease patient discomfort, Harrison *et al.* investigated the feasibility of taking blood when the patient is anesthetized [174]. To this end, two serum samples were prepared from each of 22 patients: (1) from the awake patient (PRE) and (2) from the patient 4 min after induction of general anesthesia (PER). When the sera were compared for their ability to support the *in vitro* proliferation of primary human chondrocytes, the PER sample supported higher growth in 2 of 22 patients, equivalent growth in another 11, and significantly lower growth in the remaining 9. Only the opiate analgesics had a significant and inhibitory effect on chondrocyte proliferation, suggesting that exposure to opiates should be avoided when blood products are collected for use in the *in vitro* culture of human cells.

6.2. Co-culture

Co-cultures with different cell types are important to induce various desired effects on cell differentiation, cell activities, cell death, and proliferative properties due to reciprocal influences. A system for the direct co-culture of ECs on SMCs was developed by Lavender *et al.* [175]. The co-culture consisted of a culture substrate, a basal adhesion protein, a layer of porcine SMCs, a medial adhesion protein, and a layer of ECs. Conditions that led to successful co-culture were a polystyrene culture substrate, a quiescent state for SMCs, subconfluent density for SMC seeding and confluent density for EC seeding, and FN for the basal adhesion protein. The EC adhesion was not enhanced by addition of FN, types I and IV collagen, or LN to the medial layer. The SMCs did not migrate over ECs and the cells were present in two distinct layers. Co-culture could not be consistently maintained for as long as 10 days. After exposure to 5 dyn/cm^2 for 7.5 h, ECs remained adherent to SMCs. Junction formed between ECs, but at a lower level than observed with EC monoculture.

Lichtenberg *et al.* developed a novel bioreactor for 3-D cell culture that allowed for a co-cultivation of cells with diametrically different physiological conditions such as mammalian and amphibian cardiomyocytes [176]. As shown in Fig. 3.80, the bioreactor consists of four transparent glass chambers generating two separate compartments. The compartments and chambers were connected with silicone tubes. Chambers were divided by a permeable polycarbonate membrane with a pore size of 12 μm, allowing a separation of co-cultures. Such a pore size allowed the filtering only of cultured cells and not of molecules that are important for an interaction between co-cultures. The surface of the membrane in each bioreactor compartment measured 4 cm^2.

Khademhosseini *et al.* developed a method for patterning cellular co-cultures that uses the layer-by-layer deposition of ionic biopolymers. The scheme is illustrated in Fig. 3.81 [177]. To generate patterned co-cultures, a glass substrate was patterned with non-biofouling HAC (HA$_c$) and subsequently stained with FN which adheres to the exposed substrate. The first cell type was then seeded and allowed to

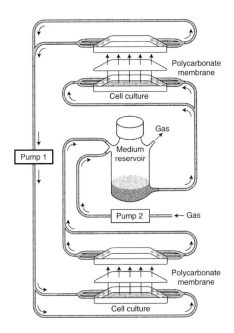

Fig. 3.80 Schematic drawing of the multifunctional membrane bioreactor for a cell co-culture.

adhere to the FN-coated regions. Subsequent ionic adsorption of poly-L-lysine (PLL) to HAC patterns was used to switch the HAC surfaces from cell repulsive to adherent thereby facilitating the adhesion of a second cell type. They demonstrated the utility of this approach to pattern co-cultures of hepatocytes or ES cells with fibroblasts.

To construct the wall of the urinary conduit consisting of an inner uroepithelium, an intermediate layer of connective tissue in the lamina propria, and an outer SMC layer, Fossum *et al.* isolated human UCs, fibroblasts, and SMCs and expanded them *in vitro* [178]. After 1 week of primary urothelial culture the cells were treated with trypsin for reseeding as a single-cell suspension. The monolayer culture of UCs was incubated for 1 week before co-cultures were initiated by seeding fibroblasts at a density of 60,000 cells/cm^2 on top of the UCs. The double layer of cells was incubated for 1 week before the three-layer culture was established by seeding SMCs on top of the fibroblasts at a density of 90,000 cells/cm^2. Both fibroblasts and SMCs were seeded as single-cell suspensions in the same manner as for ECs. During the monolayer primary cultures the three different types of cells had a morphological appearance and growth pattern typical of each different phenotype. Each cell type attached well when seeded on either the bottom of the culture flask (UCs) or another cell layer (fibroblasts and SMCs). After 3 weeks of culture in a three-layered system the tissue formed sheets that were thick enough for careful handling and lifting from the well.

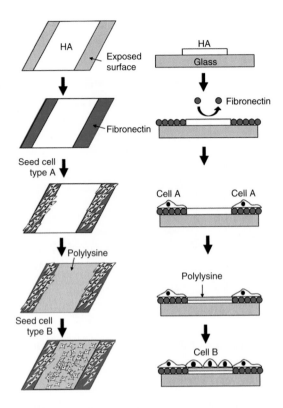

Fig. 3.81   Schematic diagram of the HAC–PLL layering approach to pattern co-cultures.

6.3. Bioreactors

Martin and Vermette gave a review on design, characterization, and recent advances of bioreactors for tissue mass culture, such as the continuous stirred tank, the hollow fiber, the Couette-Taylor, the airlift, and the rotating-wall reactors [179]. In addition, they examined the role of the uterus, the reactor built by Nature and reported the environment provided to a growing embryo, yielding possible paths for further reactor development. Kino-Oka *et al.* designed a novel bioreactor system to perform a series of batchwise cultures of anchorage-dependent cells by means of automated operations of medium change and passage for cell transfer [180]. The experimental data on contamination frequency ensured the biological cleanliness in the bioreactor system, which facilitated the operations in a closed environment, as compared with that in flask culture system with manual handlings. In addition, the tools for growth prediction (based on growth kinetics) and real-time growth monitoring by measurement of medium components (based on small-volume analyzing machinery) were installed into the bioreactor system to schedule the operations of medium change and passage and to confirm that culture proceeds as scheduled, respectively. The successive culture of anchorage-dependent cells was conducted with the bioreactor running in an automated way, realizing 79-fold cell expansion for 169 h.

6.3.1. Spinner Flask Reactor

Vunjak-Novakovic *et al.* found that cartilage constructs from mixing cultures at 50 rpm were more regular in shape and contained up to 70% more cells, 60% more sulfated GAG, and 125% more total collagen compared with static cultures [181]. Their compositions were approximately higher than most bioreactor results. Freed *et al.* examined the mixing effect on the production of GAG and collagen [182]. They found that mixing did not increase the amount of GAG in a cell–polymer construct relative to static levels, but total collagen did increase by 50%. However, the outer capsule was composed mainly by type I collagen, which would account for the increase in the total collagen composition of the construct. Gooch *et al.* found that constructs exposed to any intensity of mixing (80–160 rpm) contained more cartilage and synthesized and released more GAG but retained lower fractions of GAG within the scaffold.

To visualize the medium flow in a spinner flask, Wang *et al.* used microspheres (20–30 μm in diameter and density of 1.2 g/cm^3) stained with methylene blue [183]. They found that different regions existed in medium (outer, inner, and middle), where overall particle movement was different. In the middle region the flow was turbulent and scaffolds were best placed.

6.3.2. Perfusion Reactor

Cartmell *et al.* examined the effects of varying continuous flow rate on cell viability, proliferation, and gene expression within cell-seeded 3-D scaffolds [184]. They produced cylindrical scaffolds measuring 6 mm in diameter and length from the trabecular bone of human femoral metaphyses. Average porosity was 82% and average pore size was 650 μm. Hydrated bone scaffolds were seeded with 2 million of MC3T3-E1 immature osteoblast-like cells suspended in α-MEM containing 10% FCS per scaffold. The cells were then allowed to adhere to the constructs over a period of 24 h before placement into the 3-D tissue system developed to facilitate controlled perfusion of a culture medium through cell-seeded constructs. For each perfusion experiment, eight cylindrical constructs were located into individual chambers of the stainless steel perfusion block. Side holes in the perfusion block allowed the entrance and exit of defined culture medium to each of the independent chambers. Medium perfusion rates of 1.0, 0.2, 0.1, and 0.01 ml/min were applied continuously to the constructs for 7 days. A live/dead fluorescent cell stain that labels viable cells "green" (calcein) and dead cells "red" (ethidium homodimer) was used to assess cell viability within the constructs. Confocal microscopy revealed a clear influence of perfusion rate on cell viability. At day 1, a high proportion of the cells were viable throughout the constructs. After 1 week in static culture, viable cells were nearly confluent on the periphery of the constructs, but only a few viable cells were found at the construct center. At 1 week, constructs for each perfusion rate were retrieved for total DNA content analysis in order to assess cell proliferation. Two freeze–thaw cycles were completed before crushing each construct and then sonicated to rupture cell membranes and release the cellular

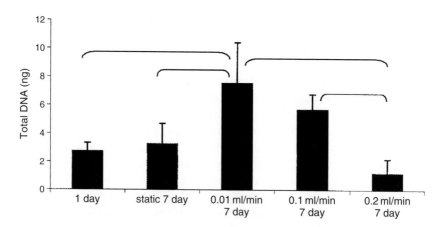

Fig. 3.82 DNA assay after 1 week of perfusion. *1 day*: scaffold seeded 1 day before perfusion. *Static 7 day*: static sample (not perfused). *0.01, 0.1. and 0.2 ml/min 7 day*: flow rates, perfused for 1 week.

DNA into solution. Figure 3.82 shows the effects of medium perfusion rate on cell proliferation, as measured by total DNA content within the constructs. Clearly, only the lowest perfusion rate of 0.01 ml/min enhanced cell proliferation relative to static controls. Real-time RT-PCR showed that all three bone-related genes (Runx 2, OCN, and ALP) analyzed were detected in cellular constructs perfused at less than or equal to 0.2 ml/min. Differences in the number of viable cells were most evident at the center of the constructs, suggesting that slow perfusion may have improved mass transport throughout the constructs. Perfusion medium at lower rates may also have allowed for autocrine effects, as soluble proteins released by cells may have been flushed away at higher flow rates before exerting an influence on neighboring cells. Expression of Runx 2, ALP, and OCN was enhanced within cell-seeded constructs perfused at 0.2 ml/min relative to lower flow rates. Runx 2 expression, but not ALP or OCN expression, was also upregulated compared with static controls.

Neves *et al.* used a suite of MRI and magnetic resonance spectroscopy (MRS) methods to monitor and optimize the performance of a perfusion bioreactor system for generating meniscal cartilage constructs with properties approaching those of the native tissue [185]. Sheep meniscal fibrochondrocytes of the fourth passage were seeded onto scaffolds at a density of 3×10^4 cells/cm^2 in spinner flasks. The scaffolds consisted of 15-μm-diameter PET fibers, with a void volume of 97% and a density of 45 mg/cm^3, and had a form of discs, 12 mm in diameter and 4 mm thick. Over a period of 3 days cells attached to the surface of the scaffolds with no significant cell loss and an adhesion yield greater than 95% and then the scaffolds were transferred to a perfusion bioreactor. To measure the construct perfusion, solution of a contrast agent gadolinium-diethyltriaminepentaacetic acid (Gd-DTPA) was added to the perfusion medium. This paramagnetic agent, which is used in clinical imaging to enhance tissue contrast and to measure perfusion, produces an increase in signal intensity in T_1-weighted images. Axial flow through and around the scaffolds was measured by a time-of-flight MRI method. The non-invasive nature of these measurements allowed a continuous assessment of bioreactor behavior and the influence of different flow rates on its performance.

To distribute cells into the central area of porous ceramics, Uemura *et al.* modified a culture technique by introducing a low pressure system and a perfusion culture system (Fig. 3.83) [186]. The low-pressure system consisted of a rotary vacuum pump (VP), a vacuum controller (VC), and a polycarbonate vacuum desiccator (D), connected to each other by rubber and silicone tubes. After subculturing bone marrow–derived osteoblastic (BMO) cells for 2 days, composites of BMO and porous β-TCP were transferred to a perfusion culture container (Minucells, and Minutissue, Bad Abbach, Germany). The container was connected with medium bottles by silicone tubes. The peristaltic pump was adjusted to deliver fresh medium at a rate of 2 ml/hr. The whole system was put on a table at room temperature, but the perfusion container was maintained at 37°C by a thermo plate. Figure 3.84 shows the pressure dependence of the ALP activity of the BMO/porous β-TCP composites at 4 and 8 weeks after implantation at subcutaneous sites in rats. The lower the pressure applied to cells and porous ceramics at subculture, the stronger the ALP activity. However, the activity level reached a maximum around 100 mmHg and decreased below 50 mmHg, indicating that there exists an appropriate pressure for the application of this low-pressure system.

Freyria *et al.* investigated two types of perfusion mode (direct and free flow) in terms of the biosynthetic activities of chondrocytes grown in collagen sponges [187]. Results demonstrated that both bovine and human-derived chondrocytes displayed a dose-dependent response to flow rate (0~1 ml/min) in terms of cell number and GAG content. This may reflect the weak adhesion of cells to the sponge scaffolds and the immature state of the constructs even after 3 weeks of proliferative culture. An optimal flow rate was defined between 0.1 and 0.3 ml/min for direct perfusion and free flow bioreactors. Using fresh bovine chondrocytes and a lower flow rate of 0.1 ml/min, a comparison was made between free flow system and

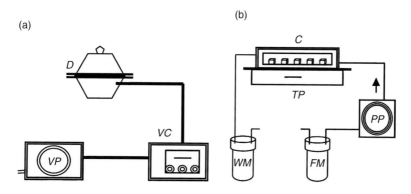

Fig. 3.83 Schematic diagram of the low-pressure system (a) and the perfusion culture system (b).

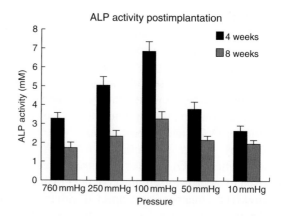

Fig. 3.84 Temporal changes in ALP activity of subcultured composite grafts treated with different pressures at 4 and 8 weeks.

direct perfusion system. In the free flow bioreactor, no cell loss was observed and higher GAG production was measured compared with static cultured controls. However, as with direct perfusion, the enhancement effect of free flow perfusion was strongly dependent on the maturation and organization of the constructs before the stimulation. An increase in culture duration of 18 days before mechanical conditioning resulted in enhanced GAG production compared with controls. Additional enhancement was found in specimens that were further subjected to a prolonged duration of perfusion (63% increase after an additional 4 days of perfusion) after prematuration. The free flow system had an advantage over the direct perfusion system, especially when using sponge scaffolds, which have lower mechanical properties; however, mass transfer of nutrients was still more optimal throughout the scaffolds in a direct perfusion system as demonstrated by histological analysis.

6.3.3. Rotating Reactor

To study whether the rotating bioreactor produces higher fractions of GAG and collagen than do mixed flasks or static culture, Martin *et al.* cultured constructs for 6 weeks and produced tissue that had compositions of GAG and total collagen that were 75 and 39%, respectively, of native cartilage levels [188]. When cultured for 7 months, the GAG content exceeded physiological levels whereas the collagen content stayed at 39%. In addition, the equilibrium moduli (950 kPa) and hydraulic permeability (5 × 10^{-15} m4/Ns) of the construct were comparable to physiological levels, due mainly to the increase in GAG. Freed *et al.* found similar results, obtaining 68% of native GAG and 33% of native type II collagen compositions [189]. The major improvement in the engineered tissue was not just the increased matrix compositions. The morphology of the constructs showed a more continuous matrix than in other bioreactors. There was less of an outer capsule, and ECM was deposited more evenly through the thickness of the scaffold, suggesting that oxygen and nutrients were reaching the middle of the construct in sufficient amounts.

A rotating-wall bioreactor culture of knee meniscal chondrocytes on an agarose hydrogel and a non-woven mesh of PGA was compared with static cultures [190]. Constructs were cultured for up to 7 weeks in static and rotating-wall bioreactor culture. Cell numbers were 22 times higher in PGA than agarose after 7 weeks in culture. Static PGA scaffolds had more than twice the amount of sulfated GAGs and three times the amount of collagen compared to static agarose constructs at week 7. The rotating-wall bioreactor was not found with increased matrix production or cell proliferation significantly over static cultures.

Detamore and Athanasiou determined the effects of a rotating bioreactor in temporomandibular joint (TMJ) disc tissue engineering [191]. Porcine TMJ disc cells were seeded at a density of 20 million cells/ml onto non-woven PGA scaffolds in spinner flasks for 1 week and then cultured either under static conditions or in a rotating bioreactor for a period of 6 weeks. Between the bioreactor and static cultures, there were marked differences in gross appearance, histological structure, and distribution of types I and II collagen. Engineered constructs from the bioreactor contracted earlier and to a greater extent, resulting in a denser matrix and cell composition. In addition, immunostaining intensity was generally uniform in static constructs, in contrast to higher intensity around the periphery of bioreactor constructs. Moreover, bioreactor constructs had higher amounts of type II collagen than did static constructs. However, differences in total matrix content and compressive stiffness were generally not significant.

6.4. Kinetics

To evaluate the role of nutrient diffusion and consumption within cell-seeded scaffolds in the absence of flow, Botchwey *et al.* developed a 1-D glucose diffusion model with the following assumptions [192]:

1. The pores within the scaffold are cylindrical with a uniform distribution of glucose consumption sources, Q g s^{-1} m^{-1}, within each of the pores. Q is given by:

$$Q = \frac{N\hat{r}}{nl}$$

(1)

   where N is the total number of cells within the scaffold, $\hat{r}$ is the single osteoblast glucose consumption rate, n is the number of cylindrical pores within the scaffold, and l is the length of the modeled pores.

2. In the absence of flow, diffusion within the scaffold is given by:

$$D\sigma = \frac{\delta^2}{\delta x^2} C(x) = -Q$$

(2)

   where D is the diffusivity of glucose, σ is the average cross-sectional area of the modeled pores, and $C(x)$ is the concentration of glucose as a function of depth within the scaffold.

3. The boundary conditions are given by:

$$C_0 = C(0) \quad \text{and} \quad \frac{\delta}{\delta x} C(l/2) = 0 \tag{3}$$

where C_0 represents media glucose concentration at the exterior boundary of the scaffold. Symmetry of glucose concentration within the pores in the absence of flow results in the 0 value of the derivative. The analytical solution to the boundary value problem is then given by:

$$C(x) = C_0 - \frac{N\hat{r}}{2Dn\sigma l}(Lx - x^2) \tag{4}$$

Botchwey *et al.* quantified also the effects of internal fluid perfusion and flow-induced nutrient transport to cells within the scaffolds during dynamic bioreactor cultivation. To facilitate the analysis, they made the following simplifying assumptions:

(a) The scaffolds are roughly cylindrical and their 3-D geometry may be represented by a network interconnecting cylindrical channels with tortuosity, τ. The Darcy's permeability constant, K, for such a construct is given by [193]:

$$K = \frac{\varepsilon \sigma^2}{96\eta\tau^2} \tag{5}$$

where ε is the pore volume fraction, σ is the pore diameter, and η is the viscosity of the fluid. These physical dimensions and 3-D geometry are assumed not to change during the time period of experiment.

(b) There is a homogeneous velocity profile, V_∞, at a large distance from the porous scaffold that creates a pressure gradient, VP, across the thickness of the construct.

(c) The normal vector of the scaffold front surface is parallel to the velocity at large distances from the disc.

(d) The scaffold thickness is small and permeable to fluid flow but it is assumed that the velocity of the fluid in the scaffold is small compared with V_∞.

(e) The pressure gradient is equal to the pressure difference per unit length along the scaffold, $\Delta P/L$, and is given by the drag force on the scaffold per unit volume as per the text by Happel and Brenner [194], where the force due to drag on a circular cylinder is given by:

$$F = \frac{4\pi\eta V_\infty R}{\ln(2R/L) - 0.72} \tag{6}$$

where R and L are the radius and length of the scaffold, respectively.

4. The rate of internal fluid perfusion within scaffolds may then be calculated as follows:

$$V = -\left(\frac{K}{\eta}\right)\frac{\Delta P}{L} \tag{7}$$

where V is the velocity of the fluid flow through the porous scaffold.

To calculate the internal fluid perfusion and flow-induced glucose flux, Botchwey *et al.* employed three culture models. To facilitate the calculation of spinner flask and perfusion culture conditions, characteristic values of exterior fluid flow rates in these systems were taken from the literature and applied to the analytical model of interflow. For spinner flasks, calculated values of 0.124 and 0.02 m/s were accepted as representative maximum and minimum external fluid velocities [195]. For direct perfusion culture, they used a value of 0.03 m/s for internal flow rate [196]. To adequately describe the hydrodynamic conditions that occur within porous scaffolds during culture in the rotating bioreactor, Botchwey *et al.* fabricated microcarrier scaffolds. They adapted double emulsion microencapsulation methods for the formation of hollow, lighter than water microspheres of PLGA (50:50). Individual microcarriers were then used to create 3-D microcarrier scaffolds by permanently fusing together microcapsules 500–860 μm in size to form 4 mm × 2.5-mm cylinders using microsphere sintering methods. Table 3.19 gives the physical and experimental data of this microcarrier [197]. Figure 3.85 shows the internal fluid perfusion rates based on a broad range of physical scaffold properties. In this calculation they chose three representative values of measured scaffold velocity of 100, 50, and 10 mm/s.

6.5. Mechanical Stimulation

Costa *et al.* tested the hypothesis that cells align along the direction of greatest local tension [198]. Figure 3.86 shows the alignment of dermal fibroblasts in a partially constrained square collagen gel. Cells aligned parallel to free edges in

Table 3.19
Nomenclature and data used in calculations

Physical and experimental data	
$U_0 = 0.098 \pm 0.005$	Velocity of scaffold motion (m/s)
$R = 0.002$	Radius of the cylindrical scaffold (m)
$L = 0.0025$	Length of the cylindrical scaffold (m)
$\varepsilon = 0.31$	Fluid volume fraction of the Scaffold
$\delta = 0.000187$	Median pore diameter of the Scaffold (m)
0.01	Volume of fluid filling the scaffold (ml)
1–3	Range of tortuosity (τ) accepted by the authors
Theoretical model nomenclature	
U	Superficial fluid velocity through the scaffold (m/s)
Q	Volume flow rate through the scaffold (m³/s)
$\Delta P/L$	Pressure gradient across the scaffold (N m)
η	Dynamic viscosity (kg m⁻¹ s⁻¹)
K	Darcy's permeability constant (m²)
σ	Shear stress (N m⁻²)
λ	Residence time (s)

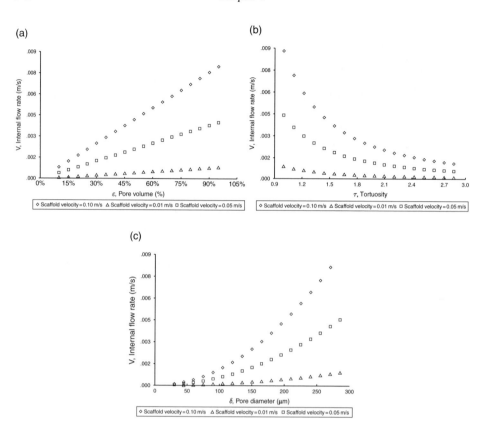

Fig. 3.85   Internal perfusion rates of fluid perfusion within 3-D scaffolds as determined by theoretical modeling.

partially constrained gels and, in contrast to the hypothesis, fibroblasts in fully constrained gels remained randomly aligned independent of geometry. Collagen orientation in the gels was determined by confocal reflectance microscopy. Collagen was randomly oriented in the gel with two parallel cut edges but no cells, whereas in the presence of fibroblast collagen became aligned more strongly near the free boundary and less strongly near the center of the gel.

To investigate the relationship between mechano-transduced anabolic and catabolic processes of cells, Berry *et al.* applied cyclic tensile strain to a fully contracted fibroblast-seeded collagen gels and determined cell viability, cell proliferation, collagen synthesis, MMP expression, and construct mechanical properties [199]. While fibroblasts deposit the collagen, MMP digests fibrous collagens. There exist tissue inhibitors of MMP (TIMP) that counterbalance the actions of MMPs by directly binding the proteases. The [³H] thymidine and [³H] proline incorporation was used as a measure of cell proliferation and collagen synthesis, respectively. The cells were preloaded at either 2 or 10 mN and the preloaded gels were subsequently subjected to a further 10% cyclic strain (0–10%) at 1 Hz. In all cases cellular viability

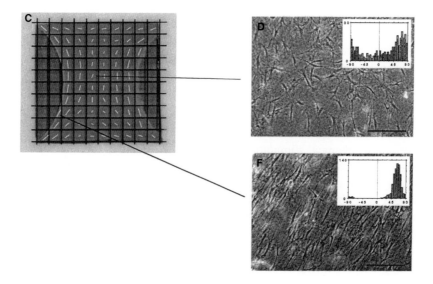

Fig. 3.86 Vector map in a partially unconstrained square gel (C) and microgram and histogram from the indicated regions (D: 71 ± 32° and F: 61 ± 13°).

was maintained during the conditioning period. Table 3.20 summarizes the results. As can be seen, cell proliferation was enhanced by the application of cyclic strain within gels preloaded at both 2 and 10 mN, whereas collagen synthesis was enhanced by cyclic strain within gels preloaded at 2 mN only. The profile of MMP expression was broadly similar in gels preloaded at 2 mN with or without the application of cyclic strain. By contrast, gels preloaded at 10 mN and subjected to cyclic strain expressed enhanced levels of staining for latent MMP-9, and both latent and active MMP-2, when compared with the other conditioning regimens. The structural stiffness of gels preloaded at 2 mN and subjected to cyclic strain was enhanced

Table 3.20

Results for fibroblast-seeded collagen gels preloaded at 2 or 10 mN and subjected to cyclic strain, compared with constructs subjected to preloading alone

	Preload of 2 mN, 10% Cyclic Strain	Preload of 10 mN, 10% Cyclic Strain
[3H] Thymidine incorporation	↑	↑
[3H] Proline incorporation	↑	—
MMP expression		
Latent	↑ 1	↑ 1, ↑ 2, ↓ 3, ↑ 9
Active		↑ 1, ↑ 2, ↑ 3
Structural stiffness	↑	—
Initial failure load	—	↓

Key: ↑, Increased; ↓, decreased; –, unchanged.

compared with control specimens, reflecting the increase in collagen synthesis. By contrast, the initial failure loads for cyclically strained gels preloaded at 10 mN were reduced, potentially because of enhanced catabolic activity.

To examine whether short periods of mechanical stimulation (minutes per day) applied over a long duration (up to 4 weeks) would improve the properties of cartilaginous tissue formed *in vitro*, cartilaginous tissue was developed from isolated chondrocytes seeded on the surface of calcium phosphate ceramic substrates [200]. *In vitro*–formed tissues were subjected to different stimulation protocols for 1 week. The optimal mechanical stimulation parameters identified in this short-term study were then applied to the cultures for up to 4 weeks. As shown in Table 3.21, long-term intermittent compressive mechanical stimulation (applied under the optimal conditions described above) significantly affected both the physical and the biochemical properties of the tissue. Mechanical stimulation applied at a 5% compressive amplitude at a frequency of 1 Hz for 400 cycles every second day resulted in the greatest increase in collagen synthesis (37% over control) while not significantly affecting proteoglycan synthesis (2% over control). This condition, applied to the chondrocyte cultures for 4 weeks, resulted in a significant increase in the amount of tissue that formed (stimulated, 2.4 mg dry wt; unstimulated, 1.6 mg dry wt). Stimulated tissues contained 40% more collagen (stimulated, 590 μg; unstimulated, 420 μg), and 30% more proteoglycans (stimulated, 393 μg; unstimulated, 302 μg) as well as displaying a 2- to 3-fold increase in compressive mechanical properties

Table 3.21

Properties of *in vitro*–formed tissues[a]

	Static culture ($n = 10$)	Stimulated for 2 weeks ($n = 9$)	Stimulated for 4 weeks ($n = 9$)
Thickness (mm)	0.82 ± 0.04	0.89 ± 0.05	1.07 ± 0.07[b]
Dry weight (mg)	1.61 ± 0.08	1.7 ± 0.1	2.4 ± 0.2[b]
Water content (%)	88.6 ± 0.1[c]	88.5 ± 0.4	87.5 ± 0.4[c]
DNA (μg/construct)	11 ± 2	11 ± 2	11 ± 1
DNA (10^6 cells)	1.5 ± 0.3	1.4 ± 0.2	1.4 ± 0.1
Proteoglycans/DNA (μg/μg)	24 ± 4[c]	31 ± 5	43 ± 7[c]
Proteoglycans/DNA (μg/10^6 cells)	184 ± 28[c]	235 ± 43	328 ± 55[c]
Collagen/DNA (μg/μg)	32 ± 5[c]	44 ± 7	59 ± 9[c]
Collagen/DNA (μg/10^6 cells)	248 ± 36[c]	328 ± 63	454 ± 71[c]
Equilibrium stress (kPa)[d]	5 ± 1	6 ± 2	10 ± 1[b]
Equilibrium modulus (kPa)[c]	24 ± 6	38 ± 8	80 ± 23[b]

[a] After 8 weeks in culture, the physical properties and matrix accumulation relative to the DNA content (or number of cells based on 7.7 μg of DNA per 10^6 cells[34]) of the *in vitro*–formed cartilaginous tissues in static culture and those subjected to either 2 or 4 weeks of stimulation were analyzed as described in Materials and Methods. Three separate experiments were performed. The results were pooled and expressed as mean ± standard error of the mean.
[b] Significantly different from other stimulation protocols ($p < 0.05$).
[c] Statistical difference between static culture and 4 week-stimulated culture ($p < 0.05$).
[d] From equilibrium testing at 25% compressive strain.

(maximal equilibrium stress: stimulated, 10 kPa; unstimulated, 5 kPa and maximum equilibrium modulus: stimulated, 80 kPa; unstimulated, 24 kPa). Only short periods of mechanical stimulation (6 min every second day) were needed to affect the quality of cartilaginous tissue formed *in vitro*.

To investigate the independent effects of cyclic flexure on *in vitro* TEHVs development, Engelmayr Jr. *et al.* designed a bioreactor with the capacity to provide cyclic three-point flexure to 12 rectangular samples of TEHV biomaterial [201]. Ovine vascular smooth muscle cells (VSMCs) were seeded for 30 h onto strips of non-woven 50:50 PGA and PLLA scaffold. After 4 days of incubation, SMC-seeded and unseeded scaffolds were either maintained under static conditions (static group), or subjected to unidirectional cyclic three-point flexure at a physiological frequency and amplitude in a bioreactor (flex group) for 3 weeks. After seeding or incubation, the effective stiffness (*E*) was measured, with SMC-seeded scaffolds further characterized by DNA, collagen, sulfated GAG(S-GAG), and elastin content, as well as by histology. The seeding period was over 90% efficient, with a significant accumulation of S-GAG, no significant change in *E*, and no collagen detected. Following 3 weeks of incubation, unseeded scaffolds exhibited no significant change in *E* in the flex or static groups. In contrast, *E* of SMC-seeded scaffolds increased 429% in the flex group and 351% in the static group, with a trend of increased *E*, a 63% increase in collagen, increased vimentin expression, and a more homogenous transmural cell distribution in the flex versus static group. Moreover, a positive linear relationship was found between the mean E and mean collagen concentration, as shown in Fig. 3.87.

To evaluate the effect of strain direction on the ECM production, Lee *et al.* seeded human ligament fibroblasts (HLFs) on parallel aligned, vertically aligned to

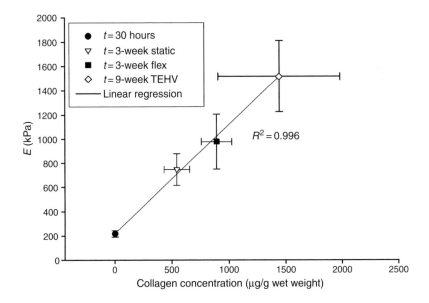

Fig. 3.87 E results for SMC-seeded scaffolds plotted versus collagen concentration.

the strain direction, and randomly oriented nanofiber sheets [202]. The nanofiber matrix was fabricated using electrospinning technique. A rotating target was used to align the nanofibers. The HLFs on the aligned nanofibers were spindle shaped and oriented in the direction of the nanofibers. After 48-h culture, 5% uniaxial strain was applied for 24 h at a frequency of 12 cycles/min. The amount of collagen produced was measured 2 days after halting the strain application. The HLFs were more sensitive to strain in the longitudinal direction. Grenier *et al.* designed to evaluate the effects of static mechanical forces on the functionality of the produced tissue constructs [203]. Living tissue sheets reconstructed by the self-assembly approach from human cells, without the addition of synthetic material or ECM, were subjected to mechanical load to induce cell and ECM alignment. In addition, the effects of alignment on the function of substitutes reconstructed from these living tissue sheets were evaluated. Results show that tissue constructs made from living tissue sheets, in which fibroblasts and ECM were aligned, presented higher mechanical resistance. This was assessed by the modulus of elasticity and ultimate strength as compared with tissue constructs in which components were randomly oriented. Moreover, tissue-engineered vascular media made from a prealigned living tissue sheet, produced with smooth muscle cells, possessed greater contractile capacity compared with those produced from living tissue sheets that were not prealigned.

Williams and Wick described a perfusion bioreactor system designed to address bioprocessing issues for tissue engineering of small diameter vascular grafts [204]. Sheets of PGA non-woven felts of 19×50 mm^2 were sutured into tubes (4.5 mm inner diameter and 50 mm long). Cylindrical bioreactors consisting of modules flanked by two head plates were made from hand-blown glass. Modules were custom made from glass flanges that had inner and outer diameters of 52 and 57 mm, respectively. Scaffolds were mounted on two glass tubes inserted through opposite sides of the module walls to provide medium perfusion through the lumen. Lumen flow allowed for mechanical stimulation (shear stress and pulsatile flow). Head plates had a 6.35-mm outer diameter glass port inserted through the face to allow medium flow around the external surface of the constructs for nutrient delivery. This external flow could also be used for cell seeding. A custom-built 2-l medium reservoir vessel was used to facilitate simultaneous perfusion of multiple constructs. The reservoir contained several 6.35-mm hose barb outlet ports for tubing connections and a male luer fitting, on which a 0.2-μm syringe filter was attached for gas exchange. A PGA scaffold was mounted between the glass tubes in each bioreactor module and held in place by cable ties. The medium reservoir was filled with 1l of complete culture medium and sealed with a lid that had four hose barb connectors. One of the connectors was used for air exchange in the incubator by attaching a 0.2-μm filter. The outer three ports were used for connections to pump medium between the reservoir and the bioreactor. To seed the polymer scaffold, an SMC suspension was pumped reciprocally through the lumen by a dual syringe pump. Simultaneously, medium was pumped continuously over the external surface of the scaffold by a peristaltic pump. After seeding, the syringe pump was replaced by a peristaltic pump to provide pulsatile medium flow through the construct lumen at a time-averaged flow rate of

40 ml/min and a frequency of 1.5 Hz. Culture medium was also pumped by a peristaltic pump across the external surface of the scaffold at 40 ml/min to provide nutrients to the external surface of the construct. During bioreactor operation, 1l of medium was replaced in the reservoir every 4–7 days in a sterile flow hood. In some experiments, ECs were perfused through the construct lumen reciprocally by a dual syringe pump at 4 ml/min 2 days before harvest. Cell proliferation was more than 3-fold after 4 days, SMCs expressed differentiated phenotype after 16 days, and collagen and elastin were distributed throughout the construct after 25 days of culture. In bioreactor experiments in which the construct lumen was seeded with ECs by perfusion after 13 days of SMC culture, EC seeding efficiency was 100%, and a confluent monolayer was observed in the lumen within 48 h.

Mauck *et al.* studied the role of growth factors in the development of engineered constructs subjected to physiologic deformational loading [205]. For this study chondrocyte-seeded agarose hydrogels and a culture medium supplemented with either TGF-β_1 or IGF-I were used. Under free-swelling conditions in control medium (C), the [proteoglycan content][collagen content][equilibrium aggregate modulus] of cell-laden 2% agarose constructs reached a peak of [0.54% wet weight][0.16%][13.4 kPa], whereas the addition of TGF-β_1 or IGF-I to the control medium led to significantly higher peaks of [1.18%][0.97%][23.6 kPa]$_{\text{TGF-}\beta}$ and [1.00%][0.63%][19.3 kPa]$_{\text{IGF-I}}$, respectively, by day 28 or 35. Under dynamic loading in control medium (L), the measured parameters were [1.10%][0.52%] [24.5 kPa], and with the addition of TGF-β_1 or IGF-I to the control medium these further increased to [1.49%][1.07%] [50.5 kPa]$_{\text{TGF-}\beta}$ and [1.48%][0.81%][46.2 kPa]$_{\text{IGF-I}}$, respectively. Type II collagen accumulated primarily in the pericellular area under free-swelling conditions, but spanned the entire tissue in dynamically loaded constructs. Applied in concert, dynamic deformational loading and TGF-β_1 or IGF-I increased the aggregate modulus of engineered constructs by 277 or 245%, respectively, an increase greater than the sum of either stimulus applied alone.

Jeong *et al.* hypothesized that a radial distention induces the phenotype of VSMCs in *in vitro*–engineered tissues to be similar to that of VSMCs in native tissues *in vivo* [206]. To test the hypothesis, rabbit aortic VSMCs were seeded onto an elastic P(LA/CL) (50:50) scaffolds. The seeded polymer scaffolds were connected to each end of sleeves coupled to pulsatile perfusion bioreactors, and incubated. The seeded scaffolds were initially placed statistically for 2 days with the medium changed every day. Then, a pulsatile flow of culture medium was applied in the perfusion system so that VSMCs on the scaffolds were subjected to radial distention for 8 weeks. The media flow rate was 130 ml/min and the pressure was 25 mmHg with a pulse of 1 Hz. The amplitude of radial distention was 5% of the initial radius. As control experiments, cells were grown under the static conditions. The pulsatile strain and shear stress enhanced the VSMC proliferation and collagen production. In addition, a significant cell alignment in a direction radial to the distending direction was observed in VSM tissues exposed to radial distention, whereas VSMCs in VSM tissues engineered in the static condition aligned randomly. The expression of SM α-actin, a differentiated phenotype of SMCs, was upregulated by 2.5-fold in

VSM tissues engineered under the mechano-active condition, compared to VSM tissues engineered under the static condition.

To examine the effect of mechanical conditioning on cell–scaffold constructs implanted in the body with the purpose of *in vivo* tissue engineering, Case *et al.* used a hydraulic bone chamber to investigate mechanical influence on *de novo* bone formation (see Fig. 2.8 in Chapter 2). The disassembled hydraulic bone chamber consisted of a hollow, titanium-threaded cylinder and a removable cap, with circular openings regularly spaced around the base of the chamber to enable cellular and vascular infiltration [207]. The chamber base with inner dimensions of 6.3 mm (both diameter and depth) was implanted securely into drilled and tapered holes into cancellous bone, while the chamber interior was exposed to an environment rich in vascularity and cellular activity. The chamber inside was subjected to a compressive mechanical stimulus by activating the hydraulic loading mechanism of the chamber with a piston positioned inside the chamber, separating the base and upper chamber areas. A barbed connector was threaded to the exterior of the chamber above the piston, and polyethylene tubing was attached to the barb. Tubing was routed subcutaneously from the chamber to an exit side on the animal back and externally connected to a solenoid-driven loading system controlled by pressure transducer feedback. Using this chamber and polymer discs seeded with articular chondrocytes, followed by culture for 4 weeks *in vitro*, Case *et al.* found that biopsies from the chamber after 4-week implantation into rabbit femoral metaphyses consisted of a complex composite of bone, cartilage, and fibrous tissue, with bone formation in direct opposition to the cartilage constructs. Application of an intermittent cyclic mechanical load was found to increase the bone volume fraction of the chamber tissue from 0.4 to 3.6% as compared with no-load control biopsies.

Grad *et al.* investigated the effect of articular motion on the gene expression of superficial zone protein (SZP) and hyaluronan synthases (HASs) and on the release of SZP and hyaluronan of chondrocytes seeded onto biodegradable scaffolds [208]. Cylindrical (8×4 mm^2) porous PU scaffolds were seeded with bovine articular chondrocytes (BACs) and subjected to static or dynamic compression, with and without articulation against a ceramic hip ball. After loading, the mRNA expression of SZP and HASs was analyzed, and SZP immunoreactivity and hyaluronan concentration of conditioned media were determined. Surface motion significantly upregulated the mRNA expression of SZP and HASs. Axial compression alone had no effect on SZP and increased HAS mRNA only at high strain amplitude. The SZP was immunodetected only in the media of constructs exposed to surface motion. The release of hyaluronan into the culture medium was significantly enhanced by surface motion.

A multi-cue bioreactor (MCB) was designed by McCulloch *et al.* to apply pressure and strain conditions to a cell-seeded scaffold constrained in an environment physiologically similar to that of the naturally occurring counterpart [209]. The major components of MCB are shown in Fig. 3.88. The device emulates the pressure and straining environment found at the aortic root. Aortic and pulmonary arteries obtained from freshly isolated porcine hearts were subjected to various loading regimens (Δpressure/flow/force). Through analyzing data acquired by the

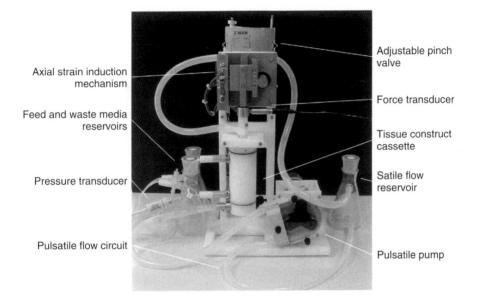

Axial strain induction mechanism

Feed and waste media reservoirs

Pressure transducer

Pulsatile flow circuit

Adjustable pinch valve

Force transducer

Tissue construct cassette

Satile flow reservoir

Pulsatile pump

Fig. 3.88 Major components of the multi-cue bioreactor (MCB).

MCB transducer array it was possible to differentiate the dynamic mechanical properties of the tissue types tested.

6.6. Cell counting and distribution in scaffolds

Stabler *et al.* investigated the utility of ^{1}H NMR spectroscopy to noninvasively quantify viable cell number in tissue-engineered substitutes *in vitro* [210]. Agarose disc-shaped constructs containing βTC3 cells were employed as the model tissue-engineered system. Two construct prototypes containing different initial cell numbers were monitored by localized, water-suppressed ^{1}H NMR spectroscopy over the course of 13 days. The ^{1}H NMR measurements of the total choline resonance at 3.2 ppm (TCho) were compared with results from the traditional cell viability assay MTT and with insulin secretion rates. Figure 3.89 shows a strong linear correlation between TCho and MTT. Ng *et al.* showed that typical cell proliferation assays did not necessarily correlate linearly with increasing cell densities or between 2-D and 3-D cultures, and were either not suitable or only rough approximations in quantifying actual cell numbers in high cell density and 3-D cultures [211].

As for autologous bone transplants, cell survival is questionable because inside large-sized grafts the survival of cells will be compromised, as the cells will be inside the harsh environment of a hemotoma, without vascularization during the first week after transplantation. To effectively monitor the cells *in vivo*, labeling of cells is required. Aandrade *et al.* showed that the fluorescent CM-DiI label was maintained within the cell membrane after dehydration for paraffin embedding

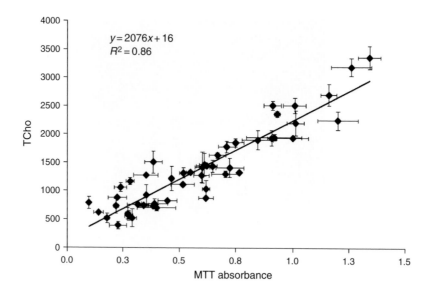

Fig. 3.89   Correlation between MTT assay absorbance values and TCho peak areas for βTC3 cells in agarose constructs for all time points, construct types, and densities tested.

[212]. For investigation of cell survival and differentiation, Kruyt *et al.* chose this CM-DiI as a label [213]. The CM-DiI solution was combined with the suspension of goat bone-marrow cells. With label concentrations above 20 μM, more than 75% of cells maintained the label for at least four population doublings, but *in vitro* transfer experiments demonstrated label transfer between vital cells when the label concentration was increased from 10 to 40 μM. Porous HAp scaffolds were seeded with labeled cells, and up to 6 weeks after implantation in nude mice cells could be treated inside tissue-engineered bone. However, transfer of the label from labeled to unlabeled cells occurred also *in vivo*. This label transfer was detected both between vital cells and between dead and living cells. *In vivo*, transfer was found 1 week after implantation of dead labeled cells. Therefore, it seems that studies of cell survival, using the CM-DiI label, are considered feasible only during the first few days after implantation.

    Washburn *et al.* applied two complementary, 3-D microscopic techniques, magnetic resonance microscopy (MRM) with a resolution on the order of tens of microns and X-ray microtomography (XMT), to map the distribution of newly formed bone in polymer scaffolds [214]. They monitored protons associated with water and generated maps of the magnetization transfer ratio (MTR) and magnetic resonance (MR) relaxation times T1 and T2, which provide information about protein deposition and mineral formation, respectively. The XMT generates 3-D maps of linear absorption coefficients (LACs) based on the attenuation of X-rays. The mineral concentration in each voxel was calculated from the known cross-sections of X-ray absorption of each element in the absorbing phase. Porous poly(ethyl methacrylate) (PEMA) scaffolds were seeded with primary chick calvarial osteoblasts and cultured under static conditions for up to 8 weeks. Bone formation within porous scaffolds was confirmed by the application of histologic stains to intact PEMA discs. Discs were treated with Alizarin red to visualize calcium deposits and with Sirius red to visualize regions of collagen deposition. The DNA analysis confirmed cell confluence on the scaffolds after 7 weeks in static culture. Quantitative MRM maps of MTR yielded maps of protein deposition, and T1 and T2 yielded maps of mineral deposition. Figure 3.90 shows the approximate location from which XMT mineral concentration and T2 profiles were extracted. The location of newly formed bone and local mineral concentrations were confirmed by XMT. By comparing MRM and XMT data from selected regions-of-interest in one sample, the inverse relationship between the MR relaxation times and mineral concentrations was validated, and calibration curves for estimating the mineral content of cell-seeded scaffolds from quantitative MRM images were developed. Ramrattan *et al.* described a histomorphometric method that examines a number of computer-defined concentric zones, based on the distance of a pixel from the scaffold edge

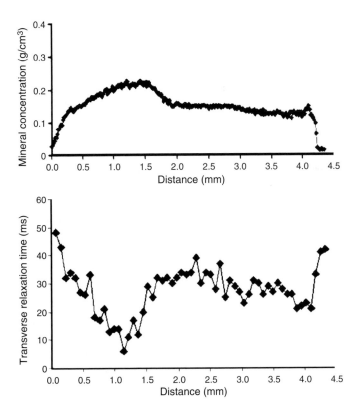

Fig. 3.90 XMT mineral concentration and T2 profiles calculated from linear absorption coefficient maps.

[215]. Each zone was automatically analyzed for tissue content, eliminating the need for user definition of a tissue ingrowth front and thus reducing errors and observer dependence.

Blum *et al.* investigated the utility of genetically modified cells expressing reporter genes for evaluating cell growth on materials *in vitro* and *in vivo* [216]. The Fisher rat fibroblastic cell line was genetically modified to express both EGFP and luciferase. They did not find particulates from the materials evaluated to interfere with the fluorescence or luminescence measurements of the reporter genes from the cellular extracts. The oligo PEG fumarate (OPF) hydrogels and titanium mesh scaffolds were used as cell carrier materials in the study. On both materials, cells were viewed microscopically via fluorescence. The Nightowl molecular light imaging equipment was used to visualize cells on the materials macroscopically both by luminescence and by fluorescence. In addition, images acquired by the molecular light imaging system were quantified. Both cells within hydrogels and on titanium mesh scaffolds were evaluated for cellularity in culture. Implants in rats containing genetically modified cells could be both qualitatively and quantitatively evaluated over 28 days.

7. EXAMPLES OF CELL CULTURE

A variety of cells have been studied for cell culture in tissue engineering. The following represents some examples.

7.1. Differentiated Cells

7.1.1. Muscular Cells

Bursac *et al.* tested the hypothesis that cardiac myocytes maintain their electrophysiological and molecular properties better if cultured on 3-D scaffold in bioreactors than if cultured as confluent monolayers [217]. The experimental design is demonstrated in Fig. 3.91. They isolated cardiomyocytes from the lower portions of ventricles of 2-day-old rats. The cardiomyocytes were cultured on discs (5 mm in diameter and 2 mm thick) of PGA mesh, coated with LN, in bioreactors (100-ml high aspect ratio vessel [HARV], Synthecon) rotating at 12 rpm for 1 week. The resulting tissue constructs were compared with confluent monolayers on LN-coated glass coverslips and with slices of native neonatal rat ventricular tissue with respect to proteins involved in cell metabolism (creatine kinase isoform-MM, CK-MM), contractile function (sarcomeric myosin heavy chain, MHC), and intercellular communication (connexin 43, Cx-43), and action potential characteristics (*e.g.*, membrane resting potential or MRP; maximum depolarization slope or MDS; action potential duration or APD) and macroscopic electrophysiological properties (maximum capture rate or MCR). As shown in Fig. 3.92, tissue constructs expressed Cx-43 at levels between those of monolayers and neonatal ventricles. Construct levels of CK-MM, MHC, and Cx-43 were 40–60% as high as those of ventricles, whereas monolayer levels were only 11–20% as high. Tissue constructs exhibited spontaneous contractile activity at the early stages

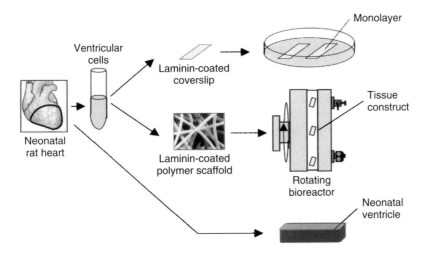

Fig. 3.91 Experimental design.

(*e.g.*, days 2–4) of culture. However, spontaneous contractions ceased by day 7. In contrast, monolayers contracted synchronously at rates ranging from 30 to 160 beats/min (brp), starting on day 1–2 and continuing through day 7. In tissue constructs and neonatal ventricles, recorded action potentials were characteristic of electrically quiescent non-pacemaker (NP) myocytes, but, in monolayers, 10–20% of the examined cells generated action potentials characteristic of electrical pacemaker (P) cells. The MRP and APD values were comparable in all three experimental groups. Tissue construct APDs were in between those of monolayers and native ventricles at all phases of repolarization.

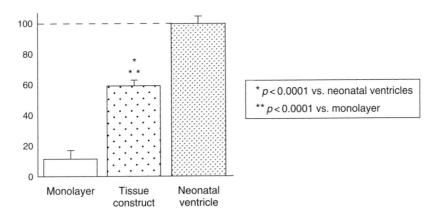

Fig. 3.92 Levels of connexin-43 assessed by scanning of Western blots (normalized per unit of total protein and expressed as a fraction of that in neonatal ventricles).

Radisic *et al.* attempted to optimize seeding of 3-D scaffolds with hypoxia-sensitive cells at physiological densities [218]. Scaffolds were discs punched from sheets of Ultrafoam™ collagen hemostat, a water-insoluble, partial HCl salt of purified bovine dermal collagen formed as a sponge with interconnected pores. Pre-wetted scaffolds were gently blotted dry, and the Matrigel/cell suspension was delivered evenly to the top surface of each scaffold using an automatic pipette. Gelation was achieved within 10 min of the transfer of inoculated scaffolds to either orbitally mixed dishes or perfused cartridges. Gel–cell-inoculated scaffolds were placed between two stainless steel screens and two silicone gaskets in polycarbonate perfusion cartridges (one scaffold per cartridge). As given in Table 3.22, medium perfusion maintained the viability of cardiac myocytes during both the seeding (1.5 and 4.5 h) and the cultivation (7 days), while the cell yield remained comparable for constructs cultured in perfusion and in dishes. For longer period (4.5 h), the decrease in viability was 3- or 9-fold higher for constructs seeded in dishes as compared to those seeded at low or high perfusion flow rate, respectively. Such profound differences were most likely the result of insufficient diffusional supply of oxygen within constructs seeded in orbital dishes as compared to those seeded in perfusion, where oxygen and nutrients are supplied by a combination of diffusion and convection. Relatively smaller decrease in viability in perfusion groups at longer seeding times (4.5 h) as compared to shorter seeding time (1.5 h) was consistent with the fact that damaged or dead cells were gradually washed away by flow leaving only healthy and viable cells on the scaffold. If not removed promptly from the tissue construct, the damaged cells could potentially send apoptosis signals to the neighboring healthy cells or release harmful intracellular compounds into the environment, which could cause further damage. Most of the cells in the constructs seeded in dishes were located in the 100- to 200-μm-thick top layer. When the cell–gel suspension was applied to pre-wetted scaffolds, about half

Table 3.22

Seeding and cultivation of cardiomyocytes in perfused cartridges[a]

Time	Seeding Set-up	Total Cell Yield	Change in Viability (%)	Live Cell Yield	L/G (mol/mol)	Glucose Consumption Rate (μmol/h)
1.5 h	Dishes (25 rpm)	1.03 ± 0.11	25.79 ± 1.24	0.70 ± 0.09	1.04 ± 0.08	2.47 ± 0.21
	Perfusion (0.5 ml/min)	0.89 ± 0.12	11.59 ± 9.05*	0.76 ± 0.05	0.81 ± 0.09*	2.22 ± 0.37
7 days	Dishes (25 rpm)	0.59 ± 0.09	33.03 ± 9.39	0.35 ± 0.05	2.26 ± 0.30	0.42 ± 0.09
	Perfusion (0.5 ml/min)	0.48 ± 0.04	−1.16 ± 5.35*	0.50 ± 0.09*	1.18 ± 0.49*	0.49 ± 0.30

* Significantly different than dish ($p < 0.05$).

[a] Cell numbers, viability, and metabolic activity measured for constructs inoculated with 12×10^6 neonatal cardiac myocytes and seeded (1.5 h) and cultured (7 days) in perfused cartridges (0.5 ml/min) or in orbitally mixed dishes (25 rpm). Data are average ± SD ($n = 3$).

of it penetrated the constructs by capillary forces, and the rest formed a layer on the top surface. In contrast, flow helped distribute the high-density cell patches evenly across the central part of the constructs seeded in perfusion. In response to electrical stimulation, perfused constructs made from neonatal rat cardiomyocytes contracted synchronously, had lower excitation thresholds (ETs), and recovered their baseline function levels of ETs and MCR after treatment with a gap junctional blocker; dish-grown constructs exhibited arrhythmic contractile patterns and failed to recover their baseline MCR levels.

Kofidis *et al.* investigated pulsatile tissue culture perfusion to manufacture full-thickness 3-D myocardial grafts [219]. Their bioreactor consisted of plain, circular chambers with transparent walls. Multiple chambers could be assembled parallel to each other, without communication, and each chamber had a closeable inlet and an outlet. A rat aorta was mounted between inlet and outlet lines at the beginning of the study. Syngeneic rat cardiomyocytes were mixed into a fibrin glue scaffold within the chamber, using a Y-shaped applicator. The vessel was supplied with perfusate by a pulsatile micropump connected to the inlet. Two weeks after continuous perfusion, the contents of every single chamber were analyzed by fluor-deoxy-glucose-positron-emmision-tomography (FDG-PET). Cardiomyocytes were isolated from neonatal Wistar rat hearts with the harvest rate over 90%. The core vessel was an abdominal aorta isolated from adult Wistar rats. In a sterile hood, two blunt needles were inserted into the input and output foramens of each chamber. The input needle was larger in diameter than the output needle (16G versus 18G). An approximately 2-cm segment of rat aorta was mounted inside the chamber on the blunt ends of the needles using Histoacryl glue and was tied with suture. A fibrin-glue applicator was used to deliver the cell/matrix mixture. One of the two syringes was filled with 2 ml of culture medium containing 10^7 cells/ml while the second syringe was filled with fibrin glue. The mixture of cells, culture medium, and fibrin glue was administered slowly into the chamber in sufficient amounts to cover the core vessel of the bioreactor, and sub-sequently it was left to consolidate. A flask filled with MEM, FCS, and BrdU was served as the supply reservoir and was connected to the input line of the bioreactor chamber by silicone tubing. After 30 min of consolidation, perfusion was initiated. Perfused and unperfused bioreactors were placed in an incubator, along with their cor-responding perfusion lines and supply flasks. Thirty-six hours after the start of perfu-sion, solid blocks of 8.5 ± 1.2 mm thickness were obtained, with mechanical stability significant enough to allow for rigorous manipulation without decomposition of the graft. Mean cellular population of each block was 10^7 cells/mm^3, when the solid graft was further perfused in the same chamber for up to 2 weeks. Cell population was higher in the vicinity of the core vessel which had the mean diameter of 1.8 ± 0.4 mm. The FDG-PET, which can assess tissue viability in engineered tissues as small as a few millimeters, maintaining the viability of the *in vitro* construct during measure-ment, revealed enhanced metabolic activity in perfused chambers.

Watzka *et al.* showed that under favorable conditions a 3-D culture of myocardial cells led to the establishment of a rudimentary capillary network within tissue aggre-gates, which presumably guaranteed sufficient tissue perfusion up to a maximum aggregate diameter of 900 μm [220]. During 3 weeks of culture of myocardial tissue from newborn mice, a mean 24% of all aggregates contracted spontaneously. The contracting aggregates displayed a tissue-like architecture with small basal and apical zones, and a large central zone. The basal and apical zone consisted of immature mesenchymal cells. The underlying shell of the aggregate contained many cardiomyocytes. Vessel-like structures were found concentrated within the aggregates. Up to 15% of the cells in the central zone of the aggregate were positive for the endothelial-specific BS-I lectin. Vessel-like structures were formed by cells, which often showed intracytoplasmatic lumena. Surrounding the neocapillaries, structures of a rudimentary basal membrane could be detected.

Hubschmid et al. investigated in vitro growth of human urinary tract SMCs under static conditions and mechanical stimulation. The cells were cultured on type I collagen—and LN-coated silicone membranes [221]. Using a Flexcell device for mechanical stimulation, a cyclic strain of 0–20% was applied in a strain–stress–time model (stretch, 104 min relaxation, 15 s), imitating physiological bladder filling and voiding. Cell proliferation and α-actin, calponin, and caldesmon phenotype marker expression were analyzed. Nonstretched cells showed significant better growth on LN during the first 8 days, thereafter becoming comparable to cells grown on type I collagen. Cyclic strain significantly reduced cell growth on both surfaces; however, better growth was observed on LN. Neither the type of surface nor mechanical stimulation influenced the expression pattern of phenotype markers; α-actin was predominantly expressed. Coating with the ECM protein LN improved in vitro growth of human urinary tract smooth muscle cells.

7.1.2. Fibroblasts

Wang et al. studied dermal tissue formation by dynamic seeding of fibroblasts in spinner flasks up to 16 scaffolds in one bioreactor, as illustrated in Fig. 3.93 [183]. This seeding system may be advantageous in the evaluation of in vitro tissue formation and circumvents the need to take samples from the same scaffold (increased group number) at different time intervals. The scaffolds were 330–380 μm thick and 1.55 cm in diameter with 70–80% porosity and pore sizes ranging from 50 to 200 μm. The scaffold disks ($n = 16$) were fixed onto 12.5-cm-long, 22-gauge needles attached to the lid of a spinner flask. Dynamic seeding of human dermal fibroblasts was performed in 125 ml of fibroblast culture medium per flask and at 20, 40, and 60 rpm. High seeding efficiencies with a homogeneous fibroblast distribution could be achieved at a stirring speed of 40 rpm and, in addition, when scaffolds were located in the middle region and with an interscaffold distance of at least 5 mm, multiple scaffolds ($n = 16$) could be identically seeded in the same bioreactor. Stirring at more than 10 rpm significantly stimulated fibroblast proliferation and GAG and collagen content in the cultured scaffolds. A stirring speed of 80 rpm showed the highest collagen deposition in the scaffold after 21 days compared with the other stirring speeds investigated.

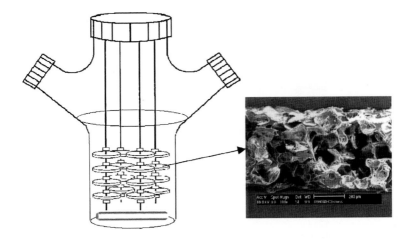

Fig. 3.93 Schematic representation of seeding system for tissue engineering of dermal substitutes in spinner flask.

Garvin *et al.* fabricated a tendon tissue using native tendon cells suspended in a type I collagen matrix that could be readily subjected to regulated, cyclic, mechanical loading [222]. Tendon internal fibroblasts isolated from the flexor digitorium profundus tendons of chickens were suspended in type I collagen mixed with growth medium and neutralized to pH 7.0, and apportioned into each well of a culture plate. Linear, tethered, 3-D cell-populated matrices were formed by placing the culture plate atop a four-place gasketed baseplate with planar-faced cylindrical posts inserted into centrally located, rectangular cutouts beneath each flexible well base, as shown in Fig. 3.94. The trough loaders had vertical holes in the floor of the rectangle through which a vacuum could be applied to deform the flexible membrane into the trough. The trough provided a space for delivery of cells and matrix. The baseplate was transferred to an incubator, where the construct was held in position under vacuum for 1.5 h until the cells and matrix formed a gelatinous material. Bioartificial tendons (BATs) were then covered with growth medium and the plates were returned to the incubator. After 24 h in culture the matrix and cell attachments to the anchor points were mechanically bonded and secured. The BATs were uniaxially loaded by placing arctangle loading posts beneath each well of the culture plates in a gasketed baseplate and applying vacuum to deform the flexible membranes downward at the east and west poles. The flexible but inelastic non-woven nylon mesh anchors deformed downward along the long sides of the arctangle loading posts thus applying uniaxial strain along the long axis of each BAT. The loading regimen was 1 h/day at 1% elongation and 1 Hz. *In vivo* data indicated that the natural strain in flexor tendons is about 1% with a normal operating limit up to 3–4%. The cell-populated collagen gel construct was broken at less than 0.7–1 mPa. The selected load duration was sufficient to stimulate gene expression changes. Analyses of tendon internal fibroblasts grown in BATs with an initial seeding density of

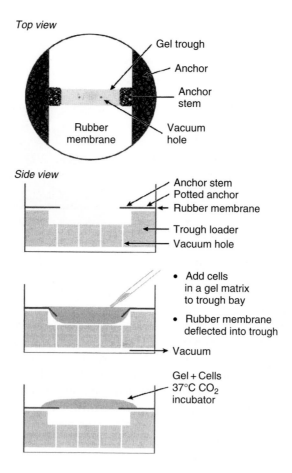

Fig. 3.94   Tissue Train six-well culture plate.

200,000 cells and of cells grown in 2-D monolayers demonstrated the typical lag, log, and stationary phases of a traditional growth curve. However, BATs with an initial seeding density of 500,000 cells did not demonstrate a typical log phase, but rather remained in a stationary phase, indicating that a comparable cell-to-matrix ratio was maintained although the initial seeding densities differed. The tenocytes dispersed in the collagen gel remodeled and contracted their matrix by an 82% reduction in area over an 8-day period, confirming that matrix contraction by fibroblasts is typically rapid in the first week of culture. As the fibroblasts exerted traction on the collagen matrix, the matrix was consolidated in the unconstrained portions of the culture. The expression patterns of genes encoding type I collagen, type III collagen, β–actin, and decorin were the same when comparing the RNA isolated from cells in BATs with that of cells in either a 2-D monolayer or native whole tendon. Expression patterns of the genes encoding for tenassin, FN, and type XII collagen were the same when compared with cells grown in either monolayer or

3-D BAT cultures. The elastic moduli of the BATs were significantly lower than that of native tendon. Moduli for various native whole tendons are reported to be 1.5 GPa for *in vitro* testing and 1.2 GPa at maximum forces *in vivo*. The biomechanical strength and moduli of the BATs were increased by applying cyclic mechanical strain *in vitro*, as shown in Table 3.23.

To generate myofibroblasts for potential use in a tissue-engineered cardiac valve replacement, Hoffman-Kim *et al.* used tissue biopsies of clinically appropriate sizes obtained from juvenile sheep [223]. Cells obtained from three tissue sources—tricuspid valve leaflet, carotid artery, and jugular vein (JV)—exhibited a myofibroblast phenotype *in vitro*, as demonstrated by their immunoreactivity with antibodies directed against vimentin, α-smooth muscle actin, FN, and chondroitin sulfate. Protein synthesis characteristics were defined for the key ECM components: collagen, GAGs, and elastin. Among the three sources, JV generated the highest numbers of cells, and JV cells produced the largest amount of collagen per cell, as shown in Fig. 3.95.

Witte and Kao developed photopolymerizable interpenetrating polymer network (IPN) systems composed of PEGDA and gelatin modified with PEG and/or polyanions. The PEG-based IPNs were utilized as a platform for the delivery of keratinocyte-active factors to study the effect of exogenous keratinocyte-active factors on the keratinocyte behavior and the keratinocyte–fibroblast paracrine relationship [224]. Adherent keratinocyte density on a polystyrene culture dish and GA-fixed gelatin hydrogels but not on IPN was significantly increased with culture time in the presence of growth supplements independent of the released KGF from the gelatin hydrogel and IPN. In the presence of fibroblasts, adherent keratinocyte density on gelatin hydrogels was higher than that without fibroblasts. This phenomenon was not observed on IPN and polycarbonate membrane. The delivered exogenous KGF (*i.e.*, released from a biomaterial matrix) operated in tandem with fibroblasts in regulating keratinocyte activation (*i.e.*, IL-1β release and adhesion) in a surface-dependent manner.

7.1.3. Chondrocytes

Malda *et al.* identified a suitable microcarrier for proliferation of chondrocytes and evaluated whether chondrocytes maintained their differentiated phenotype while proliferating on the microcarriers [225]. Dextran-based carriers (Cytodex) showed overall better cell attachment in comparison with plastic-based carriers, except for Cytodex 2 and Plus Coated carriers. Naumann *et al.* explored the use of a new

Table 3.23
Comparison of modulus of elasticity and ultimate tensile strength results for mechanically conditioned and control specimens on day 7

	Load	No Load
Day 7 elasticity (MPa)	1.80 ± 1.82	0.49 ± 0.24
Day 7 ultimate tensile strength (kPa)	327.65 ± 172.03	112.20 ± 6.07

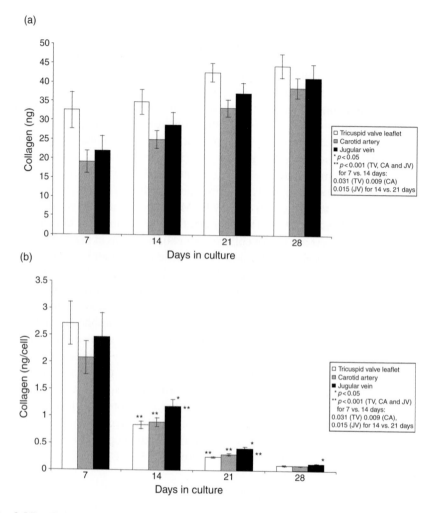

Fig. 3.95 Cells from all three sources produced collagen (a), while collagen production per cell from isolated cells decreased with time in culture (b).

culture system, which allowed isolated and expanded human chondrocytes to be redifferentiated for engineering of cartilage grafts without the use of biomaterial scaffolds [226]. A 3-D *in vitro* macroaggregate culture system with isolated human nasoseptal and auricular chondrocytes was developed. The cells were enzymatically isolated and amplified in conventional monolayer culture before the cells were seeded into a cell culture insert with a track-etched membrane. After 3 weeks of *in vitro* culture, nasoseptal and auricular chondrocytes synthesized new cartilage with the typical appearance of hyaline nasal cartilage and elastic auricular cartilage. Immunohistochemical staining of cartilage samples showed a characteristic pattern of staining for collagen antibodies that varied in location and intensity. In all samples, intense staining for cartilage-specific, types I, II, and X collagen was observed. The measurement of total GAG content demonstrated higher GAG content for reformed nasoseptal cartilage compared with elastic auricular cartilage. However, the total GAG content of engineered macroaggregates was lower than that of native cartilage. Under the tare load, the surface deformed at a decreasing rate, but sample displacement never came to equilibrium, although auricular macroaggregates demonstrated good stability and elasticity. Therefore, it was not possible to obtain valid measurement of mechanical properties such as aggregate modulus, Poisson's ratio, and permeability of native cartilage.

With aging, the capacity of chondrocytes to synthesize some types of proteoglycans and their response to stimuli, including growth factors, decrease. These age-related changes may limit the ability of the cells to maintain the tissue, and thereby contribute to the development of degeneration of the articular cartilage. Marlovits *et al.* studied chondrogenesis by aged human articular chondrocytes *in vitro* [227]. Articular chondrocytes isolated from 10 aged patients (median age, 84 years) were increased in monolayer culture. A single-cell suspension of dedifferentiated chondrocytes was inoculated in a rotating wall vessel, without the use of any scaffold or supporting gel material. After 90 days of cultivation, a 3-D cartilage-like tissue was formed, encapsulated by fibrous tissue resembling a perichondrial membrane. Morphological examination revealed differentiated chondrocytes ordered in clusters within a continuous dense cartilaginous matrix demonstrating a strong positive staining with monoclonal antibodies against type II collagen and articular proteoglycan. The surrounding fibrous membrane consisted of fibroblast-like cells, and showed a clear distinction from the cartilaginous areas when stained against type I collagen. The TEM revealed differentiated and highly metabolically active chondrocytes, producing an ECM consisting of a fine network of randomly distributed cross-banded collagen fibrils. Dorotka *et al.* evaluated the morphologic and biochemical behavior and activity of human chondrocytes taken from nonarthritic and osteoarthritic cartilage and seeded on a 3-D matrix consisting of types I, II, and III collagen [228]. Human articular chondrocytes were isolated from either nonarthritic or osteoarthritic cartilage of elderly subjects, and from nonarthritic cartilage of an adolescent subject, seeded on collagen matrices, and cultured for 12 h, 7 days, and 14 days. Chondrocytes of nonarthritic cartilage revealed a larger number of spherical cells, consistent with a chondrocytic phenotype. The biochemical assay showed a net increase in GAG content in nonarthritic chondrocytes, whereas almost no GAGs were seen in osteoarthritic cells. The DNA results suggest that more osteoarthritic cells than chondrocytes from nonarthritic cartilage attached to the matrix within the first week.

Veilleux *et al.* evaluated the effects of passage number on the proliferative, biosynthetic, and contractile activity of adult canine articular chondrocytes grown in types I and II collagen–GAG matrices that were crosslinked by dehydrothermal/CDI treatment [229]. As shown in Fig. 3.96, P0 (freshly isolated), P1 (passage 1), and P2 (passage 2) cells seeded on the type II matrices continued to proliferate over a 4-week period, but thereafter the P0 and P1 cells continued to increase in number and the P2

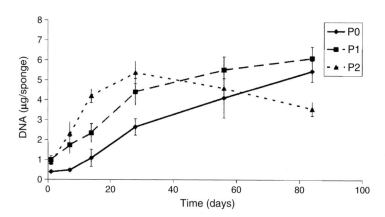

Fig. 3.96   DNA content of seeded type II collagen–GAG matrices.

cells decreased. At 4 weeks the DNA contents of the types I and II matrices seeded with P1 and P2 cells were comparable, and higher than the values for matrices seeded with freshly isolated chondrocytes. The rates of protein and GAG synthesis by the P1 and P2 cells were comparable, and higher than the rates for the P0 chondrocytes, after 1 week, and the rates were generally higher in the type II than in the type I collagen scaffolds. Western blot analysis demonstrated the presence of newly synthesized type II collagen in type II matrices in which P1 and P2 cells were grown.

To develop a method for rapid expansion of human auricular chondrocytes, Kamil *et al.* quantified the ability of *in vitro* chondrocytes multiplied by a recycled supernatant technique to generate neocartilage *in vivo*, because they had observed that enhanced cell replication could be achieved by recycling some of the used cell medium [230]. They obtained auricular chondrocytes from patients and created two groups. Group A chondrocyte number was increased by repeated passaging. Group B cells were grown from floating culture medium and their number was increased both by passaging and by repeated recycling of the culture medium. Chondrocytes from both groups were implanted in nude mice for 8 weeks to generate tissue-engineered cartilage. Flow cytometry studies performed on both groups confirmed the presence of two distinct populations of structures as the source of chondrocytes from the recycled medium.

Giannoni *et al.* assessed whether the extent of dedifferentiation of chondrocytes cultured *in vitro* varies between species and they hypothesized that the level of chondrocyte phenotype stability during expansion may contribute to the maintenance of their chondrogenic commitment and dedifferentiation potential [231]. Condyle chondrocytes were harvested from sheep, dog, and human, and expanded for 1, 6, or 12 cell duplications. At each interval, cell phenotype was monitored (morphology and biosynthesis of cartilage markers) and dedifferentiation was assessed by an *in vitro* assay of chondrogenesis in micromass pellet and an *in vivo* assay of ectopic cartilage formation in immunodeficient mice. During culture, the sheep chondrocyte phenotype was maintained better than that of human chondrocytes, which in turn dedifferentiated to a lesser extent than dog chondrocytes. Darling and Athanasiou cultured zonal articular chondrocytes on tissue culture plastic, collagen II-coated polystyrene, and aggrecan-coated polystyrene in an effort to find a surface that can either prevent or slow the loss of phenotype [232]. In addition, they encapsulated passaged cells in agarose to examine the effect of 3-D culture on dedifferentiating zonal chondrocytes. Tissue culture plastic and the collagen II–coated surface induced rapid loss of phenotype in zonal articular chondrocytes. The aggrecan-coated surface had a less detrimental effect on the chondrocytic phenotype of seeded cells, inducing gene expression characteristics comparable to those of agarose-encapsulated cells. Furthermore, when chondrocytes that had been previously passaged on a type II collagen surface were placed on an aggrecan surface, the zonal cells showed a dramatic change in gene expression from fibroblastic to chondrocytic.

Mandl *et al.* performed a comparative study to gain insight into the effect of seeding density and passaging on the capacity of chondrocytes to dedifferentiate [233]. As a guideline, they calculated that, at minimum, 20-fold multiplication is needed to fill an average cartilage defect of 4 cm^2 with the amount of donor chondrocytes they obtained. They used ear chondrocytes isolated from five children. Four different seeding densities in monolayer culture were used, ranging from 3500 to 30,000 cells/cm^2. Both passaging and decreasing seeding density yielded an increase in expanded chondrocytes, but at the same time decreased the dedifferentiation capacity. In addition, with lower seeding densities sufficient multiplication (20 times) was found to attain in less time and with less passaging than at higher seeding densities.

At the molecular level cyclin-dependent kinase inhibitors (CDKIs) are involved in mediating growth arrest in the G_1 phase of the cell cycle. Using ribonuclease protection assays and immunocytochemical staining methods, Loewenheim *et al.* analyzed expression profiles of G_1 cell cycle inhibitors at the mRNA and protein levels [234]. Analysis was carried out in proliferating, quiescent, and senescent states of primary cultures of adult human nasoseptal chondrocytes. The inhibitors included the CDKIs of the CIP/KIP family (p21^{CIP1}, p27^{KIP1}, and p57^{KIP2}) and the INK4 family (p15^{INK4b}, p16^{INK4a}, p18^{INK4c}, and p19^{INK4d}) as well as the retinoblastoma protein-family (pRb, p107, and p130) and the tumor suppressor p53. The most pronounced effect between cultures in proliferation and cultures in growth arrest was an increased expression of the CDKIs p57^{KIP2} and p15^{INK4b} for quiescent growth arrest, and of p16^{INK4a}, p15^{INK4b}, and p57^{KIP2} for senescent growth arrest. Thus, these cell cycle inhibitors represent potential candidates for selective intervention to promote cellular multiplication of chondrocytes undergoing *in vitro* expansion for tissue engineering applications.

To evaluate the degree of cellular dedifferentiation, Kino-Oka *et al.* conducted subculture of chondrocytes on a surface coated with type I collagen at a density of 1.05 mg/cm^2 [235]. In the primary culture, most of the cells were round in shape on the collagen substrate, whereas fibroblastic and partially extended cells were dominant on the polystyrene plastic (PS) substrate. The round-shaped cells on the collagen substrate were hemispherical with nebulous and punctuated F-actin filaments, whereas the fibroblastic cells on the PS substrate were flattened with fully developed stress fibers. Although serial passages of chondrocytes through sub-cultures on the collagen and PS substrates caused a decrease in the number of round-shaped cells, the morphological change was appreciably suppressed on the collagen substrate, as compared with that on the PS substrate. It was found that only round-shaped cells formed type II collagen, which supports the view that cellular dedifferentiation can be suppressed to some extent on the collagen substrate. Three-dimensional cultures in collagen gel were performed with cells isolated freshly and passaged on the collagen or PS substrate. Cell density at 21 days in the culture of cells passaged on the collagen substrate was comparable to that in the culture of freshly isolated cells, in spite of a significant reduction in cell density observed in the culture of cells passaged on the PS substrate. In addition, the expression of GAGs and type II collagen was of significance in the collagen gel with cells pas-saged on the collagen substrate, and likewise in the gel with freshly isolated cells.

Pangborn and Athanasiou studied ECM component uptake by meniscal fibro-chondrocytes when stimulated with PDGF AB, TGF-β_1, insulin-like growth factor type I, and FGF-2 at various concentrations (low, medium, and high levels for each) [236]. Growth factors were applied to monolayer cultures for 3 weeks in a soluble form as part of the culture medium. And TGF-β_1 was the only growth factor that increased the uptake of both components, showing the most consistent behavior and the highest response. There was no conclusive evidence whether the high concentra-tion of TGF-β_1 (100 ng/ml) was better than the medium concentration (10 ng/ml).

7.1.4. Bone Cells

Hofmann *et al.* cultivated human osteoblasts on three different biomaterials includ-ing human autoclaved cancellous bone (AB), human demineralized cancellous bone (DMB), and HAp ceramic cylinders (HA) [237]. The cells were cultured in a perfu-sion chamber to guarantee a constant supply of nutrients. They studied cell prolifer-ation and the expression of OCN, OP, BMP-2A, ALP, and VEGF as parameters for osteoblast function and viability. The MTT assay showed that the three different biomaterials did not alter the proliferation of the osteoblasts. In contrast, gene expression of OCN, OP, BMP-2A, and VEGF was significantly higher for DMB plasma-sterilized at 55°C, as used as biomaterial, compared with the autoclaved AB and HA matrix. Expression of bone-specific ALP was very low in osteoblasts grown on highly porous HA. The mRNA expression of VEGF was similar in all three groups without significant differences. Osteoblasts cultured on bone matrix contain-ing biologically active proteins using the perfusion chamber produced bone pieces of 1.0×0.75 cm^2 size within ten days.

Arnold *et al.* studied the influence of three growth factors, factor XIII, TGF-β, and bFGF, on the proliferation and osteogenic differentiation of porcine periosteal cells in beads made from fibrin, alginate, and HAp [238]. The beads incorporating growth factors were prepared by dropping cell suspension in fibrinogen, alginate, HAp powders, and EDTA into a mixture of CaCl$_2$ and thrombin. The cells were

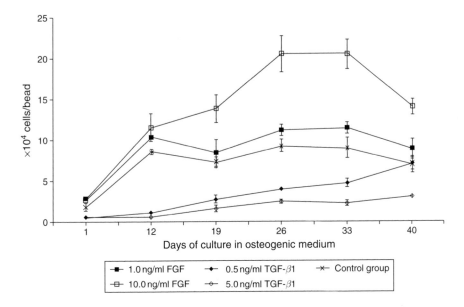

Fig. 3.97 Cell counts per bead depending on cultivation period and added growth factor, measured at days 1, 12, 19, 26, 33, and 40.

cultured in a 3-D carrier matrix (bead) in the presence of growth factors. As shown in Fig. 3.97, cell proliferation was accelerated by the presence of bFGF and TGF-β1. With regard to ALP activity, factor XIII led to significantly higher values, while bFGF and TGF-β1 resulted in lower activities. The OCN content was significantly increased by the application of bFGF. There was no universally applicable growth factor for optimizing cell growth *in vitro* and directing the osteogenic differentiation of periosteal cells.

Jones *et al.* illustrated the utility of micro-computed tomography (micro-CT) to study the process of tissue-engineered bone growth [239]. The facility was capable of acquiring 3-D images made up of 2000^3 voxels on specimens up to 60 mm in extent with resolutions down to 2 μm. This allowed the 3-D structure of tissue-engineered materials to be imaged across three orders of magnitude of detail. The capabilities of micro-CT were demonstrated by imaging bone ingrowth into a porous scaffold. This is shown in Fig. 3.98.

Bone remodeling that plays an important role in bone function is regulated by bone resorption and formation, ensuring the mechanical integrity of the skeleton. Because remodeling requires the presence of osteoblasts and osteoclasts, Nakagawa *et al.* conducted a study to assess osteoclast differentiation and attachment on tissue engineering bone [240]. According to the experimental procedure shown in Fig. 3.99, porcine bone marrow–derived mesenchymal stem cells (pMSCs) and hematopoietic cells were isolated from the bone marrow of Yucatan mini-pigs and cultured separately. The pSMCs were differentiated into osteoblasts, seeded on porous PLGA

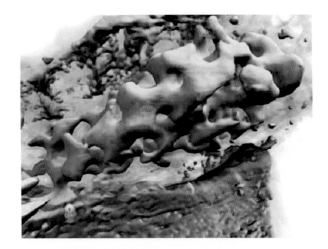

Fig. 3.98   Image from 3-D data set showing non-conjunction of new bone in scaffold, to old bone (lower left).

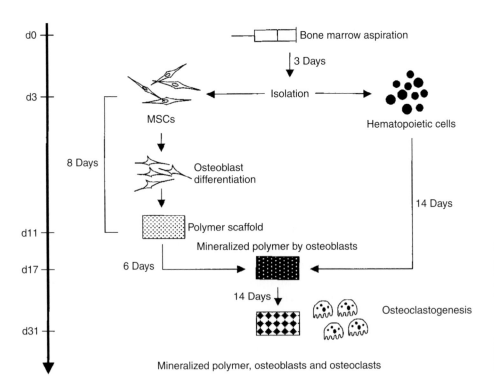

Fig. 3.99   Experimental overview.

foams, and cultured in a rotating oxygen-permeable bioreactor system. Once the cell–polymer constructs had started to mineralize, the hematopoietic cells were added and co-cultured to trigger osteoclastogenesis. It was found that osteoblasts and osteoclasts were successfully differentiated from bone marrow on the scaffolds.

Lim *et al.* investigated integrin expression by the human fetal osteoblastic cell line, hFOB 1.19, as a function of substratum surface wettability [241]. The influence of surface wettability on bone cell phenotype was also examined. Plasma-treated quartz (PTQ) and plasma-treated glass (PTG) (hydrophilic, contact angles of 0°), octadecyltrichlorosilane-treated quartz (STQ) and octadecyltrichlorosilane-treated glass (STG) (hydrophobic, contact angles above 100°), and tissue culture polystyrene were used for cell culture. The hFOB cells cultured on hydrophilic substrata displayed well-developed actin stress fibers relative to cells on hydrophobic substrata. Western blot analysis revealed that hFOB cells cultured on hydrophobic substrata (STQ or STG) expressed lower levels of α_v and β_3 integrin subunits than did cells on hydrophilic substrata (PTQ or PTG). This effect was more pronounced in cells on STQ than on STG. These variations in integrin expression were lessened by extended culture time. Double-labeled integrin/actin immunofluorescence confirmed Western blot results; that is, cells cultured on PTQ displayed distinct, large plaques of α_v and β_3 subunits and integrin $\alpha_v\beta_3$, as well as their colocalization with actin stress fiber ends, whereas cells on STQ did not display integrin plaques after 24 h and displayed only minimal plaque formation after 3 days. Vinculin, a focal adhesion protein that mediates binding between the integrin and the actin cytoskeleton, appeared in Western blots to mimic the variations of α_v and β_3 expression with respect to surface wettability. Real-time RT-PCR analysis showed that hFOB cultured on hydrophobic substrata, which had downregulated α_v and β_3 integrin subunits, displayed greater steady state mRNA levels of OP, an ECM protein containing the RGD integrin recognition sequence, than did cells cultured on hydrophilic substrata.

7.1.5. Vascular Cells

In an attempt to form a vascular tube, Nasseri *et al.* constructed tubule-shaped scaffolds from a combination of PGA/poly-4-hydroxybutyrate (P4HB) followed by coating with collagen [242]. The culture vessels containing ovine vascular myofibroblasts were rotated around the central axis at 5 rpm with a rotation radius of 9 cm. Since the constructs were not fixed inside the bioreactors, the conventional computational fluid dynamic methods were not feasible for simulation of flow patterns around the constructs. Therefore, they used first-order approximation to calculate the flow characteristics. Because of the partial rotation of freely moving constructs, there was a fairly even distribution of shear stress and medium velocity around the inner and outer surface area of the constructs. The results obtained indicated that 93% of the cells could not be detected in the medium and just 7% remained unattached to polymer scaffolds after 24 h. After 5 days the inner and outer surfaces of the tubular constructs were fully covered with myofibroblasts. They had aligned in the direction of flow by day 7. Multiple spindle-shaped cells were observed infiltrating the polymer mesh. No major cell death was observed and constructs maintained their tubular shape over the entire observation time.

To further the development of vascular tissue analogs, Seliktar *et al.* explored the use of mechanical stimulation [243]. To this end, human aortic smooth muscle cell (HASM), rat aortic smooth muscle cells (RASM), and human dermal fibroblast (HDF) were used. The cells were independently suspended in collagen solution, with the final collagen concentration of 2 mg/ml in the gels. Tubular constructs were molded over a silicone support sleeve for proper fitting into the cyclic strain bioreactor. The cell-seeded tubular constructs mounted inside a bioreactor were exposed to strain by applying stimulation to the underlying silicone support sleeves (10% radial strain) for up to 8 days. Figure 3.100 shows the ultimate stress and material modulus of strained and unstrained RASM, HASM, and HDF constructs. Whereas the mechanical properties continued to be enhanced in the RASM-seeded constructs throughout the time course of conditioning examined (*i.e.*, up to 8 days), the properties of HDF-seeded and HASM-seeded constructs were observed to

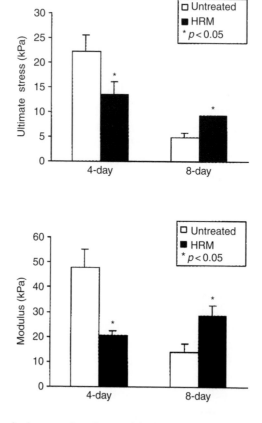

Fig. 3.100 Mechanical properties observed in the presence and absence of high-ribose medium (HRM).

deteriorate as a result of prolonged conditioning (8 days). The drop in ultimate stress and modulus did not appear to be influenced by the collagen compaction, which continued to increase between day 4 and day 8 as measured by the decrease in construct volume. Because mechanical properties of engineered constructs are dependent on the integrity of the collagenous matrix, Seliktar *et al.* thought that the deterioration of the HASM constructs could be explained by the actions of cell-secreted specialized enzymes (MMPs), which have been shown to facilitate strain-stimulated restructuring. The comparison between 4 days and 8 days of MMPs production by cyclic strain stimulation of HDF and HASM constructs is summarized in Table 3.24. As can be seen, HASM constructs accumulated active MMP-2 and also exhibited significant deterioration in the ultimate stress and modulus after 8 days. The role of MMP-2 was twofold; (1) MMP-2 facilitated a remolding response associated with cyclic strain stimulation; (2) MMP-2 accumulation contributed to the overall deterioration in mechanical integrity of the constructs when presented in large amounts. The considerably less amount of active MMP-2 accumulation in the HDF constructs after 8 days than in the HASM constructs seems to explain why the HDF constructs do not demonstrate deteriorating mechanical properties.

Dvorin *et al.* tested bFGF and VEGF for ability to stimulate proliferation of two different clonal populations of aortic valve ECs [244]. The EPCs isolated from ovine peripheral blood were grown in endothelial basal medium (EBM) with 10% heat-inactivated FCS, bFGF (2 ng/ml), and GPS (glutamine, penicillin G, and streptomycin sulfate) on 1% gelatin-coated dishes. This growth medium was referred to as EBM-B. For ovine aortic valve clone av 10, endothelial-to-mesenchymal trans-differentiation was induced by incubating the cells in EBM-B with TGF-β1 (1 ng/ml) for 5–8 days. For ovine aortic valve clone av 17, endothelial-to-mesenchymal trans-differentiation was induced by incubating cells in a reduced medium consisting of EBM, 1% FCS, and 1 $\times$ GPS for 5–8 days. The EPC colonies 780, 680, and 23, each isolated from a different lamb, were induced to transdifferentiate with TGF-β1 (10 ng/ml) for 8 days. As shown in Fig. 3.101a, in av 10 cells, bFGF induced a

Table 3.24

Percent differences in MMP-2 enzyme levels between strain-stimulated and static control constructs[a]

Sample	72 kDa (Latent)	66 kDa (Active)	Ratio (66 kDa/72 kDa)
	4-Day Conditioning		
HASM	501[b]	484[b]	61[b]
HDF	138[b]	179[b]	79[b]
HASM (high ribose)	−7	9	1
	8-Day Conditioning		
HASM	354[b]	1868[b]	313[b]
HDF	181[b]	223[b]	22
HASM (high ribose)	−37.5	−27	3

[a] Data are shown for human smooth muscle cell (HASM) constructs, human dermal fibroblast (HDF) constructs, and HASM constructs cultured with high-ribose medium.

[b] Statistically significant difference between strain-stimulated and static constructs ($n \geq 6$, $p < 0.05$).

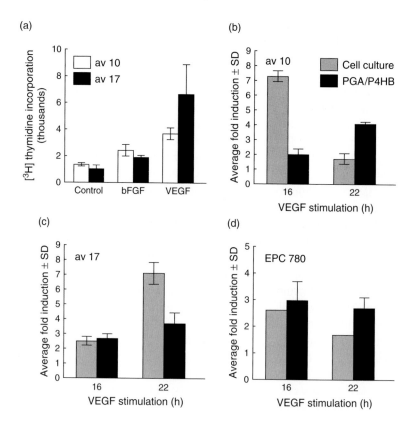

Fig. 3.101 Endothelial cell proliferation on PGA/P4HA compared with 2-D cell culture.

1.7-fold increase in [³H]thymidine incorporation compared with nonstimulated cells, whereas VEGF induced a 2.6-fold increase. In av 17 cells, bFGF induced a 1.8-fold increase whereas VEGF induced a 6.4-fold increase in [³H]thymidine in corporation. When cells were stimulated with VEGF for different periods of time so that peak entry into S phase would be detected, VEGF-induced proliferation of clone av 10 was 7-fold at 16 h and 2-fold at 22 h in 2-D cell culture, as shown in Fig. 3.101b. In contrast, av 10 cells seeded on PGA/P4HB showed 2-fold induction after 16 h compared with a 4-fold induction after 22 h. Figures 3.101c and d represent average fold induction when av 17 cells and EPC clone 780 cells were plated either on gelatin-coated cell culture wells or on PGA/P4HB and treated with VEGF for 16 or 22 h.

Griffith *et al.* described a system for generating capillary-like networks within a thick fibrin matrix [245]. Human umbilical vein ECs, growing on the surface of microcarrier beads, were embedded in fibrin gels at a known distance ($\Delta = 1.8$–4.5 mm) from a monolayer of human dermal fibroblasts. The distance of the growth medium, which contained VEGF and bFGF, from the beads, C, was varied from 2.7

to 7.2 mm. Capillaries with visible lumens sprouted in 2–3 days, reaching lengths that exceeded 500 μm within 6–8 days. On day 7, capillary network formation was largely independent of C; however, a strong inverse correlation with Δ was observed, with the maximum network formation at $\Delta = 1.8$ mm. However, the thickness of the gel was not a limiting factor for oxygen diffusion as these tissue constructs retained a relatively high oxygen tension of >125 mmHg. Fidkowski *et al.* microfabricated capillary networks with poly(glycerol sebacate) (PGS) [246]. They etched capillary patterns onto silicon wafers by standard microelectromechanical systems (MEMS) techniques. The resultant silicon wafers served as micromolds for the devices. The patterned PGS film was bonded with a flat film to create capillary networks that were perfused with a syringe pump at a physiological flow rate. During the first 10 days of culture, the cells proliferated as indicated by an increase in number and density of the cells within the device. The cell expansion slowed after 10 days and parts of the capillary networks were covered by nearly confluent ECs within 14 days as shown in Fig. 3.102. The devices were endothelialized under flow conditions, and part of the lumens reached confluence within 14 days of culture.

Gruber *et al.* investigated *in vitro* models with isolated ECs to mimic the various steps of angiogenesis [247]. In the *in vivo*–like model of the chick embryo chorioallantoic membrane assay, they observed blood vessel ingrowth into collagen sponges containing conditioned medium from undifferentiated BMSCs. In the Boyden chamber assay, the conditioned medium was chemotactic for human umbilical vascular ECs and human uterus microvascular ECs. When cells were placed on Matrigel-coated culture dishes, formation of tubular structures was enhanced. The presence of vascular endothelial growth factor–neutralizing antibodies did not affect the outcome of the two *in vitro* assays. The BMSC-conditioned medium had no effect on proliferation of ECs and on MMP-2 expression.

Afting *et al.* evaluated the use of retroviral transfection of GFP as a means of efficient and stable gene transfer into primary ovine ECs for discrimination of donor cells in cardiovascular tissue engineering [248]. Primary ECs harvested from the common carotid artery of lambs were seeded to 30% confluency on six-well plates. And 2 ml of supernatant containing replication-incompetent GFP-expressing retrovirus was mixed with Polybrene (hexadimethylene bromide) and added to each well. Polybrene is expected to make virus attachment easier because of antagonization of the negative charge of the cell surface. Cells were centrifuged for 2 h at 2000 rpm and incubated overnight. Transfection without centrifugation and Polybrene resulted in significantly lower efficiency in a control study using established cells. The GFP signal appeared 24–48 h after transfection and remained stable for up to 25 subsequent passages. Transfection efficiency 48 h after transfection was 33.4% as confirmed by FACS. This fraction did not change during the subsequent 25 passages.

Fields *et al.* developed a vector capable of stably delivering a transgene of interest to vascular ECs to enable overexpression of anticoagulant proteins on the EC surface. They derived ECs from porcine aortic ECs (PAECs), HUVECs, human microvascular ECs (HMECs), and NIH 3TS cells [249]. The PA317 and PG13

(a)

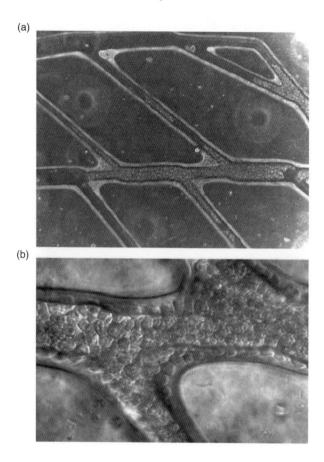

(b)

Fig. 3.102 Photomicrographs of endothelialized capillary network: Endothelial cells reached confluence at various portions of the PGS capillary within 14 days: (a) ×10; (b) ×40.

retroviral packaging cell lines stably expressing the pRet2.EGFP retroviral expression vector were utilized. The pRet2.EGFP is a modified version of the retroviral backbone MFG retroviral vector designed to optimize gene expression in primary cell lines, and expressing EGFP. Retroviral constructs derived from both PA317 and PG13 packaging cell lines demonstrated excellent ability to infect ECs *in vitro*. The EGFP expression was visualized as early as day 4 via fluorescence microscopy, and was maximal on day 7. Retroviral vectors derived from both packaging lines infected all types of ECs studied with high efficiency, apart from NIH3TS cells. Fields *et al.* studied also the long-term stability of transgene products in PAECs that were transfected with Adeno-β-gal and PG13-derived retroviral construct. Cells infected with PG13-derived retrovirus retained EGFP with 85% efficiency after 30 days, whereas cells infected with adenovirus retained either EGFP or β-galactosidase at significantly lower levels. When continued in culture past 30 days, GFP expression was maintained but PAEC replication substantially diminished, possibly reflecting the known finite life span of cultured ECs *in vitro*. To verify the applicability of this approach to engineered grafts, Fields *et al.* seeded SMCs onto PGA scaffolds and cultured in glass bioreactors under pulsatile conditions at 90 beats/min for 7 weeks. After endothelialization with EGFP-expressing PAECs, vessels were harvested from bioreactors and prepared for frozen sectioning. A dense layer of GFP-expressing cells lined the luminal vessel surface.

7.1.6. Hepatocytes

To investigate the potential of substrate-presented, acellular E-cadherin to modulate hepatocellular self-assembly and functional fate, Semler *et al.* cultured rat hepatocytes at sparse densities on surfaces designed to display recombinant E-cadherin/Fc chimeras [250]. On these substrates, hepatocytes were observed to recognize microdisplayed E-cadherin/Fc and responded by modulating the spatial distribution of the intracellular cadherin complexing protein β-catenin. Substrate-presented E-cadherin/Fc was also found to markedly alter patterns of hepatocyte morphogenesis, as cellular spreading and 2-D reorganization were significantly inhibited under these conditions, leading to multicellular aggregates that were considerably more 3-D in nature. Increasing cadherin exposure was also associated with elevated levels of albumin and urea secretion, two markers of hepatocyte differentiation, over control cultures. This suggested that cell substrate cadherin engagement established more functionally competent hepatocellular phenotypes, coinciding with the notion that E-cadherin is a differentiation-inducing ligand for these cells. To profile gene expression dynamics during interleukin 6–stimulated inflammation in hepatocytes maintained in a stable, collagen gel *in vitro* model system, Jayaraman *et al.* used DNA microarrays [251]. The observed expression profile was also compared with that obtained from rat liver tissue after burn injury to determine the extent and nature of responses captured by the *in vitro* system. Several aspects of the *in vivo* hepatic inflammatory response could be captured by the *in vitro* system at the molecular systems level. Statistical analysis of the mRNA profiles was also used to characterize the temporal response in each model system and demonstrate similar behavior.

7.1.7. Oral Cells

Marty-Roix *et al.* investigated the growth kinetics and morphologic features of porcine enamel-, dentin-, and cementum-derived cells grown in monolayer culture, and their synthesis of selected matrix molecules when seeded on a collagen–GAG matrix as a regeneration template [252]. The hypothesis they assumed was that these dental tissue cells could express the gene for a contractile actin isoform, α-smooth muscle actin (SMA), and could display contractile activity. Such contraction would collapse the pores of the scaffold in which the cells were being grown, and distort its overall shape, thus affecting the performance of the cell-seeded scaffold. In this regard the distribution of SMA-containing ameloblasts and odontoblasts *in vivo* was also assessed. Ameloblasts, odontoblasts, and cementoblasts expressed the gene for SMA, and could contract a collagen–GAG scaffold. The cell mediated contraction of collagen–GAG matrices in which enamel-, dentin-, and cementum-derived cells were growing. A large amount of SMA was found in the odontoblasts after the first passage, and SMA expression in the enamel- and cementum-derived cells appeared to increase with time in culture and with passage number.

Rat dental pulp cells were isolated from maxillary incisors of male rats and their differentiation ability was evaluated by Zhang *et al.* [253]. Immunochemistry by stem cell marker STRO-1 proved the existence of stem cells or progenitors in the isolated cell population. The dissociated cells were then cultured both on smooth surfaces and on 3-D scaffold materials in medium supplemented with β-glycerophosphate, dexamethasone, and L-ascorbic acid. These cells showed the ability to differentiate into odontoblast-like cells and produced calcified nodules, which had components similar to dentin. In addition, the "odontogenic" properties of the isolated cells were supported by 3-D calcium phosphate and titanium scaffolds equally well.

Therapeutic irradiation for cancer of the head and neck, and the autoimmune disorder Sjögren's syndrome are the two main causes of irreversible salivary gland hypofunction. In such patients, both the quantity and quality of saliva are greatly altered, leading to considerable morbidity. A major problem in this endeavor has been the difficulty in obtaining a suitable autologous cellular component. Tran *et al.* described a method of culturing and expanding primary human salivary cells that can form a polarized tight epithelial barrier [254]. Human submandibular glands (huSMGs) cells showed polarization and appropriately localized tight junction proteins. The TEM micrographs showed an absence of dense core granules, but confirmed the presence of tight and intermediate junctions and desmosomes between the cells. Functional assays showed that huSMG cells had high transepithelial electrical resistance and low rates of paracellular fluid movement. Additionally, huSMG cells showed a normal karyotype without any morphological or numerical abnormalities, and most closely resembled striated and excretory duct cells in appearance.

7.1.8. Neuronal Cells

After injury to the CNS, the anatomical organization of the tissue is disrupted, posing a barrier to the regeneration of axons. Meningeal cells, a central participant in the CNS tissue response to injury, migrate into the core of the wound site in an unorganized fashion and deposit a disorganized ECM that produces a nonpermissive environment. Walsh *et al.* provided nanometer-scale topographic cues to meningeal cells and examined the ability of the composite construct to influence dorsal root ganglion regeneration *in vitro* [255]. When grown on control surfaces of meningeal cells lacking underlying topographic cues, there was no bias in neurite outgrowth. In contrast, when grown on monolayers of meningeal cells with underlying nanometer-scale topography, neurite outgrowth length was greater and was directed parallel to the underlying surface topography even though there existed an intervening meningeal cell layer. The observed outgrowth was significantly longer than on LN-coated surfaces, which are considered to be the optimal substrata for promoting outgrowth of dorsal root ganglion neurons in culture. Shany *et al.* evaluated the feasibility of using a 3-D aragonite biomatrix as a support for neuronal culture [256]. Cultures were maintained *in vitro* for up to 5 weeks. Some portions of the cell population acquired the morphological characteristics of hippocampal pyramidal or granule neurons with axons and dendrites extending in a 3-D manner along the surfaces of the crystalline biomatrix. Neurons that usually grow on a sheet of glial cells adhered to the crystalline material and regenerated mature synaptic connections, with presynaptic sites expressing the synaptic vesicle protein 2 and postsynaptic sites having the shape of dendritic spines and expressing type 1 glutamate receptors, as these cells do under conventional culture conditions.

7.1.9. Retinal Cells

The retina has a complex, multilayered architecture that is polarized with respect to the photoreceptors. Outer retinal degeneration involves the loss of photoreceptor cells and may also involve the cells of the inner nuclear layer which connect to and support the photoreceptors. Neural progenitor cells (NPCs) have been shown to integrate morphologically into the inner nuclear layer in injury models but have not been seen to replace photoreceptors. Retinal progenitor cells (RPCs) have been isolated from the mature eye and developing retina and these cells may replace photoreceptors. Lavik *et al.* fabricated highly porous scaffolds from blends of PLLA and PLGA to produce pores oriented normal to the plane of the scaffold [257]. The RPCs were seeded on the polymer scaffolds and cultured for 14 days. Seeded scaffolds were then either fixed for characterization or used in an explant or *in vivo* rat model. The scaffolds were fully covered by RPCs in 3 days. Attachment of RPCs to the polymer scaffold was associated with downregulation of immature markers and upregulation of markers of differentiation, suggesting that the scaffold might promote differentiation of RPCs. The seeded cells elaborated cellular processes and aligned in the scaffold in conjunction with degenerating retinal explants. The cells also exhibited morphologies consistent with photoreceptors including a high degree of polarization of the cells. Implantation of the seeded scaffold into the rat eye was associated with increased RPC survival.

7.2. Stem Cells

In an attempt to develop a simple method for parallel analysis of phenotype-specific surface antigens, Ko *et al.* prepared cellulose membrane–based antibody microarrays with a characteristic size of a centimeter [258]. Figure 3.103 shows the scheme of the parallel analysis for correlating cell surface antigens with the cell phenotype. In order to avoid cross-contamination between antibodies, they covalently immobilized antibodies onto an activated substrate. A binding assay was performed with neural cells obtained from the neurosphere culture of the rat fetal striatum on a microarray spotted with eight kinds of antibodies and four different proteins, followed by immunocytochemical staining of cells bound to the

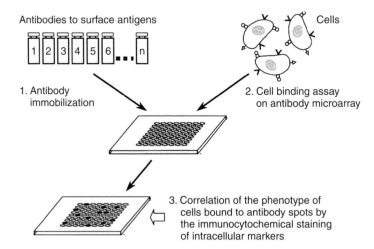

Antibodies to surface antigens

1 2 3 4 5 6 ... n

Cells

1. Antibody
   immobilization

2. Cell binding assay
   on antibody microarray

3. Correlation of the phenotype of
   cells bound to antibody spots by
   the immunocytochemical staining
   of intracellular markers

Fig. 3.103 Schematic illustration of the parallel analysis of surface markers by means of an antibody microarray.

microarray using antibodies to the intracellular markers of immature (nestin and vimentin) and mature (β-tubulin III and glial fibrillary acidic protein) neural cells. As a result, the phenotype of bound cells could be correlated to surface antigen expression, which illustrated the potential of the solid-phase cytometry developed here for the identification of surface markers. In the future their technique may be applicable to the isolation of NSCs.

7.2.1. MSCs

Despite the enormous potential benefits of MSCs for tissue engineering applications, conventional tissue culture methods limit the clinical utility of these cells because of the gradual loss of both their proliferative and differentiation potential during *ex vivo* expansion. To overcome these limitations, Mauney *et al.* reviewed basic concepts and current research involving two approaches to bone-tissue engineering; MSC combination *ex vivo* with 3-D porous biomaterials, followed by implantation and systemic or intrabone infusion of MSCs into patients with various degenerative bone conditions [259]. They discussed strategies including cultivation in the presence of FGF-2, induction of ectopotic telomerase expression, and *ex vivo* expansion on various collagenous biomaterials. In addition, they outlined mechanistic theories on the potential role of MSC–ECM interactions in mediating the retention of MSC proliferative and differentiation capacity after *ex vivo* expansion on collagenous biomaterials. Caplan also presented a review on MSCs [260].

Xia *et al.* transfected human MSCs using four retroviral pseudotypes, amphotropic murine leukemia virus 4070 (MuLV-10Al), a modification of amphotropic pseudotype 4073 (*A71G, Q74K, V139M*), gibbon ape leukemia virus (GaLV), or feline endogenous virus (RD114) encoding the neomycin resistance

(Neo) gene and EGFP as genetic markers [261]. They observed that the MuLV4073 was the most efficient pseudotype for hMSC transfection. The proliferation and differentiation characteristics of EGFP-labeled hMSCs were not significantly different from control hMSCs. The G418-selected EGFP-labeled cells were cultured for 3 weeks on two porous calcium phosphate bioceramics, a synthetic HAp and a deproteinized bone, before implantation into NOD/SCID mice for up to 4 weeks. The EGFP-labeled hMSCs could be readily visualized by their intense green fluorescence both *in vitro* and *in vivo*. In synthetic HAp implants the cells remained in a monolayer, whereas in deproteinized bone implants mineralized tissue were detected by histology, SEM, and energy-dispersive X-ray spectrometry.

Kotobuki *et al.* studied the viability of cryopreserved human mesenchymal cells and compared osteogenic potential between noncryopreserved and cryopreserved human mesenchymal cells with MSC-like characteristics, derived from the bone marrow of 28 subjects [262]. The viability of cryopreserved mesenchymal cells was approximately 90% regardless of the storage term (0.3–37 months). The cell surface antigens of both noncryopreserved and cryopreserved mesenchymal cells were negative for hematopoietic cell markers such as CD14, CD34, CD45, and HLA-DR but positive for mesenchymal characteristics such as CD29 and CD105. No difference in osteogenic potential was found between cells with or without cryopreservation treatment. In addition, cells undergoing long-term cryopreservation (about 3 years) maintained high osteogenic potential, as shown in Fig. 3.104.

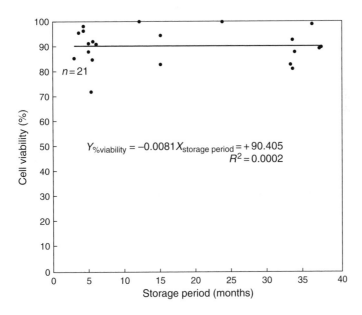

Fig. 3.104 Viability of cryopreserved human mesenchymal cells relative to storage period.

Cell Expansion

Sekiya *et al.* defined improved conditions for obtaining standardized preparations of human MSCs [263]. To isolate the cells, they took bone-marrow aspirates from the iliac crest of normal adult donors. Nucleated cells were isolated with a density gradient (Ficoll-Plaque) and resuspended in complete culture medium: α-MEM, 20% FCS, penicillin, streptomycin, and glutamine. All of the nucleated cells were plated in medium in a culture dish and incubated at 37°C with 5% CO_2. After 24 h, nonadherent cells were discarded, and adherent cells were washed with PBS. The cells were incubated for 4–11 days, harvested with trypsin and EDTA, and replated at 3–50 cells/cm². After 7–12 days, the cells were harvested with trypsin/EDTA, suspended at 1×10^6 cells/ml in 5% DMSO and 30% FCS, and frozen in liquid nitrogen (passage 1 cells). To study the effect of plating density on expansion of MSCs in culture, MSCs were cultured at 10, 50, 100, and 1000 cells/cm² in 60-cm² dishes. As shown in Fig. 3.105a, passage 3 cells plated at a density of 10 cells/cm² expanded 500-fold in 12 days, whereas the same cells plated at 1000 cells/cm² expanded 30-fold. However, cells plated at 1000 cells/cm² yielded 1.6 million cells per 60-cm² dish, whereas cells plated at 10 cells/cm² yielded only 0.3 million cells, as shown in Fig. 3.105b. As shown in Fig. 3.106, the doubling rate per day was less in cells plated at 100 or 1000 cells per cm², but the peak rate was observed on day 4, regardless of the initial cell plating density. The potential for the cells to generate single-cell-derived colonies was higher for cells expanded by initial plating at lower densities. After plating the cells at 1–1000 cells/cm², the cultures underwent a time-dependent transition from early progenitors defined as thin, spindle-shaped cells (RS-1A) to wider, spindle-shaped cells (RS-1B). To define the adipogenic potential of the expanded MSCs, cells were plated at 50 or 1000 cells/cm² in complete culture medium and expanded for 4, 7, or 12 days before plating at 5000 cells/cm² in adipogenic medium (0.5 μM dexamethasone, 0.5 mM isobutylmethylxanthine, and 50 μM indomethacin), and culturing for 4 days generated more adipocytes (assayed by the amount of Oil Red-O absorption) than cells expanded for 7 and 12 days. The number of adipocytes per culture was lower when the cells were expanded in complete medium for 12 days instead of 4 or 7 days. Cells initially plated at a density of 1000 cells/cm² produced less adipocytes than those plated at 50 cells/cm², regardless of how long they were expanded. These results indicated that the greater number of adipocytes was generated from cultures that were plated at lower densities for shorter time periods and that were enriched for RS-1A cells. To assay the chondrogenic potential of the cells, MSCs were plated at 50 cells/cm² and expanded for 4, 7, or 12 days. For chondrocyte differentiation, a micromass culture system was used [264]. Approximately 200,000 MSCs were placed in a 14-ml tube and pelleted into micromasses by centrifugation at 450 g for 10 min. The pellet was cultured for 21 days in chondrogenic media that contained 500 ng/ml BMP-6 in addition to high-glucose (25 mM) DMEM supplemented with 10 ng/ml TGF-β3, 10^{-3} M dexamethasone, 50 g/ml ascorbate-2-phosphate, 40 μg/ml proline, 100 μg/ml pyruvate, and 50 mg/ml ITS + Premix. The cultures that were expanded for 7 days and contained RS-1B cells

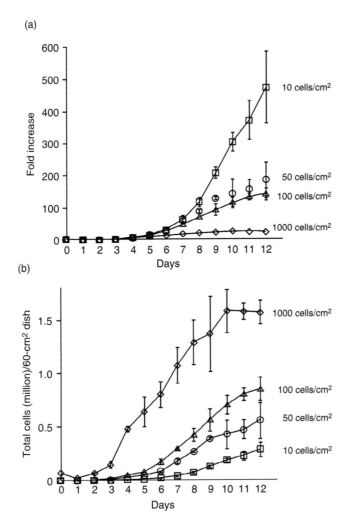

(a)

(b)

Fig. 3.105 Relation between initial plating density and expansion of passage 3 MSCs.

formed larger cartilage pellets than both the cultures that were expanded for 4 days and enriched for RS-1A cells and those that were expanded for 12 days and contained RS-1C (still wider, spindle-shaped cells). Also, the cultures that were expanded for 12 days and contained RS-1C cells formed larger cartilage pellets than cultures that were expanded for 4 days and contained RS-1A cells. Pellets derived from 7-day precultures were the heaviest among the three groups, and the weight of pellets derived from 4-day preculture was the lightest.

Tsutsumi *et al.* attempted to identify a growth factor(s) involved in self-renewal of MSC and maintenance of their multilineage differentiation potential [265]. They found that FGF-2 markedly increased the growth rate and the life span of rabbit,

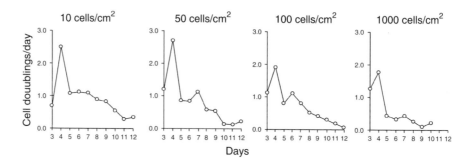

Fig. 3.106   Plating density alters cell doublings per day.

canine, and human bone-marrow MSC in monolayer cultures, as shown in Fig. 3.107. This effect of FGF-2 was more prominent in low-density cultures than in high-density cultures. In addition, all MSCs expanded *in vitro* with FGF-2, but not without FGF-2, differentiated to chondrocytes in pellet cultures. The FGF($+$) MSC also retained the osteogenic and adipogenic potential throughout many mitotic divisions. They further reported that the growth rate and the proliferative life span of MSC markedly increased when tissue culture dishes were coated with a basement membrane–like EMC [266]. The MSC that expanded 10^6-fold on the basement membrane–like ECM retained its osteogenic, chondrogenic, and adipogenic potential.

Mauney *et al.* assessed the role of a denatured collagen (DC) matrix in influencing the aging and differential potential of human BMSCs [267]. Growth of BMSCs on a DC matrix versus tissue culture polystyrene (TCPs) significantly reduced one of the main manifestations of cellular aging; the attenuation of the ability to express a major protective stress response component, HSP70, increased the proliferation capacity of *ex vivo*–expanded BMSCs, reduced the rate of morphological changes, and resulted in an increase in the retention of the potential to express osteogenic-specific functions and markers upon treatment with osteogenic stimulants. Figure 3.108 shows elevations in the level of ALP activity measured at day 7 of osteogenic stimulation (OS) as an indicator of osteogenic differentiation. Early passage (EP) cells cultured with OS treatment displayed significantly higher levels of ALP activity compared to untreated controls on TCPs and the DC matrix. In addition, EP OS-treated cells exhibited similar levels of ALP activity when either cultivated on TCPs or maintained on the DC matrix. Cultivation on the DC matrix alone, without OS treatment, did not significantly increase ALP activity. On the other hand, in late passage (LP) cells cultivated on TCPs, the extent of ALP induction by OS treatment was reduced to 15% of that seen in EP OS-treated cells. In contrast, levels of ALP activity in OS-treated LP cells cultivated on DC matrix were maintained at 69% of levels observed with EP controls. Lawson *et al.* compared the attachment and proliferation of both rat and human BMSCs on calcium-crosslinked alginate gels [268]. They found that, in contrast to rat cells, human cells did not readily attach or proliferate on both unmodified

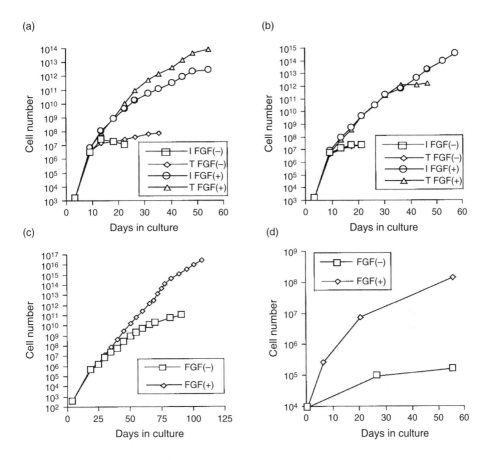

Fig. 3.107 Culture lifetime of rabbit (a, b) and human (c, d) in monolayer cultures in the presence and absence of FGF-2.

alginates and collagen-incorporating gels. However, alginate gels containing both type I collagen and β-TCP were found to enhance human cell adherence and proliferation. Furthermore, interactions between the collagen and β-TCP prevented loss of the protein from the hydrogels.

Cell Differentiation
Li *et al.* tested a 3-D nanofibrous scaffold fabricated from PCL for its ability to support and maintain multilineage differentiation of marrow-derived hMSCs *in vitro* [269]. The hMSCs were seeded onto prefabricated nanofibrous scaffolds and were induced to differentiate along adipogenic, chondrogenic, or osteogenic lineages by culturing in specific differentiation media. Gene expression analysis and immuno-histochemical observations confirmed the formation of 3-D constructs containing cells differentiated into the specified cell types. Indrawattana *et al.* studied *in vitro* chondrogenesis induction on human bone-marrow MSCs (hBMSCs) by cycling

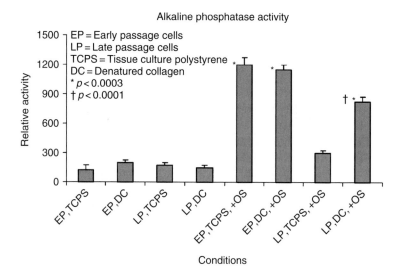

Fig. 3.108 ALP activity in response to OS treatment.

growth factors [270]. TGF-β3, BMP-6, and IGF-1 were used in combination, and pellet cultures of hBMSCs were prepared. The combined growth factors TGF-β3 and BMP-6 or TGF-β3 and IGF-1 were more effective for chondrogenesis induction, as shown in Fig. 3.109. To obtain information on the sequence of cellular and molecular events during *in vitro* chondrogenic differentiation, Lisignoli *et al.* analyzed MSCs on an HAc biomaterial (Hyaff™-11) [271]. The Hyaff™-11 was derived by the total esterification with benzyl alcohol of the carboxyl groups along the polymeric backbone of sodium hyaluronate. The scaffold used was configured as a non-woven mesh of 20-μm fibers. The materials swelled to about double their original diameter and degradation occurred by spontaneous hydrolysis of the ester bonds. Cellular differentiation was induced using two different concentrations of TGF β1 (10 and 20 ng/ml). Without TGF-β1, MSCs did not survive while in the presence of TGF-β1 the cells significantly proliferated from day 7 until day 28, and TGF-β1 at 20 ng/ml better induced the formation of cartilage-like tissue.

Alhadlaq *et al.* investigated the possibility of human-shaped articular condyle formation by rat bone marrow–derived MSCs encapsulated in a PEG-based hydrogel [272]. Rat MSC–derived chondrogenic and osteogenic cells were loaded in hydrogel in a human condyle mold. Harvested articular condyles from 4-week *in vivo* implantation demonstrated stratified layers of chondrogenesis and osteogenesis. Parallel *in vitro* experiments using goat and rat MSCs corroborated *in vivo* data by demonstrating the expression of chondrogenic and osteogenic markers. *Ex vivo*–incubated goat MSC–derived chondral constructs contained cartilage-related GAGs and collagen. By contrast, goat MSC–derived osteogenic constructs expressed ALP and osteonectin genes, and showed escalating calcium content over

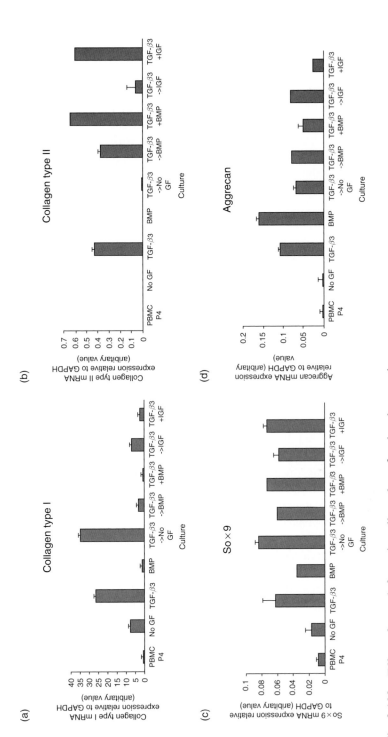

Fig. 3.109 Effect of growth factor in pellet culture for chondrogenesis.

time. Rat MSC–derived osteogenic constructs were stiffer than rat MSC–derived chondrogenic constructs upon nanoindentation with atomic force microscopy. Hirose *et al.* examined the viability, proliferation, and calcein uptake into extracellular regions of cryopreserved/thawed hBMSCs in culture to assess whether the hBMSCs after freezing and thawing could retain their bone-forming capability [273]. The viability of cells immediately after thawing was 98%. When the cells were cultured in the presence or absence of dexamethasone, the cells proliferated to reach confluency at culture day 7. The cells treated with dexamethasone formed abundant mineralized nodules at their extracellular regions after culture day 14, while the cultures without dexamethasone did not exhibit bone formation.

Williams *et al.* demonstrated that a PEG-based hydrogel provided a suitable environment for the chondrogenic differentiation of MSCs in the presence of TGF-β1 *in vitro* [274]. They encapsulated MSCs from goats in a photopolymerizable PEG-based hydrogel and cultured them with or without TGF-β1 to study the potential for chondrogenesis in a hydrogel scaffold system amendable to minimally invasive implantation. Bone marrow from the femurs of goats was aspirated and centrifuged twice (1000 rpm for 10 min) in MSC growth medium. For preparation of hydrogels, a photoinitiator and TGF were added to PEGDA solution. Immediately before photoencapsulation, MSCs were resuspended in the polymer solution. Cell–polymer–photoinitiator suspension was transferred into cylindrical molds and exposed to UV light. The hydrogels were incubated in 5% CO_2 on an orbital rocker in chondrogenic medium with or without TGF-β1. MSCs proliferated in the hydrogels with TGF-β1. The GAG and total collagen content of the hydrogels increased to 3.5% dry weight and 5.0% dry weight, respectively, in constructs cultured for 6 weeks in chondrogenic medium with TGF-β_1. Immunohistochemistry revealed the presence of aggrecan, link protein, and type II collagen. Upregulation of aggrecan and type II collagen gene expression compared with monolayer MSCs was demonstrated. Type I collagen gene expression decreased from 3 to 6 weeks in the presence of TGF-β1.

Sekiya *et al.* isolated human MSCs from bone-marrow aspirates with a density gradient. For chondrogenic differentiation, 200,000 MSCs were centrifugated, and the pellets were cultured in control media, with TGF-β3 and dexamethasone, and with TGF-β3, dexamethasone, and BMP-6 [275]. The result is shown in Fig. 3.110. As can be seen, the addition of BMP-6 (500 ng/ml) to medium containing TGF-β3 and dexamethasone further increased the gross size of the pellets (Fig. 3.110a) and increased the weight of the pellets about 10-fold (Fig. 3.110b). In addition, the presence of BMP-6 enhanced the safranin-O staining in the pellets, indicating that there was an increase in proteoglycan synthesis. However, BMP-6 without TGF-β3 or dexamethasone had no discernible effect. In contrast to the report by Johnstone *et al.* [264], TGF-β and dexamethasone were essential for chondrogenesis, but they were not sufficient in themselves in promoting differentiation into chondrocytes under the conditions used by Sekiya *et al.* The difference may be explained by the varying cell densities used to expand the cultures. Johnstone *et al.* expanded the cells under high-density plating conditions (several thousand cells/cm^2), whereas Sekiya *et al.* expanded the cells under

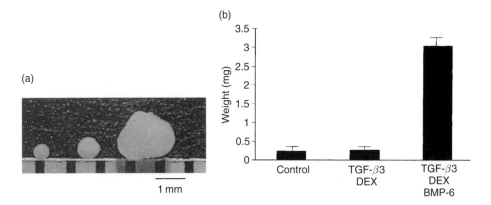

Fig. 3.110 (a) Macroscopic pictures after 21 days of pellet culture in control media (left), with TGF-β3 and dexamethasone (DEX) (center), and with TGF-β3, DEX, and BMP-6 (right). (b) Wet weight of pellets after 21 days of culture.

low-density plating conditions (several cells/cm^2). When the cells were plated at several cells/cm^2, the cells expanded several hundred-fold (eight or nine doublings) in 14 days. However, when the cells were plated at several thousand cells/cm^2, the cells increased only 2- or 4-fold (one or two doublings). Furthermore, Johnstone *et al.* reported that the chondrogenic potential of human MSCs was maintained throughout 20 passages in monolayer cultures under high-density plating conditions [264]. On the other hand, Sekiya *et al.* found that the chondrogenic potential of human MSCs decreased after each consecutive passage under their culture conditions. The potential for chondrogenesis may decrease after every passage because the number of cells which approach senescence increases.

Kitamura *et al.* compared osteogenic differentiation of hBMSCs cultured on alumina ceramics and TCPS dishes fabricated from crystal-grade polystyrene modified by a vacuum-gas plasma process [276]. The TCPS surface exhibited a slightly hydrophilic nature since the plasma treatment introduced carboxyl and hydroxyl groups onto the surface. They obtained hBMSCs from four donors (Table 3.25) and cultured the cells for two weeks in the presence of β-glycerophosphate, AA, and dexamethasone (Dex). The cells showed extensive ALP staining and mineralization. As shown in Fig. 3.111, all of the cases exhibited significantly higher levels of the

Table 3.25
Characteristics of marrow donors

Cell	Age	Sex
1	62	Male
2	58	Male
3	25	Male
4	75	Female

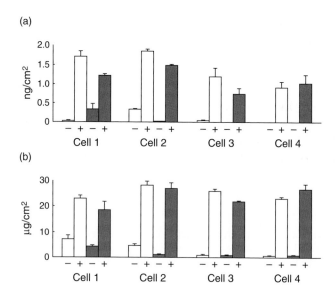

Fig. 3.111 Osteocalcin (a) and calcium (b) content of hBMSCs subcultured for two weeks on P-alumina discs (■) and control TCPS dishes (□).

factors in the Dex (+) cells than in the Dex (−) cells, although the similarity of the relative levels of these factors in the cultured cells on the alumina disc to those on the TCPS dishes seemed to depend on cellular individuality. The hBMSCs from elderly persons (62 and 75 years old), a middle-aged person (58 years old), and a young person (25 years old) showed similar high levels of ALP activity, OCN content, and calcium content.

Meinel *et al.* cultured human bone-marrow-derived MSC on three different protein scaffolds (unmodified and crosslinked collagen and silk) in three different hydrodynamic environments (static dish, spinner flask, and perfused cartridge) [277]. The objective was to determine the effects of scaffold degradation (fast for unmodified collagen, slow for crosslinked collagen and silk) and the regime of fluid flow (static medium, convective flow around constructs, and interstitial flow through constructs) on bone formation. Whole human bone marrow (25 cm³) was diluted and cells were separated by density gradient centrifugation at 800 g. Cells were seeded in flasks and the adherent cells were allowed to reach 80% confluence for the first passage, and passage 2 (P2) cells (80% confluence after 7 days) were used for the experiments. The MSC nature of the cells was determined on the basis of the expression of CD105/endoglin (a putative marker of MSCs), and CD71 (a receptor expressed in proliferating cells) as well as the lack of expression of CD31 and CD34 (markers for cells that are of endothelial or hematopoietic origin). Expanded cells could be induced to undergo either chondrogenic or osteogenic differentiation via medium supplementation with chondrogenic or osteogenic factors, respectively, without notable differences in cell differentiation capacity over three passages in culture. Constructs were cultured in dishes, spinner flasks, or perfusion bioreactors. Hydrodynamic environments of the three systems are described in Table 3.26. In static dishes, MSCs were cultured without exposure to hydrodynamic shear, and with mass transfer by molecular diffusion only. In static culture, calcium deposition was similar for MSCs grown on collagen scaffolds and films. Under medium flow, MSCs on collagen scaffolds deposited more calcium and had a higher ALP activity than MSCs on collagen films. The amounts of DNA were markedly higher in constructs based on slowly degrading (modified collagen and silk) scaffolds than on fast degrading (unmodified collagen) scaffolds. In spinner flasks, medium flow around constructs resulted in the formation of bone rods within

Table 3.26
Hydrodynamic conditions

Cultivation Vessel	Static Dish	Spinner Flask	Perfused Cartridge
Number of constructs	1	8	1
Operating volume	8 ml	120 ml	1.5 ml (cartridge) 15 ml (recirculation loop)
Mixing mechanism	None	Magnetic stirring of medium Stirring rate: 50 rpm (0.83 s^{-1}); Impeller Re number: 1370	By medium recirculation
Fluid flow	None	Turbulent flow around constructs (generated by stirring)	Laminar flow through construct Flow rate 0.2 ml/min Interstitial velocity: 35 μm/s
Mass transport	Molecular diffusion	Convection (construct periphery); molecular diffusion (construct interiors)	Convection and diffusion (throughout construct volume)
Hydrodynamic shear	None	Turbulent (construct periphery); none (construct interiors); Integrated shear factor: 8.37 s^{-1} Size of the smallest turbulent eddies: 250 μm Velocity of the smallest turbulent eddies: 0.4 cm/s	Laminar (throughout construct volume) <0.05 dyn/cm^2

the peripheral region that were interconnected and perpendicular to the construct sur-face, whereas in perfused constructs, individual bone rods oriented in the direction of fluid flow formed throughout the construct volume. On collagen scaffolds cultured in spinner flasks, the deposition of mineralized matrix was accompanied by a substantial loss in the scaffold weight and in DNA content, both of which decreased to 25% of initial over 4 weeks of culture. The comparison with the corresponding silk-based constructs, which maintained both their weight and their DNA content (80% of initial after 4 weeks), suggested that the maintenance (or loss) of DNA correlated with the maintenance (or loss) in scaffold weight because of degradation.

To examine the osteoblastic differentiation of rMSCs on non-woven PLLGA (90% LLA and 10% GA) meshes under TGF-β_1 treatment, Lieb *et al.* performed an ALP activity assay, immunohistochemistry for non-collagenous bone markers such as bone sialoprotein and osteonectin, an immunoassay, and the gene expression of these proteins as well as of OCN [278]. And TGF-β_1 was added according to the following dosing regimen: single dose of TGF-β_1 (1, 10, or 20 ng/ml), multiple doses of TGF-β_1 (1 ng/ml, added with every fourth medium change once a week), and control (complete medium supplemented with the dilution buffer for TGF-β_1). Whereas bone sialoprotein appeared to be increased depending on dose in the immunochemical stainings after supplementation with TGF-β_1, osteonectin remained unchanged. Both ALP activity and OCN were suppressed by high doses of TGF-β_1, such as single doses of 10 ng/ml or four doses of 1 ng/ml added once a week. Considering the effects of TGF-β_1 both on differentiation and on matrix for-mation and mineralization, TGF-β_1 at 1 ng/ml, added once a week in the first 1–2 weeks, was selected as an effective dose to improve bone-like tissue formation *in vitro*. Lee *et al.* examined whether MSCs were able to differentiate into functional hepatocyte-like cells *in vitro* [279]. The MSCs were isolated from human bone-marrow and umbilical cord blood, and the surface phenotype and the mesodermal multilineage differentiation potentials of these cells were characterized and tested. To effectively induce hepatic differentiation, they designed a two-step protocol with the use of hepatocyte growth factor and oncostatin M. After 4 weeks of induction, cuboidal morphology, which is characteristic of hepatocytes, was observed, and cells also expressed marker genes specific of liver cells in a time-dependent manner. Differentiated cells further demonstrated *in vitro* functions characteristic of liver cells, including albumin production, glycogen storage, urea secretion, uptake of low-density lipoprotein, and Phenobarbital-inducible cytochrome P450 activity.

Although BMSCs are easy to isolate and expand rapidly from patients without leading to major ethical and technical problems, practical application to human muscle degenerative diseases depends on the ability to control their differentiation into functional skeletal muscle cells with high efficiency and purity. Dezawa *et al.* reported a method to systematically induce skeletal muscle lineage cells with high purity from a large population of adherent MSCs, rather than from a rare subpopula-tion of myogenic stem cells contained in the bone marrow [280]. The induced popu-lation effectively differentiated into mature myotubes with some cells persisting as Pax7-positive satellite cells that continued to function in host muscle to restore

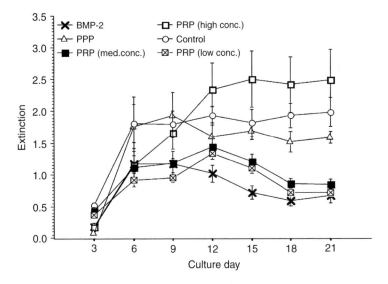

Fig. 3.112 Schematic diagram of the induction process of skeletal muscle lineage cells.

degenerating muscles in the absence of repeated transplantation. The induction procedure is shown in Fig. 3.112. Human and rat MSCs plated at a set cell density were treated with bFGF, forskolin (FSK; known to upregulate intracellular cyclic adenosine 3′,5′-monophosphate), PDGF, and neuregulin for 3 days (cells at this stage are referred to as C-MSCs). The C-MSCs were then transfected with an NICD expression plasmid by lipofection followed by G418 selection and allowed to recover to 100% confluency (referred to as CN-MSCs). Although MyoD expression was detected in CN-MSCs, the frequency of spontaneous cell fusion (the fusion index) was very low in both rat and human CN-MSCs 5 days after cells reached 100% confluency. When cells were supplied with either 2% horse serum or insulin–transferrin–selenium (ITS) serum-free medium, both of which promoted differentiation of myoblasts to myotubes, the fusion index was ~24% at 5 days after administration of 2% horse serum of 12% by ITS serum-free serum.

Hankemeier *et al.* analyzed the effect of low-dose FGF-2 (3 ng/ml) and high-dose FGF-2 (30 ng/ml) on proliferation, differentiation, and apoptosis of human BMSCs, and compared the results with those of a control group without FGF-2 [281]. Low-dose FGF-2 triggered a biphasic BMSC response: on day 7, cell proliferation reached its maximum and was significantly higher compared with the other groups. On days 14 or 28, types I and III collagen, FN, and α-smooth muscle actin mRNA expression was significantly enhanced in the presence of low-dose FGF-2. In contrast, high-dose FGF-2 did not stimulate differentiation or proliferation. Vimentin mRNA was expressed only in cultures with low-dose and high-dose FGF-2 after 14 and 28 days. Cell density was significantly higher in cultures with low-dose FGF-2 compared with the group with high-dose FGF-2 on days 7, 14, and 28. The apoptosis rate remained stable, at a rather high level, in all groups. Microscopic investigation of the cell cultures with low-dose FGF-2 showed more homogeneous, dense, fibroblast-like, spindle-shaped cells with long cell processes compared with cultures with high-dose, or no FGF-2. Arpornmaeklong *et al.* used a cell culture system to compare PRP and rhBMP-2 on osteogenic differentiation of rat BMSCs *in vitro* [282]. Table 3.27 gives the characterization result of the blood. Marrow-derived bone-forming cells from rats were seeded on porous collagenous carriers and exposed to different concentrations of PRP, rhBMP-2, or platelet-poor plasma (PPP). Cultures without additional supplements were used as controls. During a culture period of 21 days cell proliferation and the ALP and calcium contents were measured in 3-day intervals. Figure 3.113 shows the cell proliferation result. The PRP showed a dose-dependent stimulation of cell proliferation, while reducing ALP activity and calcium deposition in the culture. And BMP-2 led to an opposite cell response and induced the highest ALP activity and mineral deposition. These data suggest that PRP inhibited osteogenic differentiation of marrow-derived pre-osteoblasts in a dose-dependent manner. They concluded that PRP is not a substitute for BMP-2 in osteogenic induction.

Shimko *et al.* made a direct, long-term, *in vitro* comparison of mineralization processes in adult, marrow-derived MSCs and ES cells from the $129/Sv+c/+p$ mouse strain. The MSCs were observed to grow at a slower rate than ES cells [283]. The MSCs expressed seven times more ALP per cell than did ES cells and immediately showed type I collagen and OCN production. The ES cells also produced collagen I and OCN, but production was delayed. Mineral deposition by ES cells was nearly 50 times higher than by MSCs. The calcium-to-phosphorous ratio (Ca:P) of the ES cell mineral (1.26:1) was significantly higher than that of the MSCs (0.29:1), still 25% lower than HAp (1.67:1). Addition of bFGF significantly inhibited ALP expression, mineral deposition, and Ca:P ratios in MSCs and had little effect on ES cells.

To investigate whether highly purified muscle-derived cells (MDCs) by pre-plate technique will express markers of stem cells and differentiate into osteogenic lineage, Sun *et al.* implanted highly purified MDCs on a porous gelatin scaffold with rhBMP-2 immobilized on the surface [284]. For BMP immobilization,

Table 3.27

Characterization of blood products

Cell Count (cells/μl)	Whole Blood	PRP	PPP
Red blood cells	$6.20 \pm 0.10 \times 10^6$	$0.20 \pm 0.09 \times 10^6$	0
White blood cells	$5.33 \pm 0.55 \times 10^3$	$7.1 \pm 0.36 \times 10^3$	0
Platelets	$0.745 \pm 0.48 \times 10^6$	$6.63 \pm 0.23 \times 10^6$	$6.93 \pm 0.82 \times 10^4$
Concentration of TGF-β1 (ng/ml)	44.8 ± 1.64	386.416 ± 81.34	8.28 ± 0.14

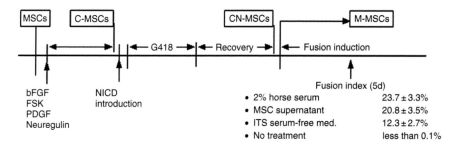

Fig. 3.113   Proliferation of cells in 3-D cell culture (WST assay).

crosslinked gelatin was soaked in EDAC solution at 4°C. After 3 min the excessive EDAC was removed and then the scaffold was further soaked with rhBMP-2 for 2 min. Primary muscle cells were isolated from newborn rat muscle and then pre-plated in collagen-coated flasks. After six serial platings, the culture was enriched with small, round cells (pp6). The cells isolated from pp6 slow adhering cells possessed round mononuclear phenotype, marked ALP stain, and matrix mineralization. The synthesis and secretion of ALP from pp6 MDCs were persistently higher than that of pp1–pp5 groups. The efficacy of rhBMP-2 immobilization on the gelatin scaffolds was manifested as the synthesis and secretion of ALP and OCN from MDCs was always significantly higher than that of the control samples.

Kaigler *et al.* tested the potential of BMSCs to modulate the growth and differentiation activities of blood vessel precursors, ECs, by their secretion of soluble angiogenic factors [285]. The growth and differentiation of cultured ECs were enhanced in response to exposure to BMSC conditioned medium (CM). Enzyme-linked immunosorbent assays demonstrated that both human and mouse BMSCs secreted significant quantities of VEGF (2.4–3.1 ng/10^6 cells per day). Visage *et al.* hypothesized that ECs combined with MSCs under conditions that mimic the natural environment of the tissue would minimize inhibitory signals and that their co-culture would thus enhance the production of an *in vitro* respiratory epithelium [286]. To test this hypothesis, they developed an *in vitro* reconstruction system for tracheal epithelium that could be useful for investigating the cellular and molecular interaction of epithelial and mesenchymal cells. In this system, a Transwell insert was used as a basement membrane on which adult bone marrow MSCs were cultured on the lower side whereas normal human bronchial epithelial (NHBE) cells were cultured on the opposite upper side. Under air–liquid interface conditions, the ECs maintained their capacity to progressively differentiate and form a functional epithelium, leading to the differentiation of mucin-producing cells between days 14 and 21. Analysis of apical secretions showed that mucin-production increased over time, with peak secretion on day 21 for NHBE cells alone, whereas mucin secretion by NHBE cells co-cultured with MSCs remained constant between days 18 and 25. This *in vitro* model of respiratory epithelium exhibited morphologic, histologic, and functional features of a tracheal mucosa.

Kihara *et al.* attempted 3-D visualization of tissue fabricated by culturing MSCs in the presence of calcein which is a fluorescent marker for bone mineralization [287]. The 3-D visualization was performed by computer-assisted confocal laser scanning microscopy. It was found that the *in vitro* tissue consisted of layers of a mineralized matrix with round cells in the matrix lacunae, an unmineralized matrix (osteoid), and osteoblastic cells on the osteoid surface.

7.2.2. Adipose-Derived Stem Cells

The procurement of MSCs from periosteum and bone marrow is tedious and gives a low yield of cells. To circumvent these problems Nathan *et al.* aimed at developing a method that would be more acceptable in the clinical setting [288]. They harvested adipose tissue from which cells are more readily obtained than from bone marrow or periosteum. *In vitro* studies were performed to assess the differentiation potential of the cells obtained from a single rabbit. Successful *in vitro* transformation into alternative mesenchymal cell lines including cardiomyocytes revealed these cells to have wide differentiation potential. Further morphological and immunohistochemical characterization and gene transfection showed features consistent with MSCs. Cultured cells were then transplanted into defects created in the left medial femoral condyle. Gross osteochondral defect reconstitution and histological grading was superior to periosteum-derived stem cell repair. Biomechanically, the repair tissue approximated intact cartilage and was superior to osteochondral autografts.

Awad *et al.* evaluated the dose-dependent effects of combinations of TGF-β1 on the chondrogenic differentiation of human adipose–derived adult stem (hADAS) cells to test the hypothesis that these factors act in a synergistic manner to promote cartilage matrix biosynthesis [289]. Human subcutaneous adipose tissue samples were obtained as waste tissue from elective surgeries with patient consent. The liposuction waste tissue was digested with type I collagenase (200 units/mg) and the floating adipocytes were separated from the precipitating stromal fraction by centrifugation. The stromal cells were then plated in tissue culture flasks in stromal medium and the primary cells (P0) were cultured for 4–5 days, and then frozen in liquid nitrogen in cryopreservation medium. Cryopreserved cells were thawed and plated in stromal medium for 5–7 days, until the cultures became confluent. Cells were harvested and then suspended in 1.2% alginate at a concentration of 4×10^6 cells/ml. Spherical beads were then created by slowly dispending droplets of the alginate cell suspension from a pipette tip into a bath of $CaCl_2$. The hADAS cells suspended within alginate beads were cultured in basal medium with insulin, transferrin, and selenious acid (ITS+) or FCS and treated with different doses and combinations of TGF-β1 and dexamethasone. The combination of ITS+ and TGF-β1 significantly increased cell proliferation. Protein synthesis rates were increased by TGF-β1 and dexamethasone in the presence of ITS+ or FCS. While TGF-β1 significantly increased proteoglycan synthesis and accumulation by 1.5- to 2-fold in the presence of FCS, such effects were suppressed by dexamethasone.

To examine whether subcutaneous adipocytes can interconvert to bone-forming cells, Justesen *et al.* investigated the bone differentiation potential of subcutaneous adipocytes and compared it with cultures of subcutaneous stromal pre-adipocytes [290]. After culture of human subcutaneous adipocytes in control medium for 1–2 weeks, the cells were incubated either in osteoblast medium (OB medium) containing various combinations of calcitriol, dexamethasone, AA, and β-glycerophosphate or in adipocyte medium (AD medium) containing HEPES, biotin, pantothenate, insulin, triiodothyronine, dexamethasone, and isobutyl-methylxanthine for 4 weeks. Cells were also implanted, mixed with HAp–TCPS powder, in the subcutaneous tissue of immunodeficient mice. One week after incubation in control medium, cells formed fusiform elongated fibroblast-like cells. In OB medium, cells stained positive for ALP and expressed mRNAs encoding Cbfa1/Runx2, ALP, and OCN. In AD medium cells re-acquired adipocyte morphology with multilocular lipid-filled cells. Also, the cells expressed adipocyte-specific mRNA markers: lipoprotein lipase and peroxisome proliferator-activated receptor γ2. Bone was formed only in the *in vivo* implants of cells incubated in OB medium.

Awad *et al.* further compared the chondrogenic differentiation of hADAS cells seeded in alginate and agarose hydrogels, and porous gelatin scaffolds in chondrogenic media containing TGF-β1 [291]. Figure 3.114 shows the DNA, sulfated GAG (S-GAG), and hydroxyproline (OHP) content of the scaffolds at different times in culture. For scaffolds grown in chondrogenic conditions, the DNA content of gelatin scaffolds was 37–51% greater than agarose and alginate scaffolds on day 14 and 28. The DNA content increased, reaching peak values of nearly 4.3 μg/scaffold on day 7 for the agarose and alginate and 6.5 μg/scaffold on day 14 for the gelatin with insignificant declines afterwards. For scaffolds grown in chondrogenic conditions, there were no significant differences in S-GAG content among the scaffold materials, whereas the OHP content in gelatin was 28–47% greater than agarose and alginate on days 14 and 28. When normalized by DNA content, S-GAG and OHP contents increased significantly between days 1 and 28 for all scaffold materials grown in chondrogenic conditions, as shown in Fig. 3.115. However, in general, there were no significant differences between the different scaffold materials.

7.2.3. NSCs

Wurmser *et al.* co-cultured mouse NSCs, which are committed to become neurons and glial cells, with human ECs, which form the lining of blood vessels [292]. They showed that in the presence of ECs 6% of the NSC population converted to cells that did not express neuronal or glial markers, but instead showed the stable expression of multiple endothelial markers and the capacity to form capillary networks. This was surprising because NSCs and ECs are believed to develop from the ectoderm and mesoderm, respectively. Experiments in which ECs were killed by fixation before co-culture with live NSCs (to prevent cell fusion) and karyotyping analyses revealed that NSCs had differentiated into endothelial-like cells independently of cell fusion.

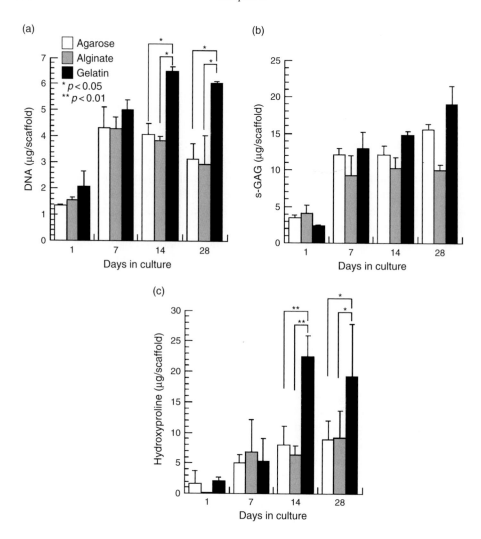

Fig. 3.114 Biochemical analysis of the scaffolds at different times in chondrogenic culture conditions.

Ko *et al.* developed an antibody microarray that permitted parallel analysis of multiple surface antigens expressed on NSCs present in a neurosphere-forming cell population [293]. As illustrated in Fig. 3.116, a microarray was prepared by micro-spotting antibodies directed to surface antigens and ligands for membrane-associated receptors onto the patterned monolayer of alkanethiols self-assembled on a gold-evaporated glass plate. Neurosphere-forming cells were subjected to a cell-binding assay on the microarray followed by immunofluorescent staining of nestin, an intra-cellular marker of NSCs. As shown in Fig. 3.117, cell binding was obviously dependent on the type of immobilized antibodies and proteins. Moreover, in most

(a)

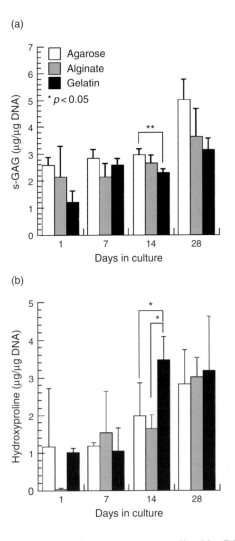

Fig. 3.115 S-GAG and hydroxyproline contents normalized by DNA content.

cases, the density of bound cells varied depending on cell passages, indicating the passage-dependent alteration of surface antigen expression. Regardless of the cell passages, almost no cells were bound to the spots with antibody against CD9, CD31, CD34, CD81, CD90, O4, FGFR, and NGFR. Such a cell-based assay facilitated to examine the specificity of surface antigens for nestin-positive NSCs. The microarray could also be used to assess the proliferation capability of cells bound to individual spots. To prepare protein microarrays that allow the functional analysis of proteins at a cellular level, Kato *et al.* utilized recombinant proteins genetically engineered to carry a fusion tag that has an affinity for metal ions, as shown in Fig. 3.118 [294].

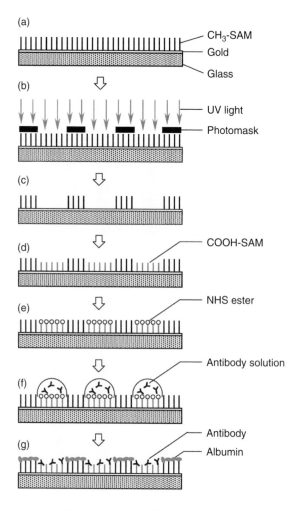

Fig. 3.116 Schematic for the preparation of an antibody microarray.

The feasibility of the method was demonstrated by culturing NSCs on the microarray that displayed oligohistidine-tagged epidermal growth factor.

7.2.4. ES Cells

Harrison *et al.* determined the ability of various poly(α-hydroxy esters) to support the *in vitro* propagation of murine ES cells in an undifferentiated state [295]. By the analysis of live and dead cell number indices and Oct-4 immunoreactivity, ES cell colonization rate during a 48-h culture period was found to be significantly greater on PLGA compared to PDLLA, PLLA, and PGA. Surface treatment of all polymers with 0.1 M potassium hydroxide revealed a significant increase in ES cell live numbers when compared to all unmodified polymers, thus revealing a correlation between polymer composition, hydrophilicity, and colonization rate.

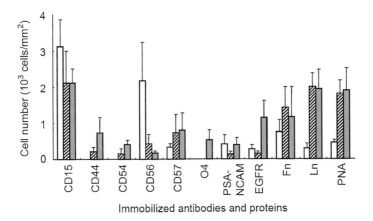

Fig. 3.117 The number of cells bound to the spot with immobilized antibodies and other proteins. Cell passage: P0 (open bar), P1 (hatched bar), and P4 (closed bar).

To control ES cell proliferation and differentiation into higher-order structures, Levenberg *et al.* examined the use of biodegradable polymer scaffolds for promoting human ES (hES) cell growth and differentiation and formation of complex 3-D structures [296]. The scaffolds consisted of a 50/50 blend of PLGA and PLLA. The sponges were fabricated by a salt-leaching process. For cell differentiation experiments, the sponges were cut into rectangular pieces of $5 \times 4 \times 1$ mm^3. For seeding on scaffolds, 0.8×10^6 cells (from undifferentiated hES or embryoid bodies [EBs] at day 8) were seeded on each scaffold by using 25 μl of a mixture containing 50% of Matrigel and the respective EB medium. After seeding the cells, scaffolds were suspended in petri dishes in their respective medium. Growth factors such as

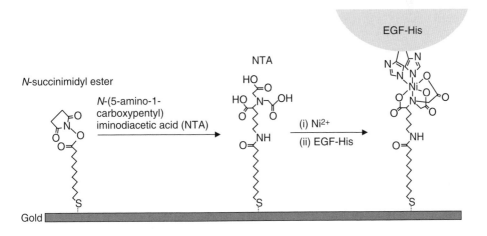

Fig. 3.118 Procedure for the immobilization of EGF-His onto an alkanethiol self-assembled monolayer.

retinoic acid, FGF-β, activin-A, or insulin-like growth factor induced differentiation into 3-D structures with characteristics of developing neural tissues, cartilage, or liver, respectively. In addition, formation of a 3-D vessel-like network was observed. When transplanted into SCID mice, the constructs continued to express specific human proteins in defined differentiated structures and appeared to recruit and anastomose with the host vasculature. Levenberg *et al.* further investigated the potential of various neurotrophins, including NGF, and neurotrophin 3 (NT-3), to induce neuronal differentiation of human ES cells on 3-D scaffolds fabricated from degradable poly(α-hydroxy esters) including PLGA and PLLA [297]. When cultured *in vitro*, neural rosette-like structures developed throughout the scaffolds with differentiation dependent on factors in the medium (*e.g.*, retinoic acid [RA], NGF, and NT-3) and the differentiation stage of the cells. Specifically, enhanced numbers of neural structures and staining of nestin and β_{III}-tubulin (indicative of neural differentiation) were observed with human ES cell–seeded polymer scaffolds when cultured with both NGF and NT-3 when compared with control medium. In addition, vascular structures were found throughout the engineered tissues when cultured with the neurotrophins, but not in the presence of RA.

After developing methods to isolate highly purified (>96%) proliferating populations of ECs from mouse ES cells, McCloskey *et al.* tested their ability to form 3-D vascular structures *in vitro* [298]. The ES cell–derived ECs were embedded in 3-D collagen gel constructs with rat tail type I collagen (2 mg/ml) at a concentration of 10^6 cells/ml of gel. The gels were observed daily with a phase-contrast microscope to analyze the time course for EC assembly. The first vessels were observed between days 3 and 5 after gel construct formation. The number and complexity of structures steadily increased, reaching a maximum before beginning to regress. By 2 weeks, all vessel-like structures had regressed back to single cells. Histology and fluorescent images of the vessel-like structures verified that tube structures were multicellular and could develop patent lumens. Co-culture of stem/progenitor cells with mature cells or tissues can drive their differentiation toward required lineages. Thus, Van Vranken *et al.* hypothesized that co-culture of murine ES cells with embryonic mesenchyme from distal lung promotes the differentiation of pneumocytes [299]. Murine ES cells were differentiated to EBs and cultured for 5 or 12 days with pulmonary mesenchyme from embryonic day 11.5 or 13.5 murine embryos, in direct contact or separated by a membrane. Controls included EBs cultured alone or with embryonic gut mesenchyme. Histology revealed epithelium-lined channels in directly co-cultured EBs, whereas EBs grown alone showed little structural organization. The lining cells expressed cytokeratin and thyroid transcription factor 1, an early developmental marker in pulmonary epithelium. Differentiation of type II pneumocytes specifically was demonstrated by the presence of surfactant protein C (SP-C) in some of the ECs. None of these markers was seen in EBs cultured alone or with embryonic gut mesenchyme. Indirect co-culture of EBs with lung mesenchyme resulted in a 14-fold increase in SP-C gene expression. Levenberg *et al.* isolated human embryonic ECs from day 13–15 EBs by using platelet EC-adhesion molecule-1 (PECAM1) antibodies and characterized their behavior *in vitro* and *in vivo* [300]. The isolated embryonic PECAM1+ cells, grown in culture, displayed characteristics similar to vessel endothelium. The cells expressed EC markers in a pattern similar to human umbilical vein ECs, their junctions were correctly organized, and they had high metabolism of acetylated low-density lipoprotein. In addition, the cells were able to differentiate and form tube-like structures when cultured on Matrigel. *In vivo*, when transplanted into SCID mice, the cells appeared to form microvessels containing mouse blood cells.

Bielby *et al.* investigated whether methods for driving osteogenic differentiation developed with murine ES cells could be applied successfully to hES cells [301]. The H1 line was propagated *in vitro* on murine feeder layers and shown to be fully pluripotent by expression of the markers Oct-4 and SSEA-4. Subsequently, differentiation was initiated via EB formation and, after 5 days in suspension culture, cells harvested from EBs were replated in a medium containing osteogenic supplements. They found that the treatment regimen previously identified as optimal for murine ES cells, and in particular the addition of dexamethasone at specific time points, also induced the greatest osteogenic response from hES cells. They identified mineralizing cells *in vitro* that immunostained positively for OCN and found an increase in expression of an essential bone transcription factor, Runx2. When implanted into SCID mice on a PDLLA scaffold, the cells had the capacity to give rise to mineralized tissue *in vivo*. After 35 days of implantation, regions of mineralized tissue could be identified within the scaffold by von Kossa staining and immunoexpression of the human form of OCN. They did not see any evidence of teratoma formation.

Induction of dopaminergic neurons from ES cells using mouse stromal cells (PA6 cells) as a feeder layer seems to be efficient to obtain dopamine-releasing cells for cell transplantation treatment of Parkinson's disease. Yamazoe *et al.* prepared stock solutions containing neural inducing factors (NIFs) by washing PA6 cells with PBS containing heparin [302]. The ES cells grew successfully in culture media supplemented with 33% NIFs stock solution, and the rate of neural differentiation of ES cell progeny increased with increasing heparin concentration in the culture solutions. The NIF-immobilized surfaces were prepared by exposing polyethyleneimine-modified surfaces to NIF stock solutions. The NIF-immobilized culture dish effectively supported cell growth as the culture medium supplemented with NIFs stock did, but its induction effect to dopaminergic neurons from ES cells was much smaller than free NIFs. Thus, NIFs stock solutions had two different activities; stimulation of cell growth and induction of ES cell differentiation to the neural fate in the presence of heparin. To achieve sustained engraftment of donor cells, Zhan *et al.* aimed to develop methods to efficiently differentiate human ES cells to hematopoietic cells, including immune-modulating leucocytes, a prerequisite of the tolerance induction strategies applying to human ES cell-mediated transplantation [303].

Bielby *et al.* examined whether inorganic soluble extracts prepared from 58S sol–gel bioactive glasses could affect the differentiation of osteoblasts in an ES-cell based system [304]. Previous work has demonstrated the ability of soluble ions released from bioactive glasses undergoing dissolution *in vitro* to stimulate gene expression characteristic of a mature phenotype in primary osteoblasts. Differentiation of ES cells into osteogenic cells was characterized by the formation of multilayered, mineralized nodules. These nodules contained cells expressing the transcription factor runx2/cbfa-1, and deposition of OCN in the ECM was detected by immunostaining. When differentiating cells were placed in an osteoblast maintenance medium supplemented with soluble extracts prepared from bioactive glass powders, increased formation of mineralized nodules (98 ± 6%, mean ± SEM) and ALP activity (56 ± 14%, mean ± SEM) were observed in a pattern characteristic of osteoblast differentiation. This effect of the glass extracts exhibited dose dependency, with ALP activity and nodule formation increasing with extract concentrations. Compared with medium supplemented with dexamethasone, which has been used to enhance osteoblast lineage derivation, the glass extracts were effective at inducing formation of mineralized nodules by murine ES cells.

Liu and Roy studied hematopoietic differentiation of ES cells within a biomimetic culture environment and specifically determined the effects of 3-D matrix scaffold alone as well as combined with spinner flask technology on the hematopoietic differentiation of mouse ES cell [305]. The ES cells differentiated on porous 3-D scaffold structures developed EBs similar to those in traditional 2-D cultures, but, unlike 2-D differentiation, these EBs integrated with the scaffold and appeared embedded in a network of ECM. The efficiency of hematopoietic precursor cells (HPC) generation on 3-D, as indicated by the expression of various HPC-specific surface markers (CD34, Sca-1, Flk-1, and c-Kit) and colony-forming cell (CFC) assays, was increased (about twofold) over their 2-D counterparts. Spinner flask technology also contributed to the higher hematopoietic differentiation efficiency of ES cells seeded on scaffolds. Continued differentiation of 3-D-derived HPCs into the myeloid lineage demonstrated increased efficiency (twofold) of generating myeloid compared with differentiation from 2-D-derived HPCs.

Human ES cells are known to derive from the inner cell mass of blastocyst. Although the embryos of other developmental stages have also been used as a source for ES cells in animal models, the feasibility of obtaining ES cell lines from human morula is not known, despite being an obvious source available through assisted reproduction and preimplantation genetic diagnosis programs. Strelchenko *et al.* described a technique for derivation of ES cells from human morula [306]. Established eight morula-derived ES cell lines were shown to have no morphological differences from the ES cells derived from blastocysts. They expressed the same ES cell–specific markers, including Oct-4, TRA-2–39, stage-specific embryonic antigens SSEA-3 and SSEA-4, and high MW glycoproteins TRA-1–60 and TRA-1–81, detected in the same colony of morula-derived ES cells showing specific ALP expression. No differences were observed in these marker expressions in the morula-derived ES cells cultured in the feeder layer–free medium. Similar to ES cells originating from blastocyst, the morula-derived ES cells were shown to spontaneously differentiate *in vitro* into a variety of cell types, including the neuron-like cells and contracting primitive cardiocyte-like cells.

The cytoplasm of an oocyte can reprogram the genome of a somatic cell to an embryonic state. Cowan *et al.* have demonstrated that human ES cells also have the

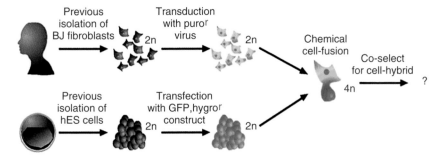

Fig. 3.119 Generation of stable hybrid cells through the fusion of hES cells and human somatic cells.

capacity to reprogram adult somatic cell chromosomes after cell fusion [307]. As illustrated in Fig. 3.119 existing somatic and human ES cell lines were stably transduced or transfected with independent drug-resistant markers and induced to undergo cell fusion in the presence of PEG. Fused cells were then grown under standard conditions for the maintenance of human ES cells in the presence of antibiotics to select for cell hybrids. This approach might lead to an alternative route for creating genetically tailored human ES cell lines for use in the treatment of human disease. However, a substantial technical barrier remains before human ES cells could be used for therapeutic purposes: the elimination of the ES cell chromosomes either before or after cell fusion. A brief profile of K. Eggan who is one of the authors of this work was given by Check [308].

REFERENCES

1. L. Ma, C. Gao, Z. Mao *et al.*, Collagen/chitosan porous scaffolds with improved biostability for skin tissue engineering, *Biomaterials*, **24**, 4833 (2003).
2. A.B. Caruso and M.G. Dunn, Functional evaluation of collagen fiber scaffolds for ACL reconstruction: Cyclic loading in proteolytic enzyme solutions, *J. Biomed. Mater. Res.*, **69A**, 164 (2004).
3. T. Taguchi, L. Xu, H. Kobayashi *et al.*, Encapsulation of chondrocytes in injectable alkali-treated collagen gels prepared using poly(ethylene glycol)-based 4-armed star polymer, *Biomaterials*, **26**, 1247 (2005).
4. J.M. Orban, L.B. Wilson, J.A. Kofroth *et al.*, Crosslinking of collagen gels by transglutaminase, *J. Biomed. Mater. Res.*, **68A**, 756 (2004).
5. P. Angele, J. Abke, R. Kujat *et al.*, Influence of different collagen species on physicochemical properties of crosslinked collagen matrices, *Biomaterials*, **25**, 2831 (2004).
6. M.J. Yost, C.F. Baicu, C.E. Stonerock *et al.*, A novel tubular scaffold for cardiovascular tissue engineering, *Tissue Engineering*, **10**, 273 (2004).
7. Y. Nahmias, A. Arneja, T.T. Tower *et al.*, Cell patterning on biological gels via cell spraying through a mask, *Tissue Engineering*, **11**, 701 (2005).
8. E. E.A. Gentleman, K.C. Nauman, G.A. Dee *et al.*, Short collagen fibers provide control of contraction and permeability in fibroblast-seeded collagen gels, *Tissue Engineering*, **10**, 421 (2004).

9. K.E. Lewus and E.A. Nauman, *In vitro* characterization of a bone marrow stem cell-seeded collagen gel composite for soft tissue grafts: Effects of fiber number and serum concentration, *Tissue Engineering*, **11**, 1015 (2005).

10. A.S. Sosnik and M.J. Sefton, Semi-synthetic collagen/poloxamine matrices for tissue engineering, *Biomaterials*, **26**, 7425 (2005).

11. Y. Liu, X.Z. Shu, S.D. Gray *et al.*, Disulfide-crosslinked hyaluronan-gelatin sponge: Growth of fibrous tissue *in vivo*, *J. Biomed. Mater. Res.*, **68A**, 142 (2004).

12. R.E. Unger, M. Wolf, K. Peters *et al.*, Growth of human cells on a non-woven silk fibroin net: A potential for use in tissue engineering, *Biomaterials*, **25**, 1069 (2004).

13. L. Meinel, S. Hofmann, V. Karageorgiou *et al.*, The inflammatory responses to silk films *in vitro* and *in vivo*, *Biomaterials*, **26**, 147 (2005).

14. H.J. Kim, U.-J. Kim, G. Vunjak-Novakovic *et al.*, Influence of macroporous protein scaffolds on bone tissue engineering from bone marrow stem cells, *Biomaterials*, **26**, 4442 (2005).

15. S. Zhang *et al.*, Fabrication of novel biomaterials through molecular self-assembly, *Nature Biotech.*, **21**, 1171 (2003).

16. E. Genove, C. Shen, S. Zhang *et al.*, The effect of functionalized self-assembling peptide scaffolds on human aortic endothelial cell function, *Biomaterials*, **26**, 3341 (2005).

17. R. Barbucci, R. Rappuoli, A. Borzacchiello *et al.*, Synthesis, chemical and rheological characterization of new hyaluronic acid-based hydrogels, *J. Biomater. Sci. Polymer Edn*, **11**, 383 (2000).

18. R. Barbucci and G. Leone, Formation of defined microporous 3D structures starting from cross-linked hydrogels, *J. Biomed. Mater. Res. Part B: Appl. Biomater.*, **68B**, 117 (2004).

19. X.Z. Shu, Y. Liu, Y. Luo *et al.*, Disulfide cross-linked hyaluronan hydrogels, *Biomacromolecules*, **3**, 1304 (2002).

20. J.B. Leach, K.A. Bivens, C.N. Collins *et al.*, Development of photocrosslinkable hyaluronic acid-polyethylene glycol-peptide composite hydrogels for soft tissue engineering, *J. Biomed. Mater. Res.*, **70A**, 74 (2004).

21. M.M. Stevens, H.F. Qanadilo, R. Langer *et al.*, A rapid-curing alginate gel system: Utility in periosteum-derived cartilage tissue engineering, *Biomaterials*, **25**, 887 (2004).

22. J.L. Drury, R.G. Dennis, and D.J. Mooney, The tensile properties of alginate hydrogels, *Biomaterials*, **25**, 3187 (2004).

23. T. Boontheekul, H.-J. Kong, and D.J. Mooney, Controlling alginate gel degradation utilizing partial oxidation and bimodal molecular weight distribution, *Biomaterials*, **26**, 2455 (2005).

24. Q. Li, C.G. Williams, D.D.N. Sun *et al.*, Photocrosslinkable polysaccharides based on chondroitin sulfate, *J. Biomed. Mater. Res.*, **68A**, 28 (2004).

25. T. Freier, R. Montenegro, H.S. Koh *et al.*, Chitin-based tubes for tissue engineering in the nervous system, *Biomaterials*, **26**, 4624 (2005).

26. C.-H. Chang, H.-C. Liu, C.-C. Lin *et al.*, Gelatin-chondroitin-hyaluronan tri-copolymer scaffold for cartilage tissue engineering, *Biomaterials*, **24**, 4853 (2003).

27. L. Ferreira, M.H. Gil, A.M.S. Cabrita *et al.*, Biocatalytic synthesis of highly ordered degradable dextran-based hydrogels, *Biomaterials*, **26**, 4707 (2005).

28. M.M. Stevens, M. Mayer, D.G. Anderson *et al.*, Direct patterning of mammalian cells onto porous tissue engineering substrates using agarose stamps, *Biomaterials*, **26**, 7636 (2005).

29. J.M. Moran, D. Pazzano, and L.J. Bonassar, Characterization of polylactic acid-polyglycolic acid composites for cartilage tissue engineering, *Tissue Engineering*, **9**, 63 (2003).

30. A. Prokop, A. Jubel, H.J. Helling *et al.*, Soft tissue reactions of different biodegradable polylactide implants, *Biomaterials*, **25**, 259 (2004).
31. K.G. Shea, R.D. Bloebaum, J.M. Avent *et al.*, Analysis of lymph nodes for polyethylene particles in patients who have had a primary joint replacement, *J. Bone Jt. Surg.*, **78A**, 497 (1996).
32. C.G. Pitt, F.I. Chasalow, Y.M. Hibionada *et al.*, Aliphatic polyesters. I. The degradation of poly(ε-caprolactone) *in vivo*, *J. Appl. Polym. Sci.*, **26**, 3779 (1981).
33. S.F. Williams, D.P. Martin, D.M. Horowitz *et al.*, PHA applications: Addressing the price performance issue: I. Tissue engineering, *Int. J. Biol. Macromol.*, **25**, 111 (1999).
34. S.J. Holland, A.M. Jolly, M. Yasin *et al.*, Polymers for biodegradable medical devices. II. Hydroxybutyrate-hydroxyvalerate copolymers: Hydrolytic degradation studies, *Biomaterials*, **8**, 289 (1987).
35. K. Zhu, X. Lin, and S. Yang, Preparation and properties of D,L-lactide and ethylene oxide copolymer: A modifying biodegradable polymeric material, *J. Polym. Sci.*, Part C, *Polym. Lett.*, **24**, 331 (1986).
36. E. Lieb, J. Tessmar, M. Hacker *et al.*, Poly(D,L-lactic acid)-poly(ethylene glycol)-monomethyl ether diblock copolymers control adhesion and osteoblastic differentiation of marrow stromal cells, *Tissue Engineering*, **9**, 71 (2003).
37. N. Saito, T. Okada, H. Horiuchi *et al.*, A biodegradable polymer as a cytokine delivery system for inducing bone formation, *Nature Biotechnology*, **19**, 332 (2001).
38. K. Kim, M. Yu, X. Zong *et al.*, Control of degradation rate and hydrophilicity in electrospun non-woven poly(D,L-lactide) nanofiber scaffolds for biomedical applications, *Biomaterials*, **24**, 4977 (2003).
39. M.-H. Huang, S. Li, D.W. Hutmacher *et al.*, Degradation and cell culture studies on block copolymers prepared by ring opening polymerization of epsilon-caprolactone in the presence of poly(ethylene glycol), *J. Biomed. Mater. Res.*, **69A**, 417 (2004).
40. M. Deng, R. Wang, G. Rong *et al.*, Synthesis of a novel structural triblock copolymer of poly(gamma-benzyl-L-glutamic acid)-b-poly (ethylene oxide)-b-poly(epsilon-caprolactone), *Biomaterials*, **25**, 3553 (2004).
41. M.F. Meek, K. Jansen, R. Steendam *et al.*, *In vitro* degradation and biocompatibility of poly(DL-lactide-epsilon-caprolactone) nerve guides, *J. Biomed. Mater. Res.*, **68A**, 43 (2004).
42. S.I. Jeong, B.-S. Kim, W. Kang *et al.*, *In vivo* biocompatibility and degradation behavior of elastic poly(L-lactide-co-epsilon-caprolactone) scaffolds, *Biomaterials*, **25**, 5939 (2004).
43. A.P. Pego, A.A. Poot, D.W. Grijpma *et al.*, Biodegradable elastomeric scaffolds for soft tissue engineering, *J. Controlled Release*, **87**, 69 (2003).
44. D.W. Grijpma, Q. Hou, and J. Feijen, Preparation of biodegradable networks by photo-crosslinking lactide, epsilon-caprolactone and trimethylene carbonate-based oligomers functionalized with fumaric acid monoethyl ester, *Biomaterials*, **26**, 2795 (2005).
45. A.P. Pego, M.J.A.Van Luyn, L.A. Brouwer *et al.*, *In vivo* behavior of poly(1,3-trimethylene carbonate) and copolymers of 1,3-trimethylene carbonate with D,L-lactide or epsilon-caprolactone: Degradation and tissue response, *J. Biomed. Mater. Res.*, **67A**, 1044 (2003).
46. A. Goraltchouk, T. Freier, and M.S. Shoichet, Synthesis of degradable poly(L-lactide-*co*-ethylene glycol) porous tubes by liquid-liquid centrifugal casting for use as nerve guidance channels, *Biomaterials*, **26**, 7555 (2005).
47. M. Mizutani and T. Matsuda, Liquid acrylate-endcapped biodegradable poly(epsilon-caprolactone-co-trimethylene carbonate). I. Preparation and visible light-induced photocuring characteristics, *J. Biomed. Mater. Res.*, **62**, 387 (2002).

48. H.M. Younes, E. Bravo-Grimaldo, and B.G. Amsden, Synthesis, characterization and *in vitro* degradation of a biodegradable elastomer, *Biomaterials*, **25**, 5261 (2004).

49. Z. Ding, J. Chen, S. Gao *et al.*, Immobilization of chitosan onto poly-L-lactic acid film surface by plasma graft polymerization to control the morphology of fibroblast and liver cells, *Biomaterials*, **25**, 1059 (2004).

50. F. Cellesi, N. Tirelli, and J.A. Hubbell, Towards a fully-synthetic substitute of alginate: Development of a new process using thermal gelation and chemical cross-linking, *Biomaterials*, **25**, 5115 (2004).

51. K.S. Masters, D.N. Shah, L.A. Leinwand *et al.*, Crosslinked hyaluronan scaffolds as a biologically active carrier for valvular interstitial cells, *Biomaterials*, **26**, 2517 (2005).

52. B. Dhariwala, E. Hunt, and T. Boland, Rapid prototyping of tissue-engineering constructs, using photopolymerizable hydrogels and stereolithography, *Tissue Engineering*, **10**, 1316 (2004).

53. V. Mironov, V. Kasyanov, X.Z. Shu *et al.*, Fabrication of tubular tissue constructs by centrifugal casting of cells suspended in an *in situ* crosslinkable hyaluronan-gelatin hydrogel, *Biomaterials*, **26**, 7628 (2005).

54. J. Guan, M.S. Sacks, E.J. Beckman *et al.*, Biodegradable poly(ether ester urethane)urea elastomers based on poly(ether ester) triblock copolymers and putrescine: Synthesis, characterization and cytocompatibility, *Biomaterials*, **25**, 85 (2004).

55. B. Saad, P. Neuenschwander, G.K. Uhlschmid *et al.*, New versatile, elastomeric, degradable polymeric materials for medicine, *Int. J. Biol. Macromol.*, **25**, 293 (1999).

56. S. Grad, L. Kupcsik, K. Gorna *et al.*, The use of biodegradable polyurethane scaffolds for cartilage tissue engineering: Potential and limitations, *Biomaterials*, **24**, 5163 (2003).

57. J.Y. Zhang, E.J. Beckman, N.P. Piesco *et al.*, A new peptide-based urethane polymer: Synthesis, biodegradation, and potential to support cell growth *in vitro*, *Biomaterials*, **21**, 1247 (2001).

58. J. Zhang, B.A. Doll, E.J. Beckman *et al.*, A biodegradable polyurethane-ascorbic acid scaffold for bone tissue engineering, *J. Biomed. Mater. Res.*, **67A**, 389 (2003).

59. J. Guan, K.L. Fujimoto, M.S. Sacks *et al.*, Preparation and characterization of highly porous, biodegradable polyurethane scaffolds for soft tissue applications, *Biomaterials*, **26**, 3961 (2005).

60. D. Dean, N.S. Topham, C. Meneghetti *et al.*, Poly(propylene fumarate) and poly(DL-lactic-co-glycolic acid) as scaffold materials for solid and foam-coated composite tissue-engineered constructs for cranial reconstruction, *Tissue Engineering*, **9**, 495 (2003).

61. E.L. Hedberg, C.K. Shih, J.J. Lemoine *et al.*, *In vitro* degradation of porous poly (propylene fumarate)/poly(DL-lactic-co-glycolic acid) composite scaffolds, *Biomaterials*, **26**, 3215 (2005).

62. S.L. Bourke, J. Kohn, and M.G. Dunn, Preliminary development of a novel resorbable synthetic polymer fiber scaffold for anterior cruciate ligament reconstruction, *Tissue Engineering*, **10**, 43 (2004).

63. Y. Cui, X. Zhao, X. Tang *et al.*, Novel micro-crosslinked poly(organophosphazenes) with improved mechanical properties and controllable degradation rate as potential biodegradable matrix, *Biomaterials*, **25**, 451 (2004).

64. A.A. Deschamps, A.A. van Apeldoorn, H. Hayen *et al.*, *In vivo* and *in vitro* degradation of poly(ether ester) block copolymers based on poly(ethylene glycol) and poly(butylene terephthalate), *Biomaterials*, **25**, 247 (2004).

65. Y. Wang, G.A. Ameer, B.J. Sheppard, and R. Langer, A tough biodegradable elastomer, *Nat. Biotechnol.*, **20**, 602 (2002).

66. H.H.K. Xu, S. Takagi, J.B. Quinn *et al.*, Fast-setting calcium phosphate scaffolds with tailored macropore formation rates for bone regeneration, *J. Biomed. Mater. Res.*, **68A**, 725 (2004).

67. J.J. Blaker, J.E. Gough, V. Maquet *et al.*, *In vitro* evaluation of novel bioactive composites based on Bioglass-filled polylactide foams for bone tissue engineering scaffolds, *J. Biomed. Mater. Res.*, **67A**, 1401 (2003).

68. M. Kikuchi, Y. Koyama, T. Yamada *et al.*, Development of guided bone regeneration membrane composed of beta-tricalcium phosphate and poly(L-lactide-co-glycolide-co-epsilon-caprolactone) composites, *Biomaterials*, **25**, 5979 (2004).

69. C. Zou, W. Weng, X. Deng *et al.*, Preparation and characterization of porous beta-tricalcium phosphate/collagen composites with an integrated structure, *Biomaterials*, **26**, 5276 (2005).

70. H.-W. Kim, H.-E. Kim, and V. Salih, Stimulation of osteoblast responses to biomimetic nanocomposites of gelatin-hydroxyapatite for tissue engineering scaffolds, *Biomaterials*, **26**, 5221 (2005).

71. F.J. O'Brien, B.A. Harley, I.V. Yannas *et al.*, Influence of freezing rate on pore structure in freeze-dried collagen-GAG scaffolds, *Biomaterials*, **25**, 1077 (2004).

72. M.-H. Ho, P.-Y. Kuo, H.-J. Hsieh *et al.*, Preparation of porous scaffolds by using freeze-extraction and freeze-gelation methods, *Biomaterials*, **25**, 129 (2004).

73. Y. Cao, M.R. Davidson, A.J. O'Connor *et al.*, Architecture control of three-dimensional polymeric scaffolds for soft tissue engineering. I. Establishment and validation of numerical models, *J. Biomed. Mater. Res.*, **71A**, 81 (2004).

74. S. Stokols and M.H. Tuszynski, The fabrication and characterization of linearly oriented nerve guidance scaffolds for spinal cord injury, *Biomaterials*, **25**, 5839 (2004).

75. Z. Ma, C. Gao, Y. Gong *et al.*, Paraffin spheres as porogen to fabricate poly(L-lactic acid) scaffolds with improved cytocompatibility for cartilage tissue engineering, *J. Biomed. Mater. Res. Part B: Applied Biomater.*, **67B**, 610 (2003).

76. M. Hacker, J. Tessmar, M. Neubauer *et al.*, Towards biomimetic scaffolds: Anhydrous scaffold fabrication from biodegradable amine-reactive diblock copolymers, *Biomaterials*, **24**, 4459 (2003).

77. D. Tadic, F. Beckmann, K. Schwarz *et al.*, A novel method to produce hydroxyapatite objects with interconnecting porosity that avoids sintering, *Biomaterials*, **25**, 3335 (2004).

78. J.-Y. Zhang, B.A. Doll, E.J. Beckman *et al.*, Three-dimensional biocompatible ascorbic acid-containing scaffold for bone tissue engineering, *Tissue Engineering*, **9**, 1143 (2003).

79. K.-U. Lewandrowski, D.D. Hile. B.M.J. Thompson *et al.*, Quantitative measures of osteoinductivity of a porous poly(propylene fumarate) bone graft extender, *Tissue Engineering*, **9**, 85 (2003).

80. T. Cao, K.-H. Ho, and S.-H. Teoh, Scaffold design and *in vitro* study of osteochondral coculture in a three-dimensional porous polycaprolactone scaffold fabricated by fused deposition modeling. *Tissue Engineering*, **9**, S-103 (2003).

81. C.M. Smith, A.L. Stone, R.L. Parkhill *et al.*, Three-dimensional bioassembly tool for generating viable tissue-engineered constructs, *Tissue Engineering*, **10**, 1566 (2004).

82. Y. Yang, S. Basu, D.L. Tomasko *et al.*, Fabrication of well-defined PLGA scaffolds using novel microembossing and carbon dioxide bonding, *Biomaterials*, **26**, 2585 (2005).

83. M. Lee, J.C.Y. Dunn, and B.M. Wu, Scaffold fabrication by indirect three-dimensional printing, *Biomaterials*, **26**, 4281 (2005).

84. J.M. Williams, A. Adewunmi, R.M. Schek *et al.*, Bone tissue engineering using polycaprolactone scaffolds fabricated via selective laser sintering, *Biomaterials*, **26**, 4817 (2005).

85. I.K. Kwon, S. Kidoaki, and T. Matsuda, Electrospun nano- to microfiber fabrics made of biodegradable copolyesters: structural characteristics, mechanical properties and cell adhesion potential, *Biomaterials*, **26**, 3929 (2005).

86. X.M. Mo, C.Y. Yu, M. Kotaki *et al.*, Electrospun P(LLA-CL) nanofiber: A biomimetic extracellular matrix for smooth muscle cell and endothelial cell proliferation, *Biomaterials*, **25**, 1883 (2004).

87. K. Fujihara, M. Kotaki, and S. Ramakrishna, Guided bone regeneration membrane made of polycaprolactone/calcium carbonate composite nano-fibers, *Biomaterials*, **26**, 4139 (2005).

88. X. Zong, H. Bien, C.-Y. Chung *et al.*, Electrospun fine-textured scaffolds for heart tissue constructs, *Biomaterials*, **26**, 5330 (2005).

89. Z. Ma, M. Kotaki, R. Inai *et al.*, Potential of nanofiber matrix as tissue-engineering scaffolds, *Tissue Engineering*, **11**, 101 (2005).

90. X. Geng, O.-H. Kwon, and J. Jang, Electrospinning of chitosan dissolved in concentrated acetic acid solution, *Biomaterials*, **26**, 5427 (2005).

91. M. Li, M.J. Mondrinos, M.R. Gandhi *et al.*, Electrospun protein fibers as matrices for tissue engineering, *Biomaterials*, **26**, 5999 (2005).

92. E. Leclerc, K.S. Furukawa, F. Miyata *et al.*, Fabrication of microstructures in photosensitive biodegradable polymers for tissue engineering applications, *Biomaterials*, **25**, 4683 (2004).

93. T.B.F. Woodfield, J. Malda, J. de Wijn *et al.*, Design of porous scaffolds for cartilage tissue engineering using a three-dimensional fiber-deposition technique, *Biomaterials*, **25**, 4149 (2004).

94. Y. Luo and M.S. Shoichet, A photolabile hydrogel for guided three-dimensional cell growth and migration, *Nature Materials*, **3**, 249 (2004).

95. W.F. Daamen, H. Th.B. van Moerkerk, T. Hafmans *et al.*, Preparation and evaluation of molecularly-defined collagen-elastin-glycosaminoglycan scaffolds for tissue engineering, *Biomaterials*, **24**, 4001 (2003).

96. W.F. Elbjeirami, E.O. Yonter, B.C. Starcher *et al.*, Enhancing mechanical properties of tissue-engineered constructs via lysyl oxidase crosslinking activity, *J. Biomed. Mater. Res.*, **66A**, 513 (2003).

97. T.W. Hudson, S.Y. Liu, and C.E. Schmidt, Engineering an improved acellular nerve graft via optimized chemical processing, *Tissue Engineering*, **10**, 1346 (2004).

98. J.F. Cartmell and M.G. Dunn, Development of cell-seeded patellar tendon allografts for anterior cruciate ligament reconstruction, *Tissue Engineering*, **10**, 1065 (2004).

99. A.L. Brown, T.T. Brook-Allred, J.E. Waddell *et al.*, Bladder acellular matrix as a substrate for studying *in vitro* bladder smooth muscle–urothelial cell interactions, *Biomaterials*, **26**, 529 (2005).

100. S.-H. Lu, M.S. Sacks, S.Y. Chung *et al.*, Biaxial mechanical properties of muscle-derived cell seeded small intestinal submucosa for bladder wall reconstitution, *Biomaterials*, **26**, 443 (2005).

101. W.F. Daamen, T. Hafmans, J.H. Veerkamp *et al.*, Isolation of intact elastin fibers devoid of microfibrils, *Tissue Engineering*, **11**, 1168 (2005).

102. J.D. Berglund, R.M. Nerem, and A. Sambanis, Incorporation of intact elastin scaffolds in tissue-engineered collagen-based vascular grafts, *Tissue Engineering*, **10**, 1526 (2004).

103. T. Walles, B. Giere, M. Hofmann *et al.*, Experimental generation of a tissue-engineered functional and vascularized trachea, *J. Thorac. Cardiovasc. Surg.*, **128**, 900 (2004).

104. J.R. Mauney, C. Jaquiery, V. Volloch *et al.*, *In vitro* and *in vivo* evaluation of differentially demineralized cancellous bone scaffolds combined with human bone marrow stromal cells for tissue engineering, *Biomaterials*, **26**, 3173 (2005).

105. D.O. Freytes, A.E. Rundell, J.V. Geest *et al.*, Analytically derived material properties of multilaminated extracellular matrix devices using the ball-burst test, *Biomaterials*, **26**, 5518 (2005).
106. A. Mol, M.I. van Lieshout, C.G. Dam-de Veen *et al.*, Fibrin as a cell carrier in cardiovascular tissue engineering applications, *Biomaterials*, **26**, 3113 (2005).
107. A.H. Zisch, S.M. Zeisberger, M. Ehrbar *et al.*, Engineered fibrin matrices for functional display of cell membrane-bound growth factor-like activities: Study of angiogenic signaling by ephrin-B2, *Biomaterials*, **25**, 3245 (2004).
108. D. Green, D. Howard, X. Yang *et al.*, Natural marine sponge fiber skeleton: A biomimetic scaffold for human osteoprogenitor cell attachment, growth, and differentiation, *Tissue Engineering*, **9**, 1159 (2003).
109. K.J. Burg, C.E. Austin, C.R. Culberson *et al.*, A novel approach to tissue engineering: Injectable composites, *Transact. 2000 World Biomaterials Cong.*, May 2000.
110. J.B. McGlohorn, L.W. Grimes, S.S. Webster *et al.*, Characterization of cellular carriers for use in injectable tissue-engineering composites, *J. Biomed. Mater. Res.*, **66A**, 441 (2003).
111. S.-H. Lee, B.-S. Kim, S.-H. Kim *et al.*, Elastic biodegradable poly(glycolide-co-caprolactone) scaffold for tissue engineering, *J. Biomed. Mater. Res.*, **66A**, 29 (2003).
112. J. Dong, T. Uemura, Y. Shirasaki *et al.*, Promotion of bone formation using highly pure porous beta-TCP combined with bone marrow-derived osteoprogenitor cells, *Biomaterials*, **23**, 4493 (2002).
113. S. Michna, W. Wu, and J.A. Lewis, Concentrated hydroxyapatite inks for direct-write assembly of 3-D periodic scaffolds, *Biomaterials*, **26**, 5632 (2005).
114. V. Maquet, A.R. Boccaccini, L. Pravata *et al.*, Porous poly(α-hydroxyacid)/Bioglass® composite scaffolds for bone tissue engineering. I: Preparation and *in vitro* characterisation, *Biomaterials*, **25**, 4185 (2004).
115. R.M. Schek, J.M. Taboas, S.J. Segvich *et al.*, Engineered osteochondral grafts using biphasic composite solid free-form fabricated scaffolds, *Tissue Engineering*, **10**, 1376 (2004).
116. H.-W. Kim, J.C. Knowles, and H.-F. Kim, Hydroxyapatite and gelatin composite foams processed via novel freeze-drying and crosslinking for use as temporary hard tissue scaffolds, *J. Biomed. Mater. Res.*, **72A**, 136 (2005).
117. Z. Li, H.R. Ramay, K.D. Hauch *et al.*, Chitosan-alginate hybrid scaffolds for bone tissue engineering, *Biomaterials*, **26**, 3919 (2005).
118. J.J. Stankus, J. Guan, and W.R. Wagner, Fabrication of biodegradable elastomeric scaffolds with sub-micron morphologies, *J. Biomed. Mater. Res.*, **70A**, 603 (2004).
119. G.E. Park, M.A. Pattison, K. Park *et al.*, Accelerated chondrocyte functions on NaOH-treated PLGA scaffolds, *Biomaterials*, **26**, 3075 (2005).
120. T.M. Fahmy, R.M. Samstein, C.C. Harness *et al.*, Surface modification of biodegradable polyesters with fatty acid conjugates for improved drug targeting, *Biomaterials*, **26**, 5727 (2005).
121. Y. Zhu, C. Gao, X. Liu *et al.*, Immobilization of biomacromolecules onto aminolyzed poly(L-lactic acid) toward acceleration of endothelium regeneration, *Tissue Engineering*, **10**, 53 (2004).
122. Z. Ma, C. Gao, Y. Gong *et al.*, Cartilage tissue engineering PLLA scaffold with surface immobilized collagen and basic fibroblast growth factor, *Biomaterials*, **26**, 1253 (2005).
123. J.J. Yoon, S.H. Song, D.S. Lee *et al.*, Immobilization of cell adhesive RGD peptide onto the surface of highly porous biodegradable polymer scaffolds fabricated by a gas foaming/salt leaching method, *Biomaterials*, **25**, 5613 (2004).

124. Y. Zhu, K.S. Chian, M.B. Chan-Park *et al.*, Protein bonding on biodegradable poly(L-lactide-*co*-caprolactone) membrane for esophageal tissue engineering, *Biomaterials*, **27**, 68 (2006).

125. F. Yang. C.G. Williams, D.-A. Wang *et al.*, The effect of incorporating RGD adhesive peptide in polyethylene glycol diacrylate hydrogel on osteogenesis of bone marrow stromal cells, *Biomaterials*, **26**, 5991 (2005).

126. H. Shin, J.S. Temenoff, G.C. Bowden *et al.*, Osteogenic differentiation of rat bone marrow stromal cells cultured on Arg–Gly–Asp modified hydrogels without dexamethasone and β-glycerol phosphate, *Biomaterials*, **26**, 3645 (2005).

127. G.C.M. Steffens, C. Yao, P. Prevel *et al.*, Modulation of angiogenic potential of collagen matrices by covalent incorporation of heparin and loading with vascular endothelial growth factor, *Tissue Engineering*, **10**, 1502 (2004).

128. T. Masuko, N. Iwasaki, S. Yamane *et al.*, Chitosan-RGDSGGC conjugate as a scaffold material for musculoskeletal tissue engineering, *Biomaterials*, **26**, 5339 (2005).

129. H.H. Lu, J.A. Cooper Jr, S. Manuel *et al.*, Anterior cruciate ligament regeneration using braided biodegradable scaffolds: *in vitro* optimization studies, *Biomaterials*, **26**, 4805 (2005).

130. R.J. Klebe, H. Caldwell, and S. Millan, Cells transmit spatial information by orienting collagen fibers, *Matrix*, **9**, 451 (1989).

131. C.M. Klapperich and C.R. Bertozzi, Global gene expression of cells attached to a tissue engineering scaffold, *Biomaterials*, **25**, 5631 (2004).

132. D.G. Anderson, D. Putnam, E.B. Lavik *et al.*, Biomaterial microarrays: Rapid, microscale screening of polymer-cell interaction, *Biomaterials*, **26**, 4892 (2005).

133. M. Matsusaki, T. Serizawa, A. Kishida *et al.*, Novel functional biodegradable polymer II: fibroblast growth factor-2 activities of poly(gamma-glutamic acid)-sulfonate, *Biomacromol.*, **6**, 400 (2005).

134. M. Ehrbar, A. Metters, P. Zammaretti *et al.*, Endothelial cell proliferation and progenitor maturation by fibrin-bound VEGF variants with differential susceptibilities to local cellular activity, *J. Controlled Release*, **101**, 93 (2005).

135. G. Schmidmaier, B. Wildermann, and A. Stemberger, Biodegradable poly(D,L-lactide) coating of implants for continuous release of growth factors, *J. Biomed. Mater. Res.*, **58**, 449 (2001).

136. E.J. Cho, Z. Tao, Y. Tang *et al.*, Tailored delivery of active keratinocyte growth factor from biodegradable polymer formulations, *J. Biomed. Mater. Res.*, **66A**, 417 (2003).

137. Y.-M. Lee, S.-H. Nam, Y.-J. Seol *et al.*, Enhanced bone augmentation by controlled release of recombinant human bone morphogenetic protein-2 from bioabsorbable membranes, *J. Periodontol.*, **74**, 865 (2003).

138. Y. Tabata, S. Hijikata, and Y. Ikada, Enhanced vascularization and tissue granulation by basic fibroblast growth factor impregnated in gelatin hydrogels, *J. Controlled Release*, **31**, 189 (1994).

139. M. Yamamoto, Y. Tabata, L. Hong *et al.*, Bone regeneration by transforming growth factor β1 released from a biodegradable hydrogel, *J. Controlled Release*, **64**, 133 (2000).

140. M. Ozeki, T. Ishii, Y. Hirano *et al.*, Controlled release of hepatocyte growth factor from gelatin hydrogels based on hydrogel degradation, *J. Drug Target*, **9**, 461 (2001).

141. M. Yamamoto, Y. Takahashi, and Y. Tabata, Controlled release by biodegradable hydrogels enhances the ectopic bone formation of bone morphogenetic protein, *Biomaterials*, **24**, 4375 (2003).

142. N. Isogai, T. Morotomi, S. Hayakawa *et al.*, Combined chondrocyte-copolymer implantation with slow release of basic fibroblast growth factor for tissue engineering an auricular cartilage construct, *J. Biomed. Mater. Res.*, **74A**, 408 (2005).

143. M. Fujita, M. Ishihara, M. Simizu *et al.*, Vascularization *in vivo* caused by the controlled release of fibroblast growth factor-2 from an injectable chitosan/non-anticoagulant heparin hydrogel, *Biomaterials*, **25**, 699 (2004).

144. R.A. Peatti, A.P. Nayate, M.A. Firpo *et al.*, Stimulation of *in vivo* angiogenesis by cytokine-loaded hyaluronic acid hydrogel implants, *Biomaterials*, **25**, 2789 (2004).

145. R. Montesano, S. Kumar, L. Orci *et al.*, Synergistic effect of hyaluronan oligosaccharides and vascular endothelial growth factor on angiogenesis *in vitro*, *Lab. Invest.*, **75**, 249 (1996).

146. J.E. Lee, K.E. Kim, I.C. Kwon *et al.*, Effects of the controlled-released TGF-beta 1 from chitosan microspheres on chondrocytes cultured in a collagen/chitosan/glycosaminoglycan scaffold, *Biomaterials*, **25**, 4163 (2004).

147. W.S.W. Shalaby, H. Yeh, E. Woo *et al.*, Absorbable microparticulate cation exchanger for immunotherapeutic delivery, *J. Biomed. Mater. Res.* Part B: *Appl. Biomater.*, **69B**, 173 (2004).

148. T.A. Holland, J.K.V. Tessmar, Y. Tabata *et al.*, Transforming growth factor-beta 1 release from oligo(poly(ethylene glycol) fumarate) hydrogels in conditions that model the cartilage wound healing environment, *J. Controlled Release*, **94**, 101 (2004).

149. Y. Takahashi, M. Yamamoto, and Y. Tabata, Enhanced osteoinduction by controlled release of bone morphogenetic protein-2 from biodegradable sponge composed of gelatin and beta-tricalcium phosphate, *Biomaterials*, **26**, 4856 (2005).

150. M.K. Smith, M. Peters, T.P. Richardson *et al.*, Locally enhanced angiogenesis promotes transplanted cell survival, *Tissue Engineering*, **10**, 63 (2004).

151. W.T. Bourque, M. Gross, and B.K. Hall, Expression of four growth factors during fracture repair, *Int. J. Dev. Biol.*, **37**, 573 (1993).

152. M.P. Bostrom, J.M. Lane, W.S. Berberian *et al.*, Immunolocalization and expression of bone morphogenetic proteins 2 and 4 in fracture healing, *J. Orthop. Res.*, **13**, 357 (1995).

153. Y. Yu, J.L. Yang, P.J. Chapman-Sheath *et al.*, TGF-beta, BMPS, and their signal transducing mediators, Smads, in rat fracture healing, *J. Biomed. Mater. Res.*, **60**, 392 (2002).

154. R.L. Vonau, M.P. Bostrom, P. Aspenberg *et al.*, Combination of growth factors inhibits bone ingrowth in the bone harvest chamber, *Clin Orthop Relat Res.*, **386**, 243 (2001).

155. U. Ripamonti, J. Crooks, J.C. Petit *et al.*, Periodontal tissue regeneration by combined applications of recombinant human osteogenic protein-1 and bone morphogenetic protein-2. A pilot study in Chacma baboons (Papio ursinus), *Eur. J. Oral Sci.*, **109**, 241 (2001).

156. L.J. Marden, R.S. Fan, G.F. Pierce *et al.*, Platelet-derived growth factor inhibits bone regeneration induced by osteogenin, a bone morphogenetic protein, in rat craniotomy defects, *J. Clin. Invest.*, **92**, 2897 (1993).

157. A.T. Raiche and D.A. Puleo, *In vitro* effects of combined and sequential delivery of two bone growth factors, *Biomaterials*, **25**, 677 (2004).

158. C.A. Simmons, E. Alsberg, S. Hsiong *et al.*, Dual growth factor delivery and controlled scaffold degradation enhance *in vivo* bone formation by transplanted bone marrow stromal cells, *Bone*, **35**, 562 (2004).

159. O. Anusaksathien, S.A. Webb, Q.-M. Jin *et al.*, Platelet-derived growth factor gene delivery stimulates *ex vivo* gingival repair, *Tissue Engineering*, **9**, 745 (2003).

160. Y.C. Huang, C. Simmons, D. Kaigler *et al.*, Bone regeneration in a rat cranial defect with delivery of PEI-condensed plasmid DNA encoding for bone morphogenetic protein-4 (BMP-4), *Gene Ther.*, **12**, 418 (2005).

161. S.R. Winn, A.L. Lindner, G. Haggett *et al.*, Polymer-encapsulated genetically modified cells continue to secrete human nerve growth factor for over one year in rat ventricles: Behavioral and anatomical consequences, *Exp. Neurol.*, **140**, 126 (1996).

162. J.V. Jimenez, D.R. Tyson, S. Dhar *et al.*, Human embryonic kidney cells (HEK-293 cells): Characterization and dose-response relationship for modulated release of nerve growth factor for nerve regeneration, *Plast. Reconstr. Surg.*, **113**, 605 (2004).

163. M.P. McConnell, S. Dhar, S. Naran *et al.*, *In vivo* induction and delivery of nerve growth factor, using HEK-293 cells, *Tissue Engineering*, **10**, 1492 (2004).

164. D.K. Lee, K.B. Choi, I.S. Oh, Continuous transforming growth factor β_1 secretion by cell-mediated gene therapy maintains chondrocyte redifferentiation, *Tissue Engineering*, **11**, 310 (2005).

165. N.J. Meilander, P.J. Cheung, D.L. Wilson *et al.*, Sustained *in vivo* gene delivery from agarose hydrogel prolongs nonviral gene expression in skin, *Tissue Engineering*, **11**, 546 (2005).

166. M. Kuriwaka, M. Ochi, Y. Uchio *et al.*, Optimum combination of monolayer and three-dimensional cultures for cartilage-like tissue engineering, *Tissue Engineering*, **9**, 41 (2003).

167. J.Y. Lim, X. Liu, E.A. Vogler *et al.*, Systematic variation in osteoblast adhesion and phenotype with substratum surface characteristics, *J. Biomed. Mater. Res.*, **68A**, 504 (2004).

168. G.B. Schneider, A. English, M. Abraham *et al.*, The effect of hydrogel charge density on cell attachment, *Biomaterials*, **25**, 3023 (2004).

169. J.W. Lee. Y.H. Kim, K.D. Park *et al.*, Importance of integrin beta1-mediated cell adhesion on biodegradable polymers under serum depletion in mesenchymal stem cells and chondrocytes, *Biomaterials*, **25**, 1901 (2004).

170. W.T. Godbey, B.S.S. Hindy, M.E. Sherman *et al.*, A novel use of centrifugal force for cell seeding into porous scaffolds, *Biomaterials*, **25**, 2799 (2004).

171. K.P. Chennazhy and L.K. Krishnan, Effect of passage number and matrix characteristics on differentiation of endothelial cells cultured for tissue engineering, *Biomaterials*, **26**, 5658 (2005).

172. A.J. Mathew, J.M. Baust, R.G. van Buskirk *et al.*, Cell preservation in reparative and regenerative medicine: evolution of individualized solution composition, *Tissue Engineering*, **10**, 1662 (2004).

173. J.D. Kisiday, B. Kurz, M.A. DiMicco *et al.*, Evaluation of medium supplemented with insulin–transferrin–selenium for culture of primary bovine calf chondrocytes in three-dimensional hydrogel scaffolds, *Tissue Engineering*, **11**, 141 (2005).

174. P.E. Harrison, P.M. Pfeifer, S.L. Turner *et al.*, Serum from patients anesthetized with opiates less effective in the support of chondrocyte growth *in vitro*, *Tissue Engineering*, **9**, 37 (2003).

175. M.D. Lavender, Z. Pang, C.S. Wallace *et al.*, A system for the direct co-culture of endothelium on smooth muscle cells, *Biomaterials*, **26**, 4642 (2005).

176. A. Lichtenberg, G. Dumlu, T. Walles *et al.*, A multifunctional bioreactor for three-dimensional cell (co)-culture, *Biomaterials*, **26**, 555 (2005).

177. A. Khademhosseini, K.Y. Suh, J.M. Yang *et al.*, Layer-by-layer deposition of hyaluronic acid and poly-L-lysine for patterned cell co-cultures, *Biomaterials*, **25**, 3583 (2004).

178. M. Fossum, A. Nordenskjold, and G. Kratz, Engineering of multilayered urinary tissue *in vitro*, *Tissue Engineering*, **10**, 175 (2004).

179. Y. Martin and P. Vermette, Bioreactors for tissue mass culture. Design, characterization, and recent advances, *Biomaterials*, **26**, 7481 (2005).

180. M. Kino-Oka, N. Ogawa, R. Umegaki *et al.*, Bioreactor design for successive culture of anchorage-dependent cells operated in an automated manner, *Tissue Engineering*, **11**, 535 (2005).

181. G. Vunjak-Novakovic, L.E. Freed, R.J. Biron *et al.*, Effects of mixing on the composition and morphology of tissue-engineered cartilage, *AIChE J.*, **42**, 850 (1996).
182. L.E. Freed, J.C. Marquis, R. Langer *et al.*, Composition of cell-polymer cartilage implants, *Biotechnol. Bioeng.*, **43**, 605 (1994).
183. H.-J. Wang, M.B. Haas, C.A. van Blitterswijk *et al.*, Engineering of a dermal equivalent: seeding and culturing fibroblasts in PEGT/PBT copolymer scaffolds, *Tissue Engineering*, **9**, 909 (2003).
184. S.H. Cartmell, B.D. Porter, A.J. Garcia *et al.*, Effects of medium perfusion rate on cell-seeded three-dimensional bone constructs *in vitro*, *Tissue Engineering*, **9**, 1197 (2003).
185. A.A. Neves, N. Medcalf, and K. Brindle, Functional assessment of tissue-engineered meniscal cartilage by magnetic resonance imaging and spectroscopy, *Tissue Engineering*, **9**, 51 (2003).
186. T. Uemura, J. Dong, Y. Wang *et al.*, Transplantation of cultured bone cells using combinations of scaffolds and culture techniques, *Biomaterials*, **24**, 2277 (2003).
187. A.-M. Freyria, Y. Yang, H. Chajra *et al.*, Optimization of dynamic culture conditions: Effects on biosynthetic activities of chondrocytes grown in collagen sponges, *Tissue Engineering*, **11**, 674 (2005).
188. I. Martin, B. Obradovic, S. Treppo *et al.*, Modulation of the mechanical properties of tissue engineered cartilage, *Biorheology*, **37**, 141 (2000).
189. L.E. Freed, A.P. Hollander, I. Martin *et al.*, Chondrogenesis in a cell-polymer-bioreactor system, *Exp. Cell Res.*, **240**, 58 (1998).
190. A.C. Aufderheide and K.A. Athanasiou, Comparison of scaffolds and culture conditions for tissue engineering of the knee meniscus, *Tissue Engineering*, **11**, 1095 (2005).
191. M.S. Detamore and K.A. Athanasiou, Use of rotating bioreactor toward tissue engineering the temporomandibular joint disc, *Tissue Engineering*, **11**, 1188 (2005).
192. E.A. Botchwey, M.A. Dupree, S.R. Pollack *et al.*, Tissue engineered bone: Measurement of nutrient transport in three-dimensional matrices, *J. Biomed. Mater. Res.*, **67A**, 357 (2003).
193. A.E. Scheidegger, *The Physics of Flow through Porous Media*. New York: Macmillan, 1960.
194. J. Happel and H. Brenner, *Mechanics of Fluid Transport Processes*. The Hague: Martinus Nijhoff Publ., 1983.
195. K.J. Gooch, J.H. Kwon, T. Blunk *et al.*, Effects of mixing intensity on tissue-engineered cartilage, *Biotechnol. Bioeng.*, **72**, 402 (2001).
196. A.S. Goldstein, T.M. Juarez, C.D. Helmke *et al.*, Effect of convection on osteoblastic cell growth and function in biodegradable polymer foam scaffolds, *Biomaterials*, **22**, 1279 (2001).
197. E.A. Botchwey, S.R. Pollack, S. El-Amin *et al.*, Human osteoblast-like cells in three-dimensional culture with fluid flow, *Biorheology*, **40**, 299 (2003).
198. K.D. Costa, E.J. Lee, and J.W. Holmes, Creating alignment and anisotropy in engineered heart tissue: Role of boundary conditions in a model three-dimensional culture system, *Tissue Engineering*, **9**, 567 (2003).
199. C.C. Berry, J.C. Shelton, D.L. Bader *et al.*, Influence of external uniaxial cyclic strain on oriented fibroblast-seeded collagen gels, *Tissue Engineering*, **9**, 613 (2003).
200. S.D. Waldman, C.G. Spiteri, M.D. Grynpas *et al.*, Long-term intermittent compressive stimulation improves the composition and mechanical properties of tissue-engineered cartilage, *Tissue Engineering*, **10**, 1323 (2004).
201. G.C. Engelmayr, E. Rabkin, F.W.H. Sutherland *et al.*, The independent role of cyclic flexure in the early *in vitro* development of an engineered heart valve tissue, *Biomaterials*, **26**, 175 (2005).

202. C.H. Lee, H.J. Shin, I.H. Cho *et al.*, Nanofiber alignment and direction of mechanical strain affect the ECM production of human ACL fibroblast, *Biomaterials*, **26**, 1261 (2005).

203. G. Grenier, M. Remy-Zolghadri, D. Larouche, Tissue reorganization in response to mechanical load increases functionality, *Tissue Engineering*, **11**, 90 (2005).

204. C. Williams and T. Wick, Perfusion bioreactor for small diameter tissue-engineered arteries, *Tissue Engineering*, **10**, 930 (2004).

205. R.L. Mauck, S.B. Nicoll, S.L. Seyhan *et al.*, Synergistic action of growth factors and dynamic loading for articular cartilage tissue engineering, *Tissue Engineering*, **9**, 597 (2003).

206. S.I. Jeong, J.H. Kwon, J.I. Lim *et al.*, Mechano-active tissue engineering of vascular smooth muscle using pulsatile perfusion bioreactors and elastic PLCL scaffolds, *Biomaterials*, **26**, 1405 (2005).

207. N.D. Case, A.O. Duty, A. Ratcliffe *et al.*, Bone formation on tissue-engineered cartilage constructs *in vivo*: effects of chondrocyte viability and mechanical loading, *Tissue Engineering*, **9**, 587 (2003).

208. S. Grad, C.R. Lee, K. Gorna *et al.*, Surface motion upregulates superficial zone protein and hyaluronan production in chondrocyte-seeded three-dimensional scaffolds, surface motion upregulates superficial zone protein and hyaluronan production in chondrocyte-seeded three-dimensional scaffolds, *Tissue Engineering*, **11**, 249 (2005).

209. A.D. McCulloch, A.B. Harris, C.E. Sarraf *et al.*, New multi-cue bioreactor for tissue engineering of tubular cardiovascular samples under physiological conditions, *Tissue Engineering*, **10**, 565 (2004).

210. C.L. Stabler, R.C. Long Jr., A. Sambanis *et al.*, Noninvasive measurement of viable cell number in tissue-engineered constructs *in vitro*, using 1H nuclear magnetic resonance spectroscopy, *Tissue Engineering*, **11**, 404 (2005).

211. K.W. Ng, D.T.W. Leong, and D.M. Hutmacher, The challenge to measure cell proliferation in two and three dimensions, *Tissue Engineering*, **11**, 182 (2005).

212. W.N. Aandrade, M.G. Johnston, and H.B. Hay, The relationship of blood lymphocytes to the recirculating lymphocyte pool, *Blood*, **91**, 1653 (1998).

213. M.C. Kruyt, J. de Bruijn, M. Veenhof *et al.*, Application and limitations of chloromethyl-benzamidodialkylcarbocyanine for tracing cells used in bone tissue engineering, *Tissue Engineering*, **9**, 105 (2003).

214. N.R. Washburn, M. Weir, P. Anderson *et al.*, Bone formation in polymeric scaffolds evaluated by proton magnetic resonance microscopy and X-ray microtomography, *J. Biomed. Mater. Res.*, **69A**, 738 (2004).

215. N.N. Ramrattan, R.G.J.C. Heijkants, T.G. Van Tienen *et al.*, Assessment of tissue ingrowth rates in polyurethane scaffolds for tissue engineering, *Tissue Engineering*, **11**, 1212 (2005).

216. J.S. Blum, J.S. Temenoff, H. Park *et al.*, Development and characterization of enhanced green fluorescent protein and luciferase expressing cell line for non-destructive evaluation of tissue engineering constructs, *Biomaterials*, **25**, 5809 (2004).

217. N. Bursac, M. Papadaki, J.A. White *et al.*, Cultivation in rotating bioreactors promotes maintenance of cardiac myocyte electrophysiology and molecular properties, *Tissue Engineering*, **9**, 1243 (2003).

218. M. Radisic, M. Euloth, L. Yang *et al.*, High-density seeding of myocyte cells for cardiac tissue engineering, *Biotechnol. Bioeng.*, **82**, 403 (2003).

219. T. Kofidis, A. Lenz, J. Boublik *et al.*, Pulsatile perfusion and cardiomyocyte viability in a solid three-dimensional matrix, *Biomaterials*, **24**, 5009 (2003).

220. S.B.C. Watzka, M. Steiner, P. Samorapoompichit *et al.*, Establishment of vessel-like structures in long-term three-dimensional tissue culture of myocardium: An electron microscopy study, *Tissue Engineering*, **10**, 1684 (2004).

221. U. Hubschmid, P.-M. Leong-Morgenthaler, and A. Basset-Dardare, *In vitro* growth of human urinary tract smooth muscle cells on laminin and collagen type I-coated membranes under static and dynamic conditions, *Tissue Engineering*, **11**, 161 (2005).

222. J. Garvin, J. Qi, M. Maloney *et al.*, Novel system for engineering bioartificial tendons and application of mechanical load, *Tissue Engineering*, **9**, 967 (2003).

223. D. Hoffman-Kim, M.S. Maish, P.M. Krueger *et al.*, Comparison of three myofibroblast cell sources for the tissue engineering of cardiac valves, *Tissue Engineering*, **11**, 288 (2005).

224. R.P. Witte and W.J. Kao, Keratinocyte-fibroblast paracrine interaction: The effects of substrate and culture condition, *Biomaterials*, **26**, 3673 (2005).

225. J. Malda, C.A. van Blitterswijk *et al.*, Expansion of bovine chondrocytes on microcarriers enhances redifferentiation, *Tissue Engineering*, **9**, 939 (2003).

226. A. Naumann, J.E. Dennis, J. Aigner *et al.*, Tissue engineering of autologous cartilage grafts in three-dimensional *in vitro* macroaggregate culture system, *Tissue Engineering*, **10**, 1695 (2004).

227. S. Marlovits, B. Tichy, M. Truppe *et al.*, Chondrogenesis of aged human articular cartilage in a scaffold-free bioreactor, *Tissue Engineering*, **9**, 1215 (2003).

228. R. Dorotka, U. Bindreiter, P. Vavken *et al.*, Behavior of human articular chondrocytes derived from nonarthritic and osteoarthritic cartilage in a collagen matrix, *Tissue Engineering*, **11**, 877 (2005).

229. N.H. Veilleux, I.V. Yannas, and M. Spector, Effect of passage number and collagen type on the proliferative, biosynthetic, and contractile activity of adult canine articular chondrocytes in type I and II collagen-glycosaminoglycan matrices *in vitro*, *Tissue Engineering*, **10**, 119 (2004).

230. S.HY. Kamil, A. Rodrigruez, C.A. Vacanti *et al.*, Expansion of the number of human auricular chondrocytes: Recycling of culture media containing floating cells, *Tissue Engineering*, **10**, 139 (2004).

231. P. Giannoni, A. Crovace, M. Malpeli *et al.*, Species variability in the differentiation potential of *in vitro*-expanded articular chondrocytes restricts predictive studies on cartilage repair using animal models, *Tissue Engineering*, **11**, 237 (2005).

232. E.M. Darling and K.A. Athanasiou, Retaining zonal chondrocyte phenotype by means of novel growth environments, *Tissue Engineering*, **11**, 395 (2005).

233. E.W. Mandl, S.W. van der Veen, J.A.N. Verhaar *et al.*, Multiplication of human chondrocytes with low seeding densities accelerates cell yield without losing redifferentiation capacity, *Tissue Engineering*, **10**, 109 (2004).

234. H. Loewenheim, J. Reichl, H. Winter *et al.*, *In vitro* expansion of human nasoseptal chondrocytes reveals distinct expression of G_1 cell cycle inhibitors for replicative, quiescent, and senescent culture stages, *Tissue Engineering*, **11**, 64 (2005). The same as 233.

235. M. Kino-Oka, S. Yashiki, Y. Ota *et al.*, Subculture of chondrocytes on a collagen type I-coated substrate with suppressed cellular dedifferentiation, *Tissue Engineering*, **11**, 597 (2005).

236. C.A. Pangborn and K.A. Athanasiou, Effects of growth factors on meniscal fibrochondrocytes, *Tissue Engineering*, **11**, 1141 (2005).

237. A. Hofmann, L. Konrad, L. Gotzen *et al.*, Bioengineered human bone tissue using autogenous osteoblasts cultured on different biomatrices, *J. Biomed. Mater. Res.*, **67**A, 191 (2003).

238. U. Arnold, S. Schweitzer, K. Lindenhayn *et al.*, Optimization of bone engineering by means of growth factors in a three-dimensional matrix, *J. Biomed. Mater. Res.*, **67**A, 260 (2003).

239. A.C. Jones, B. Milthorpe, H. Averdunk *et al.*, Analysis of 3D bone ingrowth into polymer scaffolds via micro-computed tomography imaging, *Biomaterials*, **25**, 4947 (2004).

240. K. Nakagawa, H. Abukawa, M.Y. Shin *et al.*, Osteoclastogenesis on tissue-engineered bone, *Tissue Engineering*, **10**, 93 (2004).

241. J.Y. Lim, A.F. Taylor, Z. Li *et al.*, Integrin expression and osteopontin regulation in human fetal osteoblastic cells mediated by substratum surface characteristics, *Tissue Engineering*, **11**, 19 (2005).

242. B.A. Nasseri, I. Pomerantseva, M.R. Kaazempur-Mofrad *et al.*, Dynamic rotational seeding and cell culture system for vascular tube formation, *Tissue Engineering*, **9**, 291 (2003).

243. D. Seliktar, R.M. Nerem, and Z.S. Galis, Mechanical strain-stimulated remodeling of tissue-engineered blood vessel constructs, *Tissue Engineering*, **9**, 657 (2003).

244. E.L. Dvorin, J. Wylie-Sears, S. Kaushal *et al.*, Quantitative evaluation of endothelial progenitors and cardiac valve endothelial cells: Proliferation and differentiation on poly-glycolic acid/poly-4-hydroxybutyrate scaffold in response to vascular endothelial growth factor and transforming growth factor beta1, *Tissue Engineering*, **9**, 487 (2003).

245. C.K. Griffith, C. Miller, R.C.A. Sainson *et al.*, Diffusion limits of an *in vitro* thick prevascularized tissue, *Tissue Engineering*, **11**, 257 (2005).

246. C. Fidkowski, M.R. Kaazempur-Mofrad, J. Borenstein *et al.*, Endothelialized microvasculature based on a biodegradable elastomer, *Tissue Engineering*, **11**, 302 (2005).

247. R. Gruber, B. Kandler, P. Holzmann *et al.*, Bone marrow stromal cells can provide a local environment that favors migration and formation of tubular structures of endothelial cells, *Tissue Engineering*, **11**, 896 (2005).

248. M. Afting, U.A. Stock, B. Nasseri *et al.*, Quantitative evaluation of endothelial progenitors and cardiac valve endothelial cells: Proliferation and differentiation on poly-glycolic acid/poly-4-hydroxybutyrate scaffold in response to vascular endothelial growth factor and transforming growth factor beta1, *Tissue Engineering*, **9**, 137 (2003).

249. R.C. Fields, A. Solan, K.T. McDonagh, Gene therapy in tissue-engineered blood vessels, *Tissue Engineering*, **9**, 1281 (2003).

250. E.J. Semler, A. Dasgupta, and P.V. Moghe, Cytomimetic engineering of hepatocyte morphogenesis and function by substrate-based presentation of acellular E-cadherin, *Tissue Engineering*, **11**, 734 (2005).

251. A. Jayaraman, M.L. Yarmush, and C.M. Roth, Evaluation of an *in vitro* model of hepatic inflammatory response by gene expression profiling, *Tissue Engineering*, **11**, 50 (2005).

252. R. Marty-Roix, J.D. Bartlett, and M. Spector, Growth of porcine enamel-, dentin-, and cementum-derived cells in collagen-glycosaminoglycan matrices *in vitro*: Expression of alpha-smooth muscle actin and contraction, *Tissue Engineering*, **9**, 175 (2003).

253. W. Zhang, X.F. Walboomers, J.G.C. Wolke *et al.*, Differentiation ability of rat postnatal dental pulp cells *in vitro, Tissue Engineering*, **11**, 357 (2005).

254. S.D. Tran, J. Wang, B.C. Bandyopadhyay *et al.*, Primary culture of polarized human salivary epithelial cells for use in developing an artificial salivary gland, *Tissue Engineering*, **11**, 172 (2005).

255. J.F. Walsh, M.E. Manwaring, and P.A. Tresco, Directional neurite outgrowth is enhanced by engineered meningeal cell-coated substrates, *Tissue Engineering*, **11**, 1085 (2005).

256. B. Shany, R. Vago, and D. Baranes, Growth of primary hippocampal neuronal tissue on an aragonite crystalline biomatrix, *Tissue Engineering*, **11**, 585 (2005).
257. E.B. Lavik, H. Klassen, K. Warfvinge *et al.*, Fabrication of degradable polymer scaffolds to direct the integration and differentiation of retinal progenitors, *Biomaterials*, **26**, 3187 (2005).
258. I.K. Ko, K. Kato, and H. Iwata, Antibody microarray for correlating cell phenotype with surface marker, *Biomaterials*, **26**, 687 (2005).
259. J.R. Mauney, V. Volloch, and D.L. Kaplan, Role of adult mesenchymal stem cells in bone tissue engineering applications: Current status and future prospect, *Tissue Engineering*, **11**, 787 (2005).
260. A.I. Caplan, *Review*: Mesenchymal stem cells—cell-based reconstructive therapy in orthopedics, *Tissue Engineering*, **11**, 1198 (2005).
261. Z. Xia, H. Ye, R.M. Locklin *et al.*, Efficient characterisation of human cell-bioceramic interactions *in vitro* and *in vivo* by using enhanced GFP-labelled mesenchymal stem cells, *Biomaterials*, **26**, 5790 (2005).
262. N. Kotobuki, M. Hirose, H. Machida *et al.*, Viability of osteogenic potential of cryopreserved human bone marrow-derived mesenchymal cells, *Tissue Engineering*, **11**, 663 (2005).
263. I. Sekiya, B.L. Larson, J.R. Smith *et al.*, Expansion of human adult stem cells from bone marrow stroma: Conditions that maximize the yields of early progenitors and evaluate their quality, *Stem Cells*, **20**, 530 (2002).
264. B. Johnstone, T.M. Hering, A.I. Caplan *et al.*, *In vitro* chondrogenesis of bone marrow-derived mesenchymal progenitor cells, *Exp. Cell Res.*, **238**, 265 (1998).
265. S. Tsutsumi, A. Shimazu, K. Miyazaki *et al.*, Retention of multilineage differentiation potential of mesenchymal cells during proliferation in response to FGF, *Biochem. Biophys. Res. Comm.*, **288**, 413 (2001).
266. T. Matsubara, S. Tsutsumi, H. Pan *et al.*, A new technique to expand human mesenchymal stem cells using basement membrane extracellular matrix, *Biochem. Biophys. Res. Comm.*, **313**, 503 (2004).
267. J.R. Mauney, D.L. Kaplan, and V. Volloch, Matrix-mediated retention of osteogenic differentiation potential by human adult bone marrow stromal cells during *ex vivo* expansion, *Biomaterials*, **25**, 3233 (2004).
268. M.A. Lawson, J.E. Barralet, L. Wang *et al.*, Adhesion and growth of bone marrow stromal cells on modified alginate hydrogels, *Tissue Engineering*, **10**, 1480 (2004).
269. W.-J. Li, R. Tuli, X. Huang *et al.*, Multilineage differentiation of human mesenchymal stem cells in a three-dimensional nanofibrous scaffold, *Biomaterials*, **26**, 5158 (2005).
270. N. Indrawattana, G. Chen, M. Tadokoro *et al.*, Growth factor combination for chondrogenic induction from human mesenchymal stem cell, *Biochem. Biophys. Res. Commun.*, **320**, 914 (2004).
271. G. Lisignoli, S. Cristino, A. Piacentrini *et al.*, Cellular and molecular events during chondrogenesis of human mesenchymal stromal cells grown in a three-dimentional Lyaluronan based scaffold, *Biomaterials*, **26**, 5677 (2005).
272. A. Alhadlaq, J.H. Elisseeff, L. Hong *et al.*, Adult stem cell driven genesis of human-shaped articular condyle, *Ann. Biomed. Eng.*, **32**, 911 (2004).
273. M. Hirose, N. Kotobuki, H. Machida *et al.*, Osteogenic potential of cryopreserved human bone marrow-derived mesenchymal cells after thawing in culture, *Mater. Sci. Eng.*, **C24**, 355 (2004).
274. C.G. Williams, T.K. Kim, A. Taboas *et al.*, *In vitro* chondrogenesis of bone marrow-derived mesenchymal stem cells in a photopolymerizing hydrogel, *Tissue Engineering*, **9**, 679 (2003).

275. I. Sekiya, D.C. Colter, and D.J. Prockop, BMP-6 enhances chondrogenesis in a sub-population of human marrow stromal cells, *Biochem. Biophys. Res. Comm.*, **284**, 411 (2001).

276. S. Kitamura, H. Ohgushi, M. Hirose *et al.*, Osteogenic differentiation of human bone marrow-derived mesenchymal cells cultured on alumina ceramics, *Artif. Organs*, **28**, 72 (2004).

277. L. Meinel, V. Karageorgiou, R. Fajardo *et al.*, Bone tissue engineering using human mesenchymal stem cells: Effects of scaffold material and medium flow, *Ann. Biomed. Eng.*, **32**, 112 (2004).

278. E. Lieb, T. Vogel, S. Milz *et al.*, Effects of transforming growth factor beta1 on bone-like tissue formation in three-dimensional cell culture. II: Osteoblastic differentiation, *Tissue Engineering*, **10**, 1414 (2004).

279. K.D. Lee, T.K. Kuo, J. Whang-Peng *et al.*, In vitro hepatic differentiation of human mesenchymal stem cells, *Hepatology*, **40**, 1275 (2004).

280. M. Dezawa, H. Ishikawa, Y. Itojazu *et al.*, Bone marrow stromal cells generate muscle cells and repair muscle degeneration, *Science*, **309**, 314 (2005).

281. S. Hankemeier, M. Keus, J. Zeichen, Modulation of proliferation and differentiation of human bone marrow stromal cells by fibroblast growth factor 2: Potential implications for tissue engineering of tendons and ligaments, *Tissue Engineering*, **11**, 41 (2005).

282. P. Arpornmaeklong, H. Kochel, R. Depprich *et al.*, Influence of platelet-rich plasma (PRP) on osteogenic differentiation of rat bone marrow stromal cells: An *in vitro* study, *Int. J. Oral Maxillofac. Surg.*, **33**, 60 (2004).

283. D.A. Shimko, C.A. Burks, K.C. Dee *et al.*, Comparison of *in vitro* mineralization by murine embryonic and adult stem cells cultured in an osteogenic medium, *Tissue Engineering*, **10**, 1386 (2004).

284. J.-S. Sun, S.Y. Wu, and F.-H. Lin, The role of muscle-derived stem cells in bone tissue engineering, *Biomaterials*, **26**, 3953 (2005).

285. D. Kaigler, P.H. Krebsbach, P.J. Polverini *et al.*, Role of vascular endothelial growth factor in bone marrow stromal cell modulation of endothelial cell, *Tissue Engineering*, **9**, 95 (2003).

286. C. Le Visage, B. Dunham, P. Flint *et al.*, Coculture of mesenchymal stem cells and respiratory epithelial cells to engineer a human composite respiratory mucosa, *Tissue Engineering*, **9**, 1426 (2004).

287. T. Kihara, A. Oshima, M. Hirose *et al.*, Three-dimensional visualization analysis of *in vitro* cultured bone fabricated by rat marrow mesenchymal stem cells, *Biochem. Biophys. Res. Commun.*, **316**, 943 (2004).

288. S. Nathan, S. Das De, A. Thambyah *et al.*, Cell-based therapy in the repair of osteochondral defects: A novel use for adipose tissue, *Tissue Engineering*, **9**, 733 (2003).

289. H.A. Awad, Y.D. Halvorsen, J.M. Gimble *et al.*, Effects of transforming growth factor beta1 and dexamethasone on the growth and chondrogenic differentiation of adipose-derived stromal cells, *Tissue Engineering*, **9**, 1301 (2003).

290. J. Justesen, S.B. Pedersen, K. Stenderup *et al.*, Subcutaneous adipocytes can differentiate into bone-forming cells *in vitro* and *in vivo*, *Tissue Engineering*, **10**, 381 (2004).

291. H.A. Awad, M.Q. Wickham, H.A. Leddy, J.M. Gimble *et al.*, Chondrogenic differentiation of adipose-derived adult stem cells in agarose, alginate, and gelatin scaffolds, *Biomaterials*, **25**, 3211 (2004).

292. A.E. Wurmser, K. Nakashima, R.G. Summers *et al.*, Cell fusion-independent differentiation of neural stem cells to the endothelial lineage, *Nature*, **430**, 350 (2004).

293. I.K. Ko, K. Kato, and H. Iwata, Parallel analysis of multiple surface markers expressed on rat neural stem cells using antibody microarrays, *Biomaterials*, **26**, 4882 (2005).

294. K. Kato, H. Sato, and H. Iwata, Immobilization of Histidine-Tagged Recombinant Proteins onto Micropatterned Surfaces for Cell-Based Functional Assays, *Langmuir*, **21**, 7071 (2005).
295. J. Harrison, S. Pattanawong, J.S. Forsythe *et al.*, Colonization and maintenance of murine embryonic stem cells on poly(alpha-hydroxy esters), *Biomaterials*, **25**, 4963 (2004).
296. S. Levenberg, N.F. Huang, E. Lavik *et al.*, Differentiation of human embryonic stem cells on three-dimensional polymer scaffolds, *Proc. Natl. Acad. Sci. USA*, **100**, 12741 (2003).
297. M. Levenberg, J.A. Burdick, T. Kraehenbuehl *et al.*, Neurotrophin-induced differentiation of human embryonic stem cells on three-dimensional polymeric scaffolds, *Tissue Engineering*, **11**, 506 (2005).
298. K.E. McCloskey, M.E. Gilroy, and R.M. Nerem, Use of embryonic stem cell-derived endothelial cells as a cell source to generate vessel structures *in vitro*, *Tissue Engineering*, **11**, 497 (2005).
299. B.E. Van Vranken, H.M. Romanska, J.M. Polak *et al.*, Coculture of embryonic stem cells with pulmonary mesenchyme: A microenvironment that promotes differentiation of pulmonary epithelium, *Tissue Engineering*, **11**, 1177 (2005).
300. S. Levenberg, J.S. Golub, M. Amit *et al.*, Endothelial cells derived from human embryonic stem cells, *Proc. Natl. Acad. Sci. USA*, **99**, 4391 (2002).
301. R.C. Bielby, R.S. Pryce, L.L. Hench *et al.*, Enhanced derivation of osteogenic cells from murine embryonic stem cells after treatment with ionic dissolution products of 58S bioactive sol–gel glass, *Tissue Engineering*, **11**, 479 (2005).
302. H. Yamazoe, Y. Murakami, K. Mizuseki *et al.*, Collection of neural inducing factors from PA6 cells using heparin solution and their immobilization on plastic culture dishes for the induction of neurons from embryonic stem cells, *Biomaterials*, **26**, 5746 (2005).
303. X. Zhan, G. Dravid, Z. Ye *et al.*, Functional antigen-presenting leucocytes derived from human embryonic stem cells *in vitro*, *Lancet*, **364**, 163 (2004).
304. R.C. Bielby, A.R. Boccaccini, J.M. Polak *et al.*, *In vitro* differentiation and *in vivo* mineralization of osteogenic cells derived from human embryonic stem cells, *Tissue Engineering*, **10**, 1518 (2004).
305. H. Liu and K. Roy, Biomimetic three-dimensional cultures significantly increase hematopoietic differentiation efficacy of embryonic stem cells, *Tissue Engineering*, **11**, 319 (2005).
306. N. Strelchenko, O. Verlinsky, V. Kukharenko *et al.*, Morula-derived human embryonic stem cells, *Reproductive Biomed.*, **9**, 623 (2004).
307. C.A. Cowan, J. Atienza, D.A. Melton *et al.*, Nuclear reprogramming of somatic cells after fusion with human embryonic stem cells, *Science*, **309**, 1369 (2005).
308. E. Check, The rocky road to success, *Nature*, **436**, 185 (2005).

Chapter 4

Challenges in Tissue Engineering

History tells us that any new medical technology has always been confronted with a number of difficulties until to be clinically approved and widely accepted. This is also true for tissue engineering. As demonstrated in previous chapters, scientists and engineers have accumulated tremendous amounts of experiences and knowledge in the field of tissue engineering by their great efforts. However, one should notice that tissue engineering has not yet matured into such a level as engineered tissues and organs are widely accepted as autografts or allografts for medical treatments, although more than 20 years have passed since the advent of tissue engineering. The promise of tissue engineering to be able to repair diseased or lost tissues is largely unrealized to date. A number of critical engineering design issues must be addressed to develop truly clinically and commercially viable strategies for tissue regeneration. Chapter 4 identifies areas where there are large gaps between current research and clinical medicine that inhibit rapid progress toward clinical applications in tissue engineering and attempts to provide solutions that will be effective for the issues.

1. PROBLEMS IN TISSUE ENGINEERING

The neotissues generated by the technique of tissue engineering and clinically used to date are composed primarily of tissues that are relatively thin, avascular, essentially noninnervated, and whose function is determined primarily by the biomechanical properties of their ECM structure. They include skin, bones, cartilages, blood vessels, and periodontal tissues. In this sense, tissue engineering is still at a very early stage of development, with a small number of tissues in current clinical applications. Recent years have seen continuous refinement and improvement of tissue engineering strategies, but a number of tough practical problems persist, including the scarcity of tissue biopsy material and the difficulty in cell expansion maintaining the phenotype. Another reason is too much focusing on basic research dealing with cells and small animals rather than on clinically driven approaches with large animals.

Before the announcement by two US research groups on the success of isolation and identification of human ES cell line in 1998, a majority of researchers had focused on tissue engineering alone. However, from 1998 there has been a great deal of both interest and concern over the use of human ES cells, resulting in the slowdown of the tissue engineering research. This is not a surprise because ES cell is thought to have the potential to give rise to virtually all cell types. Along with the activity increase in ES cell research, recent studies have focused also on stem cells such as neuronal, pancreatic, and hematopoietic stem cells. These single-cell suspensions will be useful for curing the corresponding diseases simply by injection without any cell scaffold. Such medical treatments are called "cell therapy" and a new term "regenerative medicine" has become popular since then.

Undoubtedly, basic research is of prime importance for tissue engineering to advance into human trials. Many "successful" reports have been published on treating cartilage defects, but most of the results have been obtained in small animal models and there are only few clinical trials on cartilages. One of the obstacles that we have to overcome before clinical application of tissue-engineered products is to prepare scaffolds that are actually effective also for large animal models. Probably, readers would have recognized that there exists a big gap between the studies described in Chapters 2 and 3. It does not seem likely that much more encouraging basic research will lead to narrowing the gap. What is required for tissue engineering research to reach the level of clinical applications is to strongly promote clinically driven approaches. In principle, experiments with large animals such as dogs and sheep, preferably with primates, are demanded to prove the safety and efficacy if a new medical treatment is intended to apply to humans. Rats and mice are too different from humans in many aspects. It should be emphasized that the size and mechanical strength of scaffolds applicable to large animal models are substantially different from those for small animals. In addition, researchers of tissue engineering will be discouraged by many tedious review and regulatory issues that large animal experiments involve.

Among many challenges in tissue engineering prior to its clinical trials is the assessment of mechanical properties of engineered tissues. It is often stressed that standardization of the methodology for characterization of engineered tissues should be established for the promotion of tissue engineering. However, a sufficient amount of engineered tissues including vessels, nerves, ligaments, cartilage, and bones are not yet available for biomechanical engineers to measure. Much larger engineered tissues cannot be harvested unless we perform experiments with larger animals.

2. SITES FOR NEOTISSUE CREATION

Tissue is either grown in a patient or grown outside the patient and transplanted. In terms of the tissue creation site, tissue engineering can be divided into two categories: *ex vivo* or *in vitro* tissue engineering (outside the body) and *in situ* or *in vivo* tissue engineering (inside the body). Further, there are intermediate variations between the two extremes, as shown in Fig. 4.1. In the simplest one ((1) in Fig. 4.1) a scaffold without cell seeding is implanted directly into the defect cavity in the body to serve as guidance for cell and tissue growth *in vivo*. This approach to tissue engineering that grows a tissue inside a patient's own body relies entirely on the body's own capacity to regenerate itself. The next (2) is an approach in that a scaffold is implanted immediately after seeding with a sufficient amount of cells but

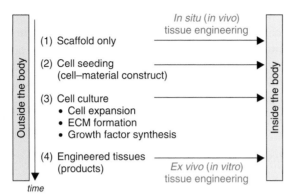

Fig. 4.1 Classification of tissue engineering based on the regeneration site.

without any cell culture *in vitro*. In the third variation, (3) a scaffold seeded with cells is first subjected to cell culture with a bioreactor for the increase in cell population, occasionally accompanied by immature ECM or tissue formation. Following the cell culture, the cell–scaffold construct is implanted in the body. The last one (4), is designed to regenerate a tissue almost to the final phase outside the body. In the fourth approach the same kind of tissues can be produced in large quantities. If the origin of the human cells used for tissue culture is the patient herself or himself, such production of many engineered tissues at once will not be necessary except for special cases. The (3) and (4) tissue engineering modalities are commonly called "*ex vivo* tissue engineering", while the approach (2) is a representative example of *in situ* tissue engineering. Attempts to grow complex tissues outside the body have limitations because artificial environments are unable to resorb and remodel the tissue, as occurs with normal tissue. At the present time, the selection of which kind of tissue engineering should be applied does not depend on the target tissue the research is aiming at, but on the thought of each researcher. Cell growth into unseeded matrices has often proven unstable, whereas implantation of a scaffold after cell seeding has met with much success. The largest advantage of scaffold implantation without cell seeding is the very simple procedure compared with the others.

The site where the targeted tissue is created is a critical factor that definitely governs the direction of tissue engineering research, especially when tissue engineering business is concerned. The possible locations are either in clean bench in factories at *ex vivo* tissue engineering or inside the body of patient at *in situ* tissue engineering. We can transplant a cell–scaffold construct in the body either immediately after cell seeding or after cell cultivation for a certain period of time using a bioreactor until cells expand and tissue forms to some extent. This semi-created tissue will become gradually matured along with remodeling inside the body after transplantation. For simplicity, this type of tissue engineering is here classified into the *ex vivo* tissue engineering, since tissue regeneration has already started *ex vivo*.

Fine-tuning of the microenvironment is typically not an option *in vivo*, whereas some technologies are available *in vitro* to recapitulate many of the key microenvironmental components. Doing so may obviate the need to artificially introduce genetic changes in cells through gene therapy and circumvent the need to use virus-mediated gene transfer and its associated risks. However, *in vitro* tissue engineering demands us to arrange all the components and factors required for tissue engineering.

2.1. *Ex vivo* Tissue Engineering

Ex vivo tissue engineering has attracted much attention, especially to American researchers. There are several reasons for this approach. One is the high possibility for the *ex vivo* approach to produce tissue in large quantities, stock, and deliver engineered tissue on demand from a medical center even in an emergency case, since a large number of tissues can be manufactured any time in clean bench in factory and stored, similar to artificial organs. This might lead to creation of a new type of biomedical industry, but *in situ* tissue engineering does not produce any tissue that can be delivered outside, even if the regenerated tissue is very compatible with patient.

Another attractive feature of *ex vivo* tissue engineering is the easy control of the cell culture in many aspects. Among them is the manageable application of cyclic mechanical stimulations to a cell–scaffold construct during culture, whereas such stress application needs a sophisticated means when a scaffold has been implanted in the human body. Sufficient supply of nutrients and oxygen to cells is difficult *in vivo* unless a scaffold is so thin that nutrients can readily diffuse in. *In vitro*, a variety of techniques have been developed for effective supply of nutrients and oxygen, especially for dynamic cell cultures such as with fluid perfusion and vessel rotation. These techniques are, in principle, not applicable to *in situ* tissue engineering.

Among tissues successfully generated by *ex vivo* tissue engineering are skin tissues. Culture of keratinocytes in monolayer at the air–medium interface yields epidermal tissue. Mostly, a 3T3 feeder layer derived from mouse fibroblasts has been used for the culture of human keratinocytes. Otherwise, the rate of cell expansion would be very low. On the other hand, creation of the dermal tissue from fibroblasts does not need such a feeder layer. Not only single epidermis or single dermis, but also a layered composite consisting of epidermis and dermis is commercially available. Both autologous and allogeneic cells are used for this tissue engineering. Although no serious side effects have been reported by the clinical use of xenogeneic 3T3 cells, research on epidermal tissue engineering without use of xenogeneic 3T3 fibroblasts has been intensively continued.

Cell culture conditions in *ex vivo* tissue engineering can be readily controlled in the initial stage, but may become difficult to control in the later cell culture. This is because the ECM produced from seeded cells into scaffold pores during cell culture may suppress free path of the nutrient fluid in the scaffold pores, resulting in

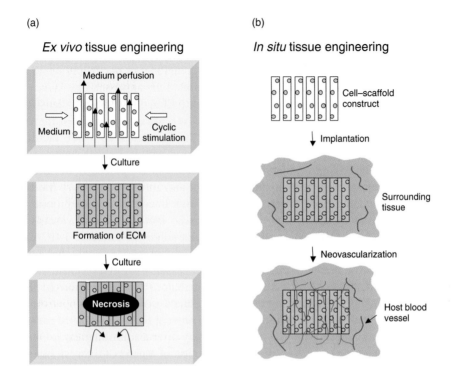

Fig. 4.2   Differences in tissue reaction between *ex vivo* and *in situ* tissue engineering.

necrosis of the cells residing in the deep core of the scaffold, as depicted in Fig. 4.2a. The absence of a microvasculature network in engineered tissues is a fatal drawback of *ex vivo* tissue engineering, especially when creation of thick tissues is the objective. Any living tissues are composed of ECM and cells which need nutrients and oxygen supply to survive. If a tissue is avascular like cartilage or very thin like epidermal tissue, culture medium would supply a sufficient amount of nutrients so far as culture medium is exchanged continuously with fresh one. Cells residing in the center of large cell aggregates tend to undergo necrosis because of low availability of nutrients to the cells and insufficient removal of metabolic wastes from the cells. In addition, all the culture conditions necessary for creating a neotissue, including serum, growth factors, trophic factors, inhibitors, optimal oxygen concentration, and mechanical stimulus, should be provided exogenously by the tissue engineers engaged in the research. Unfortunately, contemporary cell biology cannot provide us with information sufficient enough for *ex vivo* tissue regeneration. Complex biological events that take place intracellularly and intercellularly still remain unveiled. As a consequence, most of the tissues created *in vitro* are much inferior to native tissues in terms of both mechanical properties and physiological functions.

Many efforts have been devoted to replacing small-calibered arteries, but most of them have failed in producing small arterial grafts with diameters around 3 mm. The critical limitation in successful fabrication of artificially designed blood vessels is the high rate of vessel thrombosis. Graft occlusion can occur anywhere from 1 day to 4 weeks after implantation. This stimulated the approach of *ex vivo* tissue engineering for small-calibered blood vessels with ECs. The adhesion, density, and phenotype of the endothelial layer in native vessels are responsible for their anti-thrombogenic potential. Molecules expressed by ECs, such as tissue factor, can actively promote coagulation, and loss of endogenous anticoagulants such as thrombomodulin can increase thrombosis risk. The ECs that are placed in culture often downregulate expression of many anticoagulant and anti-inflammatory molecules that are expressed on native, quiescent endothelium. In contrast, many procoagulant and proinflammatory molecules, including tissue factors such as vascular cell adhesion molecule 1 (VCAM-1), and intercellular adhesion molecule 1 (ICAM-1), are upregulated in culture in response to a variety of factors. It is likely that cultured ECs seeded onto the lumen of engineered blood vessels exhibit a proinflammatory, procoagulant phenotype, which ultimately contributes to graft thrombosis. Hence, thrombosis remains a key limiting factor in the success of not only *in vivo*–but also *ex vivo*–engineered small-calibered vessels. To prevent the vessel failure caused by thrombosis, modulation of the EC phenotype may offer a solution means. One way to alter the phenotype of ECs is by the introduction of a gene of interest, but stable transfection of ECs has often been problematic. Viral transfection efficiencies with ECs are reportedly low, and long-term expression in ECs seems difficult to retain.

The replacement of diseased myocardium with *ex vivo*–engineered heart tissue also has been intensively studied. However, the *ex vivo*–engineered myocardium-like tissue also requires further improvements to enable clinical use. Potential drawbacks of the engineered 3-D constructs are as follows [1]: (1) obtained 3-D constructs are too small to allow surgical implantation into infarction areas; (2) cellular distribution and viability are not homogeneous; (3) myocyte function is limited by various factors, such as impaired nutrient availability and defective culture geometry; (4) 3-D constructs lack adequate plasticity and mechanical stability for implantation purposes; (5) used materials, especially matrix proteins, are not exactly defined or clinically approved; and (6) production costs are high because of expensive additives and collagen compounds, such as Matrigel.

2.2. *In situ* Tissue Engineering

In contrast with *ex vivo* tissue engineering, *in situ* tissue engineering seems easier to perform. This is because large part of biological conditions necessary for regenerating a tissue is spontaneously and endogenously provided inside the body, where every tissue under regeneration undergoes remodeling. However, for successful tissue engineering the implanted scaffold should help provide cells with an environment suitable for tissue regeneration. Some cases have been reported, where scaffold does not promote but disturbs tissue regeneration. A large advantage of *in situ* tissue engineering is the possible formation of vasculature network connecting the host tissue and the engineered tissue, as depicted in Fig. 4.2b. If angiogenic factors are adequately incorporated in scaffold and slowly released, they will facilitate the formation of vasculature network in the body. Vascularization is extremely important for *in situ* tissue engineering, as it is too difficult to exogenously supply nutrients and oxygen to implanted cell–scaffold constructs.

In situ tissue engineering requires no consideration of the issues of immunorejection and ethical concern, but the opportunity for business entrepreneur to participate in the *ex vivo* tissue engineering may be small, being limited to scaffold fabrication and processing for cell expansion. Such an example of *ex vivo* cell processing in industry is the harvesting of the patient's own chondrocytes from cartilage in the knee. The harvested cells are sent to company from medical center and the expanded cells are sent back to the medical center, where the cells will be transplanted in the injured site of the patient's cartilage with or without scaffold.

Our body acts like an ideal bioreactor working with a wonderful vitality under the best physiological conditions. Moreover, engineered tissues likely have a potential to adapt to their mechanical environment in the body. Given a low level of understanding in biological systems that is crucial for success in the *ex vivo* tissue engineering, rapid progress toward clinical application appears to be made by the *in situ* tissue engineering that relies mainly on the biological capabilities themselves for tissue building and remodeling. Without the fundamental knowledge that is essential to create a living cellular implant, the *ex vivo* tissue engineering would be forced to practice in a black box. Of the work done, results show that better integration between native tissue and engineered constructs occurs at an earlier rather than later stage of morphogenesis. This implies that only a threshold level of matrix production may be needed before the construct can be implanted. It may be possible to get an insight into the mechanism of tissue regeneration in the body by varying scaffold structure, cell parameters, factors, and conditions fixed in the *in situ* tissue engineering.

3. AUTOLOGOUS OR ALLOGENEIC CELLS

A source for cells, a way to cultivate them, a way to direct proper differentiation of the cells to achieve function are all key elements that enable us to succeed in developing useful therapies from tissue engineering. Pivotal part of tissue engineering is the cell source in addition to the means whereby sufficient numbers of viable undifferentiated or differentiated cells can be obtained. There are a variety of choices for cells, depending on the application. The choice of cell source influences time to clinical implementation, government regulation, and commercial strategy. Three types of cells are available for tissue engineering in terms of cell origin. They are autologous (from host's own body), allogeneic (from another human donor), and xenogeneic (sourced from animals). The last two are genetically different from the cells of patients themselves, so that immunorejective response will be induced more or less in the host when these cells come into the host body. In addition, xenogeneic cells might have a risk containing endogenous retrovirus, as exemplified by porcine endogenous retrovirus (PERV) reported in 2000. As a result, the number of research groups that have been using xenogeneic cells for tissue engineering dramatically declined from 2000, although any endogenous retrovirus has not been found in the host who might have received retrovirus by contact with xenogeneic cells. Allogeneic products are amenable to large-scale manufacturing at a single central place, while autologous tissue engineering will likely lead to more of a service industry, with a heavy emphasis on local or regional expansion. However, the use of allogeneic materials raises issues of ownership, donation, and consent not to be found with respect to autologous tissues.

3.1. Allogeneic Cells

Cell therapies such as pancreatic islet transplantation and cellular cardiomyoplasty have already shown promise in treating diabetes and cardiac infarct, respectively. Most of researchers are employing allogeneic cells in the experimental stage of tissue engineering research, although these cells should be either rendered nonimmunogenic or used in conjunction with immunosuppressants when applied to patients. Probably, researchers are anticipating the early development of solutions to circumvent the immunorejection. The most widely accepted method for this is to administer immunosuppressants to patients, similar to current organ transplantation, and to wait until immune tolerance is established in the host. Another is to select allogeneic cells that have the major antigens histocompatible with those of patient's cells, as in bone-marrow transplantation for hematopoietic diseases. Both the approaches may be applicable to tissue engineering, but the use of strong immunosuppressants should be limited to treatments for life-threatening diseases because of evoking serious side effects such as reduced immunopotency. It will require much further studies until to be able to stock a sufficient amount of engineered tissues with various major histocompatible antigens (MHAs). Human ES cells and their differentiated progeny have highly polymorphic MHC molecules that serve as major graft rejection antigens to the immune system of allogeneic hosts. An alternative to overcome the immunorejection is to remove the immunorejection-related antigens from the allogeneic cell surface using gene technology or to make antigens matching with those of patients to express, but such approaches also seem to take a long time to succeed in creating clinically applicable, allogeneic cells. The induction of tolerance by infusion of BMCs is a strategy for both xenogeneic and allogeneic tissues, although there is still much to be learned.

As an exception, allogeneic cells have been clinically applied in tissue engineering for epidermis to treat burn injuries without immunosuppressant. The tissue-engineered epidermal sheet has been proven to be effective in healing the injuries by covering and protecting the injured site until new epidermal tissue is regenerated below the sheet. This tissue-engineered sheet is superior to conventional, artificial wound-covers because the living cells present in the tissue-engineered tissue contribute to killing microorganisms invading from the outside, secrete growth factors which promote wound healing, and control the water exuding from the inside. The allogeneic tissue sheet is not integrated into the host skin, but finally detaches as a result of rejection from the host. Unfortunately, our body has few sites that allow for allogeneic cells to contribute to tissue healing as in the case of skin. Fetal cells harvested from abortional sacrifice seem to be low in immunogenicity in comparison with adult cells and hence have been applied to treatments of Parkinson's disease by cell transplantation. Apart from the medical effectiveness of this treatment, the use of about 10 fetuses per patient will invoke an ethical concern. The intracranial space is thought to be immune-privileged (low in immunological surveillance), but any clear conclusion is not yet presented concerning the feasibility of using allogeneic fetal cells for the Parkinson's disease. Further studies are needed to prove whether the intracranial space is really immunologically privileged. In the past, the intraocular space was thought to be immuno-privileged, but currently the intraocular space is known to have the system of immunological surveillance. Cartilage tissue in general has a limited immune response *in vivo* because it lacks a lymphatic and vascular presence in the tissue itself.

3.2. Autologous Cells

The so-called "self-cell therapy" has recently demonstrated clinical potential for regeneration of tissues and treatment of tissue lesions. This suggests that clinical trials of tissue engineering will be performed with autologous cells for the time being. Obviously, cells that are free of both immunological and ethical concerns are only autologous cells, but they are associated with other problems. One of them is the cell source, especially when the patient is too much aged or injured to harvest a sufficient amount of cells. Also in the case of emergency care, treatments with patient's own cells are practically impossible and autologous cells can be used only when medical treatments are applicable in an elective manner. One can say that tissue engineering with autologous cells is a sort of tailored medicine that is a model of future medicine. A disadvantage of this tailored medicine is high cost per treatment as compared with conventional medical treatments. If a large number of tissues can be produced from a single cell source in clean bench, the cost for each engineered-tissue will not be as high as that for a single tailored-made tissue. In the future, long-term, basic research will solve the immunological problems associated with the use of allogeneic cells, but an autologous source for cells might circumvent regulatory hurdles, enabling the therapeutic concept to be studied in humans earlier, and, if promising, to be made available to patients earlier. As expansion of autologous cells is for a single patient, limitations in expansion capabilities are less of an issue compared to an allogeneic source that must serve many individuals to be feasible. The economical issue related to the use of autologous cells will be solved if tissue engineering approaches using autologous cells is proven extremely effective. This will be most likely driven only by increased clinical needs. However, much more mission-oriented research must be performed to convince patients and medical

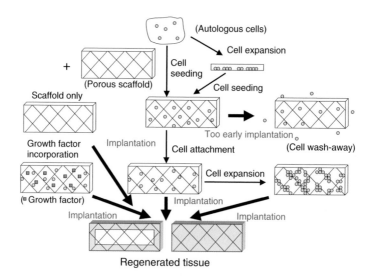

Fig. 4.3   Tissue engineering procedures starting with autologous cells.

doctors that tissue engineering with autologous cells can offer a cost-effective technique for the replacement of lost tissues. There are several approaches to tissue regeneration even from patient's own cells, as illustrated in Fig. 4.3.

4. CELL TYPES

A critical issue to be addressed in human tissue repair is to define the appropriate differentiated state of cell sources. In this respect, autologous cells include ES cells, adult somatic (or tissue) stem cells, progenitor cells committed to a definite lineage, and almost differentiated cells. It seems unclear whether there are any distinct boundaries among these cell types. It has been reported that pluripotent stem cells, almost similar to ES cells, are present in some adult tissues such as fat. Each organ or tissue appears to have its own stem cell like hematopoietic stem cell, epidermal stem cell, hair stem cell, hepatic stem cell, and pancreatic stem cell. Stem cells have the capability to multiply by themselves, although not unlimited, and to differentiate into final lineages. The destination of differentiation of stem cells *in vitro* depends on the culture conditions such as the concentration of BMP, TGF-β, and dexamethasone while differentiation *in vivo* seems to depend on the niche environment where they reside. The MSCs that differentiate into bone, cartilage, muscle, ligament, tendon, blood vessel, dermis, and fat are a kind of adult stem cells. As regards to the nature of stem cells, there are many controversial issues including their identification and transdifferentiation between different stem cells. Evidence has been presented that bone marrow contains cells that are not just multipotent but pluripotent. There is also evidence that adult-derived stem cells may be capable of giving rise to tissues that span the traditional ectodermal, mesodermal, and endodermal embryonic lineage. Multipotent cells, partially committed to a particular cell lineage, may have a practical advantage in this area. Successful tissue engineering hinges on the ability to isolate the appropriate progenitors among the complex array of cell types and cultivate them for proliferation and differentiation toward an appropriate phenotype or function.

The MSCs are multipotent and able to replicate as undifferentiated cells maintaining the potential to differentiate and produce mesenchymal tissues. This cell source has received widespread attention because of the high potential use in tissue engineering, but there remain many unsolved problems regarding the cell. Nobody has succeeded in identifying the MSC. If the MSC used in tissue engineering research is really genuine as a stem cell, this cell should regenerate the cardiac muscle, tendon, and ligament that are all mesenchymal tissues. However, no studies have clearly revealed the regeneration of these mesenchymal tissues from the MSC originated from bone marrow. Another issue that should be addressed in connection to clinical applications of stem cells is to establish technologies for isolation, identification, and characterization of other adult stem cells in addition to the MSC, if they are present at all. It is extremely difficult to harvest a sufficient amount of cells from patients for tissue engineering of the cardiac muscle, esophagus, and trachea, when these tissues have been severely damaged. Combination of the expanded adult stem cells corresponding to the target tissue with a reliable scaffold will upgrade the tissue engineering technology close to the level of human trials. It will take by far longer time to enable the differentiation of allogeneic ES cells to the desired somatic stem or progenitor cells and then to apply to its clinical use under an immunorejection-free condition than to obtain a sufficient amount of somatic stem cells from patients.

The cell source which has attracted the greatest attention in the recent clinical tissue engineering may be the bone marrow that contains various stem cells including MSCs. However, it should be noted that almost matured cells have extensively been used not only in the beginning of clinical applications of tissue engineering but also at the present time. They include keratinocytes for epidermis, fibroblasts for dermis and ligaments, osteoblasts for bones, chondrocytes for cartilages, and vascular cells (endothelial, smooth-muscle, and fibroblast cells) for blood vessels. In principle, these differentiated cells cannot commit to another lineage any more, but have still the capability to proliferate and synthesize ECM once a certain signal is delivered to them. A major clinical problem involved with these cells is the insufficient supply from patients, especially when they are aged or a target tissue is severely damaged. If the cells responsible for the target tissue regeneration can multiply within a short time of cell culture, the problem is no more critical, but such a case is not common. One possible way to circumvent this situation is to make use of less differentiated cells such as progenitor and adult stem cells, since they can multiply much more readily than the blastic cells, although not as unlimited as ES cells.

Two approaches are available to obtain the bone marrow: bone marrow puncture and surgical harvest from the iliac bone. The bone marrow puncture is generally followed by centrifugation for isolation of MNCs and used mostly for regeneration of blood vessels, while the mixture of BMCs and cancellous bone harvested from the iliac bone has been used for bone tissue engineering without purification. It should be noted that these BMCs have been clinically applied to tissue engineering without any previous differentiation into the targeted tissue lineage. Interestingly, a mesenchymal tissue is formed, exactly corresponding to the tissue residing at the site where a scaffold has been implanted after seeding with BMCs. For instance, when a tubular scaffold seeded with BMCs is anastomosed to the porcine bile duct which has partially been excised, the excised part can be replaced by a bile duct tissue regenerated from the undifferentiated BMCs. Articular cartilage has been also regenerated in a similar manner. This suggests that the biological environment at the location of excised tissue may deliver specific signals to the BMCs in the scaffold to differentiate into the cells that will finally produce the target tissue. Further studies will unveil the complicated sequence of biological events occurring during this *in situ* regeneration process. It is a great advantage of the bone marrow that the cells in the bone marrow can readily multiply *in vivo* by themselves to allow repeated harvests, but seeding cells onto a scaffold as many as possible is critical for the success of this tissue engineering because the amount of the BMCs that can be harvested from a patient is limited at present. While MSCs can be used in autologous applications and retain their multilineage potential even when harvested from elderly individuals, the ability to proliferate *in vitro* seems to decrease with age. Significant progress will be required in obtaining enough cells for the use in therapy. Harvesting autologous MSCs is also not a completely risk-free procedure.

The ES cells have no limits in terms of their proliferative capacity, allowing large banks of various cells to be established that can be HLA-matched to patients. The Korean report on a pluripotent cell line derived from a cloned human blastocyst may mean that autologous ES-derived cells may be an option in the future. Because of the possibility that all the tissues and organs in our body can be produced from ES cells, they have the highest popularity among stem cells. Some scientists believe that ES cells may have a greater ability to produce healthier tissue. Although ES cells have been shown to have the potential to turn into virtually any cell type found within the body, no studies have yet demonstrated the controlled generation of a uniform cell type. If the treatments of ischemic myocardial lesions involve the cellular replenishment of devitalized myocardial areas by transfer of healthy cardiomyocytes, which have been previously differentiated from embryonic or bone marrow–derived stem cells, these procedures involve the risks of development of teratoma from ES cells and the possibility of tumor formation from engrafted bone marrow stem cells, in addition to ethical concerns about adoption of ES cells. It is a key feature to find efficient and reproducible methods to derive therapeutically useful and safe cells from an ES cell source. The management of the material process of proliferation and differentiation from the embryonic stage to produce only the specific tissues is required. It should be pointed out that even MSCs may not be as risk-free as believed when maintained in prolonged culture. Rubio *et al*. have shown that after long-term *in vitro* expansion (4–5 months), adipose tissue–derived human MSC populations can immortalize and transform spontaneously [2]. They proposed a two-step sequential model of spontaneous MSC immortalization and tumor transformation. The first step involves senescence bypass, which is infrequent in human cells; strikingly, all MSC samples bypassed this phase. In the second step, transformed mesenchymal cell cultures spontaneously acquire tumorigenic potential, although to date premalignant phenotypes have been described only for human adipose–derived MSC and other cell types after artificial immortalization by telomerase overexpression. Spontaneous transformation has not been previously reported for any human cell type. The findings by Rubio *et al.* indicate the importance of biosafety studies of MSC biology to efficiently exploit their full clinical therapeutic potential.

It has been accepted that ES cells derived by SCNT is not rejected by the recipient who is the donor of the somatic cells. However, human ES cells express an immunogenic nonhuman sialic acid. When a combination of two animal-derived products, FCS to nourish the cells and a feeder layer of fetal mouse fibroblasts that inhibit differentiation into a variety of cell types, was used for human ES cell culture, Martin *et al.* identified a substance on the surface of cultured human ES cells, *N*-glycolylneuraminic acid (Neu5Gc), that is taken up from animal products [3]. The Neu5Gc would probably cause ES cells to be rejected if transplanted into a patient because most normal healthy humans have circulating antibodies specific for Neu5Gc. It was reported that when the culture medium for human ES cells was not conditioned by mouse cells, it promoted stem cell differentiation, mimicking the activity of BMP. This means that there must be molecules in the feeder cells that suppress BMP activity. Xu *et al.* reportedly found that high doses of FGF2 can do what mouse feeder cells do: sustain stem cells in an undifferentiated state [4]. Maitra *et al.* assessed the genomic fidelity of paired early- and late-passage hES cell lines in the course of tissue culture. Relative to early-passage lines, eight of nine late-passage hES cell lines had one or more genomic alterations commonly observed in human cancers, including aberrations in copy number (45%), mitochondrial DNA sequence (22%), and gene promoter methylation (90%). The observation that hES cell lines maintained *in vitro* develop genetic and epigenetic alterations implies that periodic monitoring of these lines using well-defined assays will be required before they are used in *in vivo* applications and that some late-passage hES cell lines may be unusable for therapeutic purposes [5]. These findings may diminish the need for and practical use of ES-derived cell sources for tissue engineering in the future with a shift in focus to progenitor cells derived from the host or human donor.

Both embryonic and adult stem cells appear to have similar potential to develop into the different cellular elements necessary for effective tissue replacement. The differences are few, but they may be significant. Allogeneic ES cells will be recognized as foreign and be rejected. There is no evidence that ES cells can be consistently driven to uniformly generate only the cell type that is needed. While they can certainly be encouraged to form any type of cell, many other types of cells are usually generated at the same time. The disadvantages of ES cells appear to overweigh the advantages. The evidence that cloned human embryos can develop to the blastocyst stage is likely to reinvigorate debates worldwide over a ban on human cloning. Many people agree that cloning to produce a human child should be banned. The Bush Administration, for example, objects to the idea that human embryos would be deliberately created and then destroyed to derive stem cells. There is a great concern that perfecting cloning techniques to improve stem cell derivation would make it easier for rogue doctors or scientists to produce a cloned human baby. The idea of using allogeneic ES cells for therapy will decline when focus has shifted toward adult stem cells, especially the active multipotent precursor cells found in bone marrow.

5. RISKS AT CELL CULTURE

Many tissue engineering strategies rely on multiplying cells from a small biopsy and directly transplanting or subsequently seeding these cells on a scaffold. It is of prime importance for cell culture to avoid contamination with other microorganisms during serum exchange and mistake in label reading. Clearly, safety considerations are required even when using expedient, autologous cells, as culture processes and reagents can alter cells, regardless of their origin. Safety considerations include not only testing and monitoring of procedures for the cells, but also testing of reagents that come in contact with the cells. When different cells originated from different patients are manually processed together in a single facility, a careless mistake in label recognition will result in serious problems. These risks can be minimized by adoption of automatic robot systems. Figure 4.4 shows a commercially available equipment for full-automated cell

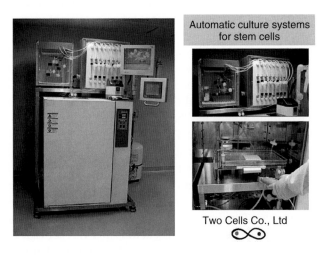

Fig. 4.4   Automatic culture systems for stem cell (Two Cells Co., Ltd).

culture. Although ancillary reagents used *ex vivo* are not intended to end up in the final product, reagent residues can have undesirable effects if they are toxic or functionally active. Published evidence shows that reagents like FCS used as cell culture medium is difficult to remove during washing and can trigger an immunologic reaction in patients who receive cell infusions. In addition, following the discovery of the first case of mad cow disease in the US at the end of 2003, bovine-derived products are expected to come under greater regulatory scrutiny [6].

Any specific medium is not necessary for isolation, fractionation, and purification of cells when procedures such as centrifugation and density gradient alone are employed. In contrast, cell expansion always requires specific media containing mitogens, nutrients, and antibiotics. The most familiar medium for cell culture is the FCS that contains very active mitogens. An urgent problem to solve for clinical tissue engineering is the limitation that xenogeneic serum should not be used for cell expansion. It would be highly desirable to use patient's own serum for cell culture, but it has been recognized that human serum is not active in cell expansion in comparison with FCS that has most widely been used for cell culture. The entire composition of this medium is unknown, and many attempts have been made to create artificial serum-free media to avoid the use of animal-derived mitogens that are antigens to human. A variety of human growth factors with a high mitogenic potential can be produced by gene recombinant technology and have been used for preparation of serum-free, cell culture media, but none of them is as active as FCS in cell expansion. A compromising way is to increase the cell number first using FCS and then to replace the FCS with the serum from the patient's blood just prior to seeding the expanded cells onto a scaffold. In this case, we need caution to completely replace the FCS with the patient's serum. Because FCS has a risk of BSE, the use of FCS for tissue engineering will be more and more severely regulated and much attention will be paid to create new cell culture media for tissue engineering.

It has been shown that even the serum derived from adult humans has a high potential to expand cells. Serum does not contain platelets, although they have high levels of growth factors inside. When a stimulus is given to platelets in fresh blood to destroy them for release of growth factors as much as possible and then remaining blood cells are removed from the blood, we may obtain serum possessing a high mitogenic potential. Figure 4.5 compares the activity of commercial FCS with the human serum prepared by this method [7]. And 200 ml taken from a healthy adult produces approximately 90 ml of serum, which will be sufficient for cell culture in the conventional tissue engineering.

Patient's own platelet-rich-plasma (PRP) has been used for bone regeneration in oral surgery. *In situ* activation of PRP releases plenty of growth factors from the platelets into the site of tissue regeneration. However, most of the released growth factors seem to scatter away from the site of tissue regeneration because an effective carrier trapping the growth factors is lacking.

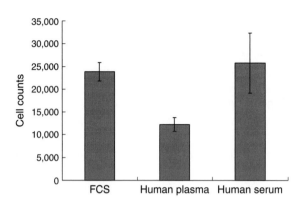

Fig. 4.5   Human MSC growth 6 days after culture in α MEM supplemented with 10% of FCS, human plasma, or human serum.

6. SCAFFOLDS FOR LARGE ANIMALS AND HUMAN TRIALS

It has been always emphasized that scaffolds play a vital role in tissue engineering because they provide a 3-D framework guiding for cells to attach, proliferate, and lay down ECM. Indeed, the biomaterials used for scaffolds represent a major area of study. The objective of scaffold is to simulate the biological microenvironment *in vitro* or *in vivo* to allow cells successfully to produce human tissues or organs. The microenvironment in living tissues comprises numerous cells, ECM proteins, and a variety of soluble and ECM-bound factors distributed in 3-D space. Great efforts have been therefore directed to provide scaffolds with optimal surface, pore structure, mechanical strength, biocompatibility, and bioabsorption kinetic, mimicking the microenvironment for the attachment, growth, morphology, and spatial organization of cells. The cell–scaffold interactions greatly affect cell proliferation, differentiation, migration, and function. These concepts demand a cell-adhesive surface and a highly porous, interconnected structure for a scaffold to ensure sufficient space for tissue development and unimpaired diffusion of nutrients, oxygen, and wastes. Materials surface has been designed by mimicking the native ECM as much as possible to provide a substrate with adhesive proteins for cell adhesion, and regulate cellular growth and function by presenting different kinds of growth factors to the cells.

However, these properties do not cover all the requirements necessary for absorbable scaffolds. The scaffolds fabricated in the studies reported to date have mostly been used for cell culture or implantation in small animals. Cell culture studies have employed either synthetic scaffolds, which have failed in approximating the physical size and mechanical strength of target tissues, or animal-derived materials, which may confound cell culture with undefined or inconsistent variables. Scaffolds from PLLA, PGA, PCL, calcium phosphates, PEG gels, alginates, or naturally derived ECMs have had only limited success. This is due in part to either their mismatch of absorption kinetics with the tissue formation rates, low nutrient diffusion rates, poor mechanical properties, or inability to create functional microenvironments for cells. Animal-derived materials including bovine, porcine, and cadaver tissues, as well as intestinal submucosa, may complicate research and therapies with their poor reproducibility, difficulty in mass production, and potential risk of unknown material contaminations.

It would be therefore difficult to overstate that a central challenge in clinical tissue engineering is to provide clinicians with appropriate scaffolds that are suited to patients and at least to large animal models. Surprisingly, very little effort has been directed toward fabricating such scaffolds that really meet the demands of clinicians. The minimal demands for scaffolds to be used for large animals and human trials include a guaranteed strength and retention of shape over time as well as a temporary structural and mechanical equivalence with human tissues, in addition to accepted non-toxicity. However, fabrication of scaffolds that meet these structural, mechanical, and biological requirements seems quite difficult to achieve so far as reported scaffold design and fabrication technologies are employed. The limited success in clinical applications of tissue engineering is largely due to these reasons, suggesting that other kinds of effort should be made to develop technologies for appropriate scaffold fabrication.

6.1. Mechanical Strength

Scaffolds that are applicable to human trials are substantially different from those used for cell culture and small animals. Scaffolds for large animals are much larger in size and stronger in mechanical strength than those for small animal studies. Preparation of tiny scaffolds will not be difficult in labs of academia unless the preparation number is small, but students may hate to manually fabricate a large number of scaffolds with complicated structures because of time-consuming and tedious work. The following list gives requirements for the scaffolds to be applied for large animals and humans.

1. should be fixed at the proper site in the body (*e.g.*, by suturing; gel is difficult to fix)
2. should be protected from compressive force if a soft scaffold is implanted (*e.g.*, by mesh cover)
3. should be processed or trimmed in the operation theater (porous β-TCP cannot be cut)
4. should be absorbed at the proper rate (PGA is absorbed too fast, whereas PLA remains too long).

It is crucial for clinical tissue engineering that scaffolds are optimized not only with porous structure and absorption kinetics, but also with mechanical strength and handling in intraoperative settings. Often, even biomaterials scientists do not have a clear perspective on the essential role of mechanical properties of scaffold in ultimate clinical application. Most of natural polymers including modified collagen, gelatin, alginate, and HAc lack in mechanical strength when used as scaffolds for large animals. Excessively brittle or limp devices may increase failure rate. The scaffolds can neither be sutured nor maintain the geometrical shape over time against the compression given from the outside. Collagen presents some disadvantages associated with its natural origin such as large batch-to-batch variation upon isolation from biological sources, as well as restricted versatility in designing devices with specific shapes and mechanical integrity. A serious problem appears when we need a large-sized scaffold which must be fixed to the apposing host tissue by some means like suturing. Most of cell–scaffold constructs used for large animals have been secured at the implanted site to prevent displacement. If a construct is of tubular type, anastomosis of the construct tube to the tubular host tissue with suture is inevitable. Most of the sponge-like materials made from amorphous polymers with high porosity will not tolerate suturing because of their low anti-tearing resistance. Sponges constructed from collagen molecules cannot be firmly fixed with suture unless a supporting assist is present.

An effective means to make the material acceptable to suturing is to reinforce the porous materials by incorporating absorbable fibers or filaments. Figure 4.6 shows the cross-section of porous scaffold reinforced with non-woven PGA fabrics. The pore structure of the main part of this scaffold was created by freeze-drying of a lactide (LA)-ε-caprolactone copolymer (P(LA/CL)) solution in which non-woven PGA fabrics had been soaked [8]. The copolymer scaffold without this reinforcement could not be fixed to the adjacent host tissue with high reliability. Such reinforcement is effective also in enhancing the burst strength of tubular scaffolds such as for blood vessel engineering. A major problem for the cartilaginous tissue engineered *in vitro* is the fixation of the engineered constructs into defects. Scaffolds have been secured by fitting into the defect and applying a fibrin sealant, but carti-

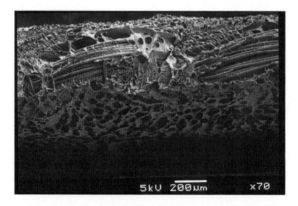

Fig. 4.6 Cross-section of porous P(LA/CL) tube reinforced with non-woven PGA fabric.

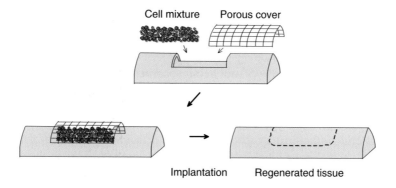

Fig. 4.7 Perforated tray (or mesh cover) to keep a cell mixture at the site of tissue regeneration.

laginous scaffolds often become detached from the defect. A rigid, mesh-like protector or an elastic stent made from absorbable material will be useful for protection of a soft scaffold from external pressure. This kind of protector is also effective in keeping a mixture of bone marrow and particulate cancellous bone at the site of bone regeneration, as illustrated in Fig. 4.7. Figure 4.8 shows an elastic, bioabsorbable stent that will prevent an engineered, tubular tissue from the stricture. The concept of using tubular structures has been studied for tissues such as the trachea, esophagus, intestine, urethra, and blood vessels. Much more emphasis must be placed on the material processing and molding to develop effective materials strategies needed for the clinical tissue engineering.

Calcium phosphate scaffolds having interconnected pores and high porosities are available, but they are brittle and unable to be trimmed with scissors so that close contact with the host tissue will not be expected. This problem would be solved if pliable composites are fabricated from calcium phosphates and soft polymers like collagen.

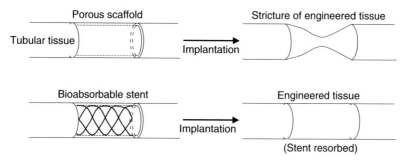

Fig. 4.8 Protection of tubular scaffold from stricture by inserted, resorbable stent.

6.2. Bioabsorption Rate

The absorption rate of scaffolds is critical for tissue engineering because any scaffold would hinder the tissue regeneration process, if the material remains long at the site of tissue regeneration without being absorbed. The scaffold should degrade at a rate that allows new tissue to receive the appropriate level of load without danger and rupture. It has been frequently mentioned that poly(α-hydroxyacid)s undergo hydrolysis accompanying acidic by-product accumulation that might evoke local or systemic inflammatory reactions. However, acidic by-products cannot accumulate if polyesters, even quickly absorbable PGA, are implanted in the site where the body liquid actively circulates as in blood vessels. Surface hydrophobicity of materials is readily reduced, for instance, by simple oxidation without significant strength loss. In many cases, absorption kinetics of PLLA is not appropriate when this polymer is used as scaffolds, since it takes several years until its complete absorption, even though the scaffold has relatively little polymer-per-unit volume. Any tissue does not need several years for the completion of tissue regeneration. The PGA is one of the common scaffold materials that have been approved by FDA for use in the body and is mechanically strong. However, this polymer loses half of the mechanical strength in a few weeks and is absorbed in a few months after implantation. In some cases, quick absorption of PGA and GA copolymers is favorable, but generally is not acceptable because of slow tissue formation. This may be the reason for few applications of PGA scaffolds to large animal experiments. Preferably, scaffolds should maintain the starting mechanical strength from several weeks to a few months, which will be required for the formation of mature ECM. Many studies have chosen PCL as a material for scaffold fabrication simply because of its good processability, but this polyester is virtually non-absorbable. The CL homopolymer needs much longer periods of time for absorption than PLLA, in marked contrast with CL copolymers with GA, LA, and TMC. These copolymers are already available and can be processed to create soft or hard structures suitable for tissue engineering. Figure 4.9 schematically represents the bioabsorption rate of the commonly used polymers.

The demand to polymers with a variety of absorption kinetics has prompted polymer scientists to synthesize new absorbable materials that will be more suited to scaffold fabrication than PLLA, PGA, and PCL. However, only few of newly synthesized biomaterials have been widely accepted, although they might exhibit better surface characteristics for cell adhesion or more flexibility in mechanical properties than existing absorbable biomaterials. Probably, for new polymers the balance between the bioabsorption rate and the mechanical strength is not superior to that of existing polymers. It is important to compare the biological and physicochemical properties of newly synthesized materials with those of existing biomaterials because regulatory agency demands to present clear evidence for safety and efficacy, and often superiority, when new materials are to be used for clinical trials. Most of existing biomaterials, especially those which have frequently been applied to medical treatments, have cleared stringent biological tests to prove

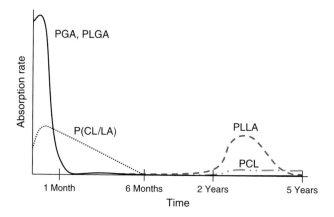

Fig. 4.9 Schematic representation of resorption rate for absorbable polyesters.

their safety. Fabricating cost is also of great concern to manufacturers. These may explain why new biomaterials have scarcely been applied to medical treatments of patients in recent years, although numerous papers have been published regarding synthesis of novel biomaterials.

6.3. Tailoring of Ultrafine Structure

The dynamic state of a tissue is regulated by a highly complex temporal and spatial coordination of many different cell–matrix and cell–cell interactions, but synthetic biomaterials represent oversimplified mimics of natural ECMs with the lack of essential, natural temporal and spatial complexity. Therefore, attention has been directed to novel design strategies for synthetic biomaterials that have biologically multifunctional networks and contain the necessary signals to recapitulate developmental processes in tissue- and organ-specific differentiation and morphogenesis, similar to natural ECMs. Based upon the paradigm of fibrin's function as a temporary matrix in tissue repair, biomimetic characteristics for synthetic materials needed to induce regeneration *in vivo* are the presence of (i) ligands for cell adhesion, (ii) a mechanism of relatively rapid and localized matrix dissolution, ideally in temporal and spatial synchrony with cell invasion, and (iii) the delivery of morphogenetic signals to attract endogenous progenitor cells and induce their differentiation to a tissue-specific pathway [9]. To this end, biomimetic materials have been tailored to serve as provisional matrices for tissue regeneration *in vivo*, but much work would remain to develop biomaterials so far as such biomolecular approaches are aiming at recapitulating the elaborate biological recognition and signaling functions of ECMs.

Along this line, designs of fine textures such as microchannel arrays have been attempted by several research groups to create in a scaffold for directing the formation of vasculature into the scaffold. However, such a structure, even if succeeded, will be soon disorganized and disappear if this structure is created on a bioabsorbable scaffold that has a reasonable absorption rate. A scaffold fabricated from very slowly absorbed biomaterials such as PLLA and PCL will not promote tissue and vasculature formation, but may conversely disturb it.

Cell adhesion to the pore surface of scaffold, which is a requisite of tissue engineering, is mediated by cell-adhesive proteins such as FN. However, this does not mean that immobilization of cell-adhesive proteins or their oligopeptides on the pore surface is indispensable. Even if a scaffold lacks in such biological cell-adhesive motifs, cells can adhere to artificial surfaces as illustrated in Fig. 4.10 [10], because any scaffold placed in serum or implanted in the body finally attracts cell-adhesive proteins present in the vicinity of the scaffold surface if the surface property is in the range of optimal hydrophilic-hydrophobic balance. Both extremely hydrophilic and hydrophobic surfaces reject protein adsorption and cell adhesion, as shown in Fig. 4.11, where serum albumin adsorption onto various non-ionic polymer surfaces is plotted against the hydrophilicity (or hydrophobicity) index [11]. Apparently, protein adsorption becomes less significant when the hydrophilicity (or hydrophobicity) shifts to the more hydrophilic (or hydrophobic) direction. The presence of negative charges on a scaffold surface enhances cell adhesion.

It is well known that a hydrophobic polymer surface can be converted to a less hydrophobic surface by oxidation such as plasma treatment that lowers water contact angle of the surface. Figure 4.12 shows the adhesion of L cell onto plasma-treated surfaces as a function of their water contact angle [12]. The initially hydrophobic polymers examined for this plasma treatment include polyethylene, PTFE, polypropylene, PET, and polystyrene. Obviously, the polymer surface with a contact angle around 70° exhibits the highest cell adhesion. The contact angles of PGA, PLLA, and PCL may be higher than 70° and alkaline hydrolysis of these

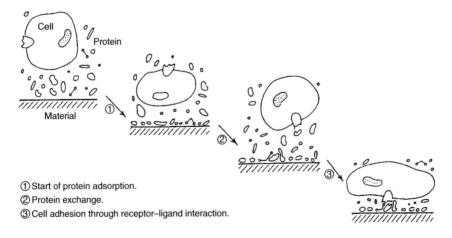

① Start of protein adsorption.
② Protein exchange.
③ Cell adhesion through receptor–ligand interaction.

Fig. 4.10 Schematic illustration of cell adhesion to protein-adsorbed surface.

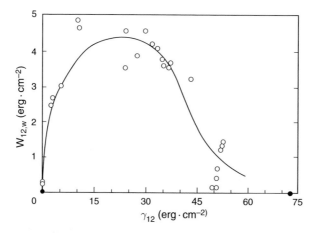

Fig. 4.11 Work of adhesion with BSA in water ($W_{12,w}$) as a function of interfacial free energy between water and polymer 1 (γ_{12}).

absorbable polyesters will reduce their contact angle, leading to higher cell adhesion. These results indicate that fixation of biological cell-adhesive proteins or oligopeptides on polymer surfaces is not always necessary to render them cell-adhesive. Cell adhesion is surely promoted by immobilization of cell-adhesive motifs, but the extent is not significantly large. Hydrogels that have the minimal capacity in cell adhesion because of their specific surface structure as illustrated in

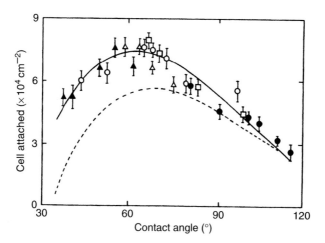

Fig. 4.12 Relationship between the contact angle and L-cell adhesion. — with plasma treatment, ---- without plasma treatment. (○) polyethylene, (●) PTFE, (△) PET, (▲) polystyrene, and (□) PP.

Grafted soluble
chain

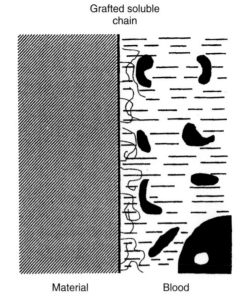

Material   Blood

Fig. 4.13 Schematic representation of the interface between the super-hydrophilic surface and blood components.

Fig. 4.13 [13] may require fixation of cell-adhesive proteins or oligopeptides if cell adhesion is required.

It is often stated that new biomaterials derived specifically for tissue engineering are needed. However, new synthesis is not always necessary because existing materials that have been already clinically accepted are, in many cases, adequate for tissue engineering. In addition, the field would not embrace development of new materials due to potential product liabilities, economical considerations, and strict regulations.

7. IMPORTANCE OF NEOVASCULARIZATION

In principle, a vascular network must be established throughout the newly formed tissue in order to support its growth. Such a functional vasculature is an essential component of any metabolically active tissue that has a thickness in excess of a few millimeters. There are three distinct mechanisms for vessel growth: vasculogenesis (*de novo* formation of blood vessels by endothelial progenitors as seen in embryogenesis), angiogenesis (sprouting of vessels from preexisting ones), and arteriogenesis (stabilization of fully formed arteries containing all three-wall layers by mural cells). Many regulating soluble signals and their receptors have been identified, including VEGFs, FGFs, TGFs, angiopoietins, ephrins, placental growth factors, and various chemokines. In angiogenesis, factors such as bFGF and VEGF induce the ECs of nearby vessels to digest their basement membranes, invade the tissues, migrate, and differentiate to form tubes. *In situ*, the developing capillary network may connect to vessels growing in from the surrounding tissue. In more detail, the process of angiogenesis is regulated by more than two dozen known cytokines that appear in complex series. Of these, the most important and thoroughly investigated are VEGF and bFGF. Neovessel formation begins with growth factor appearance is followed by the emergence of MMP enzymes. The protease activity leads to breakdown of the parent capillary basement membrane and ECM, along with release of further ECM-bound growth factors including VEGF. Proliferating ECs are then able to migrate into the matrix space and form a capillary sprout. Sprouts grow by continuing endothelial proliferation, gradually forming a lumen as they organize into tubules. New microvessels become mature when contiguous tubules anastomose with each other and a new basement membrane is formed. Each of these steps is regulated by a balance between stimulatory and inhibitory cytokines. In addition, ECM components including collagen, LN, other glycoproteins, and GAGs have been shown to have substantial roles in microvessel formation.

For the successful neovascular network formation the porous structure of scaffold is crucial, as the survival of transplanted cells is dependent on diffusion of nutrients and waste products through the pore structure. Interrupted or too narrow channels in a porous scaffold do not facilitate the vasculature formation which mediates the interaction of transplanted cells with the surrounding host tissue. For instance, surviving clusters of hepatocytes within a polymer construct are usually centered around or adjacent to large blood vessels. Most of tissues will necessitate vasculature within 2–4 weeks. Inclusion of large, interconnected pores in scaffolds may promote invasion of host fibrovascular tissues, but fibrovascular ingrowth into the scaffolds occurs at a rate less than 1 mm/day and typically takes 1–2 weeks to completely penetrate even relatively thin (*e.g.*, 3 mm thick) scaffolds.

Among a variety of growth factors that promote the formation of a new vasculature, VEGF shows promise as a molecule that may enhance the vascularization of engineered tissue, as it has been shown to act most specifically on ECs. Currently, VEGF is being employed to examine its effects on revascularization of ischemic tissues and musculocutaneous flaps. If the concentration of growth factors is too low at the site of action, these factors can be additionally provided by means of the technology of drug delivery systems (DDS). The routes of administration of VEGF vary from systemic administration to localized delivery. Additional problems related to growth factors include high cost and the difficulty for receiving approval from regulatory agencies. Probably, VEGF alone is not sufficient to create mature and stable vasculature [14]. In a study using injection of adenoviral vectors to express VEGF in normal tissue, newly formed human blood vessels that were CD31-positive were found to be disorganized, leaky, and hemorrhagic. Furthermore, VEGF may potentiate inflammation by increasing the expression of adhesion molecules on ECs or the release of chemokines. Such new blood vessels may not be adequate to the tissue's metabolic demand.

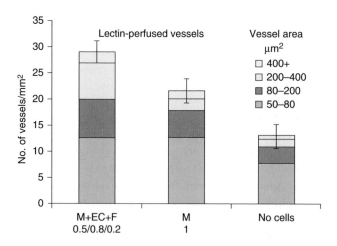

Fig. 4.14   Functional, lectin-perfused vessels in tissue-engineered muscle implants. Vessel formation was compared between tri-culture constructs seeded with myoblasts (M), endothelial cells (EC), and embryonic fibroblasts (F), and constructs seeded with myoblasts alone or without cells (no cells).

In this connection, Levenberg *et al.* have developed an interesting approach that enables formation and stabilization of endothelial vessel networks *in vitro* in 3-D engineered skeletal muscle tissue [15]. First, they induced endothelial vessel networks in engineered skeletal muscle tissue constructs using a 3-D multiculture system consisting of myoblasts, embryonic fibroblasts, and ECs coseeded on highly porous PLGA scaffolds. Then, to study the survival and vascularization of the engineered muscle implants *in vivo*, they injected labeled lectin into the mouse tail vein 2 weeks after implantation and counted the perfused vessels. The results indicated that 41% of the human CD31-positive vessels (implant-derived vessels) were perfused with lectin. Quantification of the total number of perfused vessels (host and implant derived) indicated that muscle constructs seeded with ECs had 30 ± 2 functional vessels per square millimeter compared with 21 ± 2 vessels in constructs seeded with muscle cells only (Fig. 4.14). The size distribution of functional vessels showed that including ECs in the scaffolds also increased the number of larger or stabilized vessels in the muscle implants.

In a review, Brey *et al.* presented a general outline of bioengineering research contributing to the development of therapeutic neovascularization strategies [16].

8. CARRIERS FOR GROWTH FACTORS

Many cellular processes involved in morphogenesis require a complex network of several signaling pathways. Tissue replacement with tissue engineering techniques also requires bioactive proteins such as growth factors in addition to scaffolds for tissue-forming cells. The major roles of growth factors in tissue engineering are to trigger neovascularization, accelerate cell expansion, and induce cell differentiation into the target tissue. In this connection, the roles of growth factors in wound healing are instructive. Wound healing—whether initiated by trauma, microbes, or foreign materials—proceeds via an overlapping pattern of events including coagulation, inflammation, epithelialization, formation of granulation tissue, and tissue remodeling. The process of repair is mediated in large part by interacting molecular signals, primarily cytokines, that motivate and orchestrate the manifold cellular activities which underscore inflammation and healing. The complex interplay between multiple cytokines, cells, and ECM is central to the initiation, progression, and resolution of wounds. The initial injury triggers coagulation and an acute local inflammatory response followed by mesenchymal cell recruitment, proliferation, and matrix synthesis. Most types of injury damage blood vessels, and coagulation is a rapid-fire response to initiate hemostasis and protect the host from excessive blood loss. With the adhesion, aggregation, and degranulation of circulating platelets within the forming fibrin clot, a plethora of mediators and cytokines are released, including TGF-β, PDGF, and VEGF, which influence tissue edema and initiate inflammation. A vascular permeability factor, VEGF, influences the extravasation of plasma proteins to create a temporary support structure upon which not only activated ECs, but also leukocytes and epithelial cells subsequently migrate. Latent TGF-β_1, released in large quantities by degranulating platelets, is activated from its latent complex by proteolytic and non-proteolytic mechanisms to influence wound healing from the initial insult and clot formation to the final phase of matrix deposition and remodeling. Active TGF-β_1 elicits the rapid chemotaxis of neutrophils and monocytes to the wound site in a dose-dependent manner through cell surface TGF-β receptors. Of the myriad of cytokines that have been investigated in terms of wound healing, TGF-β_1 has undoubtedly the broadest effects. Despite the vast number of reports documenting the actions of TGF-β in this context, both *in vitro* and *in vivo*, controversy remains as to its endogenous role.

The response to damage is the formation of collagen-rich scar tissue following an acute inflammatory response. Although rapid scar tissue formation represents a powerful defense mechanism against infection, it may severely compromise tissue function, as in spinal cord injuries and myocardial infarction. Functional tissue recovery may occur when (i) regeneration-competent progenitor and stem cells are present, and (ii) when these cells are conducted into a regeneration pathway by the presence of relevant morphogenetic signals, or alternatively (iii) when the regenerative process is not suppressed by signals that give way to rapid scar formation.

When a growth factor is given to the site of tissue regeneration, it may recruit the cells that are responsible for their function from the surrounding environment. It is likely that endogenous growth factors will be delivered from the cells themselves residing at the site of action or in the vicinity. However, the concentration of the secreted growth factors is often less than the necessary level. In such a case a sufficient amount of growth factors for the tissue formation should be exogenously supplied. This is the reasoning for the use of growth factors in tissue engineering. The growth factors that have frequently been used in tissue engineering research include BMP for bone formation, bFGF for neovascularization and cell expansion, and TGF-β for tissue regeneration. The BMP-2 may be the strongest osteoinductive factor administered therapeutically to restore the form and function of bone among the plethora of bioactive factors available. Interestingly, inspection of recent publications reveals that some studies have concluded growth factors to be useful for tissue engineering, whereas others have not supported the growth factor application for tissue engineering. This controversial conclusion may result from the different administration routes of growth factors. Many researchers give a growth factor to the site of tissue regeneration only once in the form of its solution (bolus administration) to find low efficiency of the growth factor. In this case the administered growth factor scatters too fast away from the site of injection. This suggests that what is important in the growth factor application in tissue engineering is to adopt the concept of the DDS that has widely been accepted in pharmaceutical therapies. The central dogma of the DDS is to maintain a proper concentration of an administered drug at the site of action for a proper time of period without scattering to other sites in the body to minimize side effects. In order to realize the primary objective of this DDS, we need material support as a drug carrier for its sustained release and guidance to the target tissue.

Delivery of BMPs has achieved good results for bone regeneration in certain instances, but exhibited even detrimental effects in other cases [17]. In cases where beneficial effects were observed, large doses of growth factor, often to more than 50 mg of growth factor per gram of carrier, were required. This supraphysiological amount of recombinant BMPs raised concerns regarding their safety, cost, and effective delivery. It is unclear why lower physiological levels are not effective, but a major reason for the low efficiency of growth factors reported by many research groups is likely to be the lack of carriers that are really effective in sustained release of growth factors. Currently used carriers for BMP-2, including collagen, fibrin gels, and ceramics, are less than ideal because of the poor capability of these biomaterials with regard to the slow release of BMP. Effective delivery systems for biologically active agents are very important. If a good carrier is employed, slow release of BMP can be achieved, as shown in Fig. 3.63 in Chapter 3. Figure 4.15 shows another example of *in vivo* release profiles of growth factor when a crosslinked gelatin is used as a carrier [18]. It is apparent that bFGF molecules entrapped in gelatin microspheres are slowly delivered into the subcutaneous tissue of mice. The profiles are quite different from the burst release observed at the bolus administration of bFGF from its aqueous solution. A plausible mechanism for slow release of growth factors from gelatin hydrogels is depicted in Fig. 4.16. It is likely that the bFGF molecules trapped in ionic hydrogels through ionic interactions will be released as a result of enzymatic degradation of gelatin hydrogel. This hypothesis seems to be ensured by the result shown in Fig. 4.17, where it is seen that basic gelatin cannot trap bFGF, in contrast with acidic gelatin [19].

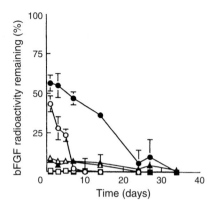

Fig. 4.15 Time course of the radioactivity remaining of ^{125}I-labeled bFGF incorporating gelatin hydrogels with water contents of 98.8% (open marks) and 96.9% (closed marks) after implantation into the back subcutis of mice: in hydrogels (o, ●), around hydrogels (△, ▲), and in the blood (□, ■).

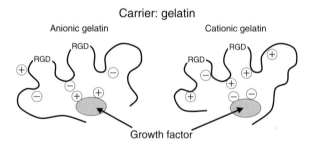

Fig. 4.16 Entrapping of growth factors by bioabsorbable gelatin.

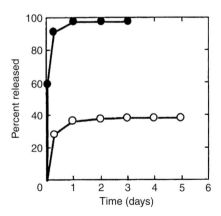

Fig. 4.17 Release profiles of bFGF from acidic and basic gelatins with the isoelectric points of (o) 4.9 and (●) 9.0, respectively.

The biomaterial to be used as a carrier of growth factors should be resorbable, similar to the scaffold for tissue engineering, but high mechanical strength is not required for the carrier. The most important requirement as a carrier of growth factors is to be able to entrap the growth factor molecules without impairing their biological activity and release them at the desired time schedule. Caution is needed when organic solvents and crosslinkers are employed for the formulation of growth factor–carrier mixtures because of possible denaturation of proteins by these reagents.

It should be pointed out that platelet has attracted attention as a means for growth factor delivery, especially in oral tissue engineering. Topically applied PRP is known to upregulate cellular activities and promote bone generation because growth factors such as PDGF and TGF-β are released from platelets upon their activation for degranulation. This blood cell stores a variety of growth factors at high concentrations, as shown in Table 4.1 [20]. The most abundant growth factor present in platelet is TGF-β. Because platelets are recruited for stopping bleeding and the subsequent wound healing, it is very likely that growth factors released from platelets will play an important role in regulating the sequence of the wound healing process. As such, wound healing is so intimately related to tissue repair that tissue engineering will also be promoted by the use of platelet. Clinically, PRP harvested from patient's own blood is given to the site of tissue regeneration along with thrombin which gives rise to the platelet destruction. However, few studies have been published to examine whether growth factors released from the degranulating platelets really remain at the site of action for a desired period of time or eventually be scattered away within very short time. In addition, the optimal amount of PRP that accelerates tissue regeneration is not yet defined. The platelet volume given to the target site in excess of the optimal might have a negative effect on tissue regeneration.

A single growth factor has been applied in most of tissue engineering studies except for the use of PRP, although the naturally occurring tissue regeneration

Table 4.1
Concentration of growth factors stored in blood   unit (pg/ml)

Growth Factor		Serum	Plasma (Succinic Acid)	Plasma (EDTA)	Plasma (Heparin)
Brain-derived neurotrophic factor	BDNF	27,793	196	428	401
Epidermal growth factor	EGF	336	6	13	
Hepatocyte growth factor	HGF	1,257	431	787	
Platelet-derived growth factor AA	PDGF-AA	4,467		148	105
Platelet-derived growth factor AB	PDGF-AB	20,126		205	308
Platelet-derived growth factor BB	PDGF-BB	3,478		60	32
Transforming growth factor-β1	TGF-β_1	48,600		2,260	
Vascular endothelial growth factor	VEGF	220		61	41

process must be regulated by a set of multiple growth factors at a determined time schedule and concentration, as seen in wound healing. In fact, time-dependent expression of cytokines, growth factors, and their receptors during embryogenesis and wound healing has been well recognized. Tissue development is controlled by the interplay of multiple signals. Engineering of complex tissue constructs is also likely to require use of multiple growth factors delivered either sequentially or with site-specific patterns. In other words, by extension from normal tissue formation and repair, important variables in tissue engineering likely include the concentration, timing, and sequence in which the bioactive molecules are introduced. Residence time is also important in which growth factor and cells interact. Extracellular FGF must be present throughout the G_1 transition period. Inappropriate growth factor combinations can stimulate multiple signaling cascades that would lead to undesirable outcomes. In the future, detailed studies are required that address this issue, especially when one aims at the regeneration of complex organs composed of multiple tissues. Recent research efforts have focused on schemes for sequential delivery of multiple growth factors and the use of biological feedback mechanism in growth factor delivery.

9. PRIMARY ROLES OF EACH PLAYER IN THE TISSUE ENGINEERING ARENA

The final objective of tissue engineering research is to develop new methods applicable to clinical treatments of patients by clinicians. The long road map on tissue engineering involving from its basic research to its clinical application may be different in each country, depending on its research and regulation systems. However, there may be no substantial difference in key technologies necessary for clinical applications of tissue engineering. Any programs to establish the clinical tissue engineering with potentially significant outcomes must combine expertise from a wide range of scientific and technical areas. Tissue engineering is the application of biological, chemical, and engineering principles toward the development of substitutes for the repair or restoration of tissue function. Successful tissue engineering depends on a number of critical steps. Few areas of technology will require more interdisciplinary research than tissue engineering or have the potential to affect more positively the quality and length of life [21]. The framework for designing and building cell–material structures should make the connection between engineered systems and biological systems. Even the development of key technologies requires collaboration of various kinds of science and technology. Molecular and cellular biology are necessary to deeply understand the inter- and intra-cellular events occurring during tissue formation. Effective isolation of cells from the cell source and their purification are crucial in tissue engineering. We should master also the techniques necessary for cell expansion with the maintenance of cell phenotype and cell differentiation into the target cell in cell culture. Designing bioreactors suitable for these purposes needs expertise in chemical and biological engineering. On the other hand, fabrication of functional, synthetic scaffolds would be virtually impossible unless profound knowledge on polymeric and ceramic biomaterials is lacking. Researchers on biomaterials in academic institutions can synthesize polymers and ceramics and fabricate scaffolds of simple shapes such as film, sheet, plate, 3-D block, and sponge, but only limited institutions may be capable to spin fibers and weave them into textiles. It would be also too difficult to produce devices possessing sophisticated 3-D structure and shape, if the institution has neither own machine shop nor specially designed facilities. In such a case, one can ask some companies equipped with such specific facilities for molding and processing. However, it would take an unexpectedly long time to receive the device fabricated as you desired unless you have a good relationship with the manufacturing company. Sometimes, such a delay, if happening in one of the research subgroups of the whole project team, would result in formation of a bottleneck that ultimately destroys the whole program, even if plenty of funds had been invested. Stem cells make them the subject of high-profile studies as agents of repair in human disease, but we need to commit the necessary time and resources to identify the best road map for translational medicine. For instance, the use of stem cell in treating cardiac-muscle diseases will need a stepwise translational pathway as shown in Table 4.2 [22].

In general, it takes a very long time for a basic, biomedical study to finally bring forth fruit with widely accepted clinical applications. In an interdisciplinary field such as tissue engineering, efforts by a single research group are not sufficient to arrive at the stage of successful application, but intimate interplays among several research groups are strongly required to reach the objective. Interdisciplinary science and technology, characteristic of tissue engineering, apparently have a high potential to pave a new way, but involve many hurdles to be overcome. In principle,

Table 4.2
From experiment to theory. A stepwise translational pathway for systematically moving experimental studies of cardiac stem cells into potential clinical therapy for heart failure

---

1. Demonstration that the progenitor/stem cell of interest can differentiate *in vitro* into fully mature cardiac muscle cells in the absence of cell fusion.
2. Documentation that the cell type can function as an authentic *in vivo* cardiac progenitor in the embryonic or postnatal heart, and can be localized to a specific region of the intact heart.
3. Identification of specific, defined molecular cues that drive the differentiation of the progenitor/stem cells into physiologically functioning cardiac muscle cells.
4. Use of genetic markers to identify cells within an intact organ/tissue; re-isolation of these cells, and documentation of differentiation (molecular and physiological) that is indistinguishable from differentiated cells from the same preparation.
5. Improvement in organ/tissue function clearly ascribed to the differentiation of the cell type of interest.
6. Documentation of stability, electrical coupling, and lack of long-term loss of specialization or transformation events.
7. Direct demonstration that results in genetically modified mice can be extrapolated to large-animal models of heart failure that better reflect human cardiac physiology.

---

interdisciplinary research comprises a set of subgroups the member of which should be top specialists in their own fields. The term "collaborative work" sounds comfortable, but many cases have been known where collaboration has not yielded a fruitful result, mostly because of a lack of steady communication among the research subgroups. The need for collaboration between scientists, engineers, and clinicians will be met most effectively when researchers from distinct backgrounds assemble into teams rather than rely on multiple collaboration between teams. What is strongly desired for the success of collaborative work is for each of the constituent subgroup member to have responsibility to develop its own technology that is an essential element for completion of the project. Establishment of a clinically applicable, new technology may be realized only when minimally essential elements are effectively integrated. If one of them is missing, any new medical treatment will be too difficult to create. Each component of the team provides critical and unique insights and resources that can greatly enhance the team's likelihood for success.

What follows focuses on the role of players working somewhere on the way from basic research to clinical application in tissue engineering. A project with the final goal of clinical trials needs several communities from specified science and technology fields. They include clinical medicine, physiology, cellular and molecular biology, gene technology, biomaterials science, chemical engineering, bioengineering, device manufacturing, and medical regulation. In contrast to cell therapy which uses single-cell suspension alone for clinical treatments, tissue engineering needs collaboration with much more specialists to treat a patient with use of a cell–scaffold construct.

9.1. Scientists and Engineers

Researchers engaged in tissue engineering should identify strategies for matching cellular processes with biomaterial processes and develop engineering of complete systems that exploit the unique performance of cell-based devices for therapeutic applications. The design rules are, however, in many cases incompletely understood. Often, highly specific polymers and materials are demanded to meet the processing conditions of some scaffolds such as 3-D printing. This will pose a significant challenge to material engineers, particularly in developing 3-D scaffolds for clinical applications. Another challenge to biomedical engineers is the biological, chemical, and physical characterizations of engineered tissues. Despite many studies on scaffold and cell culture, there has been little effort to assess the mechanical properties of engineered tissues, which experience a mechanical environment different from that of native tissues. For instance, deflection bending is a deformation mode that is typically seen in soft tissues, but it has not been studied in detail due to the shortage of engineered specimens for the measurement.

Even experienced specialists of biomaterials are not capable of fabricating scaffolds with any absorption rates and mechanical properties at the request of biologists or medical doctors. Although great efforts have been made on synthesis of absorbable biomaterials, those that have been synthesized to date do not cover the whole spectra with respect to biological, chemical, and mechanical properties. The PGA, which likely has been most frequently used for scaffold fabrication, may degrade more rapidly than desired in most cases. In contrast, PLLA and PCL, both of which are excellent in mechanical properties and processability because of controllable crystallinity, toughness, and thermal transition, are absorbed so slowly that the scaffold made from these polymers rather tends to disturb tissue regeneration. Most GA–LA copolymers degrade too quickly, similar to PGA, or yield scaffolds with poor mechanical properties. Therefore, instead of synthetic scaffolds, a variety of natural scaffolds of different types have been derived from allogeneic and xenogeneic tissues such as SIS and porcine heart valves. It is at the moment unclear whether these naturally derived scaffolds will be widely used for patients in the future, similar to biological valves that have been extensively used as xenografts to replace severely diseased heart valves.

In cell culture the modulation of cell phenotype between the synthetic and the quiescent state is important. The modulation is based on biochemical or environmental cues. For instance, a key issue in vessel culture is to balance the competing goals of SMC proliferation and ECM deposition (synthetic state or dedifferentiation), and the contractile phenotype associated with differentiation and maturation. For culture of engineered vessels *de novo*, an increased synthetic state is required, whereas at the conclusion of vessel culture, a minimally proliferative, quiescent phenotype is desired. This is another challenge to biomedical engineers. *Ex vivo* tissue engineering with cells on scaffolds often leads to minor matrix production or only thin layer formation. Large matrix formation on *ex vivo*–cultivated cell–polymer constructs is a main challenge in tissue engineering. As the engineering of biomaterials improves, the limiting factor in tissue engineering will likely become the biology.

9.2. Clinicians

Physicians who treat patients play a particularly important role in tissue engineering because they are the real reviewers and also the end-users of tissue-engineered products and have responsibility for consulting with patients. Any tissue engineering that does not attract the attention of medical doctors will be of no value. Medical doctors are strongly requested to point out the problems of current tissue engineering and to suggest possible means by which they will be solved. It is physicians who are aware of disadvantages of the medical treatments that are currently applied to patients. Physicians should provide biomedical scientists and engineers with information about what kinds of technique are needed for treatments of the disease the physicians have trouble with. This will stimulate scientists and engineers to initiate research to address the problems raised within medical centers. It is necessary to adopt a common language when researchers, engineers, and clinicians from disparate areas work together. One might insist that there is no need for discussion with clinicians because so many databases, monographs, and symposia are currently available and offer a vast amount of information on the problems related to tissue engineering. This is in part true, but a lack of direct assessments and comments from clinical sites will greatly retard the progress in tissue engineering, even if great efforts are made by scientists and engineers. Otherwise, their efforts would come to an end with "paper for paper" work. This would be a great loss in many aspects. To avoid this loss, medical doctors should more frequently participate in discussions with specialists enthusiastic in promoting clinical applications of tissue engineering.

9.3. Manufacturers

Relatively small-sized firms are currently engaged in the tissue engineering business. They attempt to sell cells, ancillaries for cell culture, bioreactors, absorbable biomaterials, or scaffolds to researchers of tissue engineering. All of the items for sale are mostly only for animal use and are strictly banned from use for human clinical trials unless approval is received from a regulatory agency. These small firms have greatly contributed to the advance in tissue engineering research, but their sales remain small in size. Those who are working in the R&D division of medical companies often complain of too rigid regulations and hope that the authorities will take into consideration the balance between the possible risks of new technology and the invaluable benefits to patients that the new technology will offer. For instance, artificial and xenograft materials for children with congenital cardiac disease have high failure rates due to the lack of growth and remodeling potential and due to postimplantation coagulation and fibrosis, whereas autologous tissues engineered using absorbable scaffolds have the capacity to grow and adapt. To release tissue-engineered products to medical markets, the manufacturer may be required to identify the compositions of engineered tissues including cellular types, collagen, chondroitin sulfate, HAc, and elastin content in addition to compressive and tensile properties of the tissues. Moreover, the destiny of the scaffold material implanted in the body should be followed until its complete bioabsorption. These issues may really be a large economical burden to small firms.

There are many possible options for a company to enter into the medical market related to tissue engineering. Companies can fabricate and sell scaffolds for human use at the request of clinics after acquisition of official approval for the products. The business to expand the cells harvested from patients and forwarded from a medical center, followed by sending back to the medical center after cell processing, creates a new type of service without manufacturing any products. Another type of tissue engineering business is to seed patient's cells onto a scaffold, subject to cell culture using bioreactors, and deliver the resultant cell–scaffold construct to the medical center that had sent the patient's cells to the service company. If a company treats allogeneic cells instead of a patient's cells, it can manufacture multiple cell–scaffold constructs and sell them to multiple medical centers with warning of possible risk of immunorejection.

The company that has an intention to commit to medical treatments of patients is expected to expand to big business in the near future. If a company plans to sell only scaffolds to medical centers, the style of business is similar to that of established medical companies that deal with conventional medical devices. However, living cells are totally different from existing devices, so that this may create a new type of medical business. Such living products must be treated with much more caution than non-living ones such as artificial organs. In this case, "high risk, high return" may be a realistic phrase.

When a company has had an interest in a certain promising product, it will first start to survey the corresponding market size to evaluate the need for the product. If the size of needs multiplied by the unit price of the product is not well matched with the total cost that will be spent for R&D, manufacturing, risk management, marketing, sales, and insurance, the company would abandon the business plan. This may be the common strategy in industries and is not exceptional for the medical industry. Obviously, medical business is more risky than non-medical, but it would be almost impossible to establish a widely applicable medical technology without collaboration of industry. Big companies having strong financial power and ample clinical trial experience can effectively support clinical trials performed by medical doctors and can expand the new technology to much broader applications. In this respect big companies will play a crucial role in establishing new medical technologies worldwide.

According to Hardingham, the application need not be for a major volume product or for a major clinical need. The target application might then be the one most likely to succeed, rather than the one most likely to make a profit. The value of not-for-profit developments is that they may give clinical success and they may provide more experience of tissue engineering applications in the clinic. Such developments may be charity funded, hospital associated, essentially local/regional activities, but this could expand the range and concept of clinical applications and would facilitate new commercial developments. These types of developments are essentially how surgical procedures have developed in the past [23].

Any scaffold to be used for patients has to be fabricated at a clean facility of GLP level and thoroughly sterilized. Companies must provide facilities that are equipped for the range of material processing and molding into specific macroscopic architectures required in scaffold fabrication. Processing approaches must be adaptable to manufacturing protocols that are cost-effective and can meet regulatory requirements for good manufacturing processes. This is one reason why translation of academic research tools into commercial products proceeds so slowly. If reinforcement of a scaffold with fibrous materials is required, it is generally too difficult to fabricate such a reinforced scaffold in labs of academia. A large amount of cell mass is needed to completely correct lost tissue function of patients. Unless collaboration with industry or at least support from industry is realized, it seems almost impossible to shift from the small-animal to the large-animal and human trial stage. This implies that tissue engineering research in the near future strongly needs active collaboration with the industry that has the unique capacity for fabrication and sterilization of large-scale materials and quality control. Industry should recognize that, as our population ages, tissue engineering and regenerative medicine will become important economic forces.

9.4. Regulatory Agencies

In marked contrast to non-medical products, medical products including drugs and medical devices need official approval for their manufacturing and sales from a regulatory organization affiliated with government. In recent years governmental regulations are more or less involved in many industries, but the involvement of regulation is much deeper in medical industry than in other non-medical ones. This is reasonable because the items dealt with medical industry are closely connected to the life of patients. The legal processes from application to final permission from government are different among countries. In common, the company that intends to manufacture or sell medical devices has to prove the non-toxicity and sterilizability of the product. It is also imperative to manufacture the product in a strictly controlled environment with the cleanness higher than a certain high level. In addition, the animal right is a great concern of researchers engaged in medical technologies. It is virtually impossible to develop a new medical technology without animal experiments that provide a proof of principle for new technology under development. Any institution that is involved in biomedical research has to establish an ethics committee to protect the right of animals when researchers of the institution have a plan to use animals for their experiment.

The highest priority of medical technologies to abide by is to guarantee the safety to patients or that there are no serious side effects that are generally controlled by regulatory authorities in many countries. In addition, any new medical technology has to prove the superiority with respect to the effectiveness and cost over existing treatments for the disease that the technology targets to cure. Tissue-engineered products substantially differ from conventional medical devices made from metal, ceramic, and plastic. In particular, it is extremely difficult to prove that a tissue-engineered product is completely free of bacteria, virus, or prions which are powerful pathogens. Conventional sterilizations with ethylene oxide gas, ionizing radiation, or dry heating are not acceptable to tissue-engineered products when they contain living cells. No other therapeutic modality comes close to approaching the complexity, fragility, or variability of human cells. And because cells represent a potential reservoir of communicable pathogens, the difficulty of detecting pathogens and sterilizing cells presents a particular headache for regulatory agencies. Human tissues used for medical purposes may be classified either as devices (as in the case of allograft heart valves and dura mater) or as biologics (as in the case of blood components and products). Engineered tissue products would be classified by certain characteristics; for instance, according to (1) the relationship between the donor and the recipient of the biological material used to produce the tissue product; (2) the degree of *ex vivo* manipulation of the cells comprising the tissue products; (3) whether the tissue product is intended for a homologous use, for metabolic or structural purposes, or to be combined with a device, drug, or biologic. The US FDA classified allogeneic living engineered skin tissue as a medical device and autologous cultured chondrocytes as a biologic device. Whether or not the clinical use of autologous cells is under regulation is different among nations.

Regulatory issues present a major challenge to the process of bringing new tissue engineering products from the academic laboratory to patients and the development of the tissue engineering industry. The approach to the regulation of products incorporating human tissues is not fully implemented even within the United States. In fact, emerging biomedical products utilizing living tissues present a new order of magnitude of complexity in their interactions with human patients. Healthcare reimbursement regulations and private insurer practices are critical components of establishing market acceptance. Ethical and immunogenic issues are always associated with the cellular products when allogeneic cells are used for the production. If undifferentiated stem cells are included in the construct to be implanted in patients, a road map to the final desired differentiation should be identified. It is therefore reasonable that regulations applied to tissue-engineered products are largely different from those for conventional medical devices. Tissue engineering is seemingly classified to the most stringent medical technology in terms of regulation, since human cells are involved in addition to animals and biomaterials. However, too stringent regulations should be avoided; otherwise, clinical applications of tissue engineering would be further delayed, although many patients for whom existing curative methods have no effect are waiting for the early establishment of new medical technologies. Some biomedical researchers say that the speed of advance in tissue engineering research has been greatly hampered by the governmental regulations and committee rules that are increasingly becoming more strict and complex. However, before protesting, researchers should make efforts to conform to the regulations. What is unfavorable to tissue engineers is the different regulations among nations. Regulations tend to be unified at least within the EU, which considers the ISO system as the unified standard, but the USA and major Asian countries have their own regulations, although they respect the ISO system. Globalization of regulations or settlement of international regulations is highly desired to promote tissue engineering as applicable to patients worldwide. If tissue engineering research continues to remain in its infancy in terms of clinical application as a result of intentional or unintentional avoidance of troublesome governmental regulations, a large number of patients eagerly hoping for treatments with tissue engineering will be forced to wait further for a long time. Efforts to develop a rational approach to the regulation of engineered tissue products and to expand globally international harmonization programs should be continued, although responses to the ethical, cultural, and legal issues on human cellular and tissue-based technologies for clinical applications are multiple and different among nations.

Finally, it should be emphasized that without any basic research we cannot expect any applications. A mature program in tissue engineering must maintain a balance between applied efforts and basic research. However, human trials of tissue engineering will further recede if much attention is biased to long-term, basic research such as on the allogeneic ES cell even though it has an enormous potential. One has to appreciate that there are many things to do even in short-term studies to realize prompt performance of human trials in tissue engineering. A reasonable balance is important between short-term and long-term research. As new results always alter the path of science, there is no definitive answer as to when engineered tissues will be widely applied to patients.

REFERENCES

1. M.R. Graham, R.K. Warrian, L.G. Girling *et al.*, Fractal or biologically variable delivery of cardioplegic solution prevents diastolic dysfunction after cardiopulmonary bypass, *J. Thorac. Cardiovasc. Surg.*, **123**, 63 (2002).
2. D. Rubio, J. Garcia-Castro, M.C. Martin *et al.*, Spontaneous human adult stem cell transformation, *Cancer Res.*, **65**, 3035 (2005).
3. M.J. Martin, A. Muotri, F. Gage *et al.*, Human embryonic stem cells express an immunogenic nonhuman sialic acid, *Nat. Med.*, **11**, 228 (2005).
4. C. Holden, Human embryonic stem cells. Getting the mice out of ES cell cultures, *Science*, **307**, 1393 (2005).
5. A. Maitra, D.E. Arking, N. Shivapurkar *et al.*, Genomic alterations in cultured human embryonic stem cells, *Nat. Genet.* (online 4 September 2005).
6. S. Louet, Reagent safety issues surface for cell/tissue therapies, *Nat. Biotechnol.*, **22**, 253 (2004).
7. H. Kawaguchi, H. Hayashi, N. Mizuno *et al.*, Periodontal tissue regeneration by transplantation of own bone marrow mesenchymal stem cell, *Regenerative Medicine* (in Japanese), **4**, 69 (2005).
8. T. Shin'oka, Y. Imai, and Y. Ikada, Transplantation of a tissue-engineered pulmonary artery, *N. Eng. J. Med.*, **344**, 532 (2001).
9. M.P. Lutolf and J.A. Hubbell, Synthetic biomaterials as instructive extracellular microenvironments for morphogenesis in tissue engineering, *Nat. Biotechnol.*, **23**, 47 (2005).
10. Y. Ikada, Interfacial biocompatibility, Chapter 3, in "ACS Symposium Series No 540 (Polymers of Biological and Biomedical Significance)", S.W. Sharaby *et al.*, eds., 1994, p. 35.
11. Y. Ikada, M. Suzuki, and Y. Tamada, Calculation of work of protein adhesion in water, *ACS Polym. Preprints*, **24**, 19 (1983).
12. Y. Tamada and Y. Ikada, Cell adhesion to plasma-treated polymer surfaces, *Polymer*, **34**, 2208 (1993).
13. Y. Ikada, Super-hydrophilic surfaces for biomedical applications, Chapter 11, in Water in *Biomaterials Surface Science*, M. Morra, ed., John Wiley & Sons, Ltd, 2001, p. 291.
14. H.M. Blau and A. Banfi, The well-tempered vessel, *Nat. Med.*, **7**, 532 (2001).
15. S. Levenberg, J. Rouwkema, M. Macdonald *et al.*, Engineering vascularized skeletal muscle tissue, *Nat. Biotechnol.*, **23**, 879 (2005).
16. E.M. Brey, S. Uriel, H.P. Greisler *et al.*, Therapeutic neovascularization: Contributions from bioengineering, *Tissue Engineering*, **11**, 567 (2005).
17. V. Karageorgiou and D. Kaplan, Porosity of 3D biomaterial scaffolds and osteogenesis, *Biomaterials*, **26**, 5474 (2005).

18. Y. Tabata, A. Nagano, and Y. Ikada, Biodegradation of hydrogel carrier incorporating fibroblast growth factor, *Tissue Engineering*, **5**, 127 (1999).
19. Y. Tabata, S. Hijikata, and Y. Ikada, Enhanced vascularization and tissue granulation by basic fibroblast growth factor impregnated in gelatin hydrogels, *J. Control. Release*, **31**, 189 (1994).
20. From the brochure of R&D Systems.
21. R. Langer and J.P. Vacanti, Tissue engineering, *Science*, **260**, 920 (1993).
22. K.R. Chien, Stem cells: Lost in translation, *Nature*, **428**, 607 (2004).
23. T. Hardingham, View from a small island, *Tissue Engineering*, **9**, 1063 (2003).

Index

1,3-Trimethylene carbonate, 18
2-D culture, 42
2-D monolayer, 362
3-D cell culture, 328, 336
3-D culture, 43
3-D matrix, 26
3T3 feeder layer, 426

Abdominal wall, 210
Absorbable, 4
Acellularization, 160
Acidic gelatin, 450
Acidosis, 64
ACL, 131, 269
ACL cell, 308
Acute liver failure, 206
ADAS, 81, 218, 396
Adipogenic medium, 382
Adipose-derived adult stem cell,
    81, 396
Adipose-derived stem cell, 396
Adipose tissue, 101
Adult stem cell, 74
Agarose, 250, 277, 290
Agarose hydrogel, 351
Aggrecan, 11, 367
ALF, 206
Alginate, 4, 12, 245, 276
Alginate gel, 384
Alkaline phosphatase, 80
Alloderm™, 32
ALP, 80
Alveolar ridge, 188
Angiogenesis, 82, 153, 446
Anoikis, 38
Anterior cruciate ligament, 130
Antibody microarray, 398
AOT, 311
Apoptosis, 38
Arteriogenesis, 446
Articular cartilage, 105

Articular joint, 61
Artificial matrix, 1
Auricular cartilage, 97
Auricular chondrocyte, 364

Basement membrane, 16
Basic fibroblast growth factor, 65
Basic gelatin, 450
bFGF, 65, 110, 153, 156, 192,
    313, 450
Bile duct, 209
Bioabsorbable, 4
Bioabsorption rate, 442
Bioadhesion, 329
Bioartificial liver, 206
Bioartificial myocardial tissue, 165
Bioartificial tendon, 361
Biochip, 309
Biodegradable, 4
Bioglass™, 272
Bioreactor, 50, 338, 391
Bladder, 213
Bladder acellular matrix, 213
Blastocyst, 404
Blend, 20, 165
BMC, 209
BM-MNC, 155
BMP, 65, 66, 450
BMP-2, 123, 171, 190, 312, 314, 322
BMP-4, 326
BMSC, 77, 389, 392
Bone, 119
Bone cell, 368
Bone marrow, 77, 433
Bone marrow and particulate cancellous
    bone, 441
Bone-marrow-mononuclear cell, 155
Bone matrix, 297
Bone morphogenetic factor, 65
Bovine spongiform encephalitis(BSE), 6
BSE, 6, 47, 437

Cadaver-derived bone, 121
Calcium carbonate, 25
Calcium phosphate, 272
Calcium phosphate cement, 272
Calcium phosphate scaffold, 441
Cambium layer, 118
Cardiac myocyte, 168, 356
Cardiomyocytes, 359
Carrier, 311, 448
Cartilage pellet, 383
CDI, 7
Cell adhesion, 37, 47, 444
Cell-adhesive protein, 444
Cell aggregate, 58
Cell carrier, 299
Cell counting, 353
Cell culture, 328, 356
Cell differentiation, 80, 385
Cell expansion, 79, 382
Cell seeding, 44, 329
Cell therapy, 81, 424
Cell transplantation, 1
Centrifugation, 333
Channel, 26, 181
Chitin, 13, 249
Chitosan, 13, 249, 288, 316
Chondrocyte, 363
Chondrogenic media, 382
Chondroitin sulfate, 13
Chondroitin sulphate, 248
Chromosomal DNA, 74
Clinical tissue engineering, 441
CNS, 378
CO_2, 284
Co-culture, 336
Collaborative work, 455
Collagen, 6, 7, 235, 292
Collagen fiber, 240, 288
Collagen fibril, 34
Collagen-GAG matrix, 377
Collagen orientation, 346
Commercial development, 458
Composite, 25, 272
Composite scaffold, 36, 301
Conjunctiva, 221
Copolymer, 20
Coral, 128
Cornea, 220
Coronary artery, 150

Corporal tissue, 217
Cryopreservation, 381, 396
CS, 13, 293

DCC, 304
DDS, 447
Dedifferentiation, 42, 366
Degradable, 4
Demineralized cancellous bone, 368
Denatured collagen, 384
Dentin, 198
Dextran, 250
DHT, 7
Dicyclohexylcarbodiimide, 304
Dimethyl sulfoxide, 28
Direct perfusion, 55
DMB, 368
DMSO, 28
DNA, 326
DNA assay, 50
DNA content, 392
Dopaminergic neuron, 403
Drug delivery system, 447
Dura mater, 219

EB, 85
EC, 217, 375, 428
ECM, 1, 14, 32, 36, 349, 438
ECM-like scaffold, 32, 292
EDAC, 7, 235
EDAC/NHS, 293
EGF, 65
EGFP, 49, 376, 381
Elastic cartilage, 97, 105
Elastic scaffold, 35, 299
Elastic stent, 441
Elastin, 8, 14, 292
Electrospinning, 30, 285
Embryoid body, 85
Embryonic stem cell, 70
Enamel, 198
Endothelial progenitor cell, 82
Engelbreth-Holm-Swarm cell, 208
Enhanced green fluorescent protein, 49
EPC, 82, 154
Epidermal growth factor, 65
Epithelial–mesenchymal cell interaction, 211
Epithel™, 92
Equine collagen, 238

ES, 70
ES cell, 84, 394, 400, 432
Esophagus, 205
Ex vivo tissue engineering, 424, 426
Extracellular matrix, 1

FCS, 46, 335, 382, 396, 435, 437
FDG-PET, 359
Fetal calf serum, 46
FGF, 67
FGF-2, 65, 313, 384, 393
Fibrin, 10
Fibrin gel, 34, 297
Fibroblast, 360
Fibrocartilage, 105
Fibronectin, 15, 37
Fluor-Deoxy-Glucose-Positron-Emmision-
    Tomography, 359
FN, 15
Freeze drying, 27, 274
Freeze-extraction, 275
Freeze-gelation, 275

GA, 7, 235
GA–ε-CL copolymer, 299
GAG, 10, 292, 342, 349, 365
Gas forming, 29, 279
GBR, 188
GBR membrane, 190
Gelatin, 8, 241, 334
Gelatin hydrogel, 314, 450
Gene transfer, 325
Genetically tailored human ES cell, 405
GFP, 50, 84, 375
Glycolic acid–trimethylene carbonate copoly-
    mer, 191
Glycosaminoglycan, 10
GPS, 373
Green fluorescent protein, 50
GRGDS, 292, 330
GRGDY, 304
Growth factor, 65, 351, 448
Growth factor-like polymer, 309
GTR, 188
Guided bone regeneration, 188
Guided tissue regeneration, 188

HAc, 10, 245
HAc–MA, 265

HAp, 25, 35, 300, 301
Heart valve, 158
HEMA, 330
Hematopoietic differentiation, 404
Heparinoid, 309, 316
Hepatocyte, 207, 377, 392
Hepatocyte growth factor, 65
Hexafluoro-2-propanol, 243
HFIP, 243
HGF, 65, 153
HLA, 85
Homopolymer, 20
Human ES cell, 423
Hyaff™, 12, 386
Hyaline cartilage, 97, 105
Hyaluronic acid, 10
Hydraulic bone chamber, 112, 352
Hydrogel, 22, 263, 279
Hydrophilic-hydrophobic balance, 444
Hydroxyapatite, 25
Hypoxia, 64

IGF-I, 65, 324, 351
Immunological surveillance, 431
Immunorejection, 429
In situ tissue engineering, 424, 428
In vitro tissue engineering, 424
In vivo tissue engineering, 424
Inferior vena cava, 146
Injectable scaffold, 35, 299
Inorganic scaffold, 35, 300
Insulin-like growth factor, 65
Insulin–transferrin–selenite, 393
Insulintransferrinselenium, 335
Integrin, 37, 371
Internal fluid perfusion, 344
Interpenetrating polymer network, 363
ITS, 335, 382, 393, 396

Joint, 140

Keratinocyte, 91, 426
Keratinocyte growth factor, 46, 65, 311
KGF, 46, 65, 311

Lactide-*p*-dipoxanone polymer, 254
Laminin, 15
Leukemia inhibitory factor, 85
LIF, 85

Ligament, 129
Liver, 206
LLA–CL copolymer, 21
LLA–ε-CL copolymer, 256
LN, 15, 360
LO, 17, 237, 294
Luciferase, 49
Lysyl oxidase, 17

Major histocompatible antigen, 430
Mandible, 199
Matrigel™, 34, 358, 401
Matrix metalloprotease, 33
Maxillary sinus augmentation, 194
Mechanical loading, 361
Mechanical stimulation, 60, 345, 360
Melanocyte, 95
MEMS, 375
Meningeal cell, 378
MHA, 430
MHC, 85
Micro-computed tomography, 369
Micro-CT, 369
Microarray, 379
Microcapsule, 345
Microcarrier, 345, 363
Microelectromechanical system, 375
Microfracture, 107
Microgravity, 56
Microtia, 97
MMP, 33, 346, 373
MNC, 33
Mononuclear cell, 33
Morula, 404
Mosaicplasty, 106
MRI, 340
MSC, 72, 77, 164, 432
MTT, 49, 353
Muscle-derived cell, 139
Myocardial infarction, 82
Myocardial tissue, 162
Myocardium, 82
Myogenic stem cell, 392

Nanofibrous scaffold, 385
Nasal augmentation, 100
Nasoseptal cartilage, 96
Nasoseptal chondrocyte, 364
Natural polymer, 6
Natural sponge, 298

Naturally derived scaffold, 31
Neovascularization, 63, 156, 446
Neovessel formation, 447
Neu5Gc, 435
Neuron, 173
NGF, 327
N-Glycolylneuraminic acid, 435
NHS, 41, 235
N-Hydroxysuccinimide, 41
Nodule, 404
Non-woven PGA fiber, 220
Non-woven PLLGA, 392
NSC, 397

OA, 115
OCD, 113
OCP, 25
Octacalcium phosphate, 25
Odontogenic, 378
o-HAc, 12, 320
OP, 80
OPF, 321
Oral cell, 377
Orbital floor, 202
Osteoblast medium, 397
Osteochondritis dissecans, 113
Osteogenic stimulation, 384
Osteogenic supplement, 307
Osteopontin, 37

P(CL/LA), 146, 147
P(LA/CL), 209, 219, 256, 315, 351, 440
P(LA/CL) fabric, 286
P4HB, 99, 371
Particulate cancellous bone and marrow, 199
Passage-dependent alteration, 399
Passage number, 365
PCBM, 199, 203
PCL, 98, 252, 442
PCL–PEG–PCL triblock copolymer, 265
PDGF, 65, 68, 325
PDLLA, 19
PDS, 202
PEEUU, 265
PEG, 23
PEG-based hydrogel, 388
PEG-diald, 319
PEI, 326
Pellet, 388
PEO, 23

PEO macromer, 248
PEODA, 248
PEOT/PBT multiblock copolymer, 269
Perfusion, 55, 359
Perfusion reactor, 339
Periodontium, 189
Periosteum, 118
Peripheral nerve, 174
PERV, 160, 430
PET, 17
PEUU, 268, 302
PGA, 18, 150, 178, 216, 321, 442
PGA mesh, 172
PGA non-woven felt, 350
PGCL, 300
PGLA, 218
PHA, 253
Phalangeal joint, 142
Phalanx, 127
Phase separation, 27
Photolithography, 290
Photosensitive polymer, 290
PLA, 18, 250
PLA-b-PEG-b-PLA triblock copolymer, 255
Plasma treatment, 444
Plasmid, 326
Platelet derived growth factor, 65
Platelet-poor plasma, 394
PLGA, 19, 303, 304
PLGA–NH$_2$, 304
PLLA, 18, 303, 442
PLLA mesh, 203
PLLA tray, 199
PLLA–LAD, 134
Pluripotent stem cell, 70
Pluronic, 24, 263
PMMA, 17
Poly(CL–TMC), 261
Poly(D,L-lactide), 19
Poly(desamino tyrosyl-tyrosine ethyl ester
     carbonate), 268
Poly(DTE carbonate), 268
Poly(EG-block-DLLA), 278
Poly(ester urethane) urea, 268
Poly(ether ester urethane) urea, 265
Poly(ethylene glycol), 23
Poly(ethylene glycol)-terephthalate-poly
     (butylene terephthalate), 290
Poly(ethylene oxide), 23
Poly(ethylene terephthalate), 17

Poly(ethyleneimine), 326
Poly(glycerol sebacate), 24, 271
Poly(glycolide-co-lactide), 18
Poly(hydroxyalkanoate), 253
Poly(LA-*co*-lysine), 22
Poly(L-lactide), 18
Poly(LLA-co-ethylene glycol), 261
Poly(L-LA-co-GA-ε-CL), 273
Poly(LLA-co-ε-caprolactone), 145
Poly(methyl methacrylate), 17
Poly(*p*-dioxanone), 202
Poly(propylene fumarate), 24, 268
Poly(TMC), 259
Poly(TMC–CL), 259
Poly(TMC–DLLA), 259
Poly(tyrosine isocyanate), 24
Poly(α-hydroxyacid), 18, 250
Poly(γ-benzyl-L-glutamic acid)-b-PEO-b-PCL,
     256
Poly(ε-caprolactone), 98, 252
Poly-4-hydroxybutyrate, 371
Polybrene, 375
Polyethylene, 17
Polyglactin™, 161
Polyglycolide, 18
Polyhydroxyoctanoate, 161
Polylactide, 18
Polyphosphazene, 24, 269
Polysaccharide, 10, 243
Polystyrene, 309, 334, 367
Polytetrafluoroethylene, 17
Polyurethane, 23
Porcine endogenous retrovirus, 160, 430
Pore size, 26
Porogen leaching, 28, 278
Porosity, 26
PPF, 24, 282
PPF-DA, 268
PPP, 394
Preimplantation, 404
Prenatal tissue, 221
Protein, 6
Protein fiber, 288
Protein microarray, 399
PRP, 194, 394, 437, 452
PS, 367
PTFE, 17
PU urea foam, 281
Pulmonary artery, 147
PuraMatrix™, 243

Rapid prototyping, 29, 282
Recombinant plasmid, 70
Redifferentiation, 42
Regenerative medicine, 424
Regulation, 459
Reporter gene, 356
Resorbable, 4
Retinal cell, 379
Retinal progenitor cell, 379
Retroviral vector, 376
RGD, 15, 37, 325
Rheumatoid arthritis, 143
Rotary wall vessel, 56
Rotating reactor, 342
Rotating wall reactor, 56
Rotator cuff, 138
RVOT, 145, 161
RWV, 56

Scaffold, 1, 438
Schwann cell, 175
SCNT, 86, 435
SDS, 295
Seeding Efficiency, 48
Self-cell therapy, 431
Sequential release, 323
Serum, 46
S-GAG, 397
Silicone, 17
Silk, 390
Silk fibroin, 10, 241
SIS, 33, 205, 214, 296
Skeletal muscle, 139
Skin, 91
Skull base, 218
SMA, 377
Small-calibered artery, 428
Small intestine, 210
SMC, 212, 216, 256
Smooth muscle cell, 212
Sodium bis(ethylhexyl) sulfosuccinate, 311
Sodium dodecyl sulphate, 295
Solid free-form fabrication, 29
Somatic cell nuclear transfer, 86
Somatic stem cell, 74
Spinal cord, 173
Spinner flask, 53, 339, 360
Stem cell, 70, 379, 432

Stereolithography, 29
Sulphated GAG, 397
Surface modification, 36

TCPS, 389
Telopeptide, 8
Template, 1
Temporomandibular joint, 195
Tendon, 129, 135
TEVA, 149
TGF-β3, 388
TGF-β1, 65, 68, 320, 327, 351, 388
Thrombosis, 428
Tissue-Derived Scaffold, 295
Tissue-engineered vascular autograft, 149
Tissue Fleece™, 165
TMC, 21
TMC-CL copolymer, 259
TMJ, 195
TMJ disc, 343
Trachea, 169
Transdifferetiation, 83
Transforming growth factor-β, 65
Transglutaminase, 237
Triton X-200, 295
Type II collagen, 105
Type III collagen, 16
Type IV collagen, 16
Type XIII collagen, 298

UC, 337
Ultrafine structure, 443
Umbilical cord blood-derived cell, 81
Ureter, 215
Urethra, 215
Urinary bladder matrix, 297
UV irradiation, 236

Vaginal tissue, 216
Vascular cell, 371
Vascular endothelial growth factor, 65
Vascularization, 316
Vasculature, 443
Vasculogenesis, 446
VEGF, 65, 67, 208, 307, 319
VIC, 265
Vicryl mesh, 18

Viral vector, 70
Vitiligo, 94

Wettability, 371
Wound healing, 449
WSC, 7

Xenogeneic cell, 429
Xenotransplantation, 160

XMT, 354
X-ray microtomography, 354

α-Gal epitope, 160
α-smooth muscle actin,
    377
β-TCP, 25, 35, 108, 274,
    300, 322
β-tricalcium phosphate, 25